Springer Series in Statistics

Springer Series in Statistics

Andersen/Borgan/Gill/Keiding: Statistical Models Based on Counting Processes.
Anderson: Continuous-Time Markov Chains: An Applications-Oriented Approach.
Andrews/Herzberg: Data: A Collection of Problems from Many Fields for the Student and Research Worker.
Anscombe: Computing in Statistical Science through APL.
Berger: Statistical Decision Theory and Bayesian Analysis, 2nd edition.
Bolfarine/Zacks: Prediction Theory for Finite Populations.
Brémaud: Point Processes and Queues: Martingale Dynamics.
Brockwell/Davis: Time Series: Theory and Methods, 2nd edition.
Choi: ARMA Model Identification.
Daley/Vere-Jones: An Introduction to the Theory of Point Processes.
Dzhaparidze: Parameter Estimation and Hypothesis Testing in Spectral Analysis of Stationary Time Series.
Farrell: Multivariate Calculation.
Federer: Statistical Design and Analysis for Intercropping Experiments.
Fienberg/Hoaglin/Kruskal/Tanur (Eds.): A Statistical Model: Frederick Mosteller's Contributions to Statistics, Science and Public Policy.
Goodman/Kruskal: Measures of Association for Cross Classifications.
Grandell: Aspects of Risk Theory.
Hall: The Bootstrap and Edgeworth Expansion.
Härdle: Smoothing Techniques: With Implementation in S.
Hartigan: Bayes Theory.
Heyer: Theory of Statistical Experiments.
Jolliffe: Principal Component Analysis.
Kotz/Johnson (Eds.): Breakthroughs in Statistics Volume I.
Kotz/Johnson (Eds.): Breakthroughs in Statistics Volume II.
Kres: Statistical Tables for Multivariate Analysis.
Leadbetter/Lindgren/Rootzén: Extremes and Related Properties of Random Sequences and Processes.
Le Cam: Asymptotic Methods in Statistical Decision Theory.
Le Cam/Yang: Asymptotics in Statistics: Some Basic Concepts.
Manoukian: Modern Concepts and Theorems of Mathematical Statistics.
Manton/Singer/Suzman: Forecasting the Health of Elderly Populations.
Miller, Jr.: Simultaneous Statistical Inference, 2nd edition.
Mosteller/Wallace: Applied Bayesian and Classical Inference: The Case of *The Federalist Papers.*
Pollard: Convergence of Stochastic Processes.
Pratt/Gibbons: Concepts of Nonparametric Theory.
Read/Cressie: Goodness-of-Fit Statistics for Discrete Multivariate Data.
Reinsel: Elements of Multivariate Time Series Analysis.
Reiss: A Course on Point Processes.
Reiss: Approximate Distributions of Order Statistics: With Applications to Nonparametric Statistics.
Ross: Nonlinear Estimation.

(continued after index)

Gregory C. Reinsel

Elements of Multivariate Time Series Analysis

With 11 Illustrations

Springer-Verlag
New York Berlin Heidelberg London Paris
Tokyo Hong Kong Barcelona Budapest

Gregory C. Reinsel
Department of Statistics
University of Wisconsin, Madison
Madison, WI 53706-1693 USA

Mathematics Subject Classifications (1991): 62-01, 62M10, 62M20, 62H20

Library of Congress Cataloging-in-Publication Data
Reinsel, Gregory C.
Elements of multivariate time series analysis / Gregory C. Reinsel.
p. cm. – (Springer series in statistics)
Includes bibliographical references and index.
ISBN 0-387-94063-4 (New York). – ISBN 3-540-94063-4 (Berlin)
1. Time-series analysis. 2. Multivariate analysis. I. Title.
II. Series.
QA280.R45 1993
519.5′5–dc20 93-13954

Printed on acid-free paper.

Production managed by Henry Krell; manufacturing supervised by Vincent Scelta.
Photocomposed copy prepared from the author's Troff files.
Printed and bound by Edwards Brothers, Inc., Ann Arbor, MI.
Printed in the United States of America.

9 8 7 6 5 4 3 2 1

ISBN 0-387-94063-4 Springer-Verlag New York Berlin Heidelberg
ISBN 3-540-94063-4 Springer-Verlag Berlin Heidelberg New York

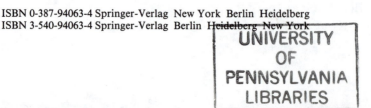

To Sandy,

and our children, Chris and Sarah

Preface

The use of methods of time series analysis in the study of multivariate time series has become of increased interest in recent years. Although the methods are rather well developed and understood for univariate time series analysis, the situation is not so complete for the multivariate case. This book is designed to introduce the basic concepts and methods that are useful in the analysis and modeling of multivariate time series, with illustrations of these basic ideas. The development includes both traditional topics such as autocovariance and autocorrelation matrices of stationary processes, properties of vector ARMA models, forecasting ARMA processes, least squares and maximum likelihood estimation techniques for vector AR and ARMA models, and model checking diagnostics for residuals, as well as topics of more recent interest for vector ARMA models such as reduced rank structure, structural indices, scalar component models, canonical correlation analyses for vector time series, multivariate unit-root models and cointegration structure, and state-space models and Kalman filtering techniques and applications. This book concentrates on the time-domain analysis of multivariate time series, and the important subject of spectral analysis is not considered here. For that topic, the reader is referred to the excellent books by Jenkins and Watts (1968), Hannan (1970), Priestley (1981), and others.

The intention of this book is to introduce topics of multivariate time series in a useful way to readers who have some background in univariate time series methods of the sort available in the book by Box and Jenkins (1976). It is also necessary for the reader to have some knowledge of matrix algebra techniques and results to completely follow all developments of the topics in the book. Appendices at the end of Chapters 1 and 4 are provided which summarize and review some basic results on matrices and the multivariate normal distribution, and results on the multivariate linear model, respectively. It is hoped that these will provide the necessary background on these topics for the reader. The book is intended to provide the basic concepts needed for an adequate understanding of the material, but elaborate and detailed mathematical developments and arguments are generally not emphasized, although substantial references are usually provided for further mathematical details. Hence, the book will be accessible to a wider audience who have a working background in univariate time series analysis and some knowledge of matrix algebra methods and who wish to become familiar with and use multivariate time series modeling techniques in applications. The book could serve as a graduate-level textbook on "multivariate time series" for a second course in time series as well as a reference book for researchers and practitioners in the area of multiple time series analysis. A set of exercise problems are included at the end of the book, which it is hoped will make the book more valuable for textbook use. Listings of the data sets used in the numerical examples in the book are also included in the Appendix on Data Sets.

I am indebted to George Tiao and Ruey Tsay for many useful and interesting discussions, in general, on the subject of multivariate time series analysis. I would like to thank Sung Ahn, Sabyasachi Basu, Sophie Yap, and others, for their helpful comments on earlier drafts of this material. I would also like to thank Lisa Ying and Eric Tam for their assistance in preparing the final figures which appear in this book. I would also like to extend my gratitude to Martin Gilchrist and others on the staff of Springer-Verlag for their interest in this book and their assistance with its preparation for publication. Finally, I express my special appreciation to my wife, Sandy, and my children, Chris and Sarah, for their help and understanding throughout this project.

Gregory Reinsel
April 1993

Contents

Preface vii

1. Vector Time Series and Model Representations 1

1.1 Stationary Multivariate Time Series and Their Properties 2

1.1.1 Covariance and Correlation Matrices for a Stationary Vector Process 2

1.1.2 Some Spectral Characteristics for a Stationary Vector Process 4

1.1.3 Some Relations for Linear Filtering of a Stationary Vector Process 5

1.2 Linear Model Representations for a Stationary Vector Process .. 7

1.2.1 Infinite Moving Average (Wold) Representation of a Stationary Vector Process 7

1.2.2 Vector Autoregressive Moving Average (ARMA) Model Representations 7

A1 Appendix: Review of Multivariate Normal Distribution and Related Topics 12

A1.1 Review of Some Basic Matrix Theory Results 12

A1.2 Expected Values and Covariance Matrices of Random Vectors 13

A1.3 The Multivariate Normal Distribution 14

A1.4 Some Basic Results on Stochastic Convergence 18

2. Vector ARMA Time Series Models and Forecasting 21

2.1 Vector Moving Average Models 21

2.1.1 Invertibility of the Vector Moving Average Model 21

2.1.2 Covariance Matrices of the Vector Moving Average Model 22

2.1.3 Features of the Vector MA(1) Model 23

2.1.4 Model Structure for Subset of Components in the Vector MA Model ... 24
2.2 Vector Autoregressive Models ... 26
2.2.1 Stationarity of the Vector Autoregressive Model ... 26
2.2.2 Yule-Walker Relations for Covariance Matrices of a Vector AR Process ... 28
2.2.3 Covariance Features of the Vector AR(1) Model ... 28
2.2.4 Univariate Model Structure Implied by Vector AR Model ... 29
2.3 Vector Mixed Autoregressive Moving Average Models ... 33
2.3.1 Stationarity and Invertibility of the Vector ARMA Model ... 33
2.3.2 Relations for the Covariance Matrices of the Vector ARMA Model ... 34
2.3.3 Some Features of the Vector ARMA(1,1) Model ... 35
2.3.4 Consideration of Parameter Identifiability for Vector ARMA Models ... 36
2.3.5 Further Aspects of Nonuniqueness of Vector ARMA Model Representations ... 39
2.4 Nonstationary Vector ARMA Models ... 40
2.4.1 Vector ARIMA Models for Nonstationary Processes ... 41
2.4.2 Cointegration in Nonstationary Vector Processes ... 42
2.4.3 The Vector IMA(1,1) Process or Exponential Smoothing Model ... 43
2.5 Prediction for Vector ARMA Models ... 45
2.5.1 Minimum Mean Squared Error Prediction ... 46
2.5.2 Forecasting for Vector ARMA Processes and Covariance Matrices of Forecast Errors ... 46
2.5.3 Computation of Forecasts for Vector ARMA Processes ... 48
2.5.4 Some Examples of Forecast Functions for Vector ARMA Models ... 49

3. Canonical Structure of Vector ARMA Models 52
3.1 Consideration of Kronecker Structure for Vector ARMA Models ... 52
3.1.1 Kronecker Indices and McMillan Degree of Vector ARMA Process ... 53
3.1.2 Echelon Form Structure of Vector ARMA Model Implied by Kronecker Indices ... 54
3.1.3 Reduced-Rank Form of Vector ARMA Model Implied by Kronecker Indices ... 56
3.2 Canonical Correlation Structure for ARMA Time Series ... 58
3.2.1 Canonical Correlations for Vector ARMA Processes ... 60
3.2.2 Relation to Scalar Component Model Structure ... 61

3.3 Partial Autoregressive and Partial Correlation Matrices 64
3.3.1 Vector Autoregressive Model Approximations and Partial Autoregression Matrices 64
3.3.2 Recursive Fitting of Vector AR Model Approximations 66
3.3.3 Partial Cross-Correlation Matrices for a Stationary Vector Process 69
3.3.4 Partial Canonical Correlations for a Stationary Vector Process 71

4. Initial Model Building and Least Squares Estimation for Vector AR Models 74

4.1 Sample Cross-Covariance and Correlation Matrices and Their Properties 74
4.1.1 Sample Estimates of Mean Vector and of Covariance and Correlation Matrices 74
4.1.2 Asymptotic Properties of Sample Correlations 76
4.2 Sample Partial AR and Partial Correlation Matrices and Their Properties 78
4.2.1 Test for Order of AR Model Based on Sample Partial Autoregression Matrices 78
4.2.2 Equivalent Test Statistics Based on Sample Partial Correlation Matrices 79
4.3 Conditional Least Squares Estimation of Vector AR Models 80
4.3.1 Least Squares Estimation for the Vector AR(1) Model 81
4.3.2 Least Squares Estimation for the Vector AR Model of General Order 83
4.3.3 Likelihood Ratio Testing for the Order of the AR Model 85
4.3.4 Derivation of the Wald Statistic for Testing the Order of the AR Model 85
4.4 Relation of LSE to Yule-Walker Estimate for Vector AR Models 89
4.5 Additional Techniques for Specification of Vector ARMA Models 91
4.5.1 Use of Order Selection Criteria for Model Specification 92
4.5.2 Sample Canonical Correlation Analysis Methods 93
4.5.3 Order Determination Using Linear LSE Methods for the Vector ARMA Model 96
A4 Appendix: Review of the General Multivariate Linear Regression Model 105
A4.1 Properties of the Maximum Likelihood Estimator of the Regression Matrix 105
A4.2 Likelihood Ratio Test of Linear Hypothesis About Regression Coefficients 107

A4.3 Asymptotically Equivalent Forms of the Test of Linear Hypothesis 108

5. Maximum Likelihood Estimation and Model Checking for Vector ARMA Models 111

5.1 Conditional Maximum Likelihood Estimation for Vector ARMA Models 111

5.1.1 Conditional Likelihood Function for the Vector ARMA Model 112

5.1.2 Likelihood Equations for Conditional ML Estimation 113

5.1.3 Iterative Computation of the Conditional MLE by GLS Estimation 115

5.1.4 Asymptotic Distribution for the MLE in the Vector ARMA Model 117

5.2 ML Estimation and LR Testing of ARMA Models Under Linear Restrictions 118

5.2.1 ML Estimation of Vector ARMA Models with Linear Constraints on the Parameters 118

5.2.2 LR Testing of the Hypothesis of the Linear Constraints 120

5.2.3 ML Estimation of Vector ARMA Models in the Echelon Canonical Form 121

5.3 Exact Likelihood Function for Vector ARMA Models 122

5.3.1 Expressions for the Exact Likelihood Function and Exact Backcasts 124

5.3.2 Special Cases of the Exact Likelihood Results 126

5.4 Innovations Form of the Exact Likelihood Function for ARMA Models 129

5.4.1 Use of Innovations Algorithm Approach for the Exact Likelihood 129

5.4.2 Prediction of Vector ARMA Processes Using the Innovations Approach 131

5.5 Overall Checking for Model Adequacy 132

5.5.1 Residual Correlation Matrices, and Overall Goodness-of-Fit Test 132

5.5.2 Asymptotic Distribution of Residual Covariances and Goodness-of-Fit Statistic 133

5.5.3 Use of the Score Test Statistic for Model Diagnostic Checking 134

5.6 Effects of Parameter Estimation Errors on Prediction Properties 138

5.6.1 Effects of Parameter Estimation Errors on Forecasting in the Vector AR(p) Model 139

5.6.2 Prediction Through Approximation by Autoregressive Model Fitting 141

5.7 Numerical Examples 142

6. Reduced-Rank and Nonstationary Co-Integrated Models 154

6.1 Nested Reduced-Rank AR Models and Partial Canonical Correlation Analysis 154

6.1.1 Specification of Ranks Through Partial Canonical Correlation Analysis 155

6.1.2 Canonical Form for the Reduced-Rank Model 157

6.1.3 Maximum Likelihood Estimation of Parameters in the Model 158

6.1.4 Relation of Reduced-Rank AR Model with Scalar Component Models and Kronecker Indices 159

6.2 Review of Estimation and Testing for Nonstationarity (Unit Roots) in Univariate ARIMA Models 162

6.2.1 Limiting Distribution Results in the AR(1) Model with a Unit Root 162

6.2.2 Unit-Root Distribution Results for General Order AR Models 163

6.3 Nonstationary (Unit-Root) Multivariate AR Models, Estimation, and Testing 165

6.3.1 Unit-Root Nonstationary Vector AR Model, and the Error-Correction Form 165

6.3.2 Asymptotic Properties of the Least Squares Estimator 166

6.3.3 Reduced-Rank Estimation of the Error-Correction Form of the Model 169

6.3.4 Likelihood Ratio Test for the Number of Unit Roots 172

6.3.5 Reduced-Rank Estimation Through Partial Canonical Correlation Analysis 174

6.3.6 Extension to Account for a Constant Term in the Estimation 175

6.3.7 Forecast Properties for the Co-integrated Model 180

6.3.8 Explicit Unit-Root Structure of the Nonstationary AR Model and Implications 181

6.3.9 Further Numerical Examples 183

6.4 Multiplicative Seasonal Vector ARMA Models 186

6.4.1 Some Special Seasonal ARMA Models for Vector Time Series 187

7. State-Space Models, Kalman Filtering, and Related Topics 192

7.1 State-Variable Models and Kalman Filtering 192

7.1.1 The Kalman Filtering Relations 193
7.1.2 Smoothing Relations in the State-Variable Model 196
7.1.3 Innovations Form of State-Space Model and Steady-State for Time-Invariant Models 197
7.2 State-Variable Representations of the Vector ARMA Model 198
7.2.1 A State-Space Form Based on the Prediction Space of Future Values 198
7.2.2 Exact Likelihood Function Through the State-Variable Approach 199
7.2.3 Alternate State-Space Forms for the Vector ARMA Model 203
7.2.4 Minimal Dimension State-Variable Representation and Kronecker Indices 207
7.2.5 (Minimal Dimension) Echelon Canonical State-Space Representation 208
7.3 Exact Likelihood Estimation for Vector ARMA Processes with Missing Values 215
7.4 Classical Approach to Smoothing and Filtering of Time Series 218
7.4.1 Smoothing for Univariate Time Series 218
7.4.2 Smoothing Relations for the Signal Plus Noise or Structural Components Model 221
7.4.3 A Simple Vector Structural Component Model for Trend 224

Appendix: Time Series Data Sets 226
Exercises and Problems 238
References 248
Index 257

CHAPTER 1

Vector Time Series and Model Representations

We study models that describe relationships among a vector of k time series variables $Y_{1t}, Y_{2t}, \ldots, Y_{kt}$ of interest. Such multivariate processes arise when several related time series processes are observed simultaneously over time, instead of observing just a single series as is the case in univariate time series analysis. Multivariate time series processes are of considerable interest in a variety of fields such as engineering, the physical sciences, particularly the earth sciences (e.g., meteorology and geophysics), and economics and business. For example, in an engineering setting, one may be interested in the study of the simultaneous behavior over time of current and voltage, or of pressure, temperature, and volume, whereas in economics, we may be interested in the variations of interest rates, money supply, unemployment, and so on, or in sales volume, prices, and advertising expenditures for a particular commodity in a business context.

In the study of multivariate processes, a framework is needed for describing not only the properties of the individual series but also the possible cross-relationships among the series. The purposes for analyzing and modeling the series jointly are to understand the dynamic relationships over time among the series and to improve accuracy of forecasts for individual series by utilizing the additional information available from the related series in the forecasts for each series. With these objectives in mind, the class of vector autoregressive moving average (ARMA) time series models is developed and its properties are examined. In later chapters, methods for empirical model building, parameter estimation, model checking, and forecasting of vector time series are discussed. The alternate state-space modeling approach is also considered, as well as special topics such as nonstationary multivariate unit-root processes and cointegration among vector series, and the presence of special simplifying structures in the ARMA model. Throughout the book, the methods will be illustrated by several examples that involve the analysis and modeling of actual multivariate time series data.

In the present introductory chapter, the concept of stationarity of a vector time series process is first introduced, and some basic covariance and correlation matrix properties and spectral properties of such a process are presented. Some features of linear filtering of a stationary time series are considered. Vector autoregressive moving average (ARMA) models are then introduced, and some of their different model representations are discussed.

1.1 Stationary Multivariate Time Series and Their Properties

Let $\boldsymbol{Y}_t = (Y_{1t}, \ldots, Y_{kt})'$, $t = 0, \pm 1, \pm 2, \ldots,$ denote a k-dimensional time series vector of random variables of interest. The choice of the univariate component time series Y_{it} that are included in $\boldsymbol{Y}_t$ will depend on the subject matter area and understanding of the system under study, but it is implicit that the component series will be interrelated both contemporaneously and across time lags. The representation and modeling of these dynamic interrelationships among the component time series will be a main interest of the multivariate time series analysis. An important concept in the representation of models and the analysis of time series, which enables useful modeling results to be obtained from a finite sample realization of the time series, is that of stationarity.

The process $\{\boldsymbol{Y}_t\}$ is *stationary* if the probability distributions of the random vectors $(\boldsymbol{Y}_{t_1}, \boldsymbol{Y}_{t_2}, \ldots, \boldsymbol{Y}_{t_n})$ and $(\boldsymbol{Y}_{t_1+l}, \boldsymbol{Y}_{t_2+l}, \ldots, \boldsymbol{Y}_{t_n+l})$ are the same for arbitrary times $t_1, t_2, \ldots, t_n$, all n, and all lags or leads $l = 0, \pm 1, \pm 2, \ldots$. Thus, the probability distribution of observations from a stationary vector process is invariant with respect to shifts in time. So, assuming finite first and second moments exist, for a stationary process we must have $E(\boldsymbol{Y}_t) = \boldsymbol{\mu}$, constant for all t, where $\boldsymbol{\mu} = (\mu_1, \mu_2, \ldots, \mu_k)'$ is the mean vector of the process. Also, the vectors $\boldsymbol{Y}_t$ must have a constant covariance matrix for all t, which we denote by $\Sigma_y \equiv \Gamma(0) = E[(\boldsymbol{Y}_t - \boldsymbol{\mu})(\boldsymbol{Y}_t - \boldsymbol{\mu})']$.

1.1.1 Covariance and Correlation Matrices for a Stationary Vector Process

In addition, for a stationary process $\{\boldsymbol{Y}_t\}$ the covariance between Y_{it} and $Y_{j,t+l}$ must depend only on the lag l, not on time t, for $i, j = 1, \ldots, k$, $l = 0, \pm 1, \pm 2, \ldots$. Hence, we let

$$\gamma_{ij}(l) = \mathrm{Cov}(Y_{it}, Y_{j,t+l}) = E[(Y_{it} - \mu_i)(Y_{j,t+l} - \mu_j)]$$

and denote the $k \times k$ matrix of *cross-covariances* at lag l as

$$\Gamma(l) = E[\,(\,\boldsymbol{Y}_t - \boldsymbol{\mu}\,)(\,\boldsymbol{Y}_{t+l} - \boldsymbol{\mu}\,)'\,] = \begin{bmatrix} \gamma_{11}(l) & \gamma_{12}(l) & . & . & . & \gamma_{1k}(l) \\ \gamma_{21}(l) & \gamma_{22}(l) & . & . & . & \gamma_{2k}(l) \\ . & . & . & . & . & . \\ . & . & . & . & . & . \\ \gamma_{k1}(l) & \gamma_{k2}(l) & . & . & . & \gamma_{kk}(l) \end{bmatrix}, \tag{1.1}$$

for $l = 0, \pm 1, \pm 2, \ldots$. Also, the corresponding *cross-correlation* matrix at lag l is denoted by

$$\rho(l) = V^{-1/2}\,\Gamma(l)\,V^{-1/2} = \begin{bmatrix} \rho_{11}(l) & \rho_{12}(l) & . & . & . & \rho_{1k}(l) \\ \rho_{21}(l) & \rho_{22}(l) & . & . & . & \rho_{2k}(l) \\ . & . & . & . & . & . \\ . & . & . & . & . & . \\ \rho_{k1}(l) & \rho_{k2}(l) & . & . & . & \rho_{kk}(l) \end{bmatrix}, \tag{1.2}$$

for $l = 0, \pm 1, \pm 2, \ldots$, where $V^{-1/2} = \text{Diag}\,\{\,\gamma_{11}(0)^{-1/2}, \ldots, \gamma_{kk}(0)^{-1/2}\,\}$, since

$$\rho_{ij}(l) = \text{Corr}(\,Y_{it},\, Y_{j,t+l}\,) = \gamma_{ij}(l)\,/\,[\,\gamma_{ii}(0)\,\gamma_{jj}(0)\,]^{1/2}$$

with $\gamma_{ii}(0) = \text{Var}(\,Y_{it}\,)$. Thus, for $i = j$, $\rho_{ii}(l) = \rho_{ii}(-l)$ denotes the autocorrelation function of the ith series Y_{it}, and for $i \neq j$, $\rho_{ij}(l) = \rho_{ji}(-l)$ denotes the cross-correlation function between the series Y_{it} and Y_{jt}. Note that $\Gamma(l)' = \Gamma(-l)$ and $\rho(l)' = \rho(-l)$, since $\gamma_{ij}(l) = \gamma_{ji}(-l)$. In addition, the cross-covariance matrices $\Gamma(l)$ and cross-correlation matrices $\rho(l)$ have the property of non-negative definiteness, in the sense that $\sum_{i=1}^{n} \sum_{j=1}^{n} \boldsymbol{b}_i'\,\Gamma(i-j)\,\boldsymbol{b}_j \geq 0$ for all positive integers n and all k-dimensional vectors $\boldsymbol{b}_1, \ldots, \boldsymbol{b}_n$, which follows since $\text{Var}(\,\sum_{i=1}^{n} \boldsymbol{b}_i'\,\boldsymbol{Y}_{t-i}\,) \geq 0$.

The definition of stationarity given above is usually referred to as strict or strong stationarity. In general, a process $\{\boldsymbol{Y}_t\}$ that possesses finite first and second moments and which satisfies the conditions that $E(\,\boldsymbol{Y}_t\,) = \boldsymbol{\mu}$ does not depend on t and $E[\,(\,\boldsymbol{Y}_t - \boldsymbol{\mu}\,)(\,\boldsymbol{Y}_{t+l} - \boldsymbol{\mu}\,)'\,]$ depends only on l is referred to as *weak, second-order* or *covariance* stationary. In this book, the term stationary will generally be used in this latter sense of weak stationarity. For a stationary vector process, the cross-covariance and cross-correlation matrix structure provides a useful summary of information on aspects of the dynamic interrelations among the components of the process. However, because of the higher dimensionality $k > 1$ of the vector process, the cross-correlation matrices can generally take on complex structures and may be much more difficult to interpret as a whole as compared with the univariate time series case. In the next section of this chapter and in Chapter 2, we will present the classes of vector moving average (MA), autoregressive (AR), and mixed autoregressive moving average (ARMA) models, and we will examine the covariance matrix structures implied for stationary processes that are generated by these models.

1.1.2 Some Spectral Characteristics for a Stationary Vector Process

Similar to the univariate case, we define the covariance-generating function (matrix) (provided $\sum_{l=-\infty}^{\infty} |\gamma_{ij}(l)| < \infty$, $i, j = 1, \ldots, k$) as $g(z) = \sum_{l=-\infty}^{\infty} \Gamma(l)\, z^{l}$, and the *spectral density* matrix of the stationary process $\{\boldsymbol{Y}_t\}$ as

$$f(\lambda) = \frac{1}{2\pi}\, g(\, e^{-i\lambda}) = \frac{1}{2\pi} \sum_{l=-\infty}^{\infty} \Gamma(l)\, e^{-il\lambda}, \qquad -\pi \le \lambda < \pi . \tag{1.3}$$

The (h, j)th element of $f(\lambda)$, denoted as $f_{hj}(\lambda)$, is $f_{hj}(\lambda) = (2\pi)^{-1} \sum_{l=-\infty}^{\infty} \gamma_{hj}(l)\, e^{-il\lambda}$. For $h = j$, $f_{jj}(\lambda)$ is the (auto)spectral density function of the series Y_{jt}, while for $h \neq j$, $f_{hj}(\lambda)$ is the cross-spectral density function of Y_{ht} and Y_{jt}. Notice that $f_{jj}(\lambda)$ is real-valued and non-negative, but since $\gamma_{hj}(l) \neq \gamma_{hj}(-l)$ for $h \neq j$, the cross-spectral density function $f_{hj}(\lambda)$ is in general complex-valued, with $f_{hj}(\lambda)$ being equal to $f_{jh}(-\lambda)$, the complex conjugate of $f_{jh}(\lambda)$. Therefore, the spectral density matrix $f(\lambda)$ is Hermitian, that is, $f(\lambda) = f(\lambda)^{*} = f(-\lambda)'$, where * denotes the complex conjugate transpose. Moreover, $f(\lambda)$ is a non-negative definite matrix in the sense that $\boldsymbol{b}' f(\lambda)\, \boldsymbol{b} \ge 0$ for any k-dimensional vector $\boldsymbol{b}$, since $\boldsymbol{b}' f(\lambda)\, \boldsymbol{b}$ is the spectral density function of the linear combination $\boldsymbol{b}' \boldsymbol{Y}_t$ and hence must be non-negative. Note also that

$$\Gamma(l) = \int_{-\pi}^{\pi} e^{il\lambda}\, f(\lambda)\, d\lambda, \qquad l = 0, \pm 1, \pm 2, \ldots, \tag{1.4}$$

that is, $\gamma_{hj}(l) = \int_{-\pi}^{\pi} e^{il\lambda}\, f_{hj}(\lambda)\, d\lambda$.

The process $\{\boldsymbol{Y}_t\}$ also has the spectral representation as

$$\boldsymbol{Y}_t = \int_{-\pi}^{\pi} e^{-i\lambda t}\, d\boldsymbol{Z}(\lambda), \qquad \text{or} \qquad Y_{jt} = \int_{-\pi}^{\pi} e^{-i\lambda t}\, dZ_j(\lambda), \quad j = 1, \ldots, k,$$

where $\boldsymbol{Z}(\lambda) = (\, Z_1(\lambda), \ldots, Z_k(\lambda)\,)'$ is a k-dimensional complex-valued continuous-parameter process defined on the continuous interval $[\, -\pi, \pi\,]$, with the property that $E[\, d\boldsymbol{Z}(\lambda_1)\, d\bar{\boldsymbol{Z}}(\lambda_2)'\,] = 0$ if $\lambda_1 \neq \lambda_2$, while $E[\, d\boldsymbol{Z}(\lambda)\, d\bar{\boldsymbol{Z}}(\lambda)'\,] = f(\lambda)\, d\lambda$. Hence, $f(\lambda)\, d\lambda$ represents the covariance matrix of $d\boldsymbol{Z}(\lambda)$, the random vector at frequency λ in the spectral representation of the vector process $\{\boldsymbol{Y}_t\}$. The (*squared*) *coherency* spectrum of a pair of series Y_{it} and Y_{jt} is defined as $K_{ij}^2(\lambda) = |f_{ij}(\lambda)|^2 / \{ f_{ii}(\lambda)\, f_{jj}(\lambda) \}$. The coherency $K_{ij}(\lambda)$ at frequency λ can be interpreted as the correlation coefficient between the random components, $dZ_i(\lambda)$ and $dZ_j(\lambda)$, at frequency λ in the spectral representations of Y_{it} and Y_{jt}. Hence, $K_{ij}(\lambda)$ as a function of λ measures the extent to which the two processes Y_{it} and Y_{jt} are linearly related in terms of the degree of linear association of their random components at different frequencies λ. When spectral relations that involve more than two time series are considered, the related concepts of partial coherency and multiple coherency are also of interest. Although spectral methods will not be emphasized in this book, detailed accounts of the spectral theory and analysis of multivariate time series may be found in Hannan (1970) and Priestley (1981).

1.1.3 Some Relations for Linear Filtering of a Stationary Vector Process

Fundamental to the study of multivariate linear systems of time series is the representation of dynamic linear relationships through the formulation of linear filters. A multivariate *linear (time-invariant) filter* relating an r-dimensional input series $\boldsymbol{X}_t$ to a k-dimensional output series $\boldsymbol{Y}_t$ is given by the form

$$\boldsymbol{Y}_t = \sum_{j=-\infty}^{\infty} \Psi_j \boldsymbol{X}_{t-j},$$

where the Ψ_j are $k \times r$ matrices. The filter is *physically realizable* or *causal* when the $\Psi_j = 0$ for $j < 0$, so that $\boldsymbol{Y}_t = \sum_{j=0}^{\infty} \Psi_j \boldsymbol{X}_{t-j}$ is expressible in terms of only present and past values of the input process $\{\boldsymbol{X}_t\}$. The filter is said to be *stable* if $\sum_{j=-\infty}^{\infty} \| \Psi_j \| < \infty$, where $\| A \|$ denotes a norm for the matrix A such as $\| A \|^2 = \text{tr}\{A' A\}$. Under the stability condition, and an assumption that the input random vectors $\{\boldsymbol{X}_t\}$ have uniformly bounded second moments, the output random vector $\boldsymbol{Y}_t$ defined by $\boldsymbol{Y}_t = \sum_{j=-\infty}^{\infty} \Psi_j \boldsymbol{X}_{t-j}$ exists uniquely and represents the limit in mean square, $\lim_{n\to\infty} \sum_{j=-n}^{n} \Psi_j \boldsymbol{X}_{t-j}$, such that $E[(\boldsymbol{Y}_t - \sum_{j=-n}^{n} \Psi_j \boldsymbol{X}_{t-j})(\boldsymbol{Y}_t - \sum_{j=-n}^{n} \Psi_j \boldsymbol{X}_{t-j})'] \to 0$ as $n \to \infty$. When the filter is stable and the input series $\boldsymbol{X}_t$ is stationary with cross-covariance matrices $\Gamma_x(l)$, the output $\boldsymbol{Y}_t = \sum_{j=-\infty}^{\infty} \Psi_j \boldsymbol{X}_{t-j}$ is a stationary process. [A proof of similar results in the univariate case is given by Fuller (1976, pp. 29–33) or Brockwell and Davis (1987, pp. 83–84), and the arguments extend directly to the vector case.] The cross-covariance matrices of the stationary process $\{\boldsymbol{Y}_t\}$ are then given by

$$\Gamma_y(l) = \text{Cov}(\boldsymbol{Y}_t, \boldsymbol{Y}_{t+l}) = \sum_{i=-\infty}^{\infty} \sum_{j=-\infty}^{\infty} \Psi_i \, \Gamma_x(l+i-j) \, \Psi_j'. \tag{1.5}$$

It also follows, from (1.3), that the spectral density matrix of the output $\boldsymbol{Y}_t$ has the representation

$$f_y(\lambda) = \Psi(e^{i\lambda}) \, f_x(\lambda) \, \Psi(e^{-i\lambda})', \tag{1.6}$$

where $f_x(\lambda)$ is the spectral density matrix of $\boldsymbol{X}_t$, and $\Psi(z) = \sum_{j=-\infty}^{\infty} \Psi_j z^j$ is the *transfer function* (matrix) of the linear filter. In addition, the cross-covariance matrices between the output $\boldsymbol{Y}_t$ and the input $\boldsymbol{X}_t$ are given by

$$\Gamma_{yx}(l) = \text{Cov}(\boldsymbol{Y}_t, \boldsymbol{X}_{t+l}) = \sum_{j=-\infty}^{\infty} \Psi_j \, \Gamma_x(l+j)$$

and the cross-spectral density matrix between $\boldsymbol{Y}_t$ and $\boldsymbol{X}_t$ is

$$f_{yx}(\lambda) = \frac{1}{2\pi} \sum_{l=-\infty}^{\infty} \Gamma_{yx}(l) \, e^{-il\lambda} = \Psi(e^{i\lambda}) \, f_x(\lambda), \tag{1.7}$$

so that the transfer function $\Psi(z)$ satisfies the relation that

$\Psi(e^{i\lambda}) = f_{yx}(\lambda) f_x(\lambda)^{-1}$. These results are most easily seen by noting that the output $\boldsymbol{Y}_t$ of the filter will have the spectral representation

$$\boldsymbol{Y}_t = \sum_{j=-\infty}^{\infty} \Psi_j \left[\int_{-\pi}^{\pi} e^{-i\lambda(t-j)} \, d\boldsymbol{Z}_x(\lambda) \right]$$

$$= \int_{-\pi}^{\pi} \left[\sum_{j=-\infty}^{\infty} \Psi_j \, e^{-i\lambda(t-j)} \right] d\boldsymbol{Z}_x(\lambda) = \int_{-\pi}^{\pi} e^{-i\lambda t} \, d\boldsymbol{Z}_y(\lambda),$$

where $d\boldsymbol{Z}_y(\lambda) = \sum_{j=-\infty}^{\infty} \Psi_j \, e^{ij\lambda} \, d\boldsymbol{Z}_x(\lambda) = \Psi(e^{i\lambda}) \, d\boldsymbol{Z}_x(\lambda)$, so that

$$E[\, d\boldsymbol{Z}_y(\lambda) \, d\bar{\boldsymbol{Z}}_y(\lambda)' \,] = \Psi(e^{i\lambda}) \, E[\, d\boldsymbol{Z}_x(\lambda) \, d\bar{\boldsymbol{Z}}_x(\lambda)' \,] \, \Psi(e^{-i\lambda})'$$

and

$$E[\, d\boldsymbol{Z}_y(\lambda) \, d\bar{\boldsymbol{Z}}_x(\lambda)' \,] = \Psi(e^{i\lambda}) \, E[\, d\boldsymbol{Z}_x(\lambda) \, d\bar{\boldsymbol{Z}}_x(\lambda)' \,].$$

Notice, for example, from (1.6) and (1.7) that if X_t and $Y_t = \sum_{j=-\infty}^{\infty} \Psi_j \, X_{t-j}$ are univariate processes, then the squared coherency between them is

$$K_{yx}^2(\lambda) = |f_{yx}(\lambda)|^2 / \{ f_x(\lambda) \, f_y(\lambda) \}$$

$$= |\Psi(e^{i\lambda})|^2 \, f_x(\lambda)^2 / \{ \, f_x(\lambda) \, |\Psi(e^{i\lambda})| \, f_x(\lambda) \, |\Psi(e^{-i\lambda})| \, \} = 1.$$

This indicates that the squared coherency between an output series Y_t formed by application of a linear filter and the original input series X_t is equal to one at all frequencies. In practice, when a causal linear filter is used to represent the relation between an observable input process $\boldsymbol{X}_t$ and an output process $\boldsymbol{Y}_t$ in a dynamic system, there will be added unobservable noise $\boldsymbol{N}_t$ in the system and a dynamic model of the form $\boldsymbol{Y}_t = \sum_{j=0}^{\infty} \Psi_j \, \boldsymbol{X}_{t-j} + \boldsymbol{N}_t$ will be useful.

For a simple example of the above linear filtering results, consider the basic vector *white noise process* $\{\varepsilon_t\}$, which is defined to be a process with the properties that $E(\varepsilon_t) = 0$, $E(\varepsilon_t \varepsilon_t') = \Sigma$ which is a $k \times k$ covariance matrix assumed to be positive-definite, and $E(\varepsilon_t \varepsilon_{t+l}') = 0$ for $l \neq 0$. Hence, ε_t has spectral density matrix $f_\varepsilon(\lambda) = (2\pi)^{-1} \Sigma$. Then the process $\boldsymbol{Y}_t = \sum_{j=0}^{\infty} \Psi_j \, \varepsilon_{t-j}$, with $\sum_{j=0}^{\infty} \| \Psi_j \| < \infty$, is stationary and has cross-covariance matrices

$$\Gamma_y(l) = \sum_{j=0}^{\infty} \Psi_j \Sigma \Psi_{j+l}'$$

and spectral density matrix

$$f_y(\lambda) = (2\pi)^{-1} \, \Psi(e^{i\lambda}) \, \Sigma \, \Psi(e^{-i\lambda})',$$

and the cross-covariance matrices between $\{\boldsymbol{Y}_t\}$ and $\{\varepsilon_t\}$ are $\Gamma_{y\varepsilon}(l) = \Psi_{-l} \, \Sigma$ for $l \leq 0$ and zero for $l > 0$.

1.2 Linear Model Representations for a Stationary Vector Process

1.2.1 Infinite Moving Average (Wold) Representation of a Stationary Vector Process

A multivariate generalization of Wold's Theorem states that if $\{\boldsymbol{Y}_t\}$ is a purely nondeterministic (i.e., $\boldsymbol{Y}_t$ does not contain any purely deterministic component process whose future values can be perfectly predicted from the past values) stationary process with (constant) mean vector $\boldsymbol{\mu}$, then $\boldsymbol{Y}_t - \boldsymbol{\mu}$ can always be represented as the output of a causal linear filter with white noise input. [For a proof of this result in the univariate case, see Anderson (1971, pp. 420–421) or Brockwell and Davis (1987, pp. 180–182); also see Hannan (1970, pp. 157–158) for the vector case.] Thus, according to this result, $\boldsymbol{Y}_t$ can be represented as an infinite vector moving average (MA) process,

$$\boldsymbol{Y}_t = \boldsymbol{\mu} + \sum_{j=0}^{\infty} \Psi_j\, \varepsilon_{t-j} = \boldsymbol{\mu} + \Psi(B)\, \varepsilon_t\,, \qquad \Psi_0 = I\,, \tag{1.8}$$

where $\Psi(B) = \sum_{j=0}^{\infty} \Psi_j B^j$ is a $k \times k$ matrix in the backshift operator B, such that $B^j\, \varepsilon_t = \varepsilon_{t-j}$, with the typical (i, j)th element of $\Psi(B)$ given by $\psi_{ij}(B) = \sum_{k=0}^{\infty} c_{ij}(k)\, B^k$, and the coefficients Ψ_j are not necessarily absolutely summable but do satisfy the weaker condition $\sum_{j=0}^{\infty} \| \Psi_j \|^2 < \infty$.

In (1.8), the ε_t form a vector white noise process with $\varepsilon_t = (\varepsilon_{1t}, \ldots, \varepsilon_{kt})'$ such that $E(\varepsilon_t) = 0$, $E(\varepsilon_t\, \varepsilon_t') = \Sigma$, and $E(\varepsilon_t\, \varepsilon_{t+l}') = 0$ for $l \neq 0$. Sometimes, as will be indicated, additional properties will be assumed for the ε_t, such as mutual independence over different times. The Wold representation in (1.8) is obtained by defining ε_t as the error of the best one-step ahead linear predictor of $\boldsymbol{Y}_t$ based on the infinite past $\boldsymbol{Y}_{t-1}, \boldsymbol{Y}_{t-2}, \ldots$, $\varepsilon_t = \boldsymbol{Y}_t - \hat{\boldsymbol{Y}}_{t-1}(1)$, where $\hat{\boldsymbol{Y}}_{t-1}(1)$ is the best linear predictor of $\boldsymbol{Y}_t$ based on $\boldsymbol{Y}_{t-1}, \boldsymbol{Y}_{t-2}, \ldots$ [$\hat{\boldsymbol{Y}}_{t-1}(1)$ is the unique projection of $\boldsymbol{Y}_t$ onto the linear space spanned by $\boldsymbol{Y}_{t-1}, \boldsymbol{Y}_{t-2}, \ldots$]. So the ε_t are mutually uncorrelated by construction, since ε_t is uncorrelated with $\boldsymbol{Y}_{t-j}$ for all $j \geq 1$ and, hence, is uncorrelated with ε_{t-j} for all $j \geq 1$, and the ε_t have a constant covariance matrix by stationarity of the process $\{\boldsymbol{Y}_t\}$. The best one-step ahead linear predictor (projection) can be expressed as

$$\hat{\boldsymbol{Y}}_{t-1}(1) = \boldsymbol{\mu} + \sum_{j=1}^{\infty} \Psi_j\, (\boldsymbol{Y}_{t-j} - \hat{\boldsymbol{Y}}_{t-j-1}(1)) = \boldsymbol{\mu} + \sum_{j=1}^{\infty} \Psi_j\, \varepsilon_{t-j}\,.$$

Consequently, as in the previous example, we note that the coefficient matrices Ψ_j in (1.8) have the interpretation of the linear regression matrices of $\boldsymbol{Y}_t$ on the ε_{t-j} in that $\Psi_j = \mathrm{Cov}(\boldsymbol{Y}_t, \varepsilon_{t-j})\, \Sigma^{-1}$.

1.2.2 Vector Autoregressive Moving Average (ARMA) Model Representations

Suppose the matrix $\Psi(B)$ in (1.8) can be represented (at least approximately, in practice) as the product of two matrices in the form $\Phi(B)^{-1}\Theta(B)$, where $\Phi(B)$ and $\Theta(B)$ are each finite order matrix polynomials in B. Then we are led to consider a class of linear models for vector time series $\boldsymbol{Y}_t$ defined by a relation of the form

$$(\boldsymbol{Y}_t - \boldsymbol{\mu}) - \sum_{j=1}^{p} \Phi_j (\boldsymbol{Y}_{t-j} - \boldsymbol{\mu}) = \varepsilon_t - \sum_{j=1}^{q} \Theta_j \varepsilon_{t-j}, \tag{1.9}$$

or $\Phi(B)(\boldsymbol{Y}_t - \boldsymbol{\mu}) = \Theta(B)\varepsilon_t$, where $\Phi(B) = I - \Phi_1 B - \Phi_2 B^2 - \cdots - \Phi_p B^p$, $\Theta(B) = I - \Theta_1 B - \Theta_2 B^2 - \cdots - \Theta_q B^q$ are matrix polynomials in B, the Φ_i and Θ_i are $k \times k$ matrices and ε_t is a vector white noise process with mean vector 0 and covariance matrix $\Sigma = E(\varepsilon_t \varepsilon_t')$. Notice that we can always take the leading matrix coefficient in (1.8) as $\Psi_0 = I$, and, hence, the leading coefficient matrices of the operators $\Phi(B)$ and $\Theta(B)$ in (1.9) can also be taken to be identity matrices without any loss of generality. This model representation (1.9) is a natural generalization of the univariate ARMA(p,q) models, and these vector ARMA(p,q) models will be found to be useful in the modeling of vector time series. A process $\{\boldsymbol{Y}_t\}$ will be referred to as a vector *ARMA(p,q) process* if it satisfies the relations in (1.9), at least for all t greater than some initial time origin, for given white noise sequence $\{\varepsilon_t\}$, regardless of whether the process $\{\boldsymbol{Y}_t\}$ is stationary or not. A stationary vector ARMA(p,q) process $\{\boldsymbol{Y}_t\}$ defined by the relations (1.9) is said to be *causal* if it can be represented as in (1.8), $\boldsymbol{Y}_t = \boldsymbol{\mu} + \sum_{j=0}^{\infty} \Psi_j \varepsilon_{t-j}$ for all $t = 0, \pm 1, \ldots,$ with $\sum_{j=0}^{\infty} \| \Psi_j \| < \infty$. A vector ARMA($p$,$q$) process is said to be *invertible* if it can be represented in the form $\boldsymbol{Y}_t - \boldsymbol{\mu} = \sum_{j=1}^{\infty} \Pi_j (\boldsymbol{Y}_{t-j} - \boldsymbol{\mu}) + \varepsilon_t$, with $\sum_{j=1}^{\infty} \| \Pi_j \| < \infty$.

It will be shown later that, if the roots of $\det\{ \Phi(B) \} = 0$ all are greater than one in absolute value, then a process $\{\boldsymbol{Y}_t\}$ that satisfies the ARMA(p,q) model relations (1.9) for all t can be determined which will be *stationary*, with $\{\boldsymbol{Y}_t\}$ possessing the causal infinite MA representation as in (1.8) and with $\Psi(B) = \Phi(B)^{-1}\Theta(B)$ representing a convergent matrix series for $| B | \le 1$. Also, if the roots of $\det\{ \Theta(B) \} = 0$ all are greater than one in absolute value, then the process is *invertible* with $\Pi(B) = \Theta(B)^{-1}\Phi(B) = I - \sum_{j=1}^{\infty} \Pi_j B^j$ representing a convergent matrix series for $| B | \le 1$ and $\{\boldsymbol{Y}_t\}$ possessing the infinite autoregressive (AR) representation $\Pi(B)(\boldsymbol{Y}_t - \boldsymbol{\mu}) = \varepsilon_t$, or

$$(\boldsymbol{Y}_t - \boldsymbol{\mu}) - \sum_{j=1}^{\infty} \Pi_j (\boldsymbol{Y}_{t-j} - \boldsymbol{\mu}) = \varepsilon_t . \tag{1.10}$$

That is, a stationary and causal solution $\boldsymbol{Y}_t = \boldsymbol{\mu} + \sum_{j=0}^{\infty} \Psi_j \varepsilon_{t-j}$ to the ARMA(p,q) model relations (1.9) exists for all t if the roots of $\det\{ \Phi(B) \} = 0$ all are greater than one in absolute value, and the process is invertible with infinite AR representation $\boldsymbol{Y}_t - \boldsymbol{\mu} = \sum_{j=1}^{\infty} \Pi_j (\boldsymbol{Y}_{t-j} - \boldsymbol{\mu}) + \varepsilon_t$ if the roots of $\det\{ \Theta(B) \} = 0$ all are greater than one in absolute value.

On occasion, it may be useful to express the vector ARMA(p,q) model (1.9) in the following slightly different form. Since $\Sigma = E(\varepsilon_t \varepsilon_t')$ is assumed to be positive-definite, there exists a lower triangular matrix $\Phi_0^{\#}$ with ones on the diagonals such that $\Phi_0^{\#} \Sigma \Phi_0^{\#'} = \Sigma^{\#}$ is a diagonal matrix with positive diagonal elements. Hence, by multiplying through equation (1.9) by $\Phi_0^{\#}$, we obtain the alternate form

$$\Phi_0^{\#}(\boldsymbol{Y}_t - \boldsymbol{\mu}) - \sum_{j=1}^{p} \Phi_j^{\#} (\boldsymbol{Y}_{t-j} - \boldsymbol{\mu}) = \boldsymbol{a}_t - \sum_{j=1}^{q} \Theta_j^{\#} \boldsymbol{a}_{t-j}, \tag{1.11}$$

where $\Phi_j^{\#} = \Phi_0^{\#} \Phi_j$, $\Theta_j^{\#} = \Phi_0^{\#} \Theta_j \Phi_0^{\#-1}$, and $\boldsymbol{a}_t = \Phi_0^{\#} \varepsilon_t$ has diagonal covariance matrix $\Sigma^{\#} = \text{Cov}(\boldsymbol{a}_t) = \text{Diag}(\sigma_{a_1}^2, \sigma_{a_2}^2, \ldots, \sigma_{a_k}^2)$. This representation explicitly displays the contemporaneous relationships among the components of the vector series $\boldsymbol{Y}_t$ through the lower triangular coefficient matrix $\Phi_0^{\#}$, with diagonal covariance matrix for the errors $\boldsymbol{a}_t$, whereas model (1.9) places the information on the contemporaneous relationships among the components of $\boldsymbol{Y}_t$ in the general covariance matrix Σ of the errors ε_t. Similarly, the model (1.9) can always be expressed in an equivalent form as

$$(\boldsymbol{Y}_t - \boldsymbol{\mu}) - \sum_{j=1}^{p} \Phi_j (\boldsymbol{Y}_{t-j} - \boldsymbol{\mu}) = \Theta_0^{+} \boldsymbol{a}_t - \sum_{j=1}^{q} \Theta_j^{+} \boldsymbol{a}_{t-j}, \tag{1.12}$$

where $\Theta_0^{+} = \Phi_0^{\#-1}$, $\Theta_j^{+} = \Theta_j \Phi_0^{\#-1}$, and $\boldsymbol{a}_t = \Phi_0^{\#} \varepsilon_t$ as in (1.11), with diagonal covariance matrix $\Sigma^{\#}$ and lower triangular coefficient matrix Θ_0^{+}.

More generally, for that matter, the vector ARMA(p,q) model (1.9) can always be expressed in the equivalent form as

$$\Phi_0^{\#}(\boldsymbol{Y}_t - \boldsymbol{\mu}) - \sum_{j=1}^{p} \Phi_j^{\#} (\boldsymbol{Y}_{t-j} - \boldsymbol{\mu}) = \Theta_0^{\#} \varepsilon_t - \sum_{j=1}^{q} \Theta_j^{\#} \varepsilon_{t-j}, \tag{1.13}$$

where $\Phi_0^{\#}$ is an arbitrary nonsingular matrix and $\Sigma = \text{Cov}(\varepsilon_t)$ is a general positive-definite matrix, with $\Theta_0^{\#} = \Phi_0^{\#}$, $\Phi_j^{\#} = \Phi_0^{\#} \Phi_j$, and $\Theta_j^{\#} = \Phi_0^{\#} \Theta_j$. Intuitively, for purposes of parsimony in the number of unknown coefficient parameters that need to be estimated in the model, we would be interested in models of the form (1.13) for the identification of the $\Phi_0^{\#}$ which leads to the simplest structure in some sense, such as in terms of the number of unknown parameters, for the coefficient matrices $\Phi_1^{\#}, \ldots, \Phi_p^{\#}$, $\Theta_1^{\#}, \ldots, \Theta_q^{\#}$. For unique identifiability of the parameters of the model in the form (1.13), it will be necessary to constrain (normalize) the form of $\Phi_0^{\#}$ at least to be lower triangular with ones on the diagonal. As will be discussed in Section 3.1, a representation of an ARMA model in a certain special form of (1.13) can prove to be more useful in certain circumstances than the "standard" or "reduced form" representation (1.9), and this special form of (1.13) will be referred to as the echelon canonical form of the ARMA model.

An alternate formulation of model for vector time series is in the form of a multiple input–multiple output "transfer function" form as

$$Y_{it} = \sum_{j \neq i} \frac{\omega_{ij}(B)}{\delta_{ij}(B)} Y_{jt} + \frac{\theta_i(B)}{\phi_i(B)} \varepsilon_{it}, \qquad i = 1, \ldots, k, \tag{1.14}$$

where the $\omega_{ij}(B)$, $\delta_{ij}(B)$, $\theta_i(B)$, and $\phi_i(B)$ are finite order polynomial operators in B. However, by multiplying through these equations by the appropriate operators $\delta_{ij}(B)$ and $\phi_i(B)$, it is easy to verify that models of this form may always be expressed in the form of a vector ARMA(p,q) model (1.9), with a different parameterization and a diagonal moving average structure. Conversely, if in the vector ARMA(p,q) model (1.9) we express the inverse of the MA operator $\Theta(B)$ as $\Theta(B)^{-1} = (1/\det\{\Theta(B)\})\,\Theta^*(B)$, where $\Theta^*(B)$ is the adjoint of the matrix $\Theta(B)$ (and hence contains elements that are finite polynomials in B), then the vector ARMA(p,q) model (1.9) can be written as

$$\Theta^*(B)\,\Phi(B)\,(\boldsymbol{Y}_t - \boldsymbol{\mu}) = \det\{\Theta(B)\}\,\varepsilon_t .$$

From these relations, equations of the multiple input–multiple output transfer function form can be obtained by dividing through each equation by the appropriate diagonal element operator of $\Theta^*(B)\,\Phi(B)$.

Now consider a circumstance where the vector process $\boldsymbol{Y}_t$ is partitioned into two groups of subcomponents of dimensions k_1 and k_2, respectively ($k_1 + k_2 = k$), as $\boldsymbol{Y}_t = (\boldsymbol{Y}_{1t}', \boldsymbol{Y}_{2t}')'$, with the white noise process $\varepsilon_t = (\varepsilon_{1t}', \varepsilon_{2t}')'$ partitioned similarly, and partition the AR and MA operators in the vector ARMA model (1.9) accordingly as

$$\Phi(B) = \begin{bmatrix} \Phi_{11}(B) & \Phi_{12}(B) \\ \Phi_{21}(B) & \Phi_{22}(B) \end{bmatrix}, \qquad \Theta(B) = \begin{bmatrix} \Theta_{11}(B) & \Theta_{12}(B) \\ \Theta_{21}(B) & \Theta_{22}(B) \end{bmatrix}.$$

Suppose in this situation that $\Phi_{12}(B)$ and $\Theta_{12}(B)$ are both identically zero, and for convenience also assume that $\Theta_{21}(B) = 0$. Then the equations for this model can be expressed in two distinct groups as

$$\Phi_{11}(B)\,\boldsymbol{Y}_{1t} = \Theta_{11}(B)\,\varepsilon_{1t} \tag{1.15a}$$

and

$$\Phi_{22}(B)\,\boldsymbol{Y}_{2t} = -\,\Phi_{21}(B)\,\boldsymbol{Y}_{1t} + \Theta_{22}(B)\,\varepsilon_{2t} . \tag{1.15b}$$

Notice that even if $\Theta_{21}(B) \neq 0$ in the above partitioned structure, we could still arrive at a model equation of a similar form to (1.15b), since the additional term in (1.15b) would be $\Theta_{21}(B)\,\varepsilon_{1t}$ and using (1.15a) this is equal to $\Theta_{21}(B)\,\Theta_{11}(B)^{-1}\Phi_{11}(B)\,\boldsymbol{Y}_{1t}$. In the terminology of causality from econometrics, under the above structure the variables $\boldsymbol{Y}_{1t}$ are said to cause $\boldsymbol{Y}_{2t}$, but $\boldsymbol{Y}_{2t}$ do not cause $\boldsymbol{Y}_{1t}$. In this circumstance, the variables $\boldsymbol{Y}_{1t}$ are referred to as *exogenous variables*, and the model equation (1.15b) is often referred to as an ARMAX model or ARMAX system (the X stands for exogenous) for the output variables $\boldsymbol{Y}_{2t}$ with $\boldsymbol{Y}_{1t}$ as the input (exogenous) variables. The implication of this model structure is that future values of the process $\boldsymbol{Y}_{1t}$ are only influenced by its own past and not by the past of $\boldsymbol{Y}_{2t}$, whereas future values of

$\boldsymbol{Y}_{2t}$ are influenced by the past of both $\boldsymbol{Y}_{1t}$ and $\boldsymbol{Y}_{2t}$. [In an econometric setting, the output variables $\boldsymbol{Y}_{2t}$ might tend to be the variables of most interest, and, hence, equation (1.15b) would represent the model structure of most interest, while equation (1.15a) may not be of direct interest.] On multiplication of equation (1.15b) on the left by $\Phi_{22}(B)^{-1}$, we obtain the representation

$$\boldsymbol{Y}_{2t} = -\,\Phi_{22}(B)^{-1}\Phi_{21}(B)\,\boldsymbol{Y}_{1t} + \Phi_{22}(B)^{-1}\Theta_{22}(B)\,\varepsilon_{2t}$$

$$\equiv \Psi_*(B)\,\boldsymbol{Y}_{1t} + \Psi_{22}(B)\,\varepsilon_{2t}\,, \tag{1.16}$$

where $\Psi_*(B) = -\,\Phi_{22}(B)^{-1}\Phi_{21}(B)$ and $\Psi_{22}(B) = \Phi_{22}(B)^{-1}\Theta_{22}(B)$. This equation may be viewed as providing a representation for the output process $\boldsymbol{Y}_{2t}$ as a causal linear filter of the input process $\boldsymbol{Y}_{1t}$, as discussed earlier in Section 1.1.3, with added unobservable noise process, that is, as $\boldsymbol{Y}_{2t} = \Psi_*(B)\,\boldsymbol{Y}_{1t} + \boldsymbol{N}_t$ where $\boldsymbol{N}_t = \Phi_{22}(B)^{-1}\Theta_{22}(B)\,\varepsilon_{2t}$ is the noise process which follows a vector ARMA model $\Phi_{22}(B)\,\boldsymbol{N}_t = \Theta_{22}(B)\,\varepsilon_{2t}$. The above ARMAX model structure will not be explicitly emphasized in this book, but it is apparent that most of the analysis methods and results for vector ARMA models readily extend to the case of vector ARMAX models.

Finally, we briefly mention here another way in which the model for a vector ARMA process $\boldsymbol{Y}_t$ can be represented. This alternate representation is in the form of a (time-invariant) *state-variable* or *state-space* model, which consists of an observation equation $\boldsymbol{Y}_t = H\,\boldsymbol{Z}_t + \boldsymbol{N}_t$ and a transition or state equation $\boldsymbol{Z}_t = \Phi\,\boldsymbol{Z}_{t-1} + \boldsymbol{a}_t$, where $\boldsymbol{Z}_t$ is an $r \times 1$ unobservable time series vector called the state vector and $\boldsymbol{N}_t$ and $\boldsymbol{a}_t$ are independent white noise processes. In this representation, the state vector $\boldsymbol{Z}_t$ conceptually contains all information from the past of the process $\boldsymbol{Y}_t$ which is relevant for the future of the process, and, hence, the dynamics of the system can be represented in the simple first-order or Markovian transition equation for the state vector. A general presentation of the state-space models will be given in Chapter 7, and in Section 7.2 it will be shown in particular that every stationary vector ARMA(p,q) process as in (1.9) can be represented in the form of the above state-space model. The above state-space model is said to be stable if all the eigenvalues of the matrix Φ are less than one in absolute value, and conversely, it can be shown that any stationary process $\boldsymbol{Y}_t$ which has a stable state-space representation of the above form can also be represented in the form of a stationary vector ARMA(p,q) model as in (1.9) [e.g., see Akaike (1974c)]. Hence, it follows that any process $\boldsymbol{Y}_t$ which satisfies a stable state-space representation can be expressed in the causal convergent infinite moving average form $\boldsymbol{Y}_t = \Psi(B)\,\varepsilon_t$ as in (1.8). The stability condition for the matrix Φ in the state-space model is equivalent to the stability condition stated in Section 1.1.3 for the matrix coefficients Ψ_j of the linear filter $\Psi(B)$, since it ensures that $\sum_{j=0}^{\infty} \|\Psi_j\| < \infty$ in the representation $\boldsymbol{Y}_t = \Psi(B)\,\varepsilon_t$.

Detailed properties of the vector ARMA(p,q) processes will be investigated in the next chapter, and more explicit relations between the different forms of models will be examined in detail for some special cases.

APPENDIX A1

Review of Multivariate Normal Distribution and Related Topics

In this appendix, we give a very brief summary and review of some basic properties of matrices, random vectors and their mean vectors and covariance matrices, the multivariate normal distribution, and stochastic convergence of sequences of random vectors, which may be useful on various occasions throughout this book. Much more detailed accounts of these and related topics may be found in books such as Graybill (1969), Searle (1982), Anderson (1984), Srivastava and Khatri (1979), and Serfling (1980).

A1.1 Review of Some Basic Matrix Theory Results

For a square $k \times k$ matrix A, the minor of the (i, j)th element a_{ij} of A is the determinant of the $(k-1) \times (k-1)$ submatrix of A obtained by deleting the ith row and jth column. The cofactor of a_{ij}, denoted as A_{ij}, say, is $(-1)^{i+j}$ times the minor of a_{ij}. The *adjoint* of A, denoted by $A^* = \text{Adj}(A)$, is the $k \times k$ matrix of cofactors whose (i, j)th element is the cofactor A_{ji} of A. The adjoint matrix of A has the property that $A\,\text{Adj}(A) = \det(A)\,I$, where $\det(A) = |A|$ denotes the determinant of the matrix A, since

$$\det(A) = \sum_{i=1}^{k} a_{ij} A_{ij} = \sum_{j=1}^{k} a_{ij} A_{ij}$$

for each $i = 1, \ldots, k$ and each $j = 1, \ldots, k$. Thus, if $\det(A) \neq 0$, then the matrix A is *nonsingular* and its inverse is given by $A^{-1} = (1/\det(A))\,\text{Adj}(A)$.

For a square $k \times k$ matrix A, an *eigenvalue* λ_i of A is a scalar such that $A\,\boldsymbol{x}_i = \lambda_i\,\boldsymbol{x}_i$ for some $k \times 1$ vector $\boldsymbol{x}_i \neq 0$, and $\boldsymbol{x}_i$ is called an *eigenvector* corresponding to the eigenvalue λ_i. Thus, since the eigenvalues λ_i and corresponding eigenvectors $\boldsymbol{x}_i$ of A must satisfy the equation $(\lambda_i I - A)\,\boldsymbol{x}_i = 0$, it follows that the eigenvalues of A are the roots of the characteristic equation $\det\{\lambda I - A\} = 0$. It is also seen that A is nonsingular (A^{-1} exists) if and only if A has no eigenvalues equal to zero or, equivalently, if and only if $\det\{A\} \neq 0$; more generally, the rank of the $k \times k$ matrix A is equal to the number of nonzero eigenvalues of A. For any symmetric matrix A, there exists an orthogonal matrix P (i.e., $P'P = I$) such that $A = P \Lambda P'$ or, equivalently, $P'A\,P = \Lambda$, where $\Lambda = \text{Diag}(\lambda_1, \ldots, \lambda_k)$ is a diagonal matrix whose diagonal elements λ_i are the corresponding eigenvalues of A. The column vectors P_i of the matrix P are the corresponding eigenvectors of A, satisfying $A\,P_i = \lambda_i\,P_i$, and the P_i form an orthonormal set of eigenvectors of A. When A is a $k \times k$ nonsymmetric matrix, a similar decomposition as $P^{-1}A\,P = \Lambda = \text{Diag}(\lambda_1, \ldots, \lambda_k)$, with P nonsingular but no longer orthogonal, exists whenever A possesses k linearly independent eigenvectors (e.g., when the eigenvalues λ_i of A are all distinct). Generally, we always have the decomposition $P^{-1} A\, P = J$, where J is the Jordan canonical form corresponding to A.

A symmetric $k \times k$ matrix A is called non-negative-definite if $\boldsymbol{x}' A\, \boldsymbol{x} \geq 0$ for all $k \times 1$ vectors $\boldsymbol{x}$, and A is *positive-definite* if $\boldsymbol{x}' A\, \boldsymbol{x} > 0$ for all $\boldsymbol{x} \neq 0$. From above, the symmetric matrix A has the decomposition such that $P' A\, P = \Lambda$, and it follows that A is non-negative-definite if and only if all its eigenvalues λ_i are non-negative, and is positive-definite if and only if all its eigenvalues are positive. In the case where A is positive-definite, $\lambda_i > 0$ for all i, and we can write $\Lambda^{1/2} = \text{Diag}(\, \lambda_1^{1/2}, \ldots, \lambda_k^{1/2}\,)$. Then we have $\Lambda^{-1/2} P' A\, P\, \Lambda^{-1/2} = I$, or $C' A\, C = I$, where $C = P\, \Lambda^{-1/2}$ is nonsingular. It follows that any positive-definite matrix A has such a representation or, equivalently, that A can be expressed as $A = B' B$ (with $B = C^{-1} = \Lambda^{1/2} P'$), where B is nonsingular.

A1.2 Expected Values and Covariance Matrices of Random Vectors

Let $\boldsymbol{Y} = (\, Y_1, \ldots, Y_k\,)'$ denote a k-dimensional random vector. If $\mu_i = E(\, Y_i\,)$ is the mean or expected value of Y_i, for $i = 1, \ldots, k$, we define the *mean vector* of $\boldsymbol{Y}$ as the $k \times 1$ vector $\boldsymbol{\mu}_y = E(\, \boldsymbol{Y}\,) = (\, \mu_1, \mu_2, \ldots, \mu_k\,)'$. In a similar way, we define the expected value of any matrix whose elements are random variables to be the matrix with each random variable replaced by its corresponding expected value. If $\boldsymbol{Y}$ and $\boldsymbol{X} = (\, X_1, \ldots, X_m\,)'$ are both random vectors, the *covariance matrix* of $\boldsymbol{Y}$ and $\boldsymbol{X}$ is

$$\Sigma_{yx} = \text{Cov}(\, \boldsymbol{Y}, \boldsymbol{X}\,) = E[\, (\, \boldsymbol{Y} - \boldsymbol{\mu}_y\,)\, (\, \boldsymbol{X} - \boldsymbol{\mu}_x\,)'\,].$$

The $(\, i, j\,)$th element of Σ_{yx} is the covariance between Y_i and X_j, $\text{Cov}(\, Y_i, X_j\,) = E[\, (\, Y_i - E(Y_i)\,)\, (\, X_j - E(X_j)\,)\,]$. In particular, the covariance matrix of $\boldsymbol{Y}$ is $\Sigma_{yy} = \text{Cov}(\, \boldsymbol{Y}\,) = E[\, (\, \boldsymbol{Y} - \boldsymbol{\mu}_y\,)\, (\, \boldsymbol{Y} - \boldsymbol{\mu}_y\,)'\,]$. If $\boldsymbol{c}$ is a k-dimensional vector of constants and A is an $n \times k$ matrix of constants, then the random vector $\boldsymbol{Z} = A\, \boldsymbol{Y} + \boldsymbol{c}$ has mean vector $E(\, \boldsymbol{Z}\,) = A\; E(\, \boldsymbol{Y}\,) + \boldsymbol{c}$ and covariance matrix $\text{Cov}(\, \boldsymbol{Z}\,) = A\; \text{Cov}(\, \boldsymbol{Y}\,)\, A'$. More generally, the covariance matrix of the random vectors $\boldsymbol{Z} = A\, \boldsymbol{Y} + \boldsymbol{c}$ and $\boldsymbol{W} = B\, \boldsymbol{X} + \boldsymbol{d}$ is $\text{Cov}(\, \boldsymbol{Z}, \boldsymbol{W}\,) = A\; \text{Cov}(\, \boldsymbol{Y}, \boldsymbol{X}\,)\, B' = A\, \Sigma_{yx}\, B'$.

The covariance matrix Σ_{yy} of a random vector $\boldsymbol{Y}$ is a symmetric and non-negative-definite matrix, since $\text{Var}(\, \boldsymbol{b}' \boldsymbol{Y}\,) = \boldsymbol{b}'\, \Sigma_{yy}\, \boldsymbol{b} \geq 0$ for all k-dimensional vectors $\boldsymbol{b}$. Therefore, Σ_{yy} can always be expressed in the form $\Sigma_{yy} = P \Lambda P'$, where P is an orthogonal matrix whose columns are an orthonormal set of eigenvectors of Σ_{yy} and $\Lambda = \text{Diag}(\, \lambda_1, \ldots, \lambda_k\,)$ is a diagonal matrix whose diagonal elements λ_i are the corresponding eigenvalues of Σ_{yy}. The matrix Σ_{yy} also possesses the *Cholesky decomposition* in the form $\Sigma_{yy} = R\, D\, R'$, where R is a lower triangular matrix with ones on the diagonal and D is a diagonal matrix with non-negative diagonal elements. Therefore, by the previous result, the random vector $\boldsymbol{Z} = R^{-1} \boldsymbol{Y}$ has diagonal covariance matrix $\text{Cov}(\, \boldsymbol{Z}\,) = R^{-1} \Sigma_{yy}\, R'^{-1} = D$.

A1.3 The Multivariate Normal Distribution

The random vector $\boldsymbol{Y}$ is said to have a (nonsingular) *multivariate normal* distribution with mean vector $\boldsymbol{\mu}_y$ and (positive-definite) covariance matrix Σ_{yy}, and we say $\boldsymbol{Y}$ is distributed as $N(\boldsymbol{\mu}_y, \Sigma_{yy})$, if $\boldsymbol{Y}$ has a joint probability density function (p.d.f.) of the form

$$f(\boldsymbol{y}) = (2\pi)^{-k/2}\, |\Sigma_{yy}|^{-1/2} \exp[\,(-1/2)(\boldsymbol{y} - \boldsymbol{\mu}_y)'\, \Sigma_{yy}^{-1} (\boldsymbol{y} - \boldsymbol{\mu}_y)\,], \qquad \text{(A1.1)}$$

where $|\Sigma_{yy}|$ denotes the determinant of Σ_{yy}. It follows that if $\boldsymbol{Y}$ is $N(\boldsymbol{\mu}_y, \Sigma_{yy})$, and if A is a $n \times k$ matrix of full rank $n \le k$, then $\boldsymbol{Z} = A\,\boldsymbol{Y} + \boldsymbol{c}$ is distributed as $N(A\,\boldsymbol{\mu}_y + \boldsymbol{c}, A\,\Sigma_{yy}\,A')$.

Marginal and Conditional Distributions for the Normal Distribution

Suppose $\boldsymbol{Y}$ is distributed as $N(\boldsymbol{\mu}_y, \Sigma_{yy})$, and we partition as $\boldsymbol{Y} = (\boldsymbol{Y}_{(1)}', \boldsymbol{Y}_{(2)}')'$, where $\boldsymbol{Y}_{(1)}$ and $\boldsymbol{Y}_{(2)}$ are $k_1 \times 1$ and $k_2 \times 1$ vectors, respectively ($k_1 + k_2 = k$). Then we have the corresponding partitions of the mean vector and covariance matrix of $\boldsymbol{Y}$ as $\boldsymbol{\mu}_y = (\boldsymbol{\mu}_{(1)}', \boldsymbol{\mu}_{(2)}')'$ where $\boldsymbol{\mu}_{(i)} = E(\boldsymbol{Y}_{(i)})$ for $i = 1, 2$, and

$$\Sigma_{yy} = \begin{bmatrix} \Sigma_{11} & \Sigma_{12} \\ \Sigma_{21} & \Sigma_{22} \end{bmatrix}, \qquad \text{(A1.2)}$$

where $\Sigma_{ij} = \text{Cov}(\boldsymbol{Y}_{(i)}, \boldsymbol{Y}_{(j)}) = E[\,(\boldsymbol{Y}_{(i)} - \boldsymbol{\mu}_{(i)})(\boldsymbol{Y}_{(j)} - \boldsymbol{\mu}_{(j)})'\,]$. Then it is known that the marginal distribution of $\boldsymbol{Y}_{(i)}$ is $N(\boldsymbol{\mu}_{(i)}, \Sigma_{ii})$, for $i = 1, 2$. Also, $\boldsymbol{Y}_{(1)}$ and $\boldsymbol{Y}_{(2)}$ are independent if and only if $\Sigma_{12} = \text{Cov}(\boldsymbol{Y}_{(1)}, \boldsymbol{Y}_{(2)}) = 0$ since then $|\Sigma_{yy}| = |\Sigma_{11}|\,|\Sigma_{22}|$ and $\Sigma_{yy}^{-1} = \text{Diag}(\Sigma_{11}^{-1}, \Sigma_{22}^{-1})$ so that the joint density in (A1.1) is readily seen to equal the product $f_1(\boldsymbol{y}_{(1)})\,f_2(\boldsymbol{y}_{(2)})$ of the marginal p.d.f.'s of $\boldsymbol{Y}_{(1)}$ and $\boldsymbol{Y}_{(2)}$.

In addition, the conditional distribution of $\boldsymbol{Y}_{(2)}$ given $\boldsymbol{Y}_{(1)}$ is multivariate normal with (conditional) mean vector

$$E(\boldsymbol{Y}_{(2)} \mid \boldsymbol{Y}_{(1)}) = \boldsymbol{\mu}_{(2)} + \Sigma_{21}\,\Sigma_{11}^{-1}(\boldsymbol{Y}_{(1)} - \boldsymbol{\mu}_{(1)}) \qquad \text{(A1.3)}$$

and (conditional) covariance matrix

$$\text{Cov}(\boldsymbol{Y}_{(2)} \mid \boldsymbol{Y}_{(1)}) = \Sigma_{22} - \Sigma_{21}\,\Sigma_{11}^{-1}\,\Sigma_{12}. \qquad \text{(A1.4)}$$

Thus, we can also see that the random vector $\boldsymbol{Z}_{(2)} = \boldsymbol{Y}_{(2)} - \boldsymbol{\mu}_{(2)} - \Sigma_{21}\,\Sigma_{11}^{-1}(\boldsymbol{Y}_{(1)} - \boldsymbol{\mu}_{(1)})$ is independent of $\boldsymbol{Y}_{(1)}$ since the covariance matrix between these two random vectors $\boldsymbol{Y}_{(1)}$ and $\boldsymbol{Z}_{(2)}$ is zero, $\text{Cov}(\boldsymbol{Z}_{(2)}, \boldsymbol{Y}_{(1)}) = \Sigma_{21} - \Sigma_{21}\,\Sigma_{11}^{-1}\,\Sigma_{11} = 0$, and $\boldsymbol{Z}_{(2)}$ has covariance matrix

$$\text{Cov}(\boldsymbol{Z}_{(2)}) = \text{Cov}(\boldsymbol{Y}_{(2)} - \Sigma_{21}\,\Sigma_{11}^{-1}\,\boldsymbol{Y}_{(1)}) = \Sigma_{22} - \Sigma_{21}\,\Sigma_{11}^{-1}\,\Sigma_{12} \equiv \Sigma_{22.1}. \qquad \text{(A1.5)}$$

The $k_2 \times k_1$ matrix $B' = \Sigma_{21}\,\Sigma_{11}^{-1}$ that occurs in the conditional distribution is the matrix of regression coefficients, or simply the *regression matrix*, of $\boldsymbol{Y}_{(2)}$ on $\boldsymbol{Y}_{(1)}$, and $\boldsymbol{\mu}_{(2)} + B'(\boldsymbol{Y}_{(1)} - \boldsymbol{\mu}_{(1)})$ is often called the (*linear*) *regression function* of $\boldsymbol{Y}_{(2)}$ on $\boldsymbol{Y}_{(1)}$. The elements of the matrix $\Sigma_{22.1}$ are called *partial covariances* (adjusted for $\boldsymbol{Y}_{(1)}$).

From the above discussion, we see that the matrix

$$C = \begin{bmatrix} I & 0 \\ -\Sigma_{21}\Sigma_{11}^{-1} & I \end{bmatrix}$$

yields a transformation from $\boldsymbol{Y}$ to $\boldsymbol{Z} = (\boldsymbol{Z}_{(1)}', \boldsymbol{Z}_{(2)}')' = C\,\boldsymbol{Y}$ such that $\boldsymbol{Z}$ has block diagonal covariance matrix

$$\Sigma_{zz} = \text{Cov}(C\,\boldsymbol{Y}) = C\,\Sigma_{yy}\,C' = \text{Diag}(\Sigma_{11}, \Sigma_{22.1}).$$

Since $|C| = 1$, from this we see that $|\Sigma_{yy}| = |\Sigma_{11}|\,|\Sigma_{22.1}|$ provides a useful expression for the determinant of the partitioned matrix Σ_{yy}. In addition, a useful expression for the inverse of the partitioned matrix Σ_{yy} in (A1.2) can be obtained from the relation $\Sigma_{yy} = C^{-1}\Sigma_{zz}\,C'^{-1}$ as $\Sigma_{yy}^{-1} = C'\,\Sigma_{zz}^{-1}\,C$, where $\Sigma_{zz}^{-1} = \text{Diag}(\Sigma_{11}^{-1}, \Sigma_{22.1}^{-1})$, that is,

$$\Sigma_{yy}^{-1} = \begin{bmatrix} I & -\Sigma_{11}^{-1}\Sigma_{12} \\ 0 & I \end{bmatrix} \begin{bmatrix} \Sigma_{11}^{-1} & 0 \\ 0 & \Sigma_{22.1}^{-1} \end{bmatrix} \begin{bmatrix} I & 0 \\ -\Sigma_{21}\Sigma_{11}^{-1} & I \end{bmatrix}. \tag{A1.6}$$

In particular, we immediately see that the lower (2, 2) block of Σ_{yy}^{-1} is equal to $\Sigma_{22.1}^{-1}$.

Relation of Conditional Expectation to Best Linear Prediction

The linear regression expression given in (A1.3) can also be viewed as the best (*minimum mean squared error*) *linear predictor* of $\boldsymbol{Y}_{(2)}$ based on linear functions of $\boldsymbol{Y}_{(1)}$, as we now discuss. We consider best linear prediction under more general distributions than normal for the random vector $\boldsymbol{Y} = (\boldsymbol{Y}_{(1)}', \boldsymbol{Y}_{(2)}')'$, where it is assumed only that $\boldsymbol{Y}$ has mean vector $\boldsymbol{\mu}_y = (\boldsymbol{\mu}_{(1)}', \boldsymbol{\mu}_{(2)}')'$ and covariance matrix Σ partitioned as in (A1.2). For any linear predictor of $\boldsymbol{Y}_{(2)}$ based on $\boldsymbol{Y}_{(1)}$, say $\hat{\boldsymbol{Y}}_{(2)}^{*} = \boldsymbol{a}_* + B_*'(\boldsymbol{Y}_{(1)} - \boldsymbol{\mu}_{(1)})$ for arbitrary $k_2 \times 1$ vector $\boldsymbol{a}_*$ and $k_2 \times k_1$ matrix B_*', the prediction error of $\hat{\boldsymbol{Y}}_{(2)}^{*}$ can be expressed as

$$\begin{aligned} \boldsymbol{e}^* &= \boldsymbol{Y}_{(2)} - [\,\boldsymbol{a}_* + B_*'(\boldsymbol{Y}_{(1)} - \boldsymbol{\mu}_{(1)})\,] \\ &= \boldsymbol{Z}_{(2)} + [\,\boldsymbol{\mu}_{(2)} - \boldsymbol{a}_* + (\Sigma_{21}\Sigma_{11}^{-1} - B_*')(\boldsymbol{Y}_{(1)} - \boldsymbol{\mu}_{(1)})\,], \end{aligned}$$

where $\boldsymbol{Z}_{(2)} = \boldsymbol{Y}_{(2)} - \boldsymbol{\mu}_{(2)} - \Sigma_{21}\Sigma_{11}^{-1}(\boldsymbol{Y}_{(1)} - \boldsymbol{\mu}_{(1)})$ which has zero mean vector and zero covariance with $\boldsymbol{Y}_{(1)}$. Hence, $\boldsymbol{Z}_{(2)}$ has zero covariance with $\boldsymbol{U}_{(2)} = \boldsymbol{\mu}_{(2)} - \boldsymbol{a}_* + (\Sigma_{21}\Sigma_{11}^{-1} - B_*')(\boldsymbol{Y}_{(1)} - \boldsymbol{\mu}_{(1)})$, and thus the mean squared error matrix of the general linear predictor is

$$\begin{aligned} E(\boldsymbol{e}^*\boldsymbol{e}^{*\prime}) &= \text{Cov}(\boldsymbol{Z}_{(2)}) + E(\boldsymbol{U}_{(2)}\boldsymbol{U}_{(2)}') \\ &= \Sigma_{22} - \Sigma_{21}\Sigma_{11}^{-1}\Sigma_{12} + (\boldsymbol{\mu}_{(2)} - \boldsymbol{a}_*)(\boldsymbol{\mu}_{(2)} - \boldsymbol{a}_*)' \\ &\quad + (\Sigma_{21}\Sigma_{11}^{-1} - B_*')\,\Sigma_{11}(\Sigma_{21}\Sigma_{11}^{-1} - B_*')'. \end{aligned} \tag{A1.7}$$

Now the last two terms in the above expression are non-negative-definite matrix terms, and a minimum is clearly obtained for the mean squared error matrix when $\boldsymbol{a}_* = \boldsymbol{\mu}_{(2)}$ and $B_*' = \Sigma_{21}\,\Sigma_{11}^{-1}$. Thus, the linear regression function

$$\hat{\boldsymbol{Y}}_{(2)} = \boldsymbol{\mu}_{(2)} + B'(\,\boldsymbol{Y}_{(1)} - \boldsymbol{\mu}_{(1)}\,) \qquad \text{with} \qquad B' = \Sigma_{21}\,\Sigma_{11}^{-1} \tag{A1.8}$$

provides the minimum mean squared error linear predictor of $\boldsymbol{Y}_{(2)}$, and from (A1.7), the mean squared error matrix of this predictor is given by

$$E(\,\boldsymbol{e}\,\boldsymbol{e}'\,) = \text{Cov}(\,\boldsymbol{e}\,) = \text{Cov}(\,\boldsymbol{Y}_{(2)} - \hat{\boldsymbol{Y}}_{(2)}\,) = \Sigma_{22} - \Sigma_{21}\,\Sigma_{11}^{-1}\,\Sigma_{12}\,.$$

Therefore, from the above developments, we see that the coefficient matrix B_*' of the best linear predictor is characterized by the property that the corresponding prediction error have zero covariance with $\boldsymbol{Y}_{(1)}$, that is, that B_*' satisfies the conditions

$$\text{Cov}(\,\boldsymbol{e}^*, \boldsymbol{Y}_{(1)}\,) = \text{Cov}(\,\boldsymbol{Y}_{(2)} - B_*'\,\boldsymbol{Y}_{(1)}, \boldsymbol{Y}_{(1)}\,) = \Sigma_{21} - B_*'\,\Sigma_{11} = 0\,,$$

leading to the solution $B_*' = \Sigma_{21}\,\Sigma_{11}^{-1}$.

Extension of Conditional Distribution Results to More Than Two Components

The results of the preceding paragraphs can be extended to the case of three or more subvector components $\boldsymbol{Y}_{(i)}$. In particular, consider $\boldsymbol{Y} = (\,\boldsymbol{Y}_{(1)}', \boldsymbol{Y}_{(2)}', \boldsymbol{Y}_{(3)}'\,)'$ and partition $\boldsymbol{\mu}_y = (\,\boldsymbol{\mu}_{(1)}', \boldsymbol{\mu}_{(2)}', \boldsymbol{\mu}_{(3)}'\,)'$ and Σ_{yy} accordingly, with $\Sigma_{ij} = \text{Cov}(\,\boldsymbol{Y}_{(i)}, \boldsymbol{Y}_{(j)}\,)$. Then, from the previous results, the conditional distribution of $\boldsymbol{Y}_{(3)}$, given $(\,\boldsymbol{Y}_{(1)}', \boldsymbol{Y}_{(2)}'\,)'$, is normal with (conditional) mean

$$\hat{\boldsymbol{Y}}_{3|12} = E(\,\boldsymbol{Y}_{(3)}\,|\,\boldsymbol{Y}_{(1)}, \boldsymbol{Y}_{(2)}) = \boldsymbol{\mu}_{(3)} + B_{13}'(\,\boldsymbol{Y}_{(1)} - \boldsymbol{\mu}_{(1)}) + B_{23}'(\,\boldsymbol{Y}_{(2)} - \boldsymbol{\mu}_{(2)}), \tag{A1.9}$$

where

$$(\,B_{13}', B_{23}'\,) = \begin{bmatrix} \Sigma_{31} & \Sigma_{32} \end{bmatrix} \begin{bmatrix} \Sigma_{11} & \Sigma_{12} \\ \Sigma_{21} & \Sigma_{22} \end{bmatrix}^{-1},$$

and covariance matrix

$$\begin{aligned} \Sigma_{33.12} &= \text{Cov}(\,\boldsymbol{Y}_{(3)} - \hat{\boldsymbol{Y}}_{3|12}\,) \\ &= \Sigma_{33} - \begin{bmatrix} \Sigma_{31} & \Sigma_{32} \end{bmatrix} \begin{bmatrix} \Sigma_{11} & \Sigma_{12} \\ \Sigma_{21} & \Sigma_{22} \end{bmatrix}^{-1} \begin{bmatrix} \Sigma_{13} \\ \Sigma_{23} \end{bmatrix}. \end{aligned} \tag{A1.10}$$

From previous results, we also have that the conditional distribution of $(\,\boldsymbol{Y}_{(2)}', \boldsymbol{Y}_{(3)}'\,)'$, given $\boldsymbol{Y}_{(1)}$, is normal with (conditional) mean vector such that $\hat{\boldsymbol{Y}}_{j|1} = E(\,\boldsymbol{Y}_{(j)}\,|\,\boldsymbol{Y}_{(1)}\,) = \boldsymbol{\mu}_{(j)} + \Sigma_{j1}\,\Sigma_{11}^{-1}(\,\boldsymbol{Y}_{(1)} - \boldsymbol{\mu}_{(1)}\,)$ and (conditional) covariance matrix such that

$$\Sigma_{jj.1} = \text{Cov}(\,\boldsymbol{Y}_{(j)} - \hat{\boldsymbol{Y}}_{j|1}\,) = \text{Cov}(\,\boldsymbol{Y}_{(j)}\,|\,\boldsymbol{Y}_{(1)}\,) = \Sigma_{jj} - \Sigma_{j1}\,\Sigma_{11}^{-1}\,\Sigma_{1j}\,,$$

for $j = 2, 3$, with

$$\Sigma_{23.1} = \text{Cov}(\boldsymbol{Y}_{(2)} - \hat{\boldsymbol{Y}}_{2|1}, \boldsymbol{Y}_{(3)} - \hat{\boldsymbol{Y}}_{3|1})$$
$$= \text{Cov}(\boldsymbol{Y}_{(2)}, \boldsymbol{Y}_{(3)} \mid \boldsymbol{Y}_{(1)}) = \Sigma_{23} - \Sigma_{21} \Sigma_{11}^{-1} \Sigma_{13} .$$

Consequently, the conditional distribution results for $\boldsymbol{Y}_{(3)}$, given $(\boldsymbol{Y}_{(1)}', \boldsymbol{Y}_{(2)}')'$, can be obtained from this conditional distribution of $(\boldsymbol{Y}_{(2)}', \boldsymbol{Y}_{(3)}')'$, given $\boldsymbol{Y}_{(1)}$, by further conditioning on $\boldsymbol{Y}_{(2)}$ in this latter distribution. Hence, it follows that the conditional distribution of $\boldsymbol{Y}_{(3)}$, given $(\boldsymbol{Y}_{(1)}', \boldsymbol{Y}_{(2)}')'$, has mean

$$\hat{\boldsymbol{Y}}_{3|12} = E(\boldsymbol{Y}_{(3)} \mid \boldsymbol{Y}_{(1)}, \boldsymbol{Y}_{(2)})$$
$$= E(\boldsymbol{Y}_{(3)} \mid \boldsymbol{Y}_{(1)}) + \Sigma_{32.1} \Sigma_{22.1}^{-1}(\boldsymbol{Y}_{(2)} - E(\boldsymbol{Y}_{(2)} \mid \boldsymbol{Y}_{(1)})) \tag{A1.11}$$
$$= \boldsymbol{\mu}_{(3)} + \Sigma_{31} \Sigma_{11}^{-1}(\boldsymbol{Y}_{(1)} - \boldsymbol{\mu}_{(1)})$$
$$+ \Sigma_{32.1} \Sigma_{22.1}^{-1}(\boldsymbol{Y}_{(2)} - \boldsymbol{\mu}_{(2)} - \Sigma_{21} \Sigma_{11}^{-1}(\boldsymbol{Y}_{(1)} - \boldsymbol{\mu}_{(1)})) \tag{A1.12}$$

and covariance matrix equal to

$$\Sigma_{33.12} = \text{Cov}(\boldsymbol{Y}_{(3)} - \hat{\boldsymbol{Y}}_{3|12}) = \Sigma_{33.1} - \Sigma_{32.1} \Sigma_{22.1}^{-1} \Sigma_{23.1} . \tag{A1.13}$$

Thus, by comparison of the coefficient matrices in (A1.9) and (A1.12), we see that

$$B_{13}' = \Sigma_{31} \Sigma_{11}^{-1} - \Sigma_{32.1} \Sigma_{22.1}^{-1} \Sigma_{21} \Sigma_{11}^{-1} \equiv \Sigma_{31} \Sigma_{11}^{-1} - \Sigma_{32.1} \Sigma_{22.1}^{-1} B_{12}' , \tag{A1.14}$$

where $B_{12}' = \Sigma_{21} \Sigma_{11}^{-1}$ is the regression matrix of $\boldsymbol{Y}_{(2)}$ on $\boldsymbol{Y}_{(1)}$, and $B_{23}' = \Sigma_{32.1} \Sigma_{22.1}^{-1}$. Note that equations (A1.11) and (A1.13) can be viewed as "updating" relations, which indicate a method by which the conditional mean and covariance matrix for $\boldsymbol{Y}_{(3)}$, given $(\boldsymbol{Y}_{(1)}', \boldsymbol{Y}_{(2)}')'$, can be obtained from the conditional mean and covariance matrix for $\boldsymbol{Y}_{(3)}$, given $\boldsymbol{Y}_{(1)}$. The linear regression expressions in (A1.9) and (A1.12) also continue to have the interpretation as the minimum mean squared error linear predictor of $\boldsymbol{Y}_{(3)}$ based on $\boldsymbol{Y}_{(1)}$ and $\boldsymbol{Y}_{(2)}$, and (A1.10) and (A1.13) provide equivalent expressions for the mean squared error matrix of this best linear predictor.

It also can be readily verified that by defining the matrix

$$C = \begin{bmatrix} I & 0 & 0 \\ -B_{12}' & I & 0 \\ -B_{13}' & -B_{23}' & I \end{bmatrix} , \tag{A1.15}$$

we can obtain the transformation $\boldsymbol{Z} = (\boldsymbol{Z}_{(1)}', \boldsymbol{Z}_{(2)}', \boldsymbol{Z}_{(3)}')' = C \boldsymbol{Y}$ such that the covariance matrix of $\boldsymbol{Z}$ is block diagonal, $\Sigma_{zz} = \text{Cov}(\boldsymbol{Z}) = C \Sigma_{yy} C'$ $= \text{Diag}(\Sigma_{11}, \Sigma_{22.1}, \Sigma_{33.12})$. That is, the components $\boldsymbol{Z}_{(1)} = \boldsymbol{Y}_{(1)}$, $\boldsymbol{Z}_{(2)} = \boldsymbol{Y}_{(2)} - B_{12}' \boldsymbol{Y}_{(1)}$, and $\boldsymbol{Z}_{(3)} = \boldsymbol{Y}_{(3)} - B_{13}' \boldsymbol{Y}_{(1)} - B_{23}' \boldsymbol{Y}_{(2)}$ are mutually uncorrelated, hence independent. The above results extend to an arbitrary number of components and arbitrary types of partitions of the vector $\boldsymbol{Y}$ and its covariance matrix Σ_{yy}. Such extension provides a statistical interpretation for the Cholesky factorization of any positive-definite matrix (i.e., covariance matrix) Σ_{yy} as $\Sigma_{yy} = R\,D\,R'$, since we will have the above decomposition $C\,\Sigma_{yy}\,C' = D$. In this decomposition, the matrix C is a lower triangular matrix having ones on the diagonal as in (A1.15) and D is a diagonal matrix with positive diagonal elements [whose jth diagonal element is interpretable as the (conditional) variance in the conditional distribution of the jth variable Y_j given the preceding $j-1$ variables $Y_1, \ldots, Y_{j-1}$].

Distribution of Quadratic Form in Normal Random Variables

Finally, the distribution of quadratic forms in multivariate normal random vectors is often of interest, and we state the following result. If $\boldsymbol{Y}$ is multivariate normal $N(\boldsymbol{\mu}_y, \Sigma_{yy})$ and A is a $k \times k$ symmetric matrix, then

> $(\boldsymbol{Y} - \boldsymbol{\mu}_y)'\,A\,(\boldsymbol{Y} - \boldsymbol{\mu}_y)$ is distributed as (central) chi-squared with degrees of freedom equal to $r = \text{rank}(A)$ if and only if $A\,\Sigma_{yy}\,A = A$.

As a particular case, we have that $(\boldsymbol{Y} - \boldsymbol{\mu}_y)'\,\Sigma_{yy}^{-1}\,(\boldsymbol{Y} - \boldsymbol{\mu}_y)$ is distributed as chi-squared with k degrees of freedom.

The general chi-squared distribution result follows from the more particular result that if $\boldsymbol{Z}$ is distributed as normal $N(\boldsymbol{\mu}_z, I)$, with identity covariance matrix, then $(\boldsymbol{Z} - \boldsymbol{\mu}_z)'\,A\,(\boldsymbol{Z} - \boldsymbol{\mu}_z)$ is distributed as (central) chi-squared with degrees of freedom equal to $r = \text{rank}(A)$ if and only if $A^2 = A$. A (symmetric) $k \times k$ matrix A with the property that $A^2 = A$ is called *idempotent*. Such a matrix has r eigenvalues equal to one and $k - r$ eigenvalues equal to zero, where $r = \text{rank}(A)$, which is also equal to $\text{tr}(A)$. Then, in the general case where $\boldsymbol{Y}$ is $N(\boldsymbol{\mu}_y, \Sigma_{yy})$, since Σ_{yy} is positive-definite there is a nonsingular matrix B $(B = \Lambda^{1/2} P')$ such that $\Sigma_{yy} = B'B$ or, equivalently, $B'^{-1}\Sigma_{yy}\,B^{-1} = I$. Hence, $\boldsymbol{Z} = B'^{-1}\boldsymbol{Y}$ is $N(B'^{-1}\boldsymbol{\mu}_y, B'^{-1}\Sigma_{yy}\,B^{-1}) \equiv N(\boldsymbol{\mu}_z, I)$, so that by the more particular result, $(\boldsymbol{Y} - \boldsymbol{\mu}_y)'\,A\,(\boldsymbol{Y} - \boldsymbol{\mu}_y)$ $= (\boldsymbol{Z} - \boldsymbol{\mu}_z)'\,(BAB')\,(\boldsymbol{Z} - \boldsymbol{\mu}_z)$ is distributed as (central) chi-squared with r degrees of freedom if and only if BAB' is idempotent. That is, $(BAB')^2 = BAB'\,BAB' = BA\,\Sigma_{yy}\,AB'$ is equal to BAB', which is equivalent to the condition that $A\,\Sigma_{yy}\,A = A$ (since B is nonsingular).

A1.4 Some Basic Results on Stochastic Convergence

We review some fundamental concepts and results on convergence of sequences of random variables and vectors, which are particularly useful in regard to asymptotic distribution theory for estimators in time series analysis. Suppose $X_1, X_2, \ldots, X_T, \ldots$ is a sequence of random variables. The sequence $\{X_T\}$ is said to *converge in probability* to a random variable X as $T \to \infty$ if for every $\varepsilon > 0$,

$$\lim_{T\to\infty} P(\,|\,X_T - X\,| > \varepsilon\,) = 0,$$

and we denote this type of convergence as $X_T \xrightarrow{P} X$ as $T \to \infty$. In applications, the limit X may often be a fixed nonstochastic constant. More generally, let $\boldsymbol{X}_t = (X_{1t}, \ldots, X_{kt})'$, $t = 1, 2, \ldots$, denote a sequence of k-dimensional random vectors, and $\boldsymbol{X} = (X_1, \ldots, X_k)'$ another random vector. We say that $\boldsymbol{X}_T \xrightarrow{P} \boldsymbol{X}$ if $X_{iT} \xrightarrow{P} X_i$ as $T \to \infty$ for each $i = 1, \ldots, k$. In practice, a convenient useful sufficient condition that implies convergence in probability of $\boldsymbol{X}_T$ to $\boldsymbol{X}$, by use of Chebychev's Inequality, is that

$$E[(\boldsymbol{X}_T - \boldsymbol{X})(\boldsymbol{X}_T - \boldsymbol{X})'] \to 0 \quad \text{as} \quad T \to \infty.$$

In addition, the sequence of random vectors $\{\boldsymbol{X}_T\}$ is said to *converge in distribution* to $\boldsymbol{X}$ as $T \to \infty$, denoted as $\boldsymbol{X}_T \xrightarrow{D} \boldsymbol{X}$ as $T \to \infty$, if $\lim_{T\to\infty} F_T(\boldsymbol{a}) = F(\boldsymbol{a})$ for all continuity points of F, where F_T represents the joint distribution function of $\boldsymbol{X}_T$, $F_T(\boldsymbol{a}) = P(\boldsymbol{X}_T \le \boldsymbol{a})$, and F represents the joint distribution function of $\boldsymbol{X}$. In practice, convergence in distribution is most often established by consideration of the limit of the sequence of characteristic functions of the $\boldsymbol{X}_T$, $\phi_T(\boldsymbol{u}) = E(e^{i\boldsymbol{u}'\boldsymbol{X}_T})$, and use of the continuity theorem which states that if the $\phi_T(\boldsymbol{u})$ converge as $T \to \infty$ to a function $\phi(\boldsymbol{u})$ which is the characteristic function of a random vector $\boldsymbol{X}$, then $\boldsymbol{X}_T \xrightarrow{D} \boldsymbol{X}$. We also mention here the implication that if $\boldsymbol{X}_T \xrightarrow{P} \boldsymbol{X}$ as $T \to \infty$, then we have $\boldsymbol{X}_T \xrightarrow{D} \boldsymbol{X}$; and as a partial converse, if $\boldsymbol{X}_T \xrightarrow{D} \boldsymbol{X}$ as $T \to \infty$ where $\boldsymbol{X} \equiv \boldsymbol{c}$ is a fixed (nonstochastic) constant vector, then we have $\boldsymbol{X}_T \xrightarrow{P} \boldsymbol{X} = \boldsymbol{c}$.

The following are a few useful facts concerning convergence in probability and in distribution. Let $g(\boldsymbol{x})$ denote a continuous function of k variables to m variables. Then

(i) $\boldsymbol{X}_T \xrightarrow{P} \boldsymbol{X}$ as $T \to \infty$ implies that $g(\boldsymbol{X}_T) \xrightarrow{P} g(\boldsymbol{X})$ as $T \to \infty$,

and

(ii) $\boldsymbol{X}_T \xrightarrow{D} \boldsymbol{X}$ as $T \to \infty$ implies that $g(\boldsymbol{X}_T) \xrightarrow{D} g(\boldsymbol{X})$ as $T \to \infty$.

For example, if $\boldsymbol{X}_T \xrightarrow{D} \boldsymbol{X}$ as $T \to \infty$ where the random vector $\boldsymbol{X}$ has a k-dimensional (nonsingular) multivariate normal distribution with zero mean vector and covariance matrix Σ, in which case we express for convenience as $\boldsymbol{X}_T \xrightarrow{D} N(0, \Sigma)$, then from the previous section and (ii) above it follows that $\boldsymbol{X}_T'\Sigma^{-1}\boldsymbol{X}_T \xrightarrow{D} \chi_k^2$, where χ_k^2 denotes a chi-squared distribution with k degrees of freedom. To state a few additional results, let A_T denote a sequence of $k \times k$ random matrices such that $A_T \xrightarrow{P} A$ as $T \to \infty$, where A denotes a $k \times k$ matrix of fixed constants. Then we have

(iii) $\boldsymbol{X}_T \xrightarrow{P} \boldsymbol{X}$ and $A_T \xrightarrow{P} A$ as $T \to \infty$ implies that $A_T\,\boldsymbol{X}_T \xrightarrow{P} A\,\boldsymbol{X}$ as $T \to \infty$,

and

(iv) $\boldsymbol{X}_T \xrightarrow{D} \boldsymbol{X}$ and $A_T \xrightarrow{P} A$ as $T \to \infty$ implies that $A_T\,\boldsymbol{X}_T \xrightarrow{D} A\,\boldsymbol{X}$ as $T \to \infty$.

As a special case of (iv), we have that if $\boldsymbol{X}_T \xrightarrow{D} \boldsymbol{X}$ and $A_T \xrightarrow{P} 0$ as $T \to \infty$, then $A_T\,\boldsymbol{X}_T \xrightarrow{D} 0$ and so also $A_T\,\boldsymbol{X}_T \xrightarrow{P} 0$. To state one additional property, let $\{\boldsymbol{Y}_T\}$ denote another sequence of k-dimensional random vectors. Then

(v) $\boldsymbol{X}_T \xrightarrow{D} \boldsymbol{X}$ and $(\boldsymbol{X}_T - \boldsymbol{Y}_T) \xrightarrow{P} 0$ as $T \to \infty$ implies that $\boldsymbol{Y}_T \xrightarrow{D} \boldsymbol{X}$ as $T \to \infty$.

The asymptotic distribution theory is most relevant in the context of the limiting distribution of estimators of parameters, based on sample size or series length T, as $T \to \infty$. Suppose that $\hat{\boldsymbol{\beta}}_T$ denotes an estimator (such as maximum likelihood or least squares) of an unknown k-dimensional parameter vector $\boldsymbol{\beta}$ based on series length T. The estimator $\hat{\boldsymbol{\beta}}_T$ is said to be *consistent* if $\hat{\boldsymbol{\beta}}_T \xrightarrow{P} \boldsymbol{\beta}$ as $T \to \infty$. Typically, the estimator $\hat{\boldsymbol{\beta}}_T$ has an asymptotic normal distribution, in the sense that $T^{1/2}(\hat{\boldsymbol{\beta}}_T - \boldsymbol{\beta})$ converges in distribution as $T \to \infty$ to a random vector with multivariate normal distribution $N(0, \Sigma)$, in which case we write $T^{1/2}(\hat{\boldsymbol{\beta}}_T - \boldsymbol{\beta}) \xrightarrow{D} N(0, \Sigma)$. Then it is common to approximate the distribution of $\hat{\boldsymbol{\beta}}_T$ by the multivariate normal distribution $N(\boldsymbol{\beta}, (1/T)\Sigma)$ for large T. A few additional useful results in this situation, which are special cases of the above facts (i)–(v), are worth mentioning. Suppose that $T^{1/2}(\hat{\boldsymbol{\beta}}_T - \boldsymbol{\beta}) \xrightarrow{D} N(0, \Sigma)$ and that $\hat{A}_T$ is a sequence of stochastic matrices such that $\hat{A}_T \xrightarrow{P} A$, a constant matrix, as $T \to \infty$. Then it follows from (iv) that

$$T^{1/2} \hat{A}_T (\hat{\boldsymbol{\beta}}_T - \boldsymbol{\beta}) \xrightarrow{D} N(0, A \Sigma A').$$

In addition, it follows from (ii) that $T(\hat{\boldsymbol{\beta}}_T - \boldsymbol{\beta})' \Sigma^{-1} (\hat{\boldsymbol{\beta}}_T - \boldsymbol{\beta}) \xrightarrow{D} \chi_k^2$ as $T \to \infty$, and from (iv) that if $\hat{\Sigma}_T \xrightarrow{P} \Sigma$, that is, $\hat{\Sigma}_T$ is a consistent estimator of Σ, then also

$$T(\hat{\boldsymbol{\beta}}_T - \boldsymbol{\beta})' \hat{\Sigma}_T^{-1} (\hat{\boldsymbol{\beta}}_T - \boldsymbol{\beta}) \xrightarrow{D} \chi_k^2 \quad \text{as} \quad T \to \infty.$$

Finally, as a general result, if $g(\boldsymbol{\beta}) = (g_1(\boldsymbol{\beta}), \ldots, g_m(\boldsymbol{\beta}))'$ is a vector-valued function which is continuously differentiable with $\partial g / \partial \boldsymbol{\beta}' \neq 0$, then by a first-order Taylor expansion argument, it follows that

$$T^{1/2} (g(\hat{\boldsymbol{\beta}}_T) - g(\boldsymbol{\beta})) \xrightarrow{D} N(0, \Sigma_g),$$

where $\Sigma_g = \{ \partial g(\boldsymbol{\beta}) / \partial \boldsymbol{\beta}' \} \Sigma \{ \partial g(\boldsymbol{\beta})' / \partial \boldsymbol{\beta} \}$.

CHAPTER 2

Vector ARMA Time Series Models and Forecasting

In this chapter, the vector autoregressive moving average (ARMA) models that were introduced in Section 1.2.2 are examined, and the stationarity and invertibility aspects of vector ARMA processes are considered. The covariance matrix structure of vector ARMA processes is considered, in general as well as for special cases such as first-order MA, AR, and ARMA models. In addition, consideration of parameter identifiability of mixed ARMA model representations is given. Nonstationary ARMA processes are also considered, and the concept of cointegration among the component series of a nonstationary process is introduced. Forecasting of vector ARMA models, including computation of forecasts and mean squared error matrix of the forecast errors, is presented.

2.1 Vector Moving Average Models

A pure *moving average* MA(q) model is obtained by setting $\Phi(B) = I$ in the general ARMA(p,q) model (1.9) of Section 1.2.2, yielding

$$Y_t = \mu + (I - \Theta_1 B - \cdots - \Theta_q B^q) \varepsilon_t = \mu + \varepsilon_t - \sum_{j=1}^{q} \Theta_j \varepsilon_{t-j} , \qquad (2.1)$$

or $Y_t = \mu + \Theta(B) \varepsilon_t$, where $\Theta(B) = I - \Theta_1 B - \cdots - \Theta_q B^q$. The finite order moving average MA(q) process defined by (2.1) is always stationary and causal because its MA representation as in (1.8) of Section 1.2.1 has $\Psi(B) = \Theta(B) = I - \Theta_1 B - \cdots - \Theta_q B^q$ which is automatically convergent.

2.1.1 Invertibility of the Vector Moving Average Model

The vector MA(q) process is *invertible* if it can be represented in the form $Y_t - \mu = \sum_{j=1}^{\infty} \Pi_j (Y_{t-j} - \mu) + \varepsilon_t$, with $\sum_{j=1}^{\infty} \| \Pi_j \| < \infty$. Let

$$d(B) = \det\{\, \Theta(B) \,\} = \det\{\, I - \Theta_1 B - \cdots - \Theta_q B^q \,\},$$

and let $\Theta^*(B) = \mathrm{Adj}\{\, \Theta(B) \,\}$ be the adjoint of the matrix $\Theta(B)$. Then, using the relation $\Theta(B)^{-1} = (\, 1/\det\{\, \Theta(B) \,\} \,)\, \mathrm{Adj}\{\, \Theta(B) \,\}$, the MA($q$) model (2.1) can be written in the inverted form as $(\, 1/d(B) \,)\, \Theta^*(B)\, (\, \boldsymbol{Y}_t - \boldsymbol{\mu} \,) = \varepsilon_t$, in the sense that if the process satisfies $\boldsymbol{Y}_t = \boldsymbol{\mu} + \Theta(B)\, \varepsilon_t$ as in (2.1), then $\{\boldsymbol{Y}_t\}$ also satisfies this inverted form relation, provided that the infinite series on the left of this expression converges. Hence, the MA(q) process $\{\boldsymbol{Y}_t\}$ in (2.1) is invertible if the infinite AR form $\Pi(B)\, (\, \boldsymbol{Y}_t - \boldsymbol{\mu} \,) = \varepsilon_t$, with $\Pi(B) = \Theta(B)^{-1} = (\, 1/d(B) \,)\, \Theta^*(B)$, is convergent, that is, the linear filter $\Pi(B) = \Theta(B)^{-1} = I - \sum_{j=1}^{\infty} \Pi_j B^j$ must be stable. Thus, for the MA(q) to be invertible and be expressible as a convergent infinite AR form, it is required that $d(B)^{-1}$ form a convergent series for $|\, B \,| \le 1$ and, hence, that all roots of $d(B) = \det\{\, \Theta(B) \,\} = 0$ be greater than one in absolute value, which we refer to as the *invertibility condition*. In this case, we can write

$$\Pi(B)\, (\, \boldsymbol{Y}_t - \boldsymbol{\mu} \,) \equiv (\, \boldsymbol{Y}_t - \boldsymbol{\mu} \,) - \sum_{j=1}^{\infty} \Pi_j\, (\, \boldsymbol{Y}_{t-j} - \boldsymbol{\mu} \,) = \varepsilon_t\,, \tag{2.2}$$

where $\Pi(B) = I - \sum_{j=1}^{\infty} \Pi_j B^j = \Theta(B)^{-1}$, with $\sum_{j=1}^{\infty} \|\, \Pi_j \,\| < \infty$. The matrix coefficients Π_j can be obtained by equating coefficients in the relation $\Theta(B)\, \Pi(B) = I$. Hence, since

$$\begin{aligned} \Theta(B)\, \Pi(B) &= (\, I - \Theta_1 B - \cdots - \Theta_q B^q \,)\, (\, I - \Pi_1 B - \Pi_2 B^2 - \cdots \,) \\ &= I - (\, \Pi_1 + \Theta_1 \,)\, B - (\, \Pi_2 - \Theta_1 \Pi_1 + \Theta_2 \,)\, B^2 - \cdots \\ &\quad - (\, \Pi_j - \Theta_1 \Pi_{j-1} - \cdots - \Theta_q \Pi_{j-q} \,)\, B^j - \cdots\,, \end{aligned}$$

by equating coefficient matrices of various powers B^j in the relation $\Theta(B)\, \Pi(B) = I$ for $j = 1, 2, \ldots,$ we have

$$\Pi_j = \Theta_1 \Pi_{j-1} + \Theta_2 \Pi_{j-2} + \cdots + \Theta_q \Pi_{j-q}\,,$$

where $\Pi_0 = -I$ and $\Pi_j = 0$ for $j < 0$.

2.1.2 Covariance Matrices of the Vector Moving Average Model

The cross-covariance matrices of an MA(q) model are easily obtained in terms of the Θ_i and Σ as

$$\begin{aligned} \Gamma(l) &= \mathrm{Cov}(\, \boldsymbol{Y}_t, \boldsymbol{Y}_{t+l} \,) \\ &= E[\, (\, \varepsilon_t - \Theta_1 \varepsilon_{t-1} - \cdots - \Theta_q \varepsilon_{t-q})(\, \varepsilon_{t+l} - \Theta_1 \varepsilon_{t+l-1} - \cdots - \Theta_q \varepsilon_{t+l-q})' \,] \\ &= -\, \Sigma\, \Theta_l' + \Theta_1 \Sigma\, \Theta_{l+1}' + \cdots + \Theta_{q-l}\, \Sigma\, \Theta_q' = \sum_{h=0}^{q-l} \Theta_h\, \Sigma\, \Theta_{h+l}'\,, \end{aligned} \tag{2.3}$$

for $l = 0, 1, \ldots, q$, with $\Theta_0 = -I$, $\Gamma(-l) = \Gamma(l)'$, and $\Gamma(l) = 0$ for $l > q$.

Thus, for the MA(q) process, we have that all cross-correlations are zero for lags greater than q.

Conversely, the equations (2.3) can be used for purposes of obtaining the MA coefficient matrices Θ_i given the $\Gamma(l)$'s of an MA(q) process. An iterative method for solving (2.3) to determine the MA coefficients Θ_i, $i = 1, \ldots, q$, and Σ from the corresponding covariance matrices $\Gamma(l)$, $l = 0, 1, \ldots, q$, has been provided by Tunnicliffe Wilson (1972). Also see Newton (1980) and Ansley and Kohn (1985c) for an alternate iterative method to determine the Θ_i and Σ that corresponds to the innovations algorithm recursions, which will be discussed in Section 5.4.1, for a vector MA(q) model. For example, in the MA(1) model $\boldsymbol{Y}_t = \boldsymbol{\mu} + \varepsilon_t - \Theta_1 \varepsilon_{t-1}$, which is considered further in the next Section 2.1.3, this last method amounts to performing the recursions

$$\Theta_{1,m} = -\Gamma(1)' \Sigma_m^{-1}, \qquad \Sigma_{m+1} = \Gamma(0) - \Theta_m \Sigma_m \Theta_m',$$

for $m = 1, 2, \ldots,$ starting with $\Sigma_1 = \Gamma(0)$, and, hence, $\Theta_{1,1} = -\Gamma(1)' \Gamma(0)^{-1}$, with $\Theta_{1,m} \to \Theta_1$ and $\Sigma_m \to \Sigma$ as $m \to \infty$. (See Sections 3.3.2 and 5.4.1 for further details.)

2.1.3 Features of the Vector MA(1) Model

Specifically, for the MA(1) process $\boldsymbol{Y}_t = \boldsymbol{\mu} + \varepsilon_t - \Theta\, \varepsilon_{t-1}$, from (2.3) we have

$$\Gamma(0) = \Sigma + \Theta \Sigma \Theta', \qquad \Gamma(1) = -\Sigma \Theta' = \Gamma(-1)',$$

and $\Gamma(l) = 0$ for $|\, l\, | > 1$. Also note that in the MA(1) case, the invertibility condition that all roots of $\det\{ I - \Theta B \} = 0$ be greater than one in absolute value is equivalent to the condition that all eigenvalues of Θ, that is, all roots λ of $\det\{ \lambda I - \Theta \} = 0$, be less than one in absolute value. More explicitly, for any arbitrary $n > 0$, by $t{+}n$ successive substitutions of the relation $\varepsilon_{t-j} = \boldsymbol{Y}_{t-j} + \Theta\, \varepsilon_{t-j-1}$, $j = 1, \ldots, t{+}n$, in the MA(1) model equation $\boldsymbol{Y}_t = \varepsilon_t - \Theta\, \varepsilon_{t-1}$ we can obtain

$$\boldsymbol{Y}_t = -\sum_{j=1}^{t+n} \Theta^j\, \boldsymbol{Y}_{t-j} + \varepsilon_t - \Theta^{t+n+1}\, \varepsilon_{-n-1}. \tag{2.4}$$

Hence, when all eigenvalues of Θ are less than one in absolute value, $\Theta^{t+n+1} \to 0$ as $n \to \infty$ and we obtain the convergent infinite AR representation as $\boldsymbol{Y}_t = -\sum_{j=1}^{\infty} \Theta^j\, \boldsymbol{Y}_{t-j} + \varepsilon_t$, with $\sum_{j=1}^{\infty} \|\Theta^j\| < \infty$. [To exhibit the convergence directly, suppose, for example, that Θ has k distinct eigenvalues $\lambda_1, \lambda_2, \ldots, \lambda_k$, so that there is a nonsingular matrix P such that $P^{-1} \Theta P = \Lambda = \mathrm{Diag}(\lambda_1, \ldots, \lambda_k)$. Then $\Theta = P \Lambda P^{-1}$ and $\Theta^j = P \Lambda^j P^{-1}$, so that $\sum_{j=1}^{\infty} \Theta^j$ will be absolutely convergent if $\sum_{j=1}^{\infty} \Lambda^j$ is, hence if all the eigenvalues λ_i of Θ satisfy $|\lambda_i| < 1$.] Thus, in the MA(1) model, we find that $\Pi(B) = (I - \Theta B)^{-1} = I + \sum_{j=1}^{\infty} \Theta^j B^j$, and the coefficients Π_j are given explicitly by $\Pi_j = -\Theta^j$, $j \geq 1$.

2.1.4 *Model Structure for Subset of Components in the Vector MA Model*

From the preceding results in (2.3), it follows that any subset of $k_1 < k$ elements of $\boldsymbol{Y}_t$ or, more generally, any collection of $k_1 \le k$ linear combinations of $\boldsymbol{Y}_t$, follows an MA(q') model where $q' \le q$, since they will have the cross-covariance matrix structure of an MA process [see Lutkepohl (1984) for further discussion of this result]. For example, let A be a $k_1 \times k$ matrix, with $k_1 \le k$, and define $\boldsymbol{Z}_t = A\, \boldsymbol{Y}_t$. Then since $\Gamma_z(l) = \text{Cov}(\, A\, \boldsymbol{Y}_t,\, A\, \boldsymbol{Y}_{t+l}\,) = A\, \Gamma(l)\, A'$, it follows that $\Gamma_z(l) = 0$ for $l > q$, and hence $\boldsymbol{Z}_t$ is MA(q') with $q' \le q$. In particular, we obtain that the individual series Y_{it} must satisfy univariate MA(q') models of the form $Y_{it} = \mu_i + \eta_i(B)\, a_{it}$, where $q' \le q$ and a_{it} is a univariate white noise process (but the a_{it} may be cross-correlated at nonzero lags).

Example 2.1. Consider the bivariate $(k = 2)$ MA(1) model $\boldsymbol{Y}_t = (\, I - \Theta\, B\,)\, \varepsilon_t$ with

$$\Theta = \begin{bmatrix} 0.8 & 0.7 \\ -0.4 & 0.6 \end{bmatrix} \quad \text{and} \quad \Sigma = \begin{bmatrix} 4 & 1 \\ 1 & 2 \end{bmatrix}.$$

The roots of $\det\{\, \lambda\, I - \Theta\,\} = \lambda^2 - 1.4\,\lambda + 0.76 = 0$ are $\lambda = 0.7 \pm 0.5196\, i$, with absolute value equal to $(\,0.76\,)^{1/2}$, and, hence, the MA(1) model is invertible. The covariance matrices at lag 0 and 1 are

$$\Gamma(0) = \Sigma + \Theta\, \Sigma\, \Theta' = \begin{bmatrix} 8.66 & 0.76 \\ 0.76 & 2.88 \end{bmatrix}, \qquad \Gamma(1) = -\,\Sigma\, \Theta' = \begin{bmatrix} -3.9 & 1.0 \\ -2.2 & -0.8 \end{bmatrix},$$

with corresponding correlation matrices

$$\rho(0) = V^{-1/2}\, \Gamma(0)\, V^{-1/2} = \begin{bmatrix} 1.000 & 0.152 \\ 0.152 & 1.000 \end{bmatrix}, \qquad \rho(1) = \begin{bmatrix} -0.450 & 0.200 \\ -0.441 & -0.278 \end{bmatrix}.$$

In addition, univariate MA(1) model representations, $Y_{it} = (\,1 - \eta_i\, B\,)\, a_{it}$, $\sigma^2_{a_i} = \text{Var}(\, a_{it}\,)$, for each component series Y_{it} can be established directly from the relations $\rho_{ii}(1) = -\,\eta_i\, /\, (\,1 + \eta_i^2\,)$, $\gamma_{ii}(0) = \sigma^2_{a_i}\, (\,1 + \eta_i^2\,)$. Solving these lead to the values $\eta_1 = 0.628$, $\sigma^2_{a_1} = 6.211$, and $\eta_2 = 0.303$, $\sigma^2_{a_2} = 2.637$. Since, from (2.21) of Section 2.5.2, the diagonal elements σ_{ii} of Σ can be interpreted as the variances of one-step ahead forecast errors from the bivariate model, whereas the $\sigma^2_{a_i}$ are variances of one-step ahead forecast errors based on the individual univariate models only, comparison of the values σ_{ii} with the $\sigma^2_{a_i}$ values from the univariate models indicate the improvement in forecast accuracy from use of the bivariate model over the individual univariate models.

Example 2.2. Relation of MA(1) to Transfer Function Model. Consider the simple case of a bivariate $(\,k = 2\,)$ MA(1) model, $\boldsymbol{Y}_t = (\,I - \Theta\, B\,)\, \varepsilon_t$, where Θ is 2×2 with elements θ_{ij}. Let

$$d(B) = \det\{ I - \Theta B \} = (1 - \theta_{11} B)(1 - \theta_{22} B) - \theta_{12} \theta_{21} B^2$$

and

$$\Theta^*(B) = \text{Adj}\{ I - \Theta B \} = \begin{bmatrix} (1 - \theta_{22} B) & \theta_{12} B \\ \theta_{21} B & (1 - \theta_{11} B) \end{bmatrix}.$$

Then we have $\Theta^*(B)\, \boldsymbol{Y}_t = d(B)\, \varepsilon_t$, or the equations

$$(1 - \theta_{22} B)\, Y_{1t} = -\theta_{12} B\, Y_{2t} + d(B)\, \varepsilon_{1t},$$

$$(1 - \theta_{11} B)\, Y_{2t} = -\theta_{21} B\, Y_{1t} + d(B)\, \varepsilon_{2t}.$$

Hence, we obtain

$$Y_{1t} = \frac{-\theta_{12} B}{1 - \theta_{22} B} Y_{2t} + \frac{d(B)}{1 - \theta_{22} B} \varepsilon_{1t}, \qquad Y_{2t} = \frac{-\theta_{21} B}{1 - \theta_{11} B} Y_{1t} + \frac{d(B)}{1 - \theta_{11} B} \varepsilon_{2t},$$

which has the general form of a "joint" transfer function structure (1.14) of Section 1.2.2. Now suppose that $\theta_{12} = 0$, so that the matrix Θ is lower triangular. Then $d(B) = (1 - \theta_{11} B)(1 - \theta_{22} B)$, and the above equations reduce to

$$Y_{1t} = (1 - \theta_{11} B)\, \varepsilon_{1t}, \qquad Y_{2t} = \frac{-\theta_{21} B}{1 - \theta_{11} B} Y_{1t} + (1 - \theta_{22} B)\, \varepsilon_{2t},$$

which has the general structure of a "unidirectional" transfer function model as studied in Box and Jenkins (1976, Chap. 11), except that the noise terms ε_{1t} and ε_{2t} may be correlated (so that the "input" Y_{1t} and the noise ε_{2t} in the transfer model equation are not independent). However, assuming normality, we can always write $\varepsilon_{2t} = \beta\, \varepsilon_{1t} + a_{2t}$, where a_{2t} is uncorrelated with and, hence, independent of ε_{1t}, with $\beta = \sigma_{21} / \sigma_{11}$. Then, substituting $\varepsilon_{2t} = \beta (1 - \theta_{11} B)^{-1} Y_{1t} + a_{2t}$, the above transfer function equation can be reexpressed as

$$\begin{aligned} Y_{2t} &= \frac{-\theta_{21} B}{1 - \theta_{11} B} Y_{1t} + \beta(1 - \theta_{22} B)\, \varepsilon_{1t} + (1 - \theta_{22} B)\, a_{2t} \\ &= \frac{\beta - (\beta\, \theta_{22} + \theta_{21}) B}{1 - \theta_{11} B} Y_{1t} + (1 - \theta_{22} B)\, a_{2t} \end{aligned}$$

which is a standard transfer function model form with the noise a_{2t} independent of ε_{1t}, hence independent of the "input" Y_{1t}. Note that the relation $Y_{2t} = [\, \beta - (\beta\, \theta_{22} + \theta_{21}) B\,]\, a_{1t} + (1 - \theta_{22} B)\, a_{2t}$ together with $Y_{1t} = (1 - \theta_{11} B)\, a_{1t}$ (with $a_{1t} \equiv \varepsilon_{1t}$), from which the standard transfer function model form above easily follows, is directly obtained from the form $\boldsymbol{Y}_t = \Theta_0^{\dagger}\, \boldsymbol{a}_t - \Theta_1^{\dagger}\, \boldsymbol{a}_{t-1}$ as in (1.12) of Section 1.2.2, with

$$\Phi_0^{\#} = \begin{bmatrix} 1 & 0 \\ -\beta & 1 \end{bmatrix}, \qquad \Theta_0^{+} = \Phi_0^{\#-1} = \begin{bmatrix} 1 & 0 \\ \beta & 1 \end{bmatrix},$$

$$\Theta_1^{+} = \Theta_1 \, \Phi_0^{\#-1} = \begin{bmatrix} \theta_{11} & 0 \\ (\beta\,\theta_{22} + \theta_{21}) & \theta_{22} \end{bmatrix}.$$

In general, for an MA(1) model in higher dimensions $k > 2$, it is easy to verify that if the matrix Θ is lower triangular, then the MA(1) model can always be reexpressed in the form of unidirectional transfer function equations. This form can also be generalized to block triangular structure for Θ and to higher-order MA models.

2.2 Vector Autoregressive Models

The vector *autoregressive* AR(p) model is given by

$$(\, \boldsymbol{Y}_t - \boldsymbol{\mu}\,) - \sum_{j=1}^{p} \Phi_j \,(\, \boldsymbol{Y}_{t-j} - \boldsymbol{\mu}\,) = \varepsilon_t\,, \tag{2.5}$$

or $\Phi(B)\,(\,\boldsymbol{Y}_t - \boldsymbol{\mu}\,) = \varepsilon_t$, where $\Phi(B) = I - \Phi_1\,B - \cdots - \Phi_p\,B^p$. For the AR process to be *stationary* and *causal*, we require it to be expressible in the convergent causal infinite MA form as $\boldsymbol{Y}_t - \boldsymbol{\mu} = \sum_{j=0}^{\infty} \Psi_j\,\varepsilon_{t-j} = \Psi(B)\,\varepsilon_t$, where $\Psi(B) = \sum_{j=0}^{\infty} \Psi_j\,B^j$, $\sum_{j=0}^{\infty} \|\Psi_j\| < \infty$.

2.2.1 *Stationarity of the Vector Autoregressive Model*

Similar to the discussion of invertibility for the MA process, from the expression $\Phi(B)^{-1} = (\,1/\det\{\,\Phi(B)\,\}\,)\,\mathrm{Adj}\{\,\Phi(B)\,\}$, we find that $\Psi(B) = \Phi(B)^{-1}$ will be a convergent series for $|\,B\,| \le 1$ if all the roots of $\det\{\,\Phi(B)\,\} = 0$ are greater than one in absolute value. Hence, the process will be stationary if this condition is satisfied, with $\boldsymbol{Y}_t = \boldsymbol{\mu} + \Psi(B)\,\varepsilon_t$, $\Psi(B) = \Phi(B)^{-1} = \sum_{j=0}^{\infty} \Psi_j\,B^j$, $\sum_{j=0}^{\infty} \|\Psi_j\| < \infty$, representing the stationary causal solution to the AR(p) relations (2.5). We will refer to the condition that all roots of $\det\{\,\Phi(B)\,\} = 0$ be greater than one in absolute value as the *stationarity condition.* (This condition is also referred to as the *stability condition*, since it implies that the process $\boldsymbol{Y}_t$ has a representation as a stable causal infinite MA filter of a white noise input ε_t.)

Under the stationary condition, the general AR(p) model can be expressed as

$$\boldsymbol{Y}_t = \boldsymbol{\mu} + \sum_{j=0}^{\infty} \Psi_j\,\varepsilon_{t-j} = \boldsymbol{\mu} + \Psi(B)\,\varepsilon_t\,, \tag{2.6}$$

with $\Psi(B) = \Phi(B)^{-1} = \sum_{j=0}^{\infty} \Psi_j\,B^j$ and $\sum_{j=0}^{\infty} \|\Psi_j\| < \infty$. The matrix weights Ψ_j can be obtained from the relation $\Phi(B)\,\Psi(B) = I$. Hence, since

$$\Phi(B)\,\Psi(B) = (\,I - \Phi_1\,B - \cdots - \Phi_p\,B^p\,)\,(\,I + \Psi_1\,B + \Psi_2\,B^2 + \cdots\,)$$
$$= I + (\,\Psi_1 - \Phi_1\,)\,B + (\,\Psi_2 - \Phi_1\,\Psi_1 - \Phi_2\,)\,B^2 + \cdots$$
$$+ (\,\Psi_j - \Phi_1\,\Psi_{j-1} - \cdots - \Phi_p\,\Psi_{j-p}\,)\,B^j + \cdots ,$$

by equating coefficient matrices of various powers B^j in the relation $\Phi(B)\,\Psi(B) = I$ for $j = 1, 2, \ldots,$ we have

$$\Psi_j = \Phi_1\,\Psi_{j-1} + \Phi_2\,\Psi_{j-2} + \cdots + \Phi_p\,\Psi_{j-p}\,,$$

where $\Psi_0 = I$ and $\Psi_j = 0$ for $j < 0$.

Note that in the special case of the AR(1) model, $\boldsymbol{Y}_t = \Phi\,\boldsymbol{Y}_{t-1} + \varepsilon_t$, we have $\det\{\,I - \Phi\,B\,\} = 0$ if and only if $\det\{\,\lambda\,I - \Phi\,\} = 0$ with $\lambda = 1/B$. Hence, the stationarity condition for the AR(1) model is equivalent to the condition that all eigenvalues of Φ, that is, all roots of $\det\{\,\lambda\,I - \Phi\,\} = 0$, be less than one in absolute value. We consider the stationarity for the AR(1) process in more detail. For arbitrary $n > 0$, by $t+n$ successive substitutions in the right-hand side of the equation $\boldsymbol{Y}_t = \Phi\,\boldsymbol{Y}_{t-1} + \varepsilon_t$ we can obtain

$$\boldsymbol{Y}_t = \sum_{j=0}^{t+n} \Phi^j\,\varepsilon_{t-j} + \Phi^{t+n+1}\,\boldsymbol{Y}_{-n-1}\,. \tag{2.7}$$

Hence, provided that all eigenvalues of Φ are less than one in absolute value, as $n \to \infty$ this will converge to the infinite moving average representation $\boldsymbol{Y}_t = \sum_{j=0}^{\infty} \Phi^j\,\varepsilon_{t-j}$, with $\sum_{j=0}^{\infty} \|\,\Phi^j\,\| < \infty$, which is stationary. So from this discussion, in the AR(1) model we find that $\Psi(B) = (\,I - \Phi\,B\,)^{-1} = I + \sum_{j=1}^{\infty} \Phi^j\,B^j$ with $\Psi_j = \Phi^j$, $j \geq 1$. In addition, it follows that the stochastic structure of the AR(1) process for all $t \geq -n$ can be deduced from the above form (2.7) once a distribution for the initial value $\boldsymbol{Y}_{-n-1}$ is specified and the additional assumption that $\boldsymbol{Y}_{-n-1}$ is independent of the $\{\varepsilon_t\}$, for all $t \geq -n$.

This discussion can directly be extended to the general AR(p) process. The most immediate way to establish the extension to the AR(p) case is by noting that a vector AR(p) process $\boldsymbol{Y}_t$ given by (2.5) can always be expressed in the form of a kp-dimensional vector AR(1) model in terms of $\mathbf{Y}_t = (\,\boldsymbol{Y}_t', \ldots, \boldsymbol{Y}_{t-p+1}'\,)'$ as $\mathbf{Y}_t = \Phi\,\mathbf{Y}_{t-1} + \mathbf{e}_t$, with $\mathbf{e}_t = (\,\varepsilon_t', 0', \ldots, 0'\,)'$ and Φ equal to the $kp \times kp$ companion matrix associated with the AR(p) operator $\Phi(B)$, that is,

$$\Phi = \begin{bmatrix} \Phi_1 & \Phi_2 & . & . & . & \Phi_p \\ I & 0 & . & . & . & 0 \\ 0 & I & . & . & . & 0 \\ . & . & . & . & . & . \\ . & . & . & . & . & . \\ 0 & 0 & . & . & I & 0 \end{bmatrix}.$$

This AR(1) representation is, in fact, the transition equation of a state-space form, as discussed in Section 1.2.2, of the vector AR(p) model with $\mathbf{Y}_t$ representing the state vector $\mathbf{Z}_t$ in this situation. Then, because $\det\{ \Phi(B) \} = \det\{ I - \Phi B \}$, the stationarity condition on the roots of $\det\{ \Phi(B) \} = 0$ in the AR(p) model is equivalent to the condition in the corresponding AR(1) representation that all eigenvalues of the $kp \times kp$ companion matrix Φ be less than one in absolute value.

2.2.2 *Yule-Walker Relations for Covariance Matrices of a Vector AR Process*

For the AR(p) process $Y_t = \sum_{j=1}^{p} \Phi_j Y_{t-j} + \varepsilon_t$, since the lagged value Y_{t-l} can be expressed in terms of past values $\varepsilon_{t-l}, \varepsilon_{t-l-1}, \ldots$, in the infinite MA form $Y_{t-l} = \Psi(B)\, \varepsilon_{t-l} = \sum_{j=0}^{\infty} \Psi_j\, \varepsilon_{t-l-j}$, it follows that $E(Y_{t-l}\, \varepsilon_t') = 0$ for $l > 0$, and $E(Y_t\, \varepsilon_t') = \Sigma$. Hence, we have the *Yule-Walker relations* satisfied by the covariance matrices $\Gamma(l)$,

$$\Gamma(l) = E(Y_{t-l}\, Y_t') = E(Y_{t-l}\, [\textstyle\sum_{j=1}^{p} \Phi_j Y_{t-j} + \varepsilon_t]') = \displaystyle\sum_{j=1}^{p} \Gamma(l-j)\, \Phi_j', \tag{2.8}$$

for $l = 1, 2, \ldots$, with $\Gamma(0) = \sum_{j=1}^{p} \Gamma(-j)\, \Phi_j' + \Sigma$. These equations for $l = 0, 1, \ldots, p$ can be used to solve for these $\Gamma(l)$ simultaneously in terms of the AR parameter matrices Φ_j and Σ.

Conversely, for general order p, the matrices $\Phi_1, \ldots, \Phi_p$ and Σ can be determined from $\Gamma(0), \Gamma(1), \ldots, \Gamma(p)$ by solving the system of matrix Yule-Walker equations

$$\sum_{j=1}^{p} \Gamma(l-j)\, \Phi_j' = \Gamma(l), \qquad \text{for} \quad l = 1, 2, \ldots, p. \tag{2.9}$$

These equations can be written in matrix form as $\Gamma_p\, \Phi = \Gamma_{(p)}$ with solution $\Phi = \Gamma_p^{-1}\, \Gamma_{(p)}$, where

$$\Phi = (\Phi_1, \Phi_2, \ldots, \Phi_p)', \qquad \Gamma_{(p)} = (\Gamma(1)', \Gamma(2)', \ldots, \Gamma(p)')',$$

and Γ_p is a $kp \times kp$ matrix with (i, j)th block of elements equal to $\Gamma(i-j)$. Once the Φ_j are determined from (2.9), Σ can be obtained as

$$\Sigma = \Gamma(0) - \sum_{j=1}^{p} \Gamma(-j)\, \Phi_j' = \Gamma(0) - \Gamma_{(p)}'\, \Phi = \Gamma(0) - \Gamma_{(p)}'\, \Gamma_p^{-1}\, \Gamma_{(p)} = \Gamma(0) - \Phi'\, \Gamma_p\, \Phi .$$

2.2.3 *Covariance Features of the Vector AR(1) Model*

As an example of the above covariance matrix relations, for the AR(1) model with $p = 1$, $Y_t = \Phi\, Y_{t-1} + \varepsilon_t$, we have $\Gamma(0) = \Gamma(-1)\, \Phi' + \Sigma = \Gamma(1)'\, \Phi' + \Sigma$, and $\Gamma(1) = \Gamma(0)\, \Phi'$, with $\Gamma(l) = \Gamma(l-1)\, \Phi'$ for $l \geq 1$. Hence, it follows that

$$\Gamma(0) = \Phi\, \Gamma(0)\, \Phi' + \Sigma, \qquad \Gamma(1) = \Gamma(0)\, \Phi', \tag{2.10}$$

and $\Gamma(l) = \Gamma(0)\,\Phi'^{l}$, $l \geq 1$. Thus, the correlation matrices have the form $\rho(l) = V^{-1/2}\,\Gamma(l)\,V^{-1/2} = \rho(0)\,\{\,V^{1/2}\,\Phi'^{l}\,V^{-1/2}\,\}$. Now the AR(1) coefficient matrix Φ' can be expressed in the Jordan canonical form as $\Phi' = P\,J\,P^{-1}$, where P is a nonsingular matrix and J is a special upper triangular matrix that has the eigenvalues $\lambda_1, \lambda_2, \ldots, \lambda_k$ of Φ' on its diagonal and (possibly) has ones in certain positions just above the diagonal. Hence, $\Phi'^{l} = P\,J^{l}\,P^{-1}$ so that $\rho(l) = \rho(0)\,\{\,V^{1/2}\,P\,J^{l}\,P^{-1}\,V^{-1/2}\,\}$, and this illustrates that even for the simple vector AR(1) model the correlations will exhibit a mixture of decaying exponential and damping sinusoidal behavior as a function of the lag l, depending on the nature (real and/or complex conjugate values with absolute value less than one) of the eigenvalues of Φ. Equations (2.10) can be used to obtain Φ and Σ from the parameter matrices $\Gamma(0)$ and $\Gamma(1)$ as

$$\Phi' = \Gamma(0)^{-1}\,\Gamma(1) \qquad \text{and} \qquad \Sigma = \Gamma(0) - \Phi\,\Gamma(0)\,\Phi' = \Gamma(0) - \Gamma(1)'\,\Gamma(0)^{-1}\,\Gamma(1).$$

Conversely, given Φ and Σ, we can determine the $\Gamma(l)$ by first solving the linear equation $\Gamma(0) = \Phi\,\Gamma(0)\,\Phi' + \Sigma$ for $\Gamma(0)$, and then recursively obtain $\Gamma(l) = \Gamma(l-1)\,\Phi'$, $l \geq 1$. An explicit solution for $\Gamma(0)$ can be given by use of the vectorizing operation "vec" which forms a vector from a matrix by stacking the columns of the matrix below one another. Then, with $\boldsymbol{\gamma} = \text{vec}(\,\Gamma(0)\,)$ and $\boldsymbol{\sigma} = \text{vec}(\,\Sigma\,)$, and using the property that $\text{vec}(\,ABC\,) = (\,C' \otimes A\,)\,\text{vec}(\,B\,)$, where $\otimes$ denotes the Kronecker product of two matrices, we have $\boldsymbol{\gamma} = (\,\Phi \otimes \Phi\,)\,\boldsymbol{\gamma} + \boldsymbol{\sigma}$, so that $\boldsymbol{\gamma} = [\,I - (\,\Phi \otimes \Phi\,)\,]^{-1}\,\boldsymbol{\sigma}$. (For more details on the vec operator and Kronecker products and some of their basic properties, see Appendix A4.)

2.2.4 Univariate Model Structure Implied by Vector AR Model

Finally, we briefly comment on the nature of the univariate models for individual series Y_{it} from the vector AR(p) process. To be specific, consider the vector AR(1) model $(\,I - \Phi\,B\,)\,\boldsymbol{Y}_t = \varepsilon_t$. Let

$$\Phi^*(B) = \text{Adj}\{\,\Phi(B)\,\} = \text{Adj}\{\,I - \Phi\,B\,\},$$

a $k \times k$ matrix whose elements are polynomials of degrees at most $k-1$, and let $d(B) = \det\{\,I - \Phi\,B\,\}$, a polynomial of degree at most k. Then from the AR model equation, we have $d(B)\,\boldsymbol{Y}_t = \Phi^*(B)\,\varepsilon_t$. Now the right-hand side of this last equation, say $\boldsymbol{W}_t = \Phi^*(B)\,\varepsilon_t = (\,I - \Phi_1^*\,B - \cdots - \Phi_{k-1}^*\,B^{k-1}\,)\,\varepsilon_t$, is in the form of a vector MA($k-1$) process, so that from the discussion of MA processes we know that the individual series W_{it} on the right-hand side will have univariate MA($k-1$) representations as $W_{it} = \eta_i(B)\,a_{it}$. Hence, we have that $d(B)\,Y_{it} = \eta_i(B)\,a_{it}$, so that the individual series Y_{it} will follow univariate ARMA($k,k-1$) models. Note that k and $k-1$ are the maximum orders for the individual ARMA models, but because of cancellation (or near cancellation in practice) of possible common factors in $d(B)$ and $\eta_i(B)$ for the individual series, the univariate models may have lower orders, and the orders need not be identical for each series. This discussion is readily extended to the higher-order vector AR(p) model, with the result that the individual series follow univariate ARMA models of orders at most kp and $(k-1)p$. It is important to note, however, that the series a_{it} are univariate white noise but they do not necessarily have zero cross-covariances at all nonzero lags, that is, the vector $(a_{1t}, a_{2t}, \ldots, a_{kt})'$ does not form a vector white noise process.

EXAMPLE 2.3. Consider the bivariate ($k = 2$) AR(1) model $(I - \Phi B) \boldsymbol{Y}_t = \varepsilon_t$ with

$$\Phi = \begin{bmatrix} 0.8 & 0.7 \\ -0.4 & 0.6 \end{bmatrix} \quad \text{and} \quad \Sigma = \begin{bmatrix} 4 & 1 \\ 1 & 2 \end{bmatrix}.$$

Results of Example 2.1 illustrate that this model is stationary, with $\det\{\lambda I - \Phi\} = 0$ having complex roots, so that the correlations of this AR(1) process will exhibit damped sinusoidal behavior. The covariance matrix $\Gamma(0)$ is obtained by solving the linear equations $\Gamma(0) - \Phi\, \Gamma(0)\, \Phi' = \Sigma$, which together with $\Gamma(l) = \Gamma(l-1)\, \Phi'$, lead to the covariance matrices

$$\Gamma(0) = \begin{bmatrix} 18.536 & -1.500 \\ -1.500 & 8.884 \end{bmatrix}, \quad \Gamma(1) = \begin{bmatrix} 13.779 & -8.315 \\ 5.019 & 5.931 \end{bmatrix},$$

$$\Gamma(2) = \begin{bmatrix} 5.203 & -10.500 \\ 8.166 & 1.551 \end{bmatrix}, \quad \Gamma(3) = \begin{bmatrix} -3.188 & -8.381 \\ 7.619 & -2.336 \end{bmatrix},$$

$$\Gamma(4) = \begin{bmatrix} -8.417 & -3.754 \\ 4.460 & -4.449 \end{bmatrix}, \quad \Gamma(5) = \begin{bmatrix} -9.361 & 1.115 \\ 0.453 & -4.453 \end{bmatrix},$$

with corresponding correlation matrices

$$\rho(0) = \begin{bmatrix} 1.000 & -0.117 \\ -0.117 & 1.000 \end{bmatrix}, \quad \rho(1) = \begin{bmatrix} 0.743 & -0.648 \\ 0.391 & 0.668 \end{bmatrix},$$

$$\rho(2) = \begin{bmatrix} 0.281 & -0.818 \\ 0.636 & 0.175 \end{bmatrix}, \quad \rho(3) = \begin{bmatrix} -0.172 & -0.653 \\ 0.594 & -0.263 \end{bmatrix},$$

$$\rho(4) = \begin{bmatrix} -0.454 & -0.293 \\ 0.348 & -0.501 \end{bmatrix}, \quad \rho(5) = \begin{bmatrix} -0.505 & 0.087 \\ 0.035 & -0.501 \end{bmatrix}.$$

The autocorrelations and cross-correlations, $\rho_{ij}(l)$, of this process are displayed, up to 18 lags, in Figure 2.1. Hence, we find that the correlation patterns are rather complex and correlations do not die out very quickly in this example.

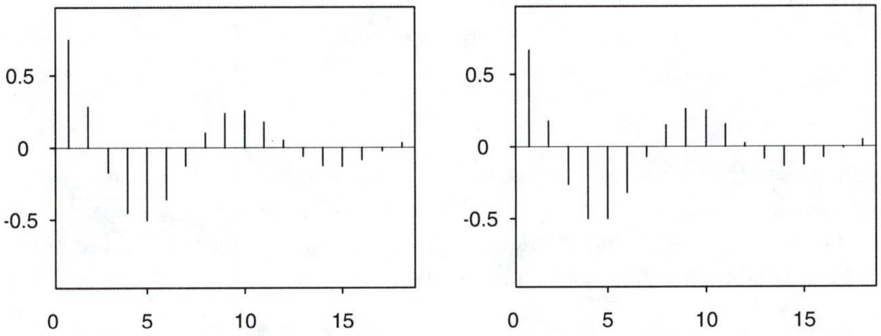

(a) Autocorrelations $\rho_{11}(l)$ and $\rho_{22}(l)$

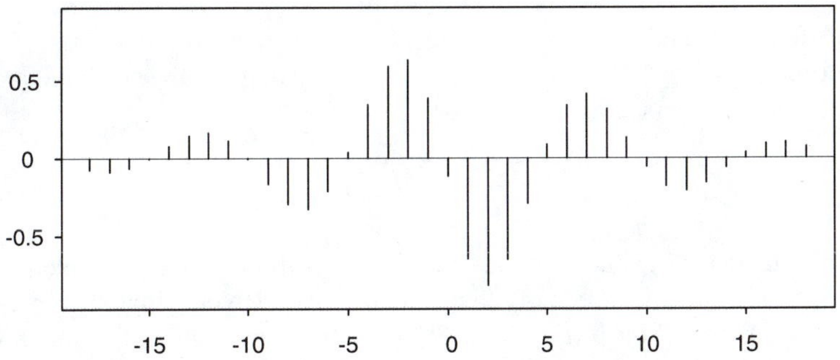

(b) Cross-correlations $\rho_{12}(l)$

Figure 2.1. Theoretical Autocorrelations and Cross-correlations, $\rho_{ij}(l)$, for the Bivariate AR(1) Process of Example 2.3

The coefficient matrices $\Psi_j = \Phi^j$, $j \geq 1$, in the infinite MA representation for this AR(1) process are

$$\Psi_1 = \begin{bmatrix} 0.80 & 0.70 \\ -0.40 & 0.60 \end{bmatrix}, \quad \Psi_2 = \begin{bmatrix} 0.36 & 0.98 \\ -0.56 & 0.08 \end{bmatrix}, \quad \Psi_3 = \begin{bmatrix} -0.10 & 0.84 \\ -0.48 & -0.34 \end{bmatrix},$$

$$\Psi_4 = \begin{bmatrix} -0.42 & 0.43 \\ -0.25 & -0.54 \end{bmatrix}, \quad \Psi_5 = \begin{bmatrix} -0.51 & -0.03 \\ 0.02 & -0.50 \end{bmatrix}, \quad \Psi_6 = \begin{bmatrix} -0.39 & -0.38 \\ 0.22 & -0.28 \end{bmatrix},$$

and so the elements of the Ψ_j matrices do not die out quickly and they exhibit damped sinusoidal behavior similar to that of the correlations.

EXAMPLE 2.4. Relation of AR(1) to Transfer Function Model. Consider the simple case of a bivariate ($k = 2$) AR(1) model, $(I - \Phi_1 B) \boldsymbol{Y}_t = \varepsilon_t$, where Φ_1 is 2×2 with elements ϕ_{ij}. When written explicitly, the AR(1) model equations are

$$Y_{1t} = \phi_{11} Y_{1,t-1} + \phi_{12} Y_{2,t-1} + \varepsilon_{1t}, \qquad Y_{2t} = \phi_{21} Y_{1,t-1} + \phi_{22} Y_{2,t-1} + \varepsilon_{2t},$$

which shows rather directly the existence of feedback between the two series. However, just as in the MA(1) model example, if $\phi_{12} = 0$, then we can show that the model can be represented in the form of a standard unidirectional transfer function model. Specifically, in this case we have

$$(1 - \phi_{11} B) Y_{1t} = \varepsilon_{1t}, \qquad (1 - \phi_{22} B) Y_{2t} = \phi_{21} Y_{1,t-1} + \varepsilon_{2t}.$$

Notice that these two equations constitute a simple example of the equations of the form of (1.15a)–(1.15b) of Section 1.2.2, with the second equation above representing a univariate ARMAX or transfer function form. As in Example 2.2 for the MA(1) model, we can again write $\varepsilon_{2t} = \beta \varepsilon_{1t} + a_{2t}$, where a_{2t} is uncorrelated with and, hence, independent of ε_{1t}, with $\beta = \sigma_{21} / \sigma_{11}$. Then, substituting $\varepsilon_{2t} = \beta (1 - \phi_{11} B) Y_{1t} + a_{2t}$, and rearranging, the second equation above can be reexpressed as

$$Y_{2t} = \frac{\beta - (\beta \phi_{11} - \phi_{21}) B}{1 - \phi_{22} B} Y_{1t} + \frac{1}{1 - \phi_{22} B} a_{2t},$$

which is a standard transfer function model form with the noise a_{2t} independent of ε_{1t}, hence independent of the "input" Y_{1t}. Note that the relation $(1 - \phi_{22} B) Y_{2t} = [\beta - (\beta \phi_{11} - \phi_{21}) B] Y_{1t} + a_{2t}$, from which the above standard transfer function model form is immediately obtained, follows directly from the model expressed in the form of (1.11) of Section 1.2.2 as $\Phi_0^{\#} \boldsymbol{Y}_t - \Phi_1^{\#} \boldsymbol{Y}_{t-1} = \boldsymbol{a}_t$, with

$$\Phi_0^{\#} = \begin{bmatrix} 1 & 0 \\ -\beta & 1 \end{bmatrix}, \qquad \Phi_1^{\#} = \Phi_0^{\#} \Phi_1 = \begin{bmatrix} \phi_{11} & 0 \\ (-\beta \phi_{11} + \phi_{21}) & \phi_{22} \end{bmatrix}.$$

In general, for an AR(1) model in higher dimensions $k > 2$, if the k series Y_{it}, $i = 1, \ldots, k$, can be arranged so that the matrix Φ is lower (or block lower) triangular, then the AR(1) model can also be reexpressed in the form of (block) unidirectional transfer function equations.

EXAMPLE 2.5. Reduced-Rank Model and Echelon Form. Consider further the bivariate AR(1) model, $(I - \Phi_1 B) \boldsymbol{Y}_t = \varepsilon_t$, as in Example 2.4. Now suppose that the matrix Φ_1 is of reduced rank one, so that $\det(\Phi_1) = \phi_{11}\phi_{22} - \phi_{21}\phi_{12} = 0$, and suppose that none of the elements ϕ_{ij} are equal to zero. (Otherwise, one of the rows or columns of Φ_1 is identically zero and a simple structure already exists.) Then define $\alpha = \phi_{21}/\phi_{11} = \phi_{22}/\phi_{12}$, and set

$$\Phi_0^\# = \begin{bmatrix} 1 & 0 \\ -\alpha & 1 \end{bmatrix}, \quad \text{with} \quad \Phi_1^\# = \Phi_0^\# \, \Phi_1 = \begin{bmatrix} \phi_{11} & \phi_{12} \\ 0 & 0 \end{bmatrix},$$

since $\phi_{21} - \alpha\,\phi_{11} = 0$ and $\phi_{22} - \alpha\,\phi_{12} = 0$. Therefore, we have the equivalent "echelon form" model structure, $\Phi_0^\# \, \boldsymbol{Y}_t = \Phi_1^\# \, \boldsymbol{Y}_{t-1} + \Phi_0^\# \, \varepsilon_t$, where the second row of $\Phi_1^\#$ is equal to zero, and $\Phi_0^\#$ is lower triangular with ones on the diagonal. (A general discussion of echelon form structure of ARMA models will be given in Section 3.1.) So we find in this example that the reduced-rank structure of Φ_1 leads to a simplified structure in the echelon form. Also found directly from this form is the interpretation that the reduced-rank structure implies that the linear combination $Y_{2t} - \alpha\, Y_{1t} = [\, -\alpha \quad 1\,]\, \boldsymbol{Y}_t$ is a white noise process.

2.3 Vector Mixed Autoregressive Moving Average Models

We now briefly consider general properties of the mixed vector *autoregressive moving average* ARMA(p,q) model, given by

$$(\, \boldsymbol{Y}_t - \boldsymbol{\mu}\,) - \sum_{j=1}^{p} \Phi_j \,(\, \boldsymbol{Y}_{t-j} - \boldsymbol{\mu}\,) = \varepsilon_t - \sum_{j=1}^{q} \Theta_j \,\varepsilon_{t-j}, \qquad (2.11)$$

or $\Phi(B)\,(\, \boldsymbol{Y}_t - \boldsymbol{\mu}\,) = \Theta(B)\,\varepsilon_t$, where $\Phi(B) = I - \Phi_1\, B - \Phi_2\, B^2 - \cdots - \Phi_p\, B^p$, $\Theta(B) = I - \Theta_1\, B - \Theta_2\, B^2 - \cdots - \Theta_q\, B^q$, and the ε_t are white noise with zero mean vector and covariance matrix Σ.

2.3.1 Stationarity and Invertibility of the Vector ARMA Model

First, the conditions for stationarity and invertibility of the vector ARMA process are clearly the same as in the pure AR and pure MA cases, respectively. Under the stationarity condition that all the roots of $\det\{\, \Phi(B)\,\} = 0$ are greater than one in absolute value, we have the convergent causal infinite MA representation as $\boldsymbol{Y}_t = \boldsymbol{\mu} + \Psi(B)\,\varepsilon_t$, where $\Psi(B) = \Phi(B)^{-1}\Theta(B) = \sum_{j=0}^{\infty} \Psi_j\, B^j$. Similar to the pure AR model situation, the coefficients Ψ_j can be determined from the relation $\Phi(B)\,\Psi(B) = \Theta(B)$, and hence by equating coefficient matrices in this relation, the Ψ_j satisfy

$$\Psi_j = \Phi_1\, \Psi_{j-1} + \Phi_2\, \Psi_{j-2} + \cdots + \Phi_p\, \Psi_{j-p} - \Theta_j, \qquad j = 1, 2, \ldots, \qquad (2.12)$$

where $\Psi_0 = I$, $\Psi_j = 0$ for $j < 0$, and $\Theta_j = 0$ for $j > q$. Conversely, under the invertibility condition that all the roots of $\det\{\, \Theta(B)\,\} = 0$ are greater than one in absolute value, $\{\boldsymbol{Y}_t\}$ has the convergent infinite AR representation as $\Pi(B)\,(\, \boldsymbol{Y}_t - \boldsymbol{\mu}\,) = \varepsilon_t$, where $\Pi(B) = \Theta(B)^{-1}\Phi(B) = I - \sum_{j=1}^{\infty} \Pi_j\, B^j$. Similar to the pure MA model situation, the weights Π_j can be determined by equating coefficient matrices in the relation $\Theta(B)\,\Pi(B) = \Phi(B)$, and, hence, they satisfy

$$\Pi_j = \Theta_1\, \Pi_{j-1} + \Theta_2\, \Pi_{j-2} + \cdots + \Theta_q\, \Pi_{j-q} + \Phi_j, \qquad j = 1, 2, \ldots, \qquad (2.13)$$

where $\Pi_0 = -I$, $\Pi_j = 0$ for $j < 0$, and $\Phi_j = 0$ for $j > p$.

For the general stationary vector ARMA(p,q) process, from the infinite MA representation $Y_t = \mu + \sum_{j=0}^{\infty} \Psi_j \varepsilon_{t-j}$, we always have the representation for the covariance matrices as $\Gamma(l) = \sum_{j=0}^{\infty} \Psi_j \Sigma \Psi'_{j+l}$. From this it follows that the covariance matrix-generating function is given by $g(z) = \sum_{l=-\infty}^{\infty} \Gamma(l) z^l = \Psi(z^{-1}) \Sigma \Psi(z)'$, and, hence, the spectral density matrix of the vector ARMA(p,q) process is given by

$$f(\lambda) = \frac{1}{2\pi} \Psi(e^{i\lambda}) \Sigma \Psi(e^{-i\lambda})',$$

where $\Psi(z) = \Phi(z)^{-1} \Theta(z)$. For example, for the vector MA(1) process, the spectral density matrix is

$$\begin{aligned} f(\lambda) &= (2\pi)^{-1} (I - \Theta e^{i\lambda}) \Sigma (I - \Theta e^{-i\lambda})' \\ &= (2\pi)^{-1} [-\Theta \Sigma e^{i\lambda} + (\Sigma + \Theta \Sigma \Theta') - \Sigma \Theta' e^{-i\lambda}]. \end{aligned}$$

2.3.2 Relations for the Covariance Matrices of the Vector ARMA Model

From the infinite MA representation $Y_{t-l} - \mu = \sum_{i=0}^{\infty} \Psi_i \varepsilon_{t-l-i}$, we find that $E[(Y_{t-l} - \mu) \varepsilon'_{t-j}] = \Psi_{j-l} \Sigma$. Thus, it is easy to determine, using (2.11), that the covariance matrices $\Gamma(l) = E[(Y_{t-l} - \mu)(Y_t - \mu)']$ satisfy the relations

$$\begin{aligned} \text{Cov}(Y_{t-l}, Y_t) &= \sum_{j=1}^{p} \text{Cov}(Y_{t-l}, Y_{t-j}) \Phi'_j \\ &\quad + \text{Cov}(Y_{t-l}, \varepsilon_t) - \sum_{j=1}^{q} \text{Cov}(Y_{t-l}, \varepsilon_{t-j}) \Theta'_j, \end{aligned}$$

and, hence,

$$\Gamma(l) = \sum_{j=1}^{p} \Gamma(l-j) \Phi'_j - \sum_{j=l}^{q} \Psi_{j-l} \Sigma \Theta'_j, \qquad l = 0, 1, \ldots, q, \tag{2.14}$$

with the convention that $\Theta_0 = -I$, and $\Gamma(l) = \sum_{j=1}^{p} \Gamma(l-j) \Phi'_j$ for $l > q$. Note that these last equations for $l = q+1, \ldots, q+p$ represent a set of linear equations in $\Phi_1, \ldots, \Phi_p$ and can be useful in determining the Φ_j from the $\Gamma(l)$. This system of linear equations $\Gamma(l) = \sum_{j=1}^{p} \Gamma(l-j) \Phi'_j$ for $l = q+1, \ldots, q+p$ will have a unique solution for the coefficient matrices Φ_j in terms of the $\Gamma(l)$ provided that the appropriate matrix $\{\Gamma(l-j)\}$, $l = q+1, \ldots, q+p$, $j = 1, \ldots, p$, be of full rank, and this will occur if certain rank conditions, including a rank identifiability condition to be discussed shortly in a subsequent subsection, hold. Specifically, for a unique solution for the Φ_j to exist requires that rank$[\Phi_p, \Theta_q] = k$ and that an additional similar rank condition be satisfied for the "backward ARMA model representation" of the process [see Hannan (1975) and An, Chen, and Hannan (1983)].

Conversely, equations (2.14) involving the $\Gamma(l)$ for $l = 0, 1, \ldots, p$ represent linear equations that can be used to determine these $\Gamma(l)$ in terms of the ARMA parameters Φ_j, Θ_j, and Σ [e.g., see Nicholls and Hall (1979), Ansley (1980), Kohn and Ansley (1982), and Mittnik (1990, 1993), for details].

2.3.3 Some Features of the Vector ARMA(1,1) Model

Consider in particular the ARMA(1,1) model, $\boldsymbol{Y}_t - \Phi\, \boldsymbol{Y}_{t-1} = \varepsilon_t - \Theta\, \varepsilon_{t-1}$. Let us consider in more detail the stationarity and invertibility for this model. As in the AR(1) model, for arbitrary $n > 0$ by $t{+}n$ successive substitutions we obtain

$$\boldsymbol{Y}_t = \varepsilon_t + \sum_{j=1}^{t+n} \Phi^{j-1} (\, \Phi - \Theta \,)\, \varepsilon_{t-j} \; + \; \Phi^{t+n} (\, \Phi\, \boldsymbol{Y}_{-n-1} - \Theta\, \varepsilon_{-n-1} \,). \qquad (2.15)$$

So a stationary (convergent) infinite moving average representation will be obtained in the above equation as $n \to \infty$ if all the eigenvalues of Φ are less than one in absolute value, the same condition as in the AR(1) model. The infinite MA form is $\boldsymbol{Y}_t = (\, I - \Phi B \,)^{-1} (\, I - \Theta B \,)\, \varepsilon_t$, with the Ψ_j satisfying $\Psi_1 = \Phi\, \Psi_0 - \Theta = \Phi - \Theta$, $\Psi_j = \Phi\, \Psi_{j-1}$, $j \ge 2$. Hence, we find that $\Psi_j = \Phi^{j-1} (\, \Phi - \Theta \,)$, $j = 1, 2, \ldots$. Similarly, we can obtain the form

$$\boldsymbol{Y}_t = \sum_{j=1}^{t+n} \Theta^{j-1} (\, \Phi - \Theta \,)\, \boldsymbol{Y}_{t-j} \; + \; \varepsilon_t \; + \; \Theta^{t+n} (\, \Phi\, \boldsymbol{Y}_{-n-1} - \Theta\, \varepsilon_{-n-1} \,). \qquad (2.16)$$

Hence, a convergent infinite autoregressive representation is obtained as $n \to \infty$, and so the process will be invertible, if all the eigenvalues of Θ are less than one in absolute value, the same condition as in the MA(1) model. Thus, we find that $\Pi_j = \Theta^{j-1} (\, \Phi - \Theta \,)$, $j \ge 1$.

Also, for the ARMA(1,1) model, equations (2.14) for the $\Gamma(l)$ take the form

$$\Gamma(0) = \Gamma(1)'\, \Phi' + \Sigma - \Psi_1\, \Sigma\, \Theta' = \Gamma(1)'\, \Phi' + \Sigma - (\, \Phi - \Theta \,)\, \Sigma\, \Theta',$$

$$\Gamma(1) = \Gamma(0)\, \Phi' - \Sigma\, \Theta',$$

and $\Gamma(l) = \Gamma(l{-}1)\, \Phi'$, $l \ge 2$. The first two equations can be used to solve for $\Gamma(0)$ and $\Gamma(1)$ in terms of Φ, Θ, and Σ, with the remaining $\Gamma(l)$ determined from $\Gamma(l) = \Gamma(l{-}1)\, \Phi' = \Gamma(1)\, \Phi'^{\,l-1}$, $l \ge 2$. Specifically, from the first two equations we have

$$\Gamma(0) - \Phi\, \Gamma(0)\, \Phi' = \Sigma - \Theta\, \Sigma\, \Phi' - (\, \Phi - \Theta \,)\, \Sigma\, \Theta' \equiv \Sigma^*,$$

and this equation can be solved for $\Gamma(0)$, equivalently, for $\boldsymbol{\gamma} = \text{vec}(\, \Gamma(0) \,)$, by the use of the vectorizing equations similar to those described in Section 2.2.3 for the case of the AR(1) model. Conversely, the first three equations above for $\Gamma(0)$, $\Gamma(1)$, and $\Gamma(2)$ can be used to solve for Φ, Θ, and Σ in terms of the $\Gamma(l)$, such as $\Phi' = \Gamma(1)^{-1}\, \Gamma(2)$, provided that the condition rank[Φ, Θ] $= k$ is satisfied so that $\Gamma(1) = \Gamma(0)\, \Phi' - \Sigma\, \Theta'$ will be nonsingular [e.g., see Hannan (1975)].

2.3.4 *Consideration of Parameter Identifiability for Vector ARMA Models*

We must note that for mixed vector ARMA(p,q) models as in (2.11) with both $p > 0$ and $q > 0$ specified, certain conditions on the matrix operators $\Phi(B)$ and $\Theta(B)$ are needed to ensure uniqueness of the parameters in the ARMA representation. Specifically, in the vector situation it is possible for two ARMA(p,q) model representations, $\Phi(B)\,\boldsymbol{Y}_t = \Theta(B)\,\varepsilon_t$ and $\Phi_*(B)\,\boldsymbol{Y}_t = \Theta_*(B)\,\varepsilon_t$, with different coefficient matrices to give rise to the same coefficients Ψ_j in the infinite MA representation $\boldsymbol{Y}_t = \Psi(B)\,\varepsilon_t = \sum_{j=0}^{\infty} \Psi_j\,\varepsilon_{t-j}$ of the process, that is, such that

$$\Psi(B) = \Phi(B)^{-1}\Theta(B) = \Phi_*(B)^{-1}\Theta_*(B).$$

Thus, the different representations also give rise to the same covariance matrix structure $\Gamma(l)$ of the process. Two ARMA(p,q) model representations with this property are said to be *observationally equivalent*. Observationally equivalent ARMA(p,q) representations can exist under some circumstances because matrix operators $\Phi_*(B)$ and $\Theta_*(B)$ could be related to $\Phi(B)$ and $\Theta(B)$, respectively, by a common left matrix factor $U(B)$ as $\Phi_*(B) = U(B)\,\Phi(B)$ and $\Theta_*(B) = U(B)\,\Theta(B)$ but such that the orders of $\Phi_*(B)$ and $\Theta_*(B)$ are not increased over those of $\Phi(B)$ and $\Theta(B)$, respectively. [This common left factor $U(B)$ would cancel when $\Phi_*(B)^{-1}\Theta_*(B)$ is formed.] The parameters of the ARMA(p,q) model $\Phi(B)\,\boldsymbol{Y}_t = \Theta(B)\,\varepsilon_t$ are *identifiable* if the Φ_j and Θ_j are uniquely determined by the impulse response matrices Ψ_j of the operator $\Psi(B)$ in the (unique) infinite MA representation $\boldsymbol{Y}_t = \Psi(B)\,\varepsilon_t = \sum_{j=0}^{\infty} \Psi_j\,\varepsilon_{t-j}$ of the process, i.e., there is no other pair of ARMA(p,q) operators ($\Phi_*(B)$, $\Theta_*(B)$) which is observationally equivalent to ($\Phi(B)$, $\Theta(B)$).

In addition to the stationarity and invertibility conditions,

(i) all roots of $\det\{\,\Phi(B)\,\} = 0$ and all roots of $\det\{\,\Theta(B)\,\} = 0$ must be greater than one in absolute value,

the following two conditions are sufficient for identifiability of the parameters in the ARMA(p,q) model (Hannan, 1969, 1975),

(ii) the matrices $\Phi(B)$ and $\Theta(B)$ have no common left factors other than unimodular ones, i.e., if $\Phi(B) = U(B)\,\Phi_1(B)$ and $\Theta(B) = U(B)\,\Theta_1(B)$, then the common factor $U(B)$ must be unimodular, that is, $\det\{\,U(B)\,\}$ is a (nonzero) constant, and

(iii) with q as small as possible and p as small as possible for that q, the matrix $[\,\Phi_p,\ \Theta_q\,]$ must be of full rank k.

When property (ii) holds, the operators $\Phi(B)$ and $\Theta(B)$ are called left-coprime, and the representation $\Psi(B) = \Phi(B)^{-1}\Theta(B)$ is said to be irreducible. It follows that if $\Phi(B)$ and $\Theta(B)$ are left-coprime and $\Phi_*(B)$ and $\Theta_*(B)$ are also left-coprime, then ($\Phi(B), \Theta(B)$) and ($\Phi_*(B), \Theta_*(B)$) are observationally equivalent if and only if there exists a (nonsingular) unimodular matrix $U(B)$ such that ($\Phi_*(B), \Theta_*(B)$) $= U(B)$ ($\Phi(B), \Theta(B)$) [see Hannan and Deistler

(1988, Sec. 2.2)]. Then, for any two observationally equivalent representations that satisfy condition (ii), condition (iii) ensures that the unimodular matrix $U(B)$ that relates them must be equal to a constant matrix [otherwise the orders of the matrix operator $U(B)\,(\,\Phi(B),\,\Theta(B)\,)$ would be increased beyond the smallest possible orders], and, hence, that $U(B) \equiv I$ since the AR and MA operators have leading coefficient matrices equal to the identity matrix.

Other more general conditions than (iii) for identifiability have been considered by Hannan (1971, 1976), Deistler, Dunsmuir, and Hannan (1978), and Hannan and Deistler (1988, Sec. 2.7), for example. Specifically, suppose that the (maximum) degrees of the ith columns of the AR and MA polynomial operators $\Phi(B)$ and $\Theta(B)$ are prescribed, denoted as p_i and q_i, respectively, for $i = 1, \ldots, k$. Let $\phi_{p_i}(i)$, $i = 1, \ldots, k$, denote the column vector of coefficients of B^{p_i} in the ith column of $\Phi(B)$, i.e., the coefficients of the highest prescribed order p_i in the ith column of $\Phi(B)$, and similarly denote $\theta_{q_i}(i)$ in terms of $\Theta(B)$. Then the parameters of the vector ARMA model with these prescribed column degrees are identifiable if the "column end matrix" $[\,\phi_{p_1}(1), \ldots, \phi_{p_k}(k), \theta_{q_1}(1), \ldots, \theta_{q_k}(k)\,]$ corresponding to the prescribed column degrees $(\,p_1, \ldots, p_k, q_1, \ldots, q_k\,)$ has full rank k. The above condition (iii) is a special case which, together with (i) and (ii), results in a situation that is referred to as *block identifiability* of the ARMA(p,q) model, and it implies that the parameters Φ_j and Θ_j of the model are identifiable simply by specification of the overall orders p and q.

However, for some ARMA(p,q) model structures condition (iii) (or its generalizations) cannot be satisfied, that is, condition (iii) is "overidentifying". In this case, further constraints on the parameter matrices must be imposed for identifiability of parameters, such as constraints which specify the ranks of certain coefficient matrices. As an alternative approach to identification, we can consider expressing the ARMA model in a certain "canonical" form such that there is one and only one unique representative model of this form for each class of observationally equivalent ARMA models, and the parameters in this (unique) representation will then necessarily be identifiable. Specifically, the ARMA model can be represented in the form of (1.13) of Section 1.2.2, $\Phi^{\#}(B)\,\boldsymbol{Y}_t = \Theta^{\#}(B)\,\varepsilon_t$, where additional order indices $K_1, \ldots, K_k$ known as the Kronecker indices or structural indices must be determined beyond the overall orders p and q. The "echelon (canonical) form" of the ARMA model in the form of (1.13) is determined as the representation such that $(\,\Phi^{\#}(B),\,\Theta^{\#}(B)\,)$ has the smallest possible row degrees, and K_i denotes the degree of the ith row of $(\,\Phi^{\#}(B),\,\Theta^{\#}(B)\,)$, that is, the maximum of the degrees of the polynomials in the ith row of $(\,\Phi^{\#}(B),\,\Theta^{\#}(B)\,)$, for $i = 1, \ldots, k$, with $p = q = \max\{\,K_1, \ldots, K_k\,\}$. The specification of these Kronecker indices or "row orders" $\{K_i\}$, which are unique for any given equivalence class of observationally equivalent ARMA models which have the (same) infinite MA operator $\Psi(B)$, then determines a unique "echelon (canonical) form" of the ARMA model (1.13) in which the unknown parameters are uniquely identifiable. In the special case where all K_i are equal to a common value p, we then have the situation of block identifiability and the echelon form has the standard ARMA model form (1.9) with $p = q$ and (Φ_p, Θ_p) of full rank k. The echelon form structure and the identifiability conditions in terms of the echelon form of the ARMA model, as in the ARMA model form (1.13), have been examined extensively by Hannan and Deistler (1988, Chap. 2) and will be discussed further in Section 3.1.

To illustrate the need for conditions such as (ii) and (iii) above, consider the k-dimensional vector ARMA(1,1) model $(I - \Phi B) \boldsymbol{Y}_t = (I - \Theta B) \varepsilon_t$, and suppose that $\text{rank}[\Phi, \Theta] = r < k$. Then there exists a $(k-r)\times k$ matrix F' such that $F' [\Phi, \Theta] = [F'\Phi, F'\Theta] = 0$. So for any arbitrary $k\times(k-r)$ matrix A, we have

$$(I - (AF') B) (I - \Phi B) \boldsymbol{Y}_t$$

$$= (I - (AF' + \Phi) B) \boldsymbol{Y}_t = (I - (AF') B) (I - \Theta B) \varepsilon_t = (I - (AF' + \Theta) B) \varepsilon_t ,$$

which is of the ARMA(1,1) form $(I - \Phi_* B) \boldsymbol{Y}_t = (I - \Theta_* B) \varepsilon_t$, where $\Phi_* = \Phi + AF'$ and $\Theta_* = \Theta + AF'$ are arbitrary to a certain extent. Hence, the parameters Φ and Θ in such an ARMA(1,1) model are not unique and would not be identifiable, unless appropriate "reduced-rank" constraints were explicitly imposed on the matrix $[\Phi, \Theta]$. That is, the parameters would not be identifiable unless the coefficient matrices Φ and Θ were explicitly parameterized to have reduced-rank forms such as $[\Phi, \Theta] = A_1 [B_1, C_1]$, where A_1 is $k \times r$ and B_1 and C_1 are each $r \times k$ matrices, and A_1, B_1, C_1 are further normalized to ensure uniqueness of this reduced-rank factorization (e.g., the matrix A_1 could be required to have zeros above the diagonal and ones on the diagonal). Equivalently, under the reduced-rank assumption, a lower triangular matrix $\Phi_0^{\#}$ with ones on the diagonal can be determined [using the $(k - r)$ rows of the matrix F' above] with the property that $\Phi_0^{\#} [\Phi, \Theta] \equiv [\Phi_1^{\#}, \Theta_1^{\#}]$ has $(k - r)$ of its rows identically equal to zero, and $\Phi_1^{\#}$ or $\Theta_1^{\#}$ may require certain additional zero coefficient constraints in some of its remaining r rows. Then we would have a unique echelon form structure $\Phi_0^{\#} \boldsymbol{Y}_t - \Phi_1^{\#} \boldsymbol{Y}_{t-1} = \Theta_0^{\#} \varepsilon_t - \Theta_1^{\#} \varepsilon_{t-1}$, with $\Theta_0^{\#} = \Phi_0^{\#}$, for the ARMA(1,1) model and this parameterization of the model in terms of the parameters of $\Phi_0^{\#}$, $\Phi_1^{\#}$, and $\Theta_1^{\#}$ would be uniquely identifiable.

EXAMPLE 2.6. For instance, consider a case of the above reduced-rank ARMA(1,1) model with $k = 3$ series and $r = \text{rank}[\Phi, \Theta] = 2$, and suppose that the third row of $[\Phi, \Theta]$ is a linear combination of the first two rows so that $[- \alpha_1, - \alpha_2, 1] [\Phi, \Theta] = 0$. Then we can define

$$\Phi_0^{\#} = \begin{bmatrix} 1 & 0 & 0 \\ 0 & 1 & 0 \\ -\alpha_1 & -\alpha_2 & 1 \end{bmatrix},$$

$$\Phi_1^{\#} = \Phi_0^{\#}\,\Phi = \begin{bmatrix} \phi_{11} & \phi_{12} & \phi_{13} \\ \phi_{21} & \phi_{22} & \phi_{23} \\ 0 & 0 & 0 \end{bmatrix}, \qquad \Theta_1^{\#} = \Phi_0^{\#}\,\Theta = \begin{bmatrix} \theta_{11} & \theta_{12} & \theta_{13} \\ \theta_{21} & \theta_{22} & \theta_{23} \\ 0 & 0 & 0 \end{bmatrix},$$

and we can obtain a model in the echelon form as in (1.13) of Section 1.2.2 with the specification that $K_1 = K_2 = 1$ and $K_3 = 0$ are the Kronecker indices. With this specification (i.e., that the third rows of $\Phi_1^{\#}$ and $\Theta_1^{\#}$ are equal to zero) and the additional constraints that either $\phi_{13} = \phi_{23} = 0$ or $\theta_{13} = \theta_{23} = 0$, the remaining parameters of the model in this form are identifiable. The need for the additional constraints is because from the third equation of the model, $Y_{3t} - \alpha_1\, Y_{1t} - \alpha_2\, Y_{2t} = \varepsilon_{3t} - \alpha_1\, \varepsilon_{1t} - \alpha_2\, \varepsilon_{2t}$, we see that $Y_{3,t-1}$ (equivalently $\varepsilon_{3,t-1}$) is a linear combination of the other variables in $\boldsymbol{Y}_{t-1}$ and ε_{t-1} and, hence, this variable must be eliminated from the first two equations of the model for identifiability of parameters. Alternatively, for this example, we could specify the coefficient matrices Φ and Θ in the standard ARMA model to have the reduced-rank form as $[\,\Phi, \Theta\,] = A_1\,[\,B_1, C_1\,]$, where the 3×2 matrix A_1 is of the form

$$A_1 = \begin{bmatrix} 1 & 0 \\ 0 & 1 \\ \alpha_1 & \alpha_2 \end{bmatrix}$$

(that is, A_1 is such that if the 3×3 matrix A is formed by appending the third column of the 3×3 identity matrix to A_1, then $A^{-1} = \Phi_0^{\#}$), and the 2×3 matrices B_1 and C_1 are of the same form as the first two rows of $\Phi_1^{\#}$ and $\Theta_1^{\#}$, respectively. Hence, we see simply that $[\,\Phi, \Theta\,] = A_1\,[\,B_1, C_1\,] = A\,[\,\Phi_1^{\#}, \Theta_1^{\#}\,] = \Phi_0^{\#-1}[\,\Phi_1^{\#}, \Theta_1^{\#}\,]$ is the representation of the (reduced-rank) coefficient matrices in the standard ARMA model form.

In the above general ARMA(1,1) model example, the ARMA(1,1) representations with different coefficient matrices are observationally equivalent since they each give rise to the same coefficients $\Psi_j = \Phi^{j-1}(\,\Phi - \Theta\,) \equiv \Phi_*^{j-1}(\,\Phi_* - \Theta_*\,)$ in the infinite MA representation $\boldsymbol{Y}_t = \sum_{j=0}^{\infty} \Psi_j\, \varepsilon_{t-j}$ of the process. Hence, identifiability of the parameters Φ and Θ is seen to require specification of additional conditions sufficient to uniquely select one parameter set from the class of equivalent structures.

2.3.5 Further Aspects of Nonuniqueness of Vector ARMA Model Representations

In addition to the complications concerning identifiability of parameters in ARMA models in the vector case, the related notion of exchangeable models also exists in the vector case because of the more complicated nature of matrix polynomials which are not present in the scalar case. Two different vector ARMA models (i.e., models of different orders) are said to be *exchangeable* if they give rise to the same covariance structure for the process $\boldsymbol{Y}_t$ or, equivalently, if they give rise to the same set of infinite MA coefficient matrices Ψ_j. (These different ARMA representations would also be called observationally equivalent within the general class of all finite order ARMA models.) The existence of exchangeable models can, in particular, be caused by the presence of a unimodular matrix factor in an AR or MA matrix polynomial operator. For example, in an MA model $\boldsymbol{Y}_t = U(B)\,\Theta(B)\,\varepsilon_t$ where $U(B)$ is unimodular, the model is equivalently representable as the finite order ARMA(p,q) model $U(B)^{-1}\boldsymbol{Y}_t = \Theta(B)\,\varepsilon_t$, since the polynomial operator $U(B)$ is a unimodular matrix if and only if $U(B)^{-1}$ is a matrix polynomial of finite order. [This last property can easily be seen from the relation $U(B)^{-1} = (\,1/\det\{\,U(B)\,\}\,)\,\mathrm{Adj}\{\,U(B)\,\}$, so that $U(B)^{-1}$ will be of finite order if and only if $\det\{\,U(B)\,\}$ is a constant.] As a simple example, the bivariate MA(1) model $\boldsymbol{Y}_t = \varepsilon_t - \Theta\,\varepsilon_{t-1}$ and the bivariate AR(1) model $\boldsymbol{Y}_t - \Phi\,\boldsymbol{Y}_{t-1} = \varepsilon_t$, where

$$\Theta = \begin{bmatrix} 0.0 & \theta_{12} \\ 0.0 & 0.0 \end{bmatrix} \quad \text{and} \quad \Phi = \begin{bmatrix} 0.0 & -\theta_{12} \\ 0.0 & 0.0 \end{bmatrix},$$

are easily seen to be exchangeable models since $(\,I - \Theta\,B\,)$ is unimodular with $(\,I - \Theta\,B\,)^{-1} = (\,I + \Theta\,B\,)$ in this example. [Note, however, that both the matrix parameter Φ in the AR(1) model and Θ in the MA(1) model representation are identifiable, and so no difficulties arise in estimation of either model.] In addition, the bivariate ARMA(1,1) model $\boldsymbol{Y}_t - \Phi_*\,\boldsymbol{Y}_{t-1} = \varepsilon_t - \Theta_*\,\varepsilon_{t-1}$, where

$$\Phi_* = \begin{bmatrix} 0.0 & \alpha \\ 0.0 & 0.0 \end{bmatrix} \quad \text{and} \quad \Theta_* = \begin{bmatrix} 0.0 & \theta_{12} + \alpha \\ 0.0 & 0.0 \end{bmatrix},$$

is obtained by left multiplication of the MA(1) model by the unimodular matrix factor $U(B) = I - \Phi_*\,B$ and is observationally equivalent to both the AR(1) and the MA(1) model for any arbitrary value of α, since $(\,I - \Phi_*\,B\,)^{-1}(\,I - \Theta_*\,B\,) = (\,I + \Phi_*\,B\,)\,(\,I - \Theta_*\,B\,) = (\,I - \Theta\,B\,)$, for example. Hence, the parameters in Φ_* and Θ_* in the ARMA(1,1) model representation are not identifiable since α is arbitrary. The topic of unique vector ARMA model representations and model specifications with uniquely identified parameters in the "echelon (canonical) form" will be discussed further in Section 3.1.

2.4 Nonstationary Vector ARMA Models

Many time series in practice exhibit nonstationary behavior, often of a homogeneous nature, that is, drifting or trending behavior such that apart from a shifting local level or local trend the series has homogeneous patterns with respect to shifts over time. Often, for example, in univariate integrated ARMA (ARIMA) time series models, a nonstationary time series can be reduced to stationarity through differencing of the series.

2.4.1 Vector ARIMA Models for Nonstationary Processes

For vector ARMA models, the condition for stationarity of the vector series $\boldsymbol{Y}_t$ is that all roots of $\det\{\,\Phi(B)\,\} = 0$ be greater than one in absolute value. To generalize this to nonstationary, but nonexplosive processes, we can consider a general form of the vector ARMA model, $\Phi(B)\,\boldsymbol{Y}_t = \Theta(B)\,\varepsilon_t$, where some of the roots of $\det\{\,\Phi(B)\,\} = 0$ are allowed to have absolute value equal to one. More specifically, because of the prominent role of the differencing operator $(\,1 - B\,)$ in univariate models, we might only allow some roots to equal one (unit roots) while the remaining roots are all greater than one in absolute value. We may consider such a general form of nonstationary vector ARMA model later, but because such models have a rather complex theory of parameter estimation and model identification, for the present we will consider only a much more restrictive class of models for nonstationary series. These models are of the form

$$\Phi_1(B)\,D(B)\,\boldsymbol{Y}_t = \Theta(B)\,\varepsilon_t\,, \tag{2.17}$$

where $D(B) = \text{Diag}[\,(\,1 - B\,)^{d_1}, (\,1 - B\,)^{d_2}, \ldots, (\,1 - B\,)^{d_k}\,]$ is a diagonal matrix, $d_1, \ldots, d_k$ are non-negative integers, and $\det\{\,\Phi_1(B)\,\} = 0$ has all roots greater than one in absolute value. Thus, this model, which will be referred to as a vector ARIMA model, simply states that after each series Y_{it} is individually differenced an appropriate number $(\,d_i\,)$ of times to reduce it to a stationary series, the resulting vector series $\boldsymbol{W}_t = D(B)\,\boldsymbol{Y}_t$ is a stationary vector ARMA(p,q) process. This model is not nearly as general as the model $\Phi(B)\,\boldsymbol{Y}_t = \Theta(B)\,\varepsilon_t$ where some roots of $\det\{\,\Phi(B)\,\} = 0$ are allowed to equal one in absolute value (or even simply allowed to equal one), and it will be found later that the more restrictive model may not be appropriate in some cases. Use of such models in these inappropriate situations will then lead to a certain type of multivariate overdifferencing and to noninvertibility of the vector MA operator $\Theta(B)$ which should be avoided because there can be associated difficulties with parameter estimation and other problems.

As a simple example of a nonstationary process, consider the nonstationary AR(1) model, $\boldsymbol{Y}_t = \Phi_1\,\boldsymbol{Y}_{t-1} + \varepsilon_t$, where some eigenvalues of Φ_1 are equal to one in absolute value. As in (2.7) in Section 2.2.1, the process can be expressed as

$$\boldsymbol{Y}_t = \sum_{j=0}^{t+n} \Phi_1^j\,\varepsilon_{t-j} + \Phi_1^{t+n+1}\,\boldsymbol{Y}_{-n-1}$$

relative to some arbitrary but fixed initial time origin $-(n+1)$. However, in the nonstationary case, because Φ_1^{t+n+1} does not converge to zero as $n \rightarrow \infty$, no convergent causal infinite MA representation is possible, and the nature of the process $\boldsymbol{Y}_t$ for $t > -(n+1)$ depends on assumptions of the initial value $\boldsymbol{Y}_{-n-1}$. For instance, it might be assumed that the initial value $\boldsymbol{Y}_{-n-1}$ has some specified distribution and is independent of the ε_t for $t > -(n+1)$, and this will determine the properties of the process $\boldsymbol{Y}_t$ for $t > -(n+1)$. As a special case, when

$\Phi_1 = I$ we have the *vector random walk process* which is generated by

$$\boldsymbol{Y}_t = \boldsymbol{Y}_{t-1} + \varepsilon_t = \sum_{j=0}^{t+n} \varepsilon_{t-j} + \boldsymbol{Y}_{-n-1} ,$$

with each component series Y_{it} following the univariate random walk $(1 - B) Y_{it} = \varepsilon_{it}$, $i = 1, \ldots, k$.

Similarly, for the general nonstationary vector ARMA model $\Phi(B) \boldsymbol{Y}_t = \Theta(B) \varepsilon_t$, such as in (2.17), relative to some arbitrary but fixed initial time origin $-(n+1)$, the process can be represented in the *truncated* random shock (finite moving average) form as

$$\boldsymbol{Y}_t = \varepsilon_t + \sum_{j=1}^{t+n} \Psi_j \, \varepsilon_{t-j} + C_{-n}(t+n) .$$

In this expression, the matrix coefficients Ψ_j are determined from the relation $\Phi(B) \Psi(B) = \Theta(B)$, as in (2.12) of Section 2.3.1, and $C_{-n}(t+n)$ represents the *complementary function* which satisfies $\Phi(B) C_{-n}(t+n) = 0$ for $t > -(n+1)$ and which embodies the "initializing" features of the process $\boldsymbol{Y}_t$ before time $-n$. The truncated random shock representation for the vector IMA(1,1) model is given in Section 2.4.3 as an additional example.

2.4.2 *Cointegration in Nonstationary Vector Processes*

The nonstationary (unit root) aspects of a vector process $\boldsymbol{Y}_t$ become more complicated in the multivariate case compared to the univariate case, due in part to the possibility of cointegration among the component series Y_{it} of a nonstationary vector process $\boldsymbol{Y}_t$. For instance, the possibility exists for each component series Y_{it} to be nonstationary with its first difference $(1 - B) Y_{it}$ stationary (in which case Y_{it} is said to be integrated of order one), but such that certain linear combinations $Z_{it} = \boldsymbol{b}_i' \boldsymbol{Y}_t$ of $\boldsymbol{Y}_t$ will be stationary. In such circumstances, the process $\boldsymbol{Y}_t$ is said to be *co-integrated* with co-integrating vectors $\boldsymbol{b}_i$ (e.g., Engle and Granger, 1987). An interpretation of co-integrated vector series $\boldsymbol{Y}_t$, particularly related to economics, is that the individual components Y_{it} share some common nonstationary components or "common trends" and, hence, they tend to have certain similar movements in their long-term behavior. A related interpretation is that the component series Y_{it}, although they may exhibit nonstationary behavior, satisfy (approximately) a long-run equilibrium relation $\boldsymbol{b}_i' \boldsymbol{Y}_t \approx 0$ such that the process $Z_{it} = \boldsymbol{b}_i' \boldsymbol{Y}_t$ which represents the deviations from the equilibrium exhibits stable behavior and so is a stationary process. A specific nonstationary ARMA model structure for which cointegration occurs is the model $\Phi(B) \boldsymbol{Y}_t = \Theta(B) \varepsilon_t$, where $\det\{ \Phi(B) \} = 0$ has $d < k$ roots equal to one and all other roots are greater than one in absolute value, and also the matrix $\Phi(1)$ has rank equal to $r = k - d$. Then for such a process, it can be established that r linearly independent vectors $\boldsymbol{b}_i$ exist such that $\boldsymbol{b}_i' \boldsymbol{Y}_t$ is stationary, and $\boldsymbol{Y}_t$ is said to have co-integrating rank r. Properties of nonstationary co-integrated systems have been investigated by Engle and Granger (1987), among others, and the estimation of co-integrated vector AR models will be considered in detail in Section 6.3.

For an example, consider further the simple bivariate ($k=2$) case of a nonstationary AR(1) model, $\boldsymbol{Y}_t = \Phi_1 \boldsymbol{Y}_{t-1} + \varepsilon_t$, and suppose Φ_1 has one eigenvalue equal to one and the remaining eigenvalue is λ such that $|\lambda| < 1$. Then we can find a nonsingular matrix $P = [P_1, P_2]$ such that $P^{-1}\Phi_1 P = J = \text{Diag}(1, \lambda)$. Letting $Q = P^{-1} = [Q_1, Q_2]'$, $\boldsymbol{Z}_t = Q\boldsymbol{Y}_t$, and $\boldsymbol{a}_t = Q\varepsilon_t$, we have $Q\boldsymbol{Y}_t = Q\Phi_1 P Q \boldsymbol{Y}_{t-1} + Q\varepsilon_t$, or $\boldsymbol{Z}_t = J\boldsymbol{Z}_{t-1} + \boldsymbol{a}_t$. Thus, $(1-B)Z_{1t} = a_{1t}$ and $(1-\lambda B)Z_{2t} = a_{2t}$. Hence, since $\boldsymbol{Y}_t = P\boldsymbol{Z}_t = P_1 Z_{1t} + P_2 Z_{2t}$, the bivariate series Y_{1t} and Y_{2t} are linear combinations of a nonstationary (random walk) component Z_{1t} and a stationary (AR(1)) component Z_{2t}. Conversely, we see that $Z_{2t} = Q_2' \boldsymbol{Y}_t$ is a linear combination of the components of the original nonstationary bivariate series $\boldsymbol{Y}_t = (Y_{1t}, Y_{2t})'$ that is stationary, and, hence, the process $\boldsymbol{Y}_t$ is co-integrated. Thus, although both component series may be unit-root nonstationary, appropriate modeling of the bivariate series $\boldsymbol{Y}_t$ does not give rise to simultaneous differencing of both component series.

2.4.3 The Vector IMA(1,1) Process or Exponential Smoothing Model

We briefly consider the nonstationary vector IMA(1,1) model given by $(1-B)\boldsymbol{Y}_t = (I - \Theta B)\varepsilon_t$, where we assume that all the eigenvalues of Θ are less than one in absolute value (that is, the MA operator is invertible). Similar to (2.15) in Section 2.3.3, we can express the process $\boldsymbol{Y}_t$ in the (finite) random shock form, relative to some arbitrary but fixed initial time origin $-(n+1)$ at which point the process is assumed to be initiated, as

$$\boldsymbol{Y}_t = \varepsilon_t + (I - \Theta)\sum_{j=1}^{t+n} \varepsilon_{t-j} + [\boldsymbol{Y}_{-n-1} - \Theta\varepsilon_{-n-1}]$$

$$\equiv \varepsilon_t + \sum_{j=1}^{t+n} \Psi_j \varepsilon_{t-j} + [\boldsymbol{Y}_{-n-1} - \Theta\varepsilon_{-n-1}],$$

where $\Psi_j = I - \Theta$ for $j \ge 1$. For convenience of notation, we will informally write this representation as

$$\boldsymbol{Y}_t = \varepsilon_t + (I - \Theta)\sum_{j=1}^{\infty} \varepsilon_{t-j} = \varepsilon_t + \sum_{j=1}^{\infty} \Psi_j \varepsilon_{t-j} = \Psi(B)\varepsilon_t,$$

where $\Psi(B) = \sum_{j=0}^{\infty} \Psi_j B^j$ is determined from the relation that $(I - B)\Psi(B) = I - \Theta B$, so that $\Psi_1 = I - \Theta$, $\Psi_j = \Psi_{j-1}$, for $j > 1$, and, hence, $\Psi_j = I - \Theta$, $j \ge 1$. Notice that the matrix weights Ψ_j do not converge in this nonstationary model and so the "infinite" MA representation is not itself meaningful, but will be used only as a convenient notational device to represent the "finite" MA representation. To properly define a nonstationary process such as the IMA(1,1) considered here, we must, in fact, make some appropriate assumptions concerning the start-up value Y_{-n-1} (or the initial "remainder term" $[\,Y_{-n-1} - \Theta\,\varepsilon_{-n-1}\,]$) at the finite past time origin $-n-1$. For example, supposing the initial term has zero mean vector and covariance matrix R_n and is independent of ε_t, for $t \ge -n$, the second-order properties of the process Y_t are determined for $t \ge -n$. Specifically, then, Y_t has covariance matrix

$$\Gamma_t(0) = E(\,Y_t\,Y_t'\,) = \Sigma + (t+n)\,(\,I-\Theta\,)\,\Sigma\,(\,I-\Theta\,)' + R_n\,,$$

for $t \ge -n$, and similarly, the cross-covariance matrix between Y_t and Y_{t+l} is

$$\Gamma_t(l) = E(\,Y_t\,Y_{t+l}'\,) = \Sigma\,(\,I-\Theta\,)' + (t+n)\,(\,I-\Theta\,)\,\Sigma\,(\,I-\Theta\,)' + R_n\,,$$

for $t \ge -n$, $l > 0$. Hence, unlike the stationary case, the covariance matrix $\Gamma_t(0)$ and cross-covariance matrices $\Gamma_t(l)$ are heavily dependent on t and do not converge as t increases.

Also, as in (2.16) in Section 2.3.3 for the vector ARMA(1,1) model, we can write

$$Y_t = \sum_{j=1}^{t+n} \Theta^{j-1}(\,I-\Theta\,)\,Y_{t-j} \;+\; \varepsilon_t \;+\; \Theta^{t+n}(\,Y_{-n-1} - \Theta\,\varepsilon_{-n-1}\,)\,.$$

This representation would be convergent as we let $n \to \infty$ under the invertibility assumption, since then $\Theta^{t+n} \to 0$. Hence, we can express the IMA(1,1) process in the inverted or infinite AR form as $\Pi(B)\,Y_t = \varepsilon_t$, where $\Pi(B) = I - \sum_{j=1}^{\infty} \Pi_j\,B^j$ satisfies the relation $(\,I - \Theta\,B\,)\,\Pi(B) = I - B$. Thus, we find that $\Pi_j = \Theta\,\Pi_{j-1}$, $j > 1$, with $\Pi_1 = I - \Theta$, so that $\Pi_j = (\,I-\Theta\,)\,\Theta^{j-1}$, $j \ge 1$. Therefore, the infinite AR representation takes the form $Y_t = (\,I-\Theta\,)\sum_{j=1}^{\infty} \Theta^{j-1}\,Y_{t-j} + \varepsilon_t$ in which the current vector value Y_t is expressible as a matrix exponentially weighted moving average (vector EWMA) of all past values Y_{t-1}, $Y_{t-2}, \ldots,$ plus the current random shock ε_t. This model is commonly referred to as the vector *exponential smoothing* model.

Example 2.7. Noninvertible IMA(1,1) Process Related to Overdifferencing. Consider the bivariate IMA(1,1) model, $(\,1 - B\,)\,Y_t = (\,I - \Theta\,B\,)\,\varepsilon_t$. Now generally, we can find a nonsingular matrix P such that $P^{-1}\Theta\,P = \Lambda = \mathrm{Diag}(\,\lambda_1, \lambda_2\,)$, where the λ_i are the eigenvalues of Θ. Consider the particular case where

$$\Theta = \begin{bmatrix} 1.5 & -0.5 \\ 1.0 & 0.0 \end{bmatrix}, \qquad \Lambda = \begin{bmatrix} 1.0 & 0.0 \\ 0.0 & 0.5 \end{bmatrix}, \qquad P = \begin{bmatrix} 1 & 1 \\ 1 & 2 \end{bmatrix},$$

so that the MA(1) operator is not invertible (one eigenvalue is equal to one). Thus, the model is

$$(\,1-B\,)\,Y_t = (\,I - P\,\Lambda\,P^{-1}B\,)\,\varepsilon_t\,, \qquad \text{or} \qquad (\,1-B\,)\,P^{-1}Y_t = (\,I - \Lambda\,B\,)\,P^{-1}\varepsilon_t\,.$$

Letting $Z_t = P^{-1}Y_t$, $a_t = P^{-1}\varepsilon_t$, we have $(\,1-B\,)\,Z_t = a_t - \Lambda\,a_{t-1}$. So we find the uncoupled model equations in terms of the variables Z_t as

$(1-B)Z_{1t} = (1-B)a_{1t}$, which implies the result $Z_{1t} = \mu_1 + a_{1t}$, and $(1-B)Z_{2t} = (1-0.5B)a_{2t}$. Thus, the transformed series $Z_{1t} = 2Y_{1t} - Y_{2t}$ is actually a (stationary) white noise series, whereas $Z_{2t} = Y_{2t} - Y_{1t}$ is a univariate (nonstationary) IMA(1,1) process. Now, after elimination of the common differencing factor $(1-B)$ in the model for Z_{1t}, the bivariate model for $\boldsymbol{Z}_t$ can be written as $(I - \phi^* B)\boldsymbol{Z}_t = \boldsymbol{\mu}^* + (I - \Lambda^* B)\boldsymbol{a}_t$, where $\boldsymbol{\mu}^* = (\mu_1, 0)'$,

$$\phi^* = \begin{bmatrix} 0 & 0 \\ 0 & 1 \end{bmatrix}, \qquad \Lambda^* = \begin{bmatrix} 0.0 & 0.0 \\ 0.0 & 0.5 \end{bmatrix}.$$

Multiplying through this equation by P, we have

$$(I - P\phi^* P^{-1} B)\boldsymbol{Y}_t = P\boldsymbol{\mu}^* + (I - P\Lambda^* P^{-1} B)\varepsilon_t,$$

or $(I - \Phi_1 B)\boldsymbol{Y}_t = \boldsymbol{\mu} + (I - \Theta_1 B)\varepsilon_t$, where $\boldsymbol{\mu} = P\boldsymbol{\mu}^*$,

$$\Phi_1 = P\phi^* P^{-1} = \begin{bmatrix} -1 & 1 \\ -2 & 2 \end{bmatrix}, \qquad \Theta_1 = P\Lambda^* P^{-1} = \begin{bmatrix} -0.5 & 0.5 \\ -1.0 & 1.0 \end{bmatrix}.$$

Note that this last model form for $\boldsymbol{Y}_t$ is of the general ARMA(1,1) form, but with only one root of $\det\{I - \Phi_1 B\} = 0$ equal to one, and the MA operator $I - \Theta_1 B$ is invertible. So in this example, we find that $Y_{1t} = Z_{1t} + Z_{2t}$ and $Y_{2t} = Z_{1t} + 2Z_{2t}$ are linear combinations of one nonstationary series Z_{2t} and one stationary (white noise) series Z_{1t}. Conversely, there is one linear combination of Y_{1t} and Y_{2t}, $Z_{1t} = 2Y_{1t} - Y_{2t}$, which is stationary so that the process $\boldsymbol{Y}_t$ is co-integrated. Consideration of the bivariate model for the first differences $(1-B)Y_{1t}$, $(1-B)Y_{2t}$ leads to an IMA(1,1) model with a noninvertible MA operator $I - \Theta B$, due to "overdifferencing" of the vector series $\boldsymbol{Y}_t$. Thus, although both series Y_{1t} and Y_{2t} exhibit nonstationary behavior, applying a first difference to both series leads to a bivariate noninvertible model, whereas the equivalent (and less troublesome) nonstationary but invertible "ARMA(1,1)" type model given above has only one unit root in its AR operator, and it also provides a better understanding of the actual nature of the nonstationarity of the bivariate series $\boldsymbol{Y}_t$ in terms of only one "common nonstationary component" series.

2.5 Prediction for Vector ARMA Models

Given a realization of a vector ARMA process up through time t, $\{\boldsymbol{Y}_s, s \leq t\}$, we consider the problem of prediction of future values $\boldsymbol{Y}_{t+l}$, $l = 1, 2, \ldots$. In this presentation, for purposes of forecasting it is assumed that the model for $\{\boldsymbol{Y}_t\}$ is known exactly, including the values of the model parameters. Although, in practice, the model must be specified and the parameters estimated from available sample data, errors due to estimation of model parameters will not have too much effect on forecast properties for sufficiently large sample sizes (e.g., see Section 5.6 for consideration of the topic of the effects of parameter estimation errors on prediction properties). Also, the practical problem of prediction of future values based on the finite past data $\boldsymbol{Y}_1, \ldots, \boldsymbol{Y}_T$ will be considered briefly in this section and in more detail in later chapters, but prediction results based on the infinite past data assumption usually provide an adequate approximation to the finite past data prediction problem.

2.5.1 Minimum Mean Squared Error Prediction

Before presenting details of forecasting for vector ARMA processes, we review some basic principles concerning prediction in a more general context. Let $\boldsymbol{Y}$ and $\mathbf{X}$ be k-dimensional and n-dimensional random vectors, respectively, and suppose we want to predict (estimate) the unknown value of $\boldsymbol{Y}$ based on a vector function of $\mathbf{X}$, say $\hat{\boldsymbol{Y}} = g(\mathbf{X})$. The mean square prediction error matrix of a predictor $\hat{\boldsymbol{Y}}$ is $E[(\boldsymbol{Y} - \hat{\boldsymbol{Y}})(\boldsymbol{Y} - \hat{\boldsymbol{Y}})']$. We refer to the minimum mean squared error (MSE) predictor of $\boldsymbol{Y}$ as that function $\hat{\boldsymbol{Y}} = g^*(\mathbf{X})$ such that, among all possible functions of $\mathbf{X}$, $\hat{\boldsymbol{Y}}$ minimizes

$$E[(\boldsymbol{b}'\boldsymbol{Y} - \boldsymbol{b}'\hat{\boldsymbol{Y}})^2] = \boldsymbol{b}' E[(\boldsymbol{Y} - \hat{\boldsymbol{Y}})(\boldsymbol{Y} - \hat{\boldsymbol{Y}})'] \boldsymbol{b}$$

for every nonzero $k \times 1$ vector of constants $\boldsymbol{b}$. Then it is well known that the *minimum mean squared error* predictor is given by $\hat{\boldsymbol{Y}} = E(\boldsymbol{Y} \mid \mathbf{X})$, the conditional expectation of $\boldsymbol{Y}$ given $\mathbf{X}$, with prediction error $\boldsymbol{e} = \boldsymbol{Y} - \hat{\boldsymbol{Y}} = \boldsymbol{Y} - E(\boldsymbol{Y} \mid \mathbf{X})$. One can also restrict prediction to consider only linear functions of $\mathbf{X}$, and, hence, consider the minimum MSE *linear* predictor as the linear function $\hat{\boldsymbol{Y}}^* = \boldsymbol{b} + B\,\mathbf{X}$ which minimizes the prediction MSE matrix among all linear functions of $\mathbf{X}$. It is well known [e.g., see (A1.8) and related results in Appendix A1] that the minimum MSE linear predictor is

$$\hat{\boldsymbol{Y}}^* = \boldsymbol{\mu}_y + \Sigma_{yx}\,\Sigma_{xx}^{-1}\,(\mathbf{X} - \boldsymbol{\mu}_x), \tag{2.18}$$

with the prediction error $\boldsymbol{e}^* = \boldsymbol{Y} - \hat{\boldsymbol{Y}}^*$ having mean zero and prediction MSE matrix (covariance matrix)

$$\text{Cov}(\boldsymbol{e}^*) = \text{Cov}(\boldsymbol{Y} - \hat{\boldsymbol{Y}}^*) = \Sigma_{yy} - \Sigma_{yx}\,\Sigma_{xx}^{-1}\,\Sigma_{xy}, \tag{2.19}$$

where $\boldsymbol{\mu}_y = E(\boldsymbol{Y})$, $\boldsymbol{\mu}_x = E(\mathbf{X})$, $\Sigma_{yy} = \text{Cov}(\boldsymbol{Y})$, $\Sigma_{yx} = \text{Cov}(\boldsymbol{Y}, \mathbf{X})$, and $\Sigma_{xx} = \text{Cov}(\mathbf{X})$. Moreover, if the prediction error $\boldsymbol{e}^* = \boldsymbol{Y} - \hat{\boldsymbol{Y}}^*$ of the best linear predictor is such that $E(\boldsymbol{e}^* \mid \mathbf{X}) = 0$ (e.g., if $\boldsymbol{e}^*$ is independent of $\mathbf{X}$), then $\hat{\boldsymbol{Y}}^*$ is also the minimum MSE predictor, i.e., $\hat{\boldsymbol{Y}}^* = \hat{\boldsymbol{Y}} = E(\boldsymbol{Y} \mid \mathbf{X})$ is a linear function of $\mathbf{X}$ and the prediction error $\boldsymbol{e} = \boldsymbol{e}^* = \boldsymbol{Y} - \hat{\boldsymbol{Y}}$ has covariance matrix $\text{Cov}(\boldsymbol{e}^*)$ as given above in (2.19).

2.5.2 Forecasting for Vector ARMA Processes and Covariance Matrices of Forecast Errors

For forecasting in the vector ARMA(p,q) model $\Phi(B)\,\boldsymbol{Y}_t = \Theta(B)\,\varepsilon_t$, we will assume that the white noise series ε_t are mutually independent random vectors. In the stationary case, the ARMA model has the "infinite" MA form $\boldsymbol{Y}_t = \Psi(B)\, \varepsilon_t$, where $\Psi(B) = \Phi(B)^{-1}\, \Theta(B) = \sum_{j=0}^{\infty} \Psi_j\, B^j$. A future value of the process at time $t+l$, relative to the forecast origin t, can be expressed as $\boldsymbol{Y}_{t+l} = \sum_{j=0}^{\infty} \Psi_j\, \varepsilon_{t+l-j}$. Now ε_{t+h}, $h > 0$, is independent of present and past values $\boldsymbol{Y}_t$, $\boldsymbol{Y}_{t-1}, \ldots$, so that $E(\, \varepsilon_{t+h} \mid \boldsymbol{Y}_t, \boldsymbol{Y}_{t-1}, \ldots) = 0$, $h > 0$. Thus, from the above results we find that the minimum mean squared error (MSE) matrix predictor of $\boldsymbol{Y}_{t+l}$ based on $\boldsymbol{Y}_t$, $\boldsymbol{Y}_{t-1}, \ldots$, can be represented as

$$\hat{\boldsymbol{Y}}_t(l) = E(\, \boldsymbol{Y}_{t+l} \mid \boldsymbol{Y}_t, \boldsymbol{Y}_{t-1}, \ldots) = \sum_{j=l}^{\infty} \Psi_j\, \varepsilon_{t+l-j}\,. \tag{2.20}$$

The forecast $\hat{\boldsymbol{Y}}_t(l)$ is also expressible as a linear function of the present and past $\boldsymbol{Y}_t$'s, since under the invertibility condition, each ε_{t+l-j} can be written as a linear function of the $\boldsymbol{Y}_t$'s using the infinite AR representation of the process.

The l-step ahead forecast error is $\boldsymbol{e}_t(l) = \boldsymbol{Y}_{t+l} - \hat{\boldsymbol{Y}}_t(l) = \sum_{j=0}^{l-1} \Psi_j\, \varepsilon_{t+l-j}$, with zero mean and covariance matrix

$$\Sigma(l) = \mathrm{Cov}(\, \boldsymbol{e}_t(l)\,) = E[\, \boldsymbol{e}_t(l)\, \boldsymbol{e}_t(l)'\,] = \sum_{j=0}^{l-1} \Psi_j\, \Sigma\, \Psi_j', \qquad \Psi_0 = I\,. \tag{2.21}$$

In particular, for $l = 1$ step ahead, we have $\boldsymbol{e}_t(1) = \boldsymbol{Y}_{t+1} - \hat{\boldsymbol{Y}}_t(1) = \varepsilon_{t+1}$ with error covariance matrix Σ, so that the white noise series ε_t can also be interpreted as a sequence of one-step ahead forecast errors for the process. It also follows from the form $\boldsymbol{e}_t(l) = \sum_{j=0}^{l-1} \Psi_j\, \varepsilon_{t+l-j}$ that forecast errors $\boldsymbol{e}_t(l)$ and $\boldsymbol{e}_t(l+i)$ at different lead times l and $l+i$, based on the same forecast origin t, will be correlated with

$$\mathrm{Cov}(\, \boldsymbol{e}_t(l), \boldsymbol{e}_t(l+i)\,) = E[\, \boldsymbol{e}_t(l)\, \boldsymbol{e}_t(l+i)'\,] = \sum_{j=0}^{l-1} \Psi_j\, \Sigma\, \Psi_{j+i}',$$

for $i \geq 1$. Note that in the ARMA(p,q) model if the ε_t are assumed only to be mutually uncorrelated (not independent), then the forecast $\hat{\boldsymbol{Y}}_t(l) = \sum_{j=l}^{\infty} \Psi_j\, \varepsilon_{t+l-j}$ is known only to be the best (minimum MSE) linear forecast of $\boldsymbol{Y}_{t+l}$, but will not be the best forecast in general unless the ε_t satisfy the condition that $E(\, \varepsilon_{t+1} \mid \boldsymbol{Y}_t, \boldsymbol{Y}_{t-1}, \ldots) = 0$ for all t, since then $E(\, \boldsymbol{e}_t(l) \mid \boldsymbol{Y}_t, \boldsymbol{Y}_{t-1}, \ldots) = 0$. For the nonstationary case, similar forecasting results as in (2.21) above can also be obtained based on use of the "finite" moving average representation for the process such as that given in Section 2.4.3 for the (nonstationary) IMA(1,1) model with known start-up values. From (2.21), we see that in the stationary case, $\Sigma(l) = \mathrm{Cov}(\, \boldsymbol{e}_t(l)\,)$ converges to $\Gamma(0) = \sum_{j=0}^{\infty} \Psi_j\, \Sigma\, \Psi_j'$ as $l \to \infty$, so that $\Sigma(l)$ approaches the covariance matrix of the process $\boldsymbol{Y}_t$ for large lead time l. However, for nonstationary series, $\Sigma(l) = \mathrm{Cov}(\, \boldsymbol{e}_t(l)\,)$ will increase without bound as l increases. Under the assumption of normality of the ε_t, the l-step ahead forecast errors $\boldsymbol{Y}_{t+l} - \hat{\boldsymbol{Y}}_t(l) = \sum_{j=0}^{l-1} \Psi_j\, \varepsilon_{t+l-j}$ then will also be normally distributed as multivariate $N(0, \Sigma(l))$ with $\Sigma(l)$ given by (2.21). Hence, it follows that the diagonal elements (forecast error variances) $\sigma_{ii}(l)$ of $\Sigma(l)$ can be used, together with the point forecasts $\hat{Y}_{it}(l)$, in the standard way to construct l-step ahead prediction interval forecasts of the future values of the component series, $Y_{i,t+l}$, for time $t+l$ [e.g., see Box and Jenkins (1976, Sec. 5.2.4)].

An alternate to expression (2.21) is obtained by writing the infinite MA representation in terms of a "standardized" white noise innovations process. Since $\Sigma = \text{Cov}(\varepsilon_t)$ is positive-definite, there is a lower triangular matrix $\Psi_0^{\#}$ with positive diagonal elements such that $\Sigma = \Psi_0^{\#} \Psi_0^{\#\prime}$, so that $\Psi_0^{\#-1} \Sigma \Psi_0^{\#\prime -1} = I$. Then the infinite MA representation can be given as $\boldsymbol{Y}_t = \boldsymbol{\mu} + \sum_{j=0}^{\infty} \Psi_j^{\#} \boldsymbol{a}_{t-j}$, where $\Psi_j^{\#} = \Psi_j \Psi_0^{\#}$ and $\boldsymbol{a}_t = \Psi_0^{\#-1} \varepsilon_t$ with $\text{Cov}(\boldsymbol{a}_t) = I$, and (2.21) becomes $\Sigma(l) = \sum_{j=0}^{l-1} \Psi_j^{\#} \Psi_j^{\#\prime}$. In this form, the elements of the matrices $\Psi_j^{\#}$, sometimes referred to as the impulse response matrices, indicate the effects of the components of the standardized shock process $\boldsymbol{a}_t$ on the components of the process $\boldsymbol{Y}_t$ at various lags. In addition, examination of the diagonal elements in the relation $\Sigma(l) = \sum_{j=0}^{l-1} \Psi_j^{\#} \Psi_j^{\#\prime}$ can be interpreted as providing a decomposition of the l-step ahead forecast error variance $\sigma_{ii}(l)$ for each component series Y_{it} into contributions from the components of the standardized innovations $\boldsymbol{a}_t$ (since the ith diagonal element of $\Psi_j^{\#} \Psi_j^{\#\prime}$ is just the sum of the squares of the elements in the ith row of $\Psi_j^{\#}$).

2.5.3 *Computation of Forecasts for Vector ARMA Processes*

From a computational point of view, forecasts for the ARMA model, $\boldsymbol{Y}_t = \sum_{j=1}^{p} \Phi_j \boldsymbol{Y}_{t-j} + \boldsymbol{\delta} + \varepsilon_t - \sum_{j=1}^{q} \Theta_j \varepsilon_{t-j}$, where $\boldsymbol{\delta} = (I - \Phi_1 - \cdots - \Phi_p)\boldsymbol{\mu}$ for stationary processes, are determined by applying conditional expectations to both sides of the ARMA(p,q) model relation, using the result that $E(\varepsilon_{t+l-j} \mid \boldsymbol{Y}_t, \boldsymbol{Y}_{t-1}, \ldots) = 0$ for $l - j > 0$. Thus, the forecasts $\hat{\boldsymbol{Y}}_t(l)$ are computed directly from the vector ARMA model difference equation recursively as

$$\hat{\boldsymbol{Y}}_t(l) = \sum_{j=1}^{p} \Phi_j \hat{\boldsymbol{Y}}_t(l-j) + \boldsymbol{\delta} - \sum_{j=l}^{q} \Theta_j \varepsilon_{t+l-j}, \qquad l = 1, 2, \ldots, q, \tag{2.22}$$

with $\hat{\boldsymbol{Y}}_t(l) = \sum_{j=1}^{p} \Phi_j \hat{\boldsymbol{Y}}_t(l-j) + \boldsymbol{\delta}$, for $l > q$, where $\hat{\boldsymbol{Y}}_t(l-j) = \boldsymbol{Y}_{t+l-j}$ for $l \le j$. Hence, the forecasts $\hat{\boldsymbol{Y}}_t(l)$ satisfy the matrix difference equation $\Phi(B)\hat{\boldsymbol{Y}}_t(l) = \boldsymbol{\delta}$, $l > q$, where B operates on the lead l, with initial values $\hat{\boldsymbol{Y}}_t(1), \ldots, \hat{\boldsymbol{Y}}_t(r)$, where $r = \max(p, q)$. Note that for pure AR models, $q = 0$, we see that

$$\hat{\boldsymbol{Y}}_t(l) = \sum_{j=1}^{p} \Phi_j \hat{\boldsymbol{Y}}_t(l-j) + \boldsymbol{\delta}$$

for all $l = 1, 2, \ldots$. So the p initial forecast values are completely determined by the last p observations $\boldsymbol{Y}_t, \boldsymbol{Y}_{t-1}, \ldots, \boldsymbol{Y}_{t-p+1}$, and, hence, for AR models all forecasts depend only on these last p observations.

For models that involve an MA term, in practice it is necessary to generate the white noise sequence ε_t recursively, using all available past data $\boldsymbol{Y}_1$, $\boldsymbol{Y}_2, \ldots, \boldsymbol{Y}_t$, as $\varepsilon_s = \boldsymbol{Y}_s - \sum_{j=1}^{p} \Phi_j \boldsymbol{Y}_{s-j} - \boldsymbol{\delta} + \sum_{j=1}^{q} \Theta_j \varepsilon_{s-j}$, $s = 1, 2, \ldots, t$, using some appropriate starting values for $\varepsilon_0, \varepsilon_{-1}, \ldots, \varepsilon_{1-q}$, and $\boldsymbol{Y}_0, \ldots, \boldsymbol{Y}_{1-p}$. For example, in the ARMA(1,1) model $\boldsymbol{Y}_t = \Phi_1 \boldsymbol{Y}_{t-1} + \varepsilon_t - \Theta_1 \varepsilon_{t-1}$, we compute $\varepsilon_s = \boldsymbol{Y}_s - \Phi_1 \boldsymbol{Y}_{s-1} + \Theta_1 \varepsilon_{s-1}$, $s = 1, 2, \ldots, t$, where a suitable starting value for $\Phi_1 \boldsymbol{Y}_0 - \Theta_1 \varepsilon_0$ is required to first compute ε_1. If the exact backcasts are used for the initial values, that is,

$$\hat{\varepsilon}_{1-j} = E(\, \varepsilon_{1-j} \mid \boldsymbol{Y}_t, \boldsymbol{Y}_{t-1}, \ldots, \boldsymbol{Y}_1 \,), \qquad j = 1, \ldots, q,$$

and

$$\hat{\boldsymbol{Y}}_{1-j} = E(\, \boldsymbol{Y}_{1-j} \mid \boldsymbol{Y}_t, \boldsymbol{Y}_{t-1}, \ldots, \boldsymbol{Y}_1 \,), \qquad j = 1, \ldots, p,$$

as will be discussed later in Section 5.3.1 in the context of exact maximum likelihood estimation for ARMA models, then it follows that the resulting forecasts $\hat{\boldsymbol{Y}}_t(l)$ as obtained through (2.22) are exactly equal to $E[\, \boldsymbol{Y}_{t+l} \mid \boldsymbol{Y}_t, \boldsymbol{Y}_{t-1}, \ldots, \boldsymbol{Y}_1 \,]$, the optimal forecasts based on the finite past history $\boldsymbol{Y}_t, \boldsymbol{Y}_{t-1}, \ldots, \boldsymbol{Y}_1$, although the above presentation of forecast properties assumed forecasts based on the infinite past history $\boldsymbol{Y}_s$, all $s \le t$. However, these two forecasts will be nearly identical for any moderate or large value of t, the number of past values available for forecasting. Alternate methods to obtain the "exact" forecasts, as well as the exact covariance matrices of the forecast errors, based on the finite sample data $\boldsymbol{Y}_1, \ldots, \boldsymbol{Y}_t$, in a convenient computational manner are through an innovations approach or through the closely related state-space model–Kalman filter approach to be discussed later in Section 5.4 and in Sections 7.1 and 7.2, respectively.

2.5.4 Some Examples of Forecast Functions for Vector ARMA Models

We now consider a few simple model examples to illustrate the above. First, consider the AR(1) model, $(\, I - \Phi B \,)(\, \boldsymbol{Y}_t - \boldsymbol{\mu} \,) = \varepsilon_t$. Then we have $\hat{\boldsymbol{Y}}_t(l) = \boldsymbol{\mu} + \Phi(\, \hat{\boldsymbol{Y}}_t(l-1) - \boldsymbol{\mu} \,)$ for $l = 1, 2, \ldots$, with $\hat{\boldsymbol{Y}}_t(1) = \boldsymbol{\mu} + \Phi(\, \boldsymbol{Y}_t - \boldsymbol{\mu} \,)$. Hence, the explicit form of the forecasts is

$$\hat{\boldsymbol{Y}}_t(l) = \boldsymbol{\mu} + \Phi^l (\, \boldsymbol{Y}_t - \boldsymbol{\mu} \,), \qquad l = 1, 2, \ldots,$$

which shows that all forecasts depend only on the last observation vector $\boldsymbol{Y}_t$ and directly exhibits the nature of the dependence of $\hat{\boldsymbol{Y}}_t(l)$ on $\boldsymbol{Y}_t$. Also we have $\Psi_j = \Phi^j$, $j \ge 1$, so that $\boldsymbol{e}_t(l) = \sum_{j=0}^{l-1} \Phi^j \varepsilon_{t+l-j}$, and

$$\Sigma(l) = \mathrm{Cov}(\, \boldsymbol{e}_t(l) \,) = \sum_{j=0}^{l-1} \Phi^j \, \Sigma \, \Phi'^j$$

is the covariance matrix of the l-step ahead forecast errors.

For the vector ARMA(1,1) model, $(\, I - \Phi B \,)(\, \boldsymbol{Y}_t - \boldsymbol{\mu} \,) = (\, I - \Theta B \,)\, \varepsilon_t$, we have $\hat{\boldsymbol{Y}}_t(1) = \boldsymbol{\mu} + \Phi\,(\,\boldsymbol{Y}_t - \boldsymbol{\mu}\,) - \Theta\,\varepsilon_t$ and

$$\hat{\boldsymbol{Y}}_t(l) = \boldsymbol{\mu} + \Phi\,(\,\hat{\boldsymbol{Y}}_t(l-1) - \boldsymbol{\mu}\,) = \boldsymbol{\mu} + \Phi^{l-1}(\,\hat{\boldsymbol{Y}}_t(1) - \boldsymbol{\mu}\,), \quad \text{for} \quad l > 1.$$

Since $\Psi_j = \Phi^{j-1}(\,\Phi - \Theta\,)$, $j \geq 1$, we have

$$\Sigma(l) = \mathrm{Cov}(\,\boldsymbol{e}_t(l)\,) = \Sigma + \sum_{j=1}^{l-1} \Phi^{j-1}(\,\Phi - \Theta\,)\,\Sigma\,(\,\Phi - \Theta\,)'\,\Phi'^{j-1}.$$

Also, from the infinite AR form, with $\Pi_j = \Theta^{j-1}(\,\Phi - \Theta\,)$, $j \geq 1$, we have the representation

$$\hat{\boldsymbol{Y}}_t(1) = \boldsymbol{\mu} + \sum_{j=1}^{\infty} \Pi_j\,(\,\boldsymbol{Y}_{t+1-j} - \boldsymbol{\mu}\,) = \boldsymbol{\mu} + \sum_{j=1}^{\infty} \Theta^{j-1}(\,\Phi - \Theta\,)\,(\,\boldsymbol{Y}_{t+1-j} - \boldsymbol{\mu}\,),$$

which shows explicitly the dependence of $\hat{\boldsymbol{Y}}_t(1)$ on the current and past values of the process $\{\boldsymbol{Y}_t\}$. More than one-step ahead forecasts $\hat{\boldsymbol{Y}}_t(l)$ can similarly be expressed in terms of current and past values explicitly through the relation $\hat{\boldsymbol{Y}}_t(l) = \boldsymbol{\mu} + \Phi^{l-1}(\,\hat{\boldsymbol{Y}}_t(1) - \boldsymbol{\mu}\,)$.

Next, consider the nonstationary vector IMA(1,1) model, $(\,I - B\,)\,\boldsymbol{Y}_t = (\,I - \Theta\,B\,)\,\varepsilon_t$, which for some purposes can be viewed as a special case of the ARMA(1,1) model with $\Phi = I$. Thus, for example, we have $\hat{\boldsymbol{Y}}_t(1) = \boldsymbol{Y}_t - \Theta\,\varepsilon_t$, $\hat{\boldsymbol{Y}}_t(l) = \hat{\boldsymbol{Y}}_t(l-1) = \hat{\boldsymbol{Y}}_t(1)$, for $l > 1$, and from the infinite AR form, $\hat{\boldsymbol{Y}}_t(1) = (\,I - \Theta\,)\sum_{j=0}^{\infty} \Theta^j\,\boldsymbol{Y}_{t-j}$, which gives the multivariate analogue of the exponential smoothing formula. Also, since $\Psi_j = I - \Theta$, $j \geq 1$, we have

$$\Sigma(l) = \mathrm{Cov}(\,\boldsymbol{e}_t(l)\,) = \Sigma + (l-1)\,(\,I - \Theta\,)\,\Sigma\,(\,I - \Theta\,)', \qquad l \geq 1.$$

In general, it follows from the infinite MA representation of the forecasts given by (2.20) that we obtain the multivariate version of the updating formula as

$$\hat{\boldsymbol{Y}}_{t+1}(l-1) = \sum_{j=l-1}^{\infty} \Psi_j\,\varepsilon_{t+l-j} = \hat{\boldsymbol{Y}}_t(l) + \Psi_{l-1}\,\varepsilon_{t+1}, \tag{2.23}$$

where $\varepsilon_{t+1} = \boldsymbol{Y}_{t+1} - \hat{\boldsymbol{Y}}_t(1)$ can also be interpreted as the one-step ahead forecast error. This provides a simple relationship to indicate how the forecast $\hat{\boldsymbol{Y}}_t(l)$ with forecast origin t is adjusted or updated to incorporate the information available from a new observation $\boldsymbol{Y}_{t+1}$ at time $t+1$. When applied to the IMA(1,1) model, we obtain the updating relations

$$\hat{\boldsymbol{Y}}_{t+1}(l-1) = \hat{\boldsymbol{Y}}_t(l) + (\,I - \Theta\,)\,\varepsilon_{t+1} = (\,I - \Theta\,)\,\boldsymbol{Y}_{t+1} + \Theta\,\hat{\boldsymbol{Y}}_t(1),$$

which gives new (updated) forecasts as a matrix "weighted average" of the old forecast $\hat{\boldsymbol{Y}}_t(l) = \hat{\boldsymbol{Y}}_t(1)$ and the new observation $\boldsymbol{Y}_{t+1}$ at time $t+1$.

From the preceding, the l-step ahead forecast error $e_{it}(l)$ for the ith component Y_{it} in the vector IMA(1,1) model has variance of the form $\mathrm{Var}(\,e_{it}(l)\,) = \sigma_{ii} + (l-1)\,b_{ii}$, $i = 1, \ldots, k$, where σ_{ii} is the ith diagonal element of Σ and b_{ii} is the ith diagonal element of $(I - \Theta) \Sigma (I - \Theta)'$. It may be of interest to compare these with the variances of l-step ahead forecast errors when the series are considered individually and forecasts are obtained using the univariate model representations for each series. From the discussion in Section 2.1.4, it follows that each series Y_{it} in the vector IMA(1,1) model has a univariate IMA(1,1) model representation given by $(1 - B) Y_{it} = (1 - \eta_i B) a_{it}$, with $\sigma_{a_i}^2 = \text{Var}(a_{it})$, $i = 1, \ldots, k$, and the parameters η_i and $\sigma_{a_i}^2$ are such that $\sigma_{a_i}^2 (1 + \eta_i^2)$ and $-\eta_i \sigma_{a_i}^2$ are the ith diagonal elements of $\Sigma + \Theta \Sigma \Theta'$ and $-\Theta \Sigma$, respectively. So if $e_{it}^*(l)$ denotes the l-step ahead forecast error for the ith component series Y_{it} based on its univariate model, then

$$\text{Var}(e_{it}^*(l)) = \sigma_{a_i}^2 + (l-1) \sigma_{a_i}^2 (1 - \eta_i)^2 \equiv \sigma_{a_i}^2 + (l-1) b_{ii}^*,$$

with $\sigma_{a_i}^2 \geq \sigma_{ii}$ and $b_{ii}^* = b_{ii}$ above, since $b_{ii}^* = \sigma_{a_i}^2 (1 - \eta_i)^2 = \sigma_{a_i}^2 (1 + \eta_i^2) - 2 \eta_i \sigma_{a_i}^2$ is seen to equal the ith diagonal element (b_{ii}) of the matrix $\Sigma + \Theta \Sigma \Theta' - \Theta \Sigma - \Sigma \Theta' = (I - \Theta) \Sigma (I - \Theta)'$. Hence, we find that variances for l-step ahead forecast errors for the ith series Y_{it} from both the vector IMA(1,1) model and the corresponding univariate model increase linearly with l at the same rate $b_{ii} = b_{ii}^*$, but with initial values $\sigma_{ii} \leq \sigma_{a_i}^2$.

We also briefly comment on the situation of a vector IMA(1,1) model with a noninvertible moving average operator, such as in Example 2.7 of Section 2.4.3, where it is assumed that the matrix Θ has $r < k$ eigenvalues equal to one and the remaining values are less than one. We also suppose that there are r linearly independent eigenvectors corresponding to the unit eigenvalues of Θ. In this situation, there exists an $r \times k$ matrix Q_2' such that $Q_2' \Theta = Q_2'$. Hence, the process $\boldsymbol{Y}_t$ satisfies $Q_2' \boldsymbol{Y}_t - Q_2' \boldsymbol{Y}_{t-1} = Q_2' \varepsilon_t - Q_2' \varepsilon_{t-1}$, which implies that $Q_2' \boldsymbol{Y}_t = \boldsymbol{\mu}_2 + Q_2' \varepsilon_t$ is actually a stationary white noise process. In such circumstances, although $\boldsymbol{Y}_t$ is nonstationary, the r linearly independent linear combinations $Q_2' \boldsymbol{Y}_t$ are stationary, and $\boldsymbol{Y}_t$ is said to be co-integrated of rank r (Engle and Granger, 1987). Also, we see that the forecasts $\hat{\boldsymbol{Y}}_t(l)$ from this model will satisfy the "co-integrating relations"

$$Q_2' \hat{\boldsymbol{Y}}_t(l) = Q_2' \hat{\boldsymbol{Y}}_t(1) = Q_2' [\boldsymbol{Y}_t - \Theta \varepsilon_t] = Q_2' \boldsymbol{Y}_t - Q_2' \varepsilon_t = \boldsymbol{\mu}_2,$$

that is, $Q_2' \hat{\boldsymbol{Y}}_t(l)$ is a fixed constant vector. In addition, the covariance matrix of the forecast errors for $Q_2' \boldsymbol{Y}_{t+l}$ is $Q_2' \Sigma(l) Q_2 = Q_2' \Sigma Q_2$, since $Q_2' (I - \Theta) = 0$, which does not increase with l as does $\Sigma(l) = \Sigma + (l-1)(I - \Theta) \Sigma (I - \Theta)'$ in general in the invertible case. Further discussion of models that involve co-integrated series and some of their properties is presented in Section 6.3.

CHAPTER 3

Canonical Structure of Vector ARMA Models

In this chapter, the canonical structure of vector ARMA model representations is briefly discussed through the introduction of the concepts of Kronecker indices and McMillan degree of a vector process $\{\boldsymbol{Y}_t\}$, and the echelon canonical form of the vector ARMA model is presented in particular. Canonical correlation structure for stationary vector ARMA processes is examined, and the relation between canonical correlation structure and the associated notion of scalar component models, introduced by Tiao and Tsay (1989) to specify simplifying structure for the vector ARMA model parameterization, is discussed. The partial correlation matrices and partial canonical correlations of a stationary vector process, and their special features for pure AR models, are also considered.

3.1 Consideration of Kronecker Structure for Vector ARMA Models

In general, the fundamental details concerning the structure of a vector ARMA(p,q) model representation,

$$\boldsymbol{Y}_t - \sum_{j=1}^{p} \Phi_j \boldsymbol{Y}_{t-j} = \varepsilon_t - \sum_{j=1}^{q} \Theta_j \varepsilon_{t-j} , \qquad (3.1)$$

beyond the mere specification of the overall orders p and q, can be characterized by use of the concepts of the Kronecker indices $K_1, \ldots, K_k$ and the McMillan degree $M = \sum_{i=1}^{k} K_i$ of a vector ARMA process. These concepts lead to specification of a canonical form of the ARMA model with certain simplifying structure in the parameterization of the AR and MA coefficients Φ_j and Θ_j, which is desirable in the search for the most appropriate and parsimonious representation for the vector ARMA process.

3.1.1 Kronecker Indices and McMillan Degree of Vector ARMA Process

For any stationary vector process $\{ \boldsymbol{Y}_t \}$ with autocovariance matrices $\Gamma(l) = \text{Cov}(\boldsymbol{Y}_t, \boldsymbol{Y}_{t+l})$, we define the infinite-dimensional (block) Hankel matrix of the autocovariances as

$$H = \begin{bmatrix} \Gamma(1)' & \Gamma(2)' & \Gamma(3)' & . & . & . \\ \Gamma(2)' & \Gamma(3)' & \Gamma(4)' & . & . & . \\ \Gamma(3)' & \Gamma(4)' & \Gamma(5)' & . & . & . \\ . & . & . & . & . & . \\ . & . & . & . & . & . \\ . & . & . & . & . & . \end{bmatrix} . \tag{3.2}$$

Then, in particular, the *McMillan degree* M of the process is defined as the rank of the Hankel matrix H. The process $\{ \boldsymbol{Y}_t \}$ follows a finite order vector ARMA(p,q) model if and only if the rank of H is finite. For an ARMA(p,q) process, it can be seen directly from the moment relations (2.14) of Section 2.3.2 that this rank, the McMillan degree M, satisfies $M \leq k\,s$, where $s = \max\{ p,\ q \}$, since, for example,

$$[\, -\boldsymbol{\Phi}_s,\ -\boldsymbol{\Phi}_{s-1}, \ldots,\ -\boldsymbol{\Phi}_1,\ I,\ 0,\ 0, \ldots \,]\ H = 0$$

with $\boldsymbol{\Phi}_j = 0$ for $j > p$. So clearly, more generally, all the $k \times k$ block rows of H beyond the sth block row ($s = \max\{ p,\ q \}$) are linearly dependent on the preceding block rows. But the McMillan degree M of a vector ARMA process could be considerably smaller than $k\,s$ due to rank deficiencies in the AR and MA coefficient matrices.

As we shall see shortly, the McMillan degree M has the interpretation as the number of linearly independent linear combinations of the present and past vectors $\boldsymbol{Y}_t, \boldsymbol{Y}_{t-1}, \ldots$ that are needed for optimal prediction of all future vectors within the ARMA structure. Note that

$$H = \text{Cov}(\mathbf{F}_{t+1}, \mathbf{P}_t) = \text{Cov}(\mathbf{F}_{t+1|t}, \mathbf{P}_t) \tag{3.3}$$

is the covariance between the collection of all present and past vectors, $\mathbf{P}_t = (\boldsymbol{Y}_t', \boldsymbol{Y}_{t-1}', \ldots)'$, and the collection of all future vectors $\mathbf{F}_{t+1} = (\boldsymbol{Y}_{t+1}', \boldsymbol{Y}_{t+2}', \ldots)'$ or the collection of predicted values of all future vectors, $\mathbf{F}_{t+1|t} = E(\mathbf{F}_{t+1} \mid \mathbf{P}_t)$. Hence, if the rank of H is equal to M, then the (linear) predictor space formed from the collection $\mathbf{F}_{t+1|t}$ of predicted values $\hat{\boldsymbol{Y}}_t(l) = E(\boldsymbol{Y}_{t+l} \mid \mathbf{P}_t)$, $l > 0$, of all future vectors is of finite dimension M. Sometimes (e.g., Hannan and Deistler, 1988, Chap. 2) the Hankel matrix H is defined in terms of the matrices $\Psi_j \equiv E(\boldsymbol{Y}_t\, \varepsilon_{t-j}')\, \Sigma^{-1}$ in the infinite moving average form $\boldsymbol{Y}_t - \boldsymbol{\mu} = \sum_{j=0}^{\infty} \Psi_j\, \varepsilon_{t-j}$ of the vector ARMA process, instead of the covariance matrices $\Gamma(j)' = E[\, (\boldsymbol{Y}_t - \boldsymbol{\mu})(\boldsymbol{Y}_{t-j} - \boldsymbol{\mu})' \,]$. Then the Hankel matrix H is closely related to the matrix $\text{Cov}(\mathbf{F}_{t+1}, \mathbf{P}_t^*)$, where $\mathbf{P}_t^* = (\varepsilon_t', \varepsilon_{t-1}', \ldots)'$, with $H = \text{Cov}(\mathbf{F}_{t+1}, \mathbf{P}_t^*)\, (I_\infty \otimes \Sigma^{-1})$, and from (2.20) of Section 2.5.2, we have

$$E(\boldsymbol{Y}_{t+l} \mid \mathbf{P}_t) = \hat{\boldsymbol{Y}}_t(l) = \sum_{j=l}^{\infty} \Psi_j \, \varepsilon_{t+l-j}$$

(assuming $\boldsymbol{\mu} = 0$), so that then $\mathbf{F}_{t+1|t} = H \, \mathbf{P}_t^*$.

In addition, the ith *Kronecker index* K_i, $i = 1, \ldots, k$, of the process $\{\boldsymbol{Y}_t\}$ is the smallest value such that the $(k K_i + i)$th row of H, that is, the ith row in the $(K_i + 1)$th block of rows of H, is linearly dependent on the previous rows of H. This also implies, through the structure of the Hankel matrix H, that all rows $k l + i$, for every $l \geq K_i$, will also be linearly dependent on the rows preceding the $(k K_i + i)$th row. The set of Kronecker indices $\{K_1, \ldots, K_k\}$ is unique for any given ARMA process, and, hence, it is not dependent on any one particular form of the observationally equivalent ARMA model representations of the process. The representation of the vector ARMA model in its equivalent minimal dimension (equal to its McMillan degree $M = \sum_{i=1}^{k} K_i$) state-space form is one way to reveal the special structure of the ARMA parameters associated with these quantities. (State-space forms of the vector ARMA model will be discussed later in Section 7.2.) We will now indicate, in particular, that the Kronecker indices have close connections with the second moment equations such as in (2.14), since these equations exhibit the row dependencies among the covariance matrices $\Gamma(j)'$ and, hence, can be used to deduce special structure among the AR and MA parameter matrices.

3.1.2 Echelon Form Structure of Vector ARMA Model Implied by Kronecker Indices

Specifically, if vector ARMA models similar to the form in (1.13) of Section 1.2.2 are considered,

$$\Phi_0^{\#} \, \boldsymbol{Y}_t - \sum_{j=1}^{p} \Phi_j^{\#} \, \boldsymbol{Y}_{t-j} = \boldsymbol{\delta} + \Theta_0^{\#} \, \varepsilon_t - \sum_{j=1}^{q} \Theta_j^{\#} \, \varepsilon_{t-j}, \tag{3.4}$$

with $\Phi_0^{\#} = \Theta_0^{\#}$ lower triangular (and having ones on the diagonal), then equations similar to (2.14) for the cross-covariance matrices $\Gamma(l)$ of the process are obtained as

$$\Phi_0^{\#} \, \Gamma(l)' - \sum_{j=1}^{p} \Phi_j^{\#} \, \Gamma(l-j)' = - \sum_{j=l}^{q} \Theta_j^{\#} \, \Sigma \, \Psi_{j-l}'. \tag{3.5}$$

So if $\phi_j(i)'$ denotes the ith row of $\Phi_j^{\#}$, then the ith Kronecker index equal to K_i implies the linear dependence in the rows of the Hankel matrix H of the form

$$\phi_0(i)' \, \Gamma(l)' - \sum_{j=1}^{K_i} \phi_j(i)' \, \Gamma(l-j)' = 0 \qquad \text{for all} \quad l \geq K_i + 1, \tag{3.6}$$

that is, $\boldsymbol{b}_i' H = 0$ with $\boldsymbol{b}_i' = (-\phi_{K_i}(i)', \ldots, -\phi_1(i)', \phi_0(i)', 0', \ldots)$. Note that by definition of the ith Kronecker index K_i, the row vector $\phi_0(i)'$ in (3.6) can be taken to have a one in the ith position and zeros for positions greater than the

ith. Therefore, a Kronecker index equal to K_i implies, in particular, that an ARMA model representation of the form (3.4) can be constructed for the process such that the ith rows of the matrices $\Phi_j^{\#}$ and $\Theta_j^{\#}$ will be zero for $j > K_i$. Notice that, through consideration of equations similar to (2.12) of Section 2.3.1, namely,

$$\Phi_0^{\#} \Psi_l - \sum_{j=1}^{p} \Phi_j^{\#} \Psi_{l-j} = -\Theta_l^{\#}, \qquad l > 0,$$

for the ARMA model in the form (3.4), the same conclusions would be obtained if the Hankel matrix H defined in terms of the infinite MA matrices Ψ_j were used.

In addition to the implications from (3.6) that the ith rows of the matrices $\Phi_j^{\#}$ and $\Theta_j^{\#}$ in (3.4) can be specified to be zero for $j > K_i$, additional zero constraints on certain elements in the ith rows of the matrices $\Phi_j^{\#}$ for $j \le K_i$ can be specified. Specifically, the lth element of the ith row $\phi_j(i)'$ can be specified to be zero whenever $j + K_l \le K_i$, because for $K_l \le K_i$ the rows $k(K_l + j) + l$, $j = 0, \ldots, (K_i - K_l)$, of the Hankel matrix H are all linearly dependent on the previous rows of H. Hence, the (i, l)th element of the AR operator $\Phi^{\#}(B) = \Phi_0^{\#} - \sum_{j=1}^{p} \Phi_j^{\#} B^j$ in model (3.4) can be specified to have nonzero coefficients only for the lags $j = K_i - K_{il} + 1, \ldots, K_i$, with zero coefficients specified for any lower lags of j (when $i \ne l$), where we define

$$K_{il} = \begin{cases} \min(K_i + 1, K_l) & \text{for } i > l \\ \min(K_i, K_l) & \text{for } i \le l \end{cases}$$

(so that whenever $K_l \le K_i$ we have $K_{il} = K_l$). So the corresponding number of unknown AR parameters in the (i, l)th element of $\Phi^{\#}(B)$ is equal to K_{il}. Thus, the AR operator $\Phi^{\#}(B)$ in model (3.4) can be specified such that the total number of unknown parameters of $\Phi^{\#}(B)$ is equal to $\sum_{i=1}^{k} \sum_{l=1}^{k} K_{il} = M + \sum \sum_{i \ne l}^{k} K_{il}$, while the number of unknown parameters in the MA operator $\Theta^{\#}(B)$ in (3.4), excluding those parameters in $\Theta_0^{\#} = \Phi_0^{\#}$, is equal to $\sum_{i=1}^{k} k K_i = k M$. The additional zero constraints mentioned for the AR coefficient matrices $\Phi_j^{\#}$ could equivalently be placed on the elements of the MA matrices $\Theta_j^{\#}$ in (3.4) instead of the AR matrices $\Phi_j^{\#}$, because of certain exchangeability of parameters between the AR and MA operators.

In summary, for a stationary linear process $\{\boldsymbol{Y}_t\}$ with specified (finite-valued) Kronecker indices $K_1, \ldots, K_k$, an ARMA representation as in (3.4) (with $p = q = \max\{K_i\}$) can be specified to describe the process, with the matrices $\Phi_j^{\#}$ and $\Theta_j^{\#}$ possessing the structure that their ith rows are zero for $j > K_i$ and the additional zero constraints structure noted above. Moreover, for a stationary vector process with given covariance matrix structure $\Gamma(l)$ (equivalently, with given infinite MA coefficients Ψ_j) and associated Kronecker indices $\{K_i\}$, Hannan and Deistler (1988, Theorem 2.5.1) [see also Dickinson, Kailath, and Morf (1974) and Forney (1975)] have shown that the ARMA model as in (3.4) with the zero constraints imposed as described above provides a unique ARMA representation for the process, with AR and MA operators $\Phi^{\#}(B)$ and $\Theta^{\#}(B)$ being left-coprime, and where all unknown parameters are identified. This (canonical) ARMA representation is referred to as a (reversed) *echelon ARMA form*. In particular, the AR coefficient matrices $\Phi_j^{\#}$ in the echelon canonical representation (3.4) are uniquely determined from the $\Gamma(l)$ by the requirement that their ith rows $\phi_j(i)'$, $j = 0, \ldots, K_i$, $i = 1, \ldots, k$, satisfy the conditions (3.6). Tsay (1991) has illustrated how specification of the Kronecker index structure for a stationary vector linear process $\{\boldsymbol{Y}_t\}$ establishes the existence of a vector ARMA representation for the process as in (3.4) with the structure for the matrices $\Phi_j^{\#}$ and $\Theta_j^{\#}$ as specified above. To establish this representation, it is noted that for each $i = 1, \ldots, k$, the process can be defined by

$$u_{i,t+1+K_i} = \phi_0(i)' \, \boldsymbol{Y}_{t+1+K_i} - \sum_{j=1}^{K_i} \phi_j(i)' \, \boldsymbol{Y}_{t+1+K_i-j} ,$$

where the $\phi_j(i)'$ are determined through (3.6), which is uncorrelated with the present and past $\mathbf{P}_t$ and, hence, it must be representable in a MA structure of order at most K_i as

$$u_{i,t+1+K_i} = \theta_0(i)' \, \varepsilon_{t+1+K_i} - \sum_{j=1}^{K_i} \theta_j(i)' \, \varepsilon_{t+1+K_i-j} .$$

From a predictive viewpoint, for each Kronecker index K_i, the above relation $\boldsymbol{b}_i' H = 0$ implies that $\text{Cov}(\boldsymbol{b}_i' \mathbf{F}_{t+1}, \mathbf{P}_t) = 0$ so that there is a (finite) linear combination $u_{i,t+1+K_i} = \boldsymbol{b}_i' \mathbf{F}_{t+1}$ of the future vector $\mathbf{F}_{t+1}$ that is uncorrelated with the present and past $\mathbf{P}_t$.

3.1.3 Reduced-Rank Form of Vector ARMA Model Implied by Kronecker Indices

It has been established that for any vector ARMA process with Kronecker indices $K_1, \ldots, K_k$, the process can be represented uniquely in the echelon canonical form as in (3.4), with $\Phi_0^{\#} = \Theta_0^{\#}$ lower triangular having ones on the diagonal, $p = q = \max\{K_i\}$, and the ith rows of the matrices $\Phi_j^{\#}$ and $\Theta_j^{\#}$ specified to be zero for $j > K_i$, $i = 1, \ldots, k$. That is, the ith row of $(\Phi^{\#}(B), \Theta^{\#}(B))$ has degree equal to K_i. Now, by multiplication of (3.4) by $\Phi_0^{\#-1}$, we obtain the vector ARMA model in the standard form as

$$\boldsymbol{Y}_t = \sum_{j=1}^{p} \Phi_0^{\#-1} \Phi_j^{\#} \, \boldsymbol{Y}_{t-j} + \varepsilon_t - \sum_{j=1}^{q} \Phi_0^{\#-1} \Theta_j^{\#} \, \varepsilon_{t-j} \equiv \sum_{j=1}^{p} \Phi_j \, \boldsymbol{Y}_{t-j} + \varepsilon_t - \sum_{j=1}^{q} \Theta_j \, \varepsilon_{t-j} . \quad (3.7)$$

In this standard ARMA model form, the AR and MA coefficient matrices $\Phi_j = \Phi_0^{\#-1} \Phi_j^{\#}$ and $\Theta_j = \Phi_0^{\#-1} \Theta_j^{\#}$ can be of a special reduced-rank form. Let the rank of the matrix $(\Phi_j, \Theta_j) = \Phi_0^{\#-1}(\Phi_j^{\#}, \Theta_j^{\#})$ be denoted as r_j, $j = 1, \ldots, p$. Then it is seen that this rank r_j is equal to the number of Kronecker indices that are greater than or equal to the lag j, since $(\Phi_j^{\#}, \Theta_j^{\#})$

has a zero row corresponding to each Kronecker index K_i for which $j > K_i$. So the ranks $r_j = \text{rank}(\, \Phi_j,\, \Theta_j \,)$ are decreasing (nonincreasing) as the lag j increases. Also, we readily see that the McMillan degree is equal to $M = \sum_{i=1}^{k} K_i = \sum_{j=1}^{p} r_j$, the sum of the ranks of the matrices $(\, \Phi_j,\, \Theta_j \,)$ in the standard ARMA model form (3.7). Moreover, for $j = 1, \ldots, p$, let D_j' denote the $r_j \times k$ submatrix of the $k \times k$ identity matrix I_k that selects the nonzero rows of $(\, \Phi_j^{\#},\, \Theta_j^{\#} \,)$, so that $(\, B_j,\, C_j \,) \equiv D_j' (\, \Phi_j^{\#},\, \Theta_j^{\#} \,)$ consists of only the r_j nonzero rows of $(\, \Phi_j^{\#},\, \Theta_j^{\#} \,)$. Then,

$$(\, \Phi_j,\, \Theta_j \,) = \Phi_0^{\#-1} D_j \,(\, B_j,\, C_j \,) \equiv A_j \,(\, B_j,\, C_j \,),$$

where $A_j = \Phi_0^{\#-1} D_j$ is $k \times r_j$, provides a reduced-rank factorization of the coefficient matrices $(\, \Phi_j,\, \Theta_j \,)$ in the standard ARMA form (3.7) of (reduced) rank r_j. The matrices A_j in this factorization have the property that $\text{range}(\, A_j \,) \supset \text{range}(\, A_{j+1} \,)$ for $j = 1, \ldots, p$, where $\text{range}(\, A \,)$ denotes the linear space generated by the columns of a matrix A. Hence, the vector ARMA model in the standard form (3.7) can be written as

$$\boldsymbol{Y}_t = \sum_{j=1}^{p} A_j \, B_j \, \boldsymbol{Y}_{t-j} + \varepsilon_t - \sum_{j=1}^{q} A_j \, C_j \, \varepsilon_{t-j} \,. \tag{3.8}$$

The vector ARMA model of this form may be referred to as a *nested reduced-rank* ARMA model representation. The special case of a pure AR nested reduced-rank model will be considered in more detail in Section 6.1.

For a few illustrative examples, consider a bivariate $(k = 2)$ process $\{\boldsymbol{Y}_t\}$. When this process possesses Kronecker indices $K_1 = K_2 = 1$, then a general ARMA(1,1) representation $\boldsymbol{Y}_t - \Phi_1 \, \boldsymbol{Y}_{t-1} = \varepsilon_t - \Theta_1 \, \varepsilon_{t-1}$ is implied for the process from the preceding discussion. However, notice that a pure AR(1) process with full rank AR matrix Φ_1 or a pure MA(1) process with full rank MA matrix Θ_1 would both also possess Kronecker indices equal to $K_1 = K_2 = 1$. This simple example thus illustrates that specification of the Kronecker indices alone does not necessarily lead to the specification of an ARMA representation where all the simplifying structure in the parameters is directly revealed. For a second case, suppose the bivariate process has Kronecker indices $K_1 = 1$ and $K_2 = 0$. Then the implied structure for the process is ARMA(1,1) as in (3.4), with either

$$\Phi_0^{\#} = \begin{bmatrix} 1 & 0 \\ X & 1 \end{bmatrix}, \qquad \Phi_1^{\#} = \begin{bmatrix} X & 0 \\ 0 & 0 \end{bmatrix}, \qquad \Theta_1^{\#} = \begin{bmatrix} X & X \\ 0 & 0 \end{bmatrix}$$

or, equivalently,

$$\Phi_0^{\#} = \begin{bmatrix} 1 & 0 \\ X & 1 \end{bmatrix}, \qquad \Phi_1^{\#} = \begin{bmatrix} X & X \\ 0 & 0 \end{bmatrix}, \qquad \Theta_1^{\#} = \begin{bmatrix} X & 0 \\ 0 & 0 \end{bmatrix},$$

where X's denote unknown parameters that need estimation and 0's indicate values that are known to be specified as zero. In either case, on multiplication of the ARMA(1,1) relation $\Phi_0^{\#} \, \boldsymbol{Y}_t - \Phi_1^{\#} \, \boldsymbol{Y}_{t-1} = \Theta_0^{\#} \, \varepsilon_t - \Theta_1^{\#} \, \varepsilon_{t-1}$ on the left by

$\Phi_0^{\#-1}$, we obtain an ARMA(1,1) representation $\boldsymbol{Y}_t - \Phi_1 \boldsymbol{Y}_{t-1} = \varepsilon_t - \Theta_1 \varepsilon_{t-1}$ in the "standard" form (2.11), but with a reduced-rank structure for the coefficient matrices such that $\text{rank}[\ \Phi_1,\ \Theta_1\ ] = 1$. For a third situation, suppose the bivariate process has Kronecker indices $K_1 = 2$ and $K_2 = 1$. Then the echelon form structure for the process is ARMA(2,2) as in (3.4), with either

$$\Phi_0^{\#} = \begin{bmatrix} 1 & 0 \\ X & 1 \end{bmatrix}, \quad \Phi_1^{\#} = \begin{bmatrix} X & 0 \\ X & X \end{bmatrix}, \quad \Phi_2^{\#} = \begin{bmatrix} X & X \\ 0 & 0 \end{bmatrix},$$

$$\Theta_1^{\#} = \begin{bmatrix} X & X \\ X & X \end{bmatrix}, \quad \Theta_2^{\#} = \begin{bmatrix} X & X \\ 0 & 0 \end{bmatrix}$$

or, equivalently,

$$\Phi_0^{\#} = \begin{bmatrix} 1 & 0 \\ X & 1 \end{bmatrix}, \quad \Phi_1^{\#} = \begin{bmatrix} X & X \\ X & X \end{bmatrix}, \quad \Phi_2^{\#} = \begin{bmatrix} X & X \\ 0 & 0 \end{bmatrix},$$

$$\Theta_1^{\#} = \begin{bmatrix} X & 0 \\ X & X \end{bmatrix}, \quad \Theta_2^{\#} = \begin{bmatrix} X & X \\ 0 & 0 \end{bmatrix}.$$

Again, in either case, on multiplication of the echelon form ARMA(2,2) relation on the left by $\Phi_0^{\#-1}$, we obtain an ARMA(2,2) representation in the "standard" form (2.11), but with a reduced-rank structure for the coefficient matrices such that $\text{rank}[\ \Phi_2,\ \Theta_2\ ] = 1$.

Excellent accounts and further illustrative examples of the Kronecker indices (also known as structural indices) and the McMillan degree of a vector ARMA process, and related topics, have been given by Solo (1986), Hannan and Kavalieris (1984), Hannan and Deistler (1988), and Tsay (1989b). Further discussion of these concepts will be presented later in Section 4.5.2 in relation to the use of canonical correlation methods between the present and past vector $\mathbf{P}_t$ and the future vector $\mathbf{F}_{t+1}$ as a method to determine the Kronecker indices K_i of the vector ARMA process, and also in Sections 7.2.4 and 7.2.5 in relation to minimal dimension state-space representations of the ARMA process.

3.2 Canonical Correlation Structure for ARMA Time Series

Canonical correlation analysis is a useful tool in the study of correlations among sets of variables in multivariate analysis, and its use in the context of multivariate time series analysis has been explored by Akaike (1976), Cooper and Wood (1982), and Tiao and Tsay (1989), among others. A canonical correlation analysis between two sets of random variables (r.v.'s), $\boldsymbol{X}_1 = (x_{11}, x_{12}, \ldots, x_{1k_1})'$ and $\boldsymbol{X}_2 = (x_{21}, x_{22}, \ldots, x_{2k_2})'$, of dimensions k_1 and k_2 (assume $k_1 \le k_2$) involves determining a set of linear combinations, $U_i = \boldsymbol{a}_i' \boldsymbol{X}_1$ and $V_i = \boldsymbol{b}_i' \boldsymbol{X}_2$, $i = 1, \ldots, k_1$, of $\boldsymbol{X}_1$ and $\boldsymbol{X}_2$ with the following properties:

1) The r.v.'s U_i and V_i are mutually uncorrelated and have unit variances, i.e., $\text{Cov}(U_i, U_j) = 1$ if $i = j$ and $= 0$ for $i \neq j$; and similarly for the V_i, and $\text{Cov}(U_i, V_j) = 0$ for $i \neq j$.

2) $\rho_1 = \text{Corr}(U_1, V_1) \geq 0$ is the largest correlation between any linear combinations of $\boldsymbol{X}_1$ and $\boldsymbol{X}_2$, that is, U_1 and V_1 are determined by the property that they are the linear combinations of $\boldsymbol{X}_1$ and $\boldsymbol{X}_2$ which possess the maximum possible correlation among all such linear combinations. In addition, U_2 and V_2 have correlation $\rho_2 = \text{Corr}(U_2, V_2) \geq 0$ and are characterized as those linear combinations with maximum correlation among all linear combinations which are uncorrelated with U_1 and V_1. Proceeding in this way, then, in general, U_i and V_i have the properties that $\rho_i = \text{Corr}(U_i, V_i) \geq 0$, with $\text{Corr}(U_i, U_j) = 0$, $\text{Corr}(V_i, V_j) = 0$, and $\text{Corr}(U_i, V_j) = 0$ for $i \neq j$, $i, j = 1, \ldots, k_1$, and U_i, V_i have maximum correlation among all linear combinations which are uncorrelated with U_j, V_j, $j = 1, \ldots, i-1$.

The variables U_i, V_i are called the canonical variables, and $\rho_1 \geq \rho_2 \geq \cdots \geq \rho_{k_1} \geq 0$ are the *canonical correlations* between $\boldsymbol{X}_1$ and $\boldsymbol{X}_2$.

If $\Omega = \text{Cov}(\boldsymbol{X})$ denotes the covariance matrix of $\boldsymbol{X} = (\boldsymbol{X}_1', \boldsymbol{X}_2')'$, with $\Omega_{ij} = \text{Cov}(\boldsymbol{X}_i, \boldsymbol{X}_j)$, then it can be shown that the values ρ_i^2 (squared canonical correlations) are the ordered eigenvalues of the matrix $\Omega_{11}^{-1}\,\Omega_{12}\,\Omega_{22}^{-1}\,\Omega_{21}$ and the vectors $\boldsymbol{a}_i$, such that $U_i = \boldsymbol{a}_i'\boldsymbol{X}_1$, are the corresponding (normalized) eigenvectors, that is, the ρ_i^2 and $\boldsymbol{a}_i$ satisfy

$$(\rho_i^2 I - \Omega_{11}^{-1}\,\Omega_{12}\,\Omega_{22}^{-1}\,\Omega_{21})\,\boldsymbol{a}_i = 0, \qquad i = 1, \ldots, k_1, \tag{3.9}$$

and $\rho_1^2 \geq \rho_2^2 \geq \cdots \geq \rho_{k_1}^2 \geq 0$ (e.g., Anderson, 1984, p. 490). Similarly, one can define the notion of *partial canonical correlations* between $\boldsymbol{X}_1$ and $\boldsymbol{X}_2$, given another set of variables $\boldsymbol{X}_3$, as the canonical correlations between $\boldsymbol{X}_1$ and $\boldsymbol{X}_2$ after they have been "adjusted" for $\boldsymbol{X}_3$ by linear regression on $\boldsymbol{X}_3$. Specifically, let $\Omega_{ij.m} = \Omega_{ij} - \Omega_{im}\,\Omega_{mm}^{-1}\,\Omega_{mj} = \text{Cov}(\boldsymbol{X}_{i.m}, \boldsymbol{X}_{j.m})$ for $i, j = 1, 2$ and $m = 3$, where $\boldsymbol{X}_{i.m} = \boldsymbol{X}_i - \Omega_{im}\,\Omega_{mm}^{-1}\,\boldsymbol{X}_m$ is $\boldsymbol{X}_i$ adjusted for linear regression on $\boldsymbol{X}_m$. Then the (squared) partial canonical correlations between $\boldsymbol{X}_1$ and $\boldsymbol{X}_2$, given $\boldsymbol{X}_3$, are the ordered eigenvalues of the matrix $\Omega_{11.3}^{-1}\,\Omega_{12.3}\,\Omega_{22.3}^{-1}\,\Omega_{21.3}$. It may be useful to note that the canonical correlations between the set of variables $(\boldsymbol{X}_1, \boldsymbol{X}_3)$ and $(\boldsymbol{X}_2, \boldsymbol{X}_3)$ that are not (trivially) equal to one are equal to the partial canonical correlations between $\boldsymbol{X}_1$ and $\boldsymbol{X}_2$, given $\boldsymbol{X}_3$.

Note that if the *i*th canonical correlation is zero, $\rho_i = 0$, then $U_i = \boldsymbol{a}_i'\boldsymbol{X}_1$ is uncorrelated with all linear combinations of $\boldsymbol{X}_2$ and conversely, because, from (3.9), $\rho_i = 0$ implies that $\Omega_{21}\,\boldsymbol{a}_i = \text{Cov}(\boldsymbol{X}_2, U_i) = 0$. More generally, a useful property to note is that if there exist (at least) s ($s \leq k_1$) linearly independent linear combinations of $\boldsymbol{X}_1$ which are completely uncorrelated with $\boldsymbol{X}_2$, say $\boldsymbol{U} = A'\boldsymbol{X}_1$ such that $\text{Cov}(\boldsymbol{X}_2, \boldsymbol{U}) = \Omega_{21}\,A = 0$, then there are (at least) s zero canonical correlations between $\boldsymbol{X}_1$ and $\boldsymbol{X}_2$. This holds since $\Omega_{21}\,A = 0$ implies that the s linearly independent columns of A satisfy $(\rho^2 I - \Omega_{11}^{-1}\,\Omega_{12}\,\Omega_{22}^{-1}\,\Omega_{21})\,A = 0$ for $\rho = 0$, and, hence, there are (at least) s

zero eigenvalues (zero canonical correlations) in (3.9). In effect, then, we see that the number s of zero canonical correlations is equal to $s = k_1 - r$, where $r = \text{rank}(\Omega_{21})$.

3.2.1 Canonical Correlations for Vector ARMA Processes

In the vector ARMA model time series context, we consider vectors such as the $k(m+1)$-dimensional vector

$$\mathbf{Y}_{m,t} = (Y_t', Y_{t-1}', \ldots, Y_{t-m}')'.$$

Then, following the approach of Tiao and Tsay (1989), we examine the canonical correlation structure between the variables $\mathbf{Y}_{m,t}$ and $\mathbf{Y}_{n,t-j-1} = (Y_{t-j-1}', Y_{t-j-2}', \ldots, Y_{t-j-1-n}')'$, for various combinations of $m = 0, 1, \ldots$, and $j = 0, 1, \ldots$, and $n \geq m$. Recall that for an ARMA(p,q) process, $Y_t - \sum_{j=1}^{p} \Phi_j Y_{t-j} = \varepsilon_t - \sum_{j=1}^{q} \Theta_j \varepsilon_{t-j}$, the variables Y_{t-k} are uncorrelated with the ε_{t-j}, $j = 0, 1, \ldots, q$, for all $k > q$. Thus, if $m \geq p$, then there are (at least) k linear combinations of $\mathbf{Y}_{m,t}$,

$$Y_t - \sum_{j=1}^{p} \Phi_j Y_{t-j} = (I, -\Phi_1, \ldots, -\Phi_p, 0, \ldots, 0)\, \mathbf{Y}_{m,t} = \varepsilon_t - \sum_{j=1}^{q} \Theta_j \varepsilon_{t-j},$$

which are uncorrelated with $\mathbf{Y}_{n,t-j-1}$ if $j \geq q$. In particular, then, for $m = n = p$ and $j = q$ there are (at least) k *zero* canonical correlations between $\mathbf{Y}_{p,t}$ and $\mathbf{Y}_{p,t-q-1}$, as well as between $\mathbf{Y}_{p,t}$ and $\mathbf{Y}_{p,t-j-1}$, $j > q$, and between $\mathbf{Y}_{m,t}$ and $\mathbf{Y}_{m,t-q-1}$, $m > p$.

However, because simplifying reduced-rank structures may exist among the coefficient matrices Φ_j and Θ_j or, equivalently, lower-order scalar component ARMA models may exist (e.g., Tiao and Tsay, 1989), there can be some zero canonical correlations between $\mathbf{Y}_{m,t}$ and $\mathbf{Y}_{n,t-j-1}$ even if $m < p$ or $j < q$. To illustrate some of the possibilities we examine the situation of a vector ARMA(2,1) model, $Y_t - \Phi_1 Y_{t-1} - \Phi_2 Y_{t-2} = \varepsilon_t - \Theta_1 \varepsilon_{t-1}$. First, from the above discussion we have that there will be (at least) k zero canonical correlations between $\mathbf{Y}_{2,t} = (Y_t', Y_{t-1}', Y_{t-2}')'$ and $\mathbf{Y}_{2,t-2} = (Y_{t-2}', Y_{t-3}', Y_{t-4}')'$ (that is, for $m = 2$ and $j = 1$), since the k linear combinations $Y_t - \Phi_1 Y_{t-1} - \Phi_2 Y_{t-2}$ are independent of $\mathbf{Y}_{2,t-2}$. But suppose that $\text{rank}(\Phi_2) = r_2 < k$. Then there exists a $(k - r_2) \times k$ matrix F_2' such that $F_2' \Phi_2 = 0$, and, hence,

$$\begin{aligned}(F_2', -F_2'\Phi_1)\, \mathbf{Y}_{1,t} &= F_2' Y_t - F_2' \Phi_1 Y_{t-1} \\ &\equiv F_2'(Y_t - \Phi_1 Y_{t-1} - \Phi_2 Y_{t-2}) = F_2'(\varepsilon_t - \Theta_1 \varepsilon_{t-1})\end{aligned}$$

is uncorrelated with $\mathbf{Y}_{1,t-2} = (Y_{t-2}', Y_{t-3}')'$. Thus, there will be $k - r_2$ zero canonical correlations between $\mathbf{Y}_{1,t} = (Y_t', Y_{t-1}')'$ and $\mathbf{Y}_{1,t-2} = (Y_{t-2}', Y_{t-3}')'$, so there are some ($< k$) zero canonical correlations when $m = 1$ and $j = 1$. If, in addition, we have $\text{rank}(\Phi_2, \Theta_1) = r_2' \geq r_2$, say, then it follows that there are

$k - r_2'$ zero canonical correlations between $\mathbf{Y}_{1,t} = (Y_t', Y_{t-1}')'$ and $\mathbf{Y}_{1,t-1} = (Y_{t-1}', Y_{t-2}')'$, with $m = 1$ and $j = 0$. Finally, if we suppose that $\text{rank}(\Phi_1, \Phi_2, \Theta_1) = r_1 \geq r_2'$, then there is a $(k - r_1) \times k$ matrix F_1' such that $F_1' Y_t$ is a $(k - r_1)$-dimensional white noise series uncorrelated with Y_{t-1}, $Y_{t-2}, \ldots$. So there will also be $k - r_1$ zero canonical correlations between $\mathbf{Y}_{0,t} \equiv Y_t$ and $\mathbf{Y}_{0,t-1} \equiv Y_{t-1}$, with $m = 0$ and $j = 0$.

Generally, many possibilities exist, but systematic examination of the number of zero canonical correlations between $\mathbf{Y}_{m,t}$ and $\mathbf{Y}_{n,t-j-1}$ for combinations of m and j can reveal the detailed structure of the nature of the ARMA(p,q) model. Arrangement of the numbers of zero canonical correlations in a 2-way table as a function of the integers $m \geq 0$ and $j \geq 0$, as suggested by Tiao and Tsay (1989), will provide the necessary detailed information. In particular, the smallest integer values of m and j for which (at least) k zero canonical correlations exist will identify the overall orders p and q of the model. We also note the special cases of the above. First, when $m = n = 0$, we are examining the canonical correlations between Y_t and Y_{t-j-1}, and these will all equal zero in an MA(q) process for $j \geq q$. Second, when $j = 0$ and $m = n$, we are examining the *partial* canonical correlations between Y_t and Y_{t-m-1}, given $Y_{t-1}, \ldots, Y_{t-m}$, and these will all equal zero in an AR(p) process for $m \geq p$.

3.2.2 Relation to Scalar Component Model Structure

The detailed information on the existence of various zero canonical correlations as discussed above can be interpreted as providing information on the presence of scalar component models (SCMs) in the structure of the vector ARMA process. Knowledge of these SCMs gives specific details on the special structure in the parameterization of the ARMA representation for the process. A *scalar component model* (Tiao and Tsay, 1989) of order (p^*, q^*), $p^* \leq p$, $q^* \leq q$, is said to exist for the process $\{Y_t\}$ if a linear combination $z_t = \boldsymbol{a}' Y_t$ exists such that

$$z_t - \sum_{j=1}^{p^*} \boldsymbol{a}' \Phi_j \, Y_{t-j} = \boldsymbol{a}' \varepsilon_t - \sum_{j=1}^{q^*} \boldsymbol{a}' \Theta_j \, \varepsilon_{t-j} , \tag{3.10}$$

so that $u_t = z_t - \sum_{j=1}^{p^*} \boldsymbol{a}' \Phi_j \, Y_{t-j}$ has the property of being uncorrelated with ε_{t-j} for $j > q^*$. A slightly more general definition is that $z_t = \boldsymbol{a}' Y_t$ follows a scalar component model structure of order (p^*, q^*) if there exists k-dimensional vectors $\boldsymbol{b}_1, \ldots, \boldsymbol{b}_{p^*}$, such that the linear combination $u_t = z_t - \sum_{j=1}^{p^*} \boldsymbol{b}_j' \, Y_{t-j}$ is uncorrelated with ε_{t-j} for $j > q^*$ but correlated with ε_{t-q^*}. Hence, from the preceding discussion we see that the presence of a scalar component model of order (p^*, q^*) implies the existence of at least one zero canonical correlation between $\mathbf{Y}_{p^*,t} = (Y_t', Y_{t-1}', \ldots, Y_{t-p^*}')'$ and $\mathbf{Y}_{n,t-j-1}$ for any $j \geq q^*$. Thus, the preceding canonical correlation analysis methods will serve to determine the presence of scalar component models of various orders.

In the vector ARMA model specification approach of Tiao and Tsay (1989) through the use of SCMs, a linearly independent set of k scalar components $z_{it} = \boldsymbol{a}_i' \boldsymbol{Y}_t$ of orders (p_i, q_i), $i = 1, \ldots, k$, is sought such that the orders $p_i + q_i$ are as small as possible. Once such a set of SCMs is obtained, the specification of an ARMA model structure for $\boldsymbol{Y}_t$ is determined through the relations

$$A\, \boldsymbol{Y}_t - \sum_{j=1}^{p} B_j\, \boldsymbol{Y}_{t-j} = A\, \varepsilon_t - \sum_{j=1}^{q} G_j\, \varepsilon_{t-j}, \tag{3.11}$$

where $A = [\boldsymbol{a}_1, \ldots, \boldsymbol{a}_k]'$, $B_j = A\, \Phi_j$, $j = 1, \ldots, p$, $G_j = A\, \Theta_j$, $j = 1, \ldots, q$, $p = \max\{p_i\}$, and $q = \max\{q_i\}$. Moreover, in (3.11) the ith row of B_j is specified to be zero for $j > p_i$ and the ith row of G_j is zero for $j > q_i$. Thus, on multiplication of (3.11) on the left by A^{-1} we are lead to a specification of an ARMA(p,q) model for $\boldsymbol{Y}_t$ in "standard" form but such that the coefficient matrices Φ_j and Θ_j are specified to have certain special reduced-rank structure. On the other hand, by inserting the factor $A^{-1}A$ in front of the $\boldsymbol{Y}_{t-j}$ and ε_{t-j} in (3.11), a vector ARMA(p,q) model in "standard" form is obtained for the *transformed* process $\boldsymbol{Z}_t = A\, \boldsymbol{Y}_t$ as

$$\boldsymbol{Z}_t - \sum_{j=1}^{p} \Phi_j^*\, \boldsymbol{Z}_{t-j} = \boldsymbol{e}_t - \sum_{j=1}^{q} \Theta_j^*\, \boldsymbol{e}_{t-j}, \tag{3.12}$$

where $\Phi_j^* = B_j A^{-1} = A\, \Phi_j A^{-1}$, $\Theta_j^* = G_j A^{-1} = A\, \Theta_j A^{-1}$, and $\boldsymbol{e}_t = A\, \varepsilon_t$. The ARMA representation (3.12) for the transformed process $\boldsymbol{Z}_t$ is such that the ith row of Φ_j^* is zero for $j > p_i$ and the ith row of Θ_j^* is zero for $j > q_i$, and some elements of the ith row of Θ_j^* for $j = 1, \ldots, q_i$ are specified to be zero because of possible redundancy of parameters in the AR and MA matrices in (3.12). The modeling approach used by Tiao and Tsay (1989) is to first identify the SCM processes $\boldsymbol{Z}_t = A\, \boldsymbol{Y}_t$, and their associated orders (p_i, q_i), through the canonical correlation methods, and then to estimate an ARMA model for the transformed vector process $\boldsymbol{Z}_t$ using the parameter structure with the zero constraints imposed as indicated in the representation (3.12).

EXAMPLE 3.1. For a specific example to illustrate the SCM structure, consider an ARMA(2,1) process $\boldsymbol{Y}_t$ of dimension $k = 4$, $\boldsymbol{Y}_t - \Phi_1\, \boldsymbol{Y}_{t-1} - \Phi_2\, \boldsymbol{Y}_{t-2} = \varepsilon_t - \Theta_1\, \varepsilon_{t-1}$. Suppose that $\text{rank}(\Phi_2) = 2$, $\text{rank}(\Theta_1) = 2$, $\text{rank}(\Phi_2, \Theta_1) = 3$, and $\text{rank}(\Phi_1, \Phi_2) = 4$, and hence also $\text{rank}(\Phi_1, \Phi_2, \Theta_1) = 4$. Then there are linearly independent vectors $\boldsymbol{a}_i$, $i = 1, 2, 3$, such that $\boldsymbol{a}_1'\, \Phi_2 = \boldsymbol{a}_1'\, \Theta_1 = 0$, $\boldsymbol{a}_2'\, \Phi_2 = 0$, and $\boldsymbol{a}_3'\, \Theta_1 = 0$, and it follows that the process $\boldsymbol{Y}_t$ has four SCMs of minimal orders equal to (1,0), (1,1), (2,0), and (2,1), with $z_{it} = \boldsymbol{a}_i'\, \boldsymbol{Y}_t$ for $i = 1, 2, 3, 4$. Canonical correlation analysis methods would identify the orders of these four SCMs as follows: (i) one zero canonical correlation would occur between $\mathbf{Y}_{1,t} = (\boldsymbol{Y}_t', \boldsymbol{Y}_{t-1}')'$ and $\mathbf{Y}_{1,t-1}$, implying the SCM of order (1,0), (ii) two zero canonical correlations would exist between $\mathbf{Y}_{1,t}$ and $\mathbf{Y}_{1,t-2}$, one due to the previous component of order (1,0) and the second one implying the SCM of order (1,1), (iii) two zero canonical correlations would occur between $\mathbf{Y}_{2,t} = (\boldsymbol{Y}_t', \boldsymbol{Y}_{t-1}', \boldsymbol{Y}_{t-2}')'$ and

$\mathbf{Y}_{2,t-1}$, one due to the previous SCM of order (1,0) and the second one implying the SCM of order (2,0), and finally, (iv) five zero canonical correlations occur between $\mathbf{Y}_{2,t}$ and $\mathbf{Y}_{2,t-2}$, four of these being due to the previous three SCMs of lower order (two are due to the same SCM of order (1,0), since both $\boldsymbol{a}_1' \boldsymbol{Y}_t - \boldsymbol{a}_1' \Phi_1 \boldsymbol{Y}_{t-1} = \boldsymbol{a}_1' \varepsilon_t$ and $\boldsymbol{a}_1' \boldsymbol{Y}_{t-1} - \boldsymbol{a}_1' \Phi_1 \boldsymbol{Y}_{t-2} = \boldsymbol{a}_1' \varepsilon_{t-1}$ are uncorrelated with $\mathbf{Y}_{2,t-2}$) and the last zero canonical correlation implies the SCM of order $(2,1)$. In addition, the transformed process $\boldsymbol{Z}_t = A\,\boldsymbol{Y}_t$ will follow an ARMA$(2,1)$ model as in (3.12) with the coefficient matrices having the structure such that

$$\Phi_1^* = \begin{bmatrix} X & X & X & X \\ X & X & X & X \\ X & X & X & X \\ X & X & X & X \end{bmatrix}, \quad \Phi_2^* = \begin{bmatrix} 0 & 0 & 0 & 0 \\ 0 & 0 & 0 & 0 \\ X & X & X & X \\ X & X & X & X \end{bmatrix}, \quad \Theta_1^* = \begin{bmatrix} 0 & 0 & 0 & 0 \\ X & X & X & X \\ 0 & 0 & 0 & 0 \\ 0 & X & X & X \end{bmatrix},$$

where X's denote unknown parameters that need estimation and 0's indicate values that are known to be specified as zero. The element in the $(4,1)$ position of Θ_1^* is specified as zero because it is redundant [to the parameter in the $(4,1)$ position of Φ_1^*] in view of the relation among the $\boldsymbol{Z}_t$ and $\boldsymbol{e}_t$ implied by the first rows of the matrices Φ_1^*, Φ_2^*, and Θ_1^* (i.e., e_{1t} is equal to a linear combination of z_{1t} and $\boldsymbol{Z}_{t-1}$).

By comparison, we consider the (canonical) structure implied for the above example through the Kronecker indices approach. Since the covariance matrices $\Gamma(l)$ for the ARMA$(2,1)$ process $\boldsymbol{Y}_t$ in the above example satisfy $\Gamma(l)' = \Phi_1\,\Gamma(l-1)' + \Phi_2\,\Gamma(l-2)'$, for $l > 1$, there are two linear combinations of the rows of $\Gamma(l)'$, $\boldsymbol{a}_1'\,\Gamma(l)'$ and $\boldsymbol{a}_2'\,\Gamma(l)'$, that are linearly dependent on the rows of $\Gamma(l-1)'$ for all $l > 1$. Thus, it follows that the Kronecker indices for this process are equal to 1, 1, 2, and 2; to be specific, say $K_1 = K_2 = 2$ and $K_3 = K_4 = 1$. Hence, on the basis of knowledge of these values alone, the process $\{\boldsymbol{Y}_t\}$ would only be known to have an echelon form ARMA$(2,2)$ representation as in (3.4), $\Phi_0^\#\,\boldsymbol{Y}_t - \Phi_1^\#\,\boldsymbol{Y}_{t-1} - \Phi_2^\#\,\boldsymbol{Y}_{t-2} = \Theta_0^\#\,\varepsilon_t - \Theta_1^\#\,\varepsilon_{t-1} - \Theta_2^\#\,\varepsilon_{t-2}$, with $\Phi_0^\# = \Theta_0^\#$, and the structure of the coefficient matrices $\Phi_j^\#$ can be specified as

$$\Phi_0^\# = \begin{bmatrix} 1 & 0 & 0 & 0 \\ 0 & 1 & 0 & 0 \\ X & X & 1 & 0 \\ X & X & 0 & 1 \end{bmatrix}, \quad \Phi_1^\# = \begin{bmatrix} X & X & 0 & 0 \\ X & X & 0 & 0 \\ X & X & X & X \\ X & X & X & X \end{bmatrix}, \quad \Phi_2^\# = \begin{bmatrix} X & X & X & X \\ X & X & X & X \\ 0 & 0 & 0 & 0 \\ 0 & 0 & 0 & 0 \end{bmatrix},$$

while all that can be specified about the structure of the MA matrices is that $\Theta_1^\#$ is a full matrix of unknown parameters and $\Theta_2^\#$ has all zeros in its third and fourth rows. Thus, in this example we see that knowledge of the Kronecker indices alone does not reveal all the special structure in the AR and MA coefficient matrices; in this case, it does not reveal the complete zero structure of $\Theta_2^\#$ nor the reduced-rank structure in $\Theta_1^\#$ because $q = 1 < p = 2$ in this ARMA model example. This more detailed structure would have to be deduced by more in depth considerations than Kronecker indices alone or, in practice, it would be determined at the stage of parameter estimation of the model.

Finally, in the general canonical correlation analysis procedures discussed above, note that when $m = j$,

$$\text{Cov}(\mathbf{Y}_{m,t}, \mathbf{Y}_{n,t-m-1}) = \begin{bmatrix} \Gamma(m+1)' & \Gamma(m+2)' & . \; . \; . & \Gamma(m+1+n)' \\ \Gamma(m)' & \Gamma(m+1)' & . \; . \; . & \Gamma(m+n)' \\ . & . & . \; . \; . & . \\ . & . & . \; . \; . & . \\ . & . & . \; . \; . & . \\ \Gamma(1)' & \Gamma(2)' & . \; . \; . & \Gamma(1+n)' \end{bmatrix} \tag{3.13}$$

is just a rearrangement of the block rows of the upper-left $k(m + 1) \times k(n + 1)$ submatrix of the Hankel matrix H in Section 3.1. Hence, for sufficiently large n, examination of the occurrence of zero canonical correlations, in a systematic fashion, associated with these matrices (that is, canonical correlation analysis between $\mathbf{Y}_{m,t}$ and $\mathbf{Y}_{n,t-m-1}$) in sequence for various values of $m = 0, 1, \ldots,$ can be useful in revealing the structure of the ARMA model in terms of its Kronecker indices and McMillan degree. In addition, in the terminology of SCMs, based on the relations from (3.4) and (3.6) a Kronecker index equal to K_i implies the existence of a SCM $z_{it} = \phi_0(i)' \boldsymbol{Y}_t$ for $\boldsymbol{Y}_t$ of order (p_i^*, q_i^*) such that $\max\{p_i^*, q_i^*\} = K_i$. Tsay (1989b) has presented a discussion which compares the canonical correlation analysis and SCM modeling approach of Tiao and Tsay (1989) with that based on consideration of Kronecker indices and the associated canonical correlation analysis approach of Akaike (1976) and Cooper and Wood (1982). Briefly, the canonical correlation analyses of Tiao and Tsay (1989), as discussed above, may be viewed as a refinement of the canonical correlation analyses associated with the determination of the Kronecker indices. As noted above, the latter approach essentially restricts consideration to canonical correlation analysis for the case with $m = j$ and, hence, it does not provide the finer information on the row structure and orders of the AR and MA coefficient matrices separately.

3.3 Partial Autoregressive and Partial Correlation Matrices

3.3.1 Vector Autoregressive Model Approximations and Partial Autoregression Matrices

For any stationary vector time series $\{\boldsymbol{Y}_t\}$ with covariance matrices $\Gamma(l)$, it may sometimes be convenient (especially for prediction purposes) to "approximate" the model for $\boldsymbol{Y}_t$ by a pure vector AR model (whether the process $\{\boldsymbol{Y}_t\}$ is a pure AR or not). Now (assume zero mean vector for the process) for any possible order m, we may determine matrix coefficients $\Phi_{1m}, \Phi_{2m}, \ldots, \Phi_{mm}$ in an AR approximation which minimize the quantity

$$\mathrm{tr}\{ E [(\boldsymbol{Y}_t - \sum_{j=1}^{m} \Phi_{jm} \boldsymbol{Y}_{t-j}) (\boldsymbol{Y}_t - \sum_{j=1}^{m} \Phi_{jm} \boldsymbol{Y}_{t-j})'] \}.$$

Notice that the expectation in the above criterion can be expressed as

$$E [(\boldsymbol{Y}_t - \Phi_{(m)}' \mathbf{Y}_{m,t-1}) (\boldsymbol{Y}_t - \Phi_{(m)}' \mathbf{Y}_{m,t-1})']$$

$$= \Gamma(0) - \Phi_{(m)}' \Gamma_{(m)} - \Gamma_{(m)}' \Phi_{(m)} + \Phi_{(m)}' \Gamma_m \Phi_{(m)} ,$$

where $\mathbf{Y}_{m,t-1} = (\boldsymbol{Y}_{t-1}', \ldots, \boldsymbol{Y}_{t-m}')'$,

$\Gamma_m = E(\mathbf{Y}_{m,t-1} \mathbf{Y}_{m,t-1}')$, $\quad \Gamma_{(m)} = E(\mathbf{Y}_{m,t-1} \boldsymbol{Y}_t') = (\Gamma(1)', \Gamma(2)', \ldots, \Gamma(m)')'$,

and $\Phi_{(m)} = (\Phi_{1m}, \Phi_{2m}, \ldots, \Phi_{mm})'$. Hence, minimization of the above criterion is seen to be a standard multivariate linear least squares regression problem, and it follows [e.g., see (A1.8) of Appendix A1] that the matrix coefficients Φ_{jm} which minimize the criterion will be the solution to the vector Yule-Walker equations of order m ,

$$\Phi_{(m)} = \{ E(\mathbf{Y}_{m,t-1} \mathbf{Y}_{m,t-1}') \}^{-1} E(\mathbf{Y}_{m,t-1} \boldsymbol{Y}_t') = \Gamma_m^{-1} \Gamma_{(m)} . \tag{3.14}$$

For this choice of the Φ_{jm} , the term $\tilde{\boldsymbol{Y}}_{t-1}(1) = \Phi_{(m)}' \mathbf{Y}_{m,t-1} = \sum_{j=1}^{m} \Phi_{jm} \boldsymbol{Y}_{t-j}$ represents the "best" (minimum mean squared error) one-step linear predictor of $\boldsymbol{Y}_t$ among all linear functions of the past m vector values $\boldsymbol{Y}_{t-1}, \ldots, \boldsymbol{Y}_{t-m}$ of the process $\{\boldsymbol{Y}_t\}$, with error covariance matrix

$$\Sigma_m = \Gamma(0) - \Phi_{(m)}' \Gamma_m \Phi_{(m)} = \Gamma(0) - \Gamma_{(m)}' \Gamma_m^{-1} \Gamma_{(m)} . \tag{3.15}$$

Thus, in this sense, this choice of the Φ_{jm} leads to the best approximation to the model for $\{\boldsymbol{Y}_t\}$ by an AR model of order m .

Similar to the univariate case, for each $m = 1, 2, \ldots,$ we can define a sequence of matrices Φ_{mm} which has been called (Tiao and Box, 1981) the *partial autoregression matrix* of order (or lag) m , as the solution for Φ_{mm} to the Yule-Walker equations of order m , $\Gamma(l) = \sum_{j=1}^{m} \Gamma(l-j) \Phi_{jm}'$, $l = 1, 2, \ldots, m$, which result from "fitting" or "approximating" an AR model of order m to the series $\boldsymbol{Y}_t$. Similar to the univariate case, it follows that the sequence of partial autoregressive matrices Φ_{mm} of order m , $m = 1, 2, \ldots,$ has the important and characteristic property that if the process $\{\boldsymbol{Y}_t\}$ is a vector AR process of order p , then $\Phi_{pp} = \Phi_p$ and, most importantly, $\Phi_{mm} = 0$ for $m > p$. Hence, the matrices Φ_{mm} have the "cutoff" property for an AR(p) model and so they can be useful in the identification and specification of the order of a pure AR structure for a vector time series. However, unlike the univariate case, note that the elements of the matrices Φ_{mm} are not partial correlations (or correlations of any kind) as for the partial autocorrelation function (PACF) ϕ_{mm} in the univariate case. They are related to the matrices of partial cross-correlations between $\boldsymbol{Y}_t$ and $\boldsymbol{Y}_{t-m}$, given $\boldsymbol{Y}_{t-1}, \ldots, \boldsymbol{Y}_{t-m+1}$, which we discuss subsequently.

3.3.2 Recursive Fitting of Vector AR Model Approximations

First, we note that for $m = 1$, we have explicitly that $\Phi_{11} = \Gamma(1)' \, \Gamma(0)^{-1}$, and for $m > 1$ the Φ_{mm} can be computed most efficiently from a recursive scheme due to Whittle (1963a). Specifically, the matrix Γ_m can be expressed in partitioned form as

$$\Gamma_m = \begin{bmatrix} \Gamma_{m-1} & \Gamma^{*}_{(m-1)} \\ \Gamma^{*\prime}_{(m-1)} & \Gamma(0) \end{bmatrix},$$

where $\Gamma^{*\prime}_{(m-1)} = (\, \Gamma(m-1), \ldots, \Gamma(1)\,)$. It then follows from a standard matrix inversion result for partitioned matrices (Rao, 1973, p. 33) that Γ_m^{-1} can be written as

$$\Gamma_m^{-1} = \begin{bmatrix} A & -\Gamma_{m-1}^{-1} \, \Gamma^{*}_{(m-1)} \, \Sigma^{*-1}_{m-1} \\ -\Sigma^{*-1}_{m-1} \, \Gamma^{*\prime}_{(m-1)} \, \Gamma^{-1}_{m-1} & \Sigma^{*-1}_{m-1} \end{bmatrix},$$

where $\Sigma^{*}_{m-1} = \Gamma(0) - \Gamma^{*\prime}_{(m-1)} \, \Gamma^{-1}_{m-1} \, \Gamma^{*}_{(m-1)}$ and

$$A = \Gamma^{-1}_{m-1} + \Gamma^{-1}_{m-1} \, \Gamma^{*}_{(m-1)} \, \Sigma^{*-1}_{m-1} \, \Gamma^{*\prime}_{(m-1)} \, \Gamma^{-1}_{m-1} \, .$$

Hence, with $\Gamma_{(m)}$ partitioned accordingly as $\Gamma_{(m)} = (\, \Gamma'_{(m-1)}, \, \Gamma(m)' \,)'$, we find that the solution $\Phi_{(m)} = \Gamma_m^{-1} \, \Gamma_{(m)}$ yields

$$\Phi'_{mm} = \Sigma^{*-1}_{m-1} \, \Gamma(m) - \Sigma^{*-1}_{m-1} \, \Gamma^{*\prime}_{(m-1)} \, \Gamma^{-1}_{m-1} \, \Gamma_{(m-1)}$$

$$= \Sigma^{*-1}_{m-1} \, \{ \Gamma(m) - \Gamma^{*\prime}_{(m-1)} \, \Phi_{(m-1)} \} = \Sigma^{*-1}_{m-1} \, \{ \Gamma(m) - \Phi^{*\prime}_{(m-1)} \, \Gamma_{(m-1)} \}, \qquad (3.16)$$

where $\Phi_{(m-1)} = \Gamma^{-1}_{m-1} \, \Gamma_{(m-1)} = (\, \Phi_{1,m-1}, \ldots, \Phi_{m-1,m-1} \,)'$ represents the Yule-Walker solution for the previous order $m - 1$, and where $\Phi^{*}_{(m-1)} = \Gamma^{-1}_{m-1} \, \Gamma^{*}_{(m-1)}$ $= (\, \Phi^{*}_{m-1,m-1}, \, \Phi^{*}_{m-2,m-1}, \ldots, \Phi^{*}_{1,m-1} \,)'$, while the solution at order m for the remaining coefficients gives the recursion

$$(\, \Phi_{1,m}, \ldots, \Phi_{m-1,m})' = \Gamma^{-1}_{m-1} \, \Gamma_{(m-1)} - \Gamma^{-1}_{m-1} \, \Gamma^{*}_{(m-1)} \, \Sigma^{*-1}_{m-1} \{ \Gamma(m) - \Gamma^{*\prime}_{(m-1)} \, \Phi_{(m-1)} \}$$

$$= \Phi_{(m-1)} - \Phi^{*}_{(m-1)} \, \Phi'_{mm} \, , \qquad (3.17)$$

that is, $\Phi'_{jm} = \Phi'_{j,m-1} - \Phi^{*\prime}_{m-j,m-1} \, \Phi'_{mm}$, $j = 1, \ldots, m-1$. In addition, using the fact from (3.15) and (3.14) that

$$\Sigma_m = \Gamma(0) - \Gamma'_{(m)} \, \Phi_{(m)} = \Gamma(0) - \textstyle\sum_{j=1}^{m-1} \Gamma(j)' \, \Phi'_{jm} - \Gamma(m)' \, \Phi'_{mm} \, ,$$

it follows from (3.16) and (3.17) that the error covariance matrix Σ_m also satisfies the recursion $\Sigma_m = \Sigma_{m-1} - \Phi_{mm} \, \Sigma^{*}_{m-1} \, \Phi'_{mm}$. Hence, the factor $\Phi_{mm} \, \Sigma^{*}_{m-1} \, \Phi'_{mm}$ is readily interpretable as the reduction in the error covariance matrix Σ_m due to inclusion of the additional predictor variables $\boldsymbol{Y}_{t-m}$ in the AR model of order m. Note that the recursion in (3.17) and that for the error covariance matrix Σ_m correspond to the relations (A1.14) and (A1.13) given in Appendix A1 for the "updating" of the linear regression function.

The coefficients $\Phi^*_{(m)} = \Gamma_m^{-1} \Gamma^*_{(m)}$ and the matrix Σ^*_m which are also needed for the recursion are computed in a similar recursive manner, starting with $\Phi^*_{11} = \Gamma(1)\,\Gamma(0)^{-1}$, as

$$\Phi^{*\prime}_{mm} = \Sigma^{-1}_{m-1}\,\{\,\Gamma(m)' - \Phi'_{(m-1)}\,\Gamma^*_{(m-1)}\,\} = \Sigma^{-1}_{m-1}\,\{\,\Gamma(m)' - \Gamma'_{(m-1)}\,\Phi^*_{(m-1)}\,\},$$

$$\Phi^{*\prime}_{jm} = \Phi^{*\prime}_{j,m-1} - \Phi'_{m-j,m-1}\,\Phi^{*\prime}_{mm}, \qquad \text{for} \quad j = 1, \ldots, m-1,$$

and $\Sigma^*_m = \Sigma^*_{m-1} - \Phi^*_{mm}\,\Sigma_{m-1}\,\Phi^{*\prime}_{mm}$. These coefficients $\Phi^*_{(m)}$ for the "backward autoregression" are discussed further in Section 3.3.3.

Remark (*Innovations representation*). For any $n \ge 1$, define the vector $\mathbf{y} = (\,Y_1', Y_2', \ldots, Y_{n+1}'\,)'$ with covariance matrix $\Gamma = \text{Cov}(\,\mathbf{y}\,)$ having $\Gamma(j-i) = \text{Cov}(\,Y_i, Y_j\,)$ in the $(\,i, j\,)$th block. Then in terms of the covariance matrix Γ, it follows from the discussion in Appendix A1 related to the matrix in (A1.15) that the sequential calculation of the AR coefficient matrices Φ_{jm} and the error covariance matrices Σ_m, for $m = 1, \ldots, n$, corresponds to the sequential calculation of the block triangular decomposition of the inverse of the covariance matrix Γ into the form $\Gamma^{-1} = \Phi' D^{-1} \Phi$, and, hence, $\Gamma = \Phi^{-1} D\, \Phi'^{-1}$, where

$$\Phi = \begin{bmatrix} I_k & 0 & 0 & . & . & . & 0 \\ -\Phi_{11} & I_k & 0 & . & . & . & 0 \\ -\Phi_{22} & -\Phi_{12} & I_k & . & . & . & 0 \\ . & . & . & . & . & . & . \\ . & . & . & . & . & . & . \\ -\Phi_{nn} & -\Phi_{n-1,n} & -\Phi_{n-2,n} & . & . & . & I_k \end{bmatrix}$$

and $D = \text{Diag}(\,\Sigma_0, \Sigma_1, \ldots, \Sigma_n\,)$, with $\Sigma_0 \equiv \Gamma(0)$. That is, Φ'^{-1} is the block upper triangular factor in the "square-root-free block" Cholesky decomposition of Γ. Since we have $\Phi\,\Gamma\,\Phi' = D$, we see that the matrix Φ affects a transformation $\mathbf{u} = \Phi\,\mathbf{y}$ from the vector $\mathbf{y}$ to the "error" vector $\mathbf{u} = (\,U_1', U_2', \ldots, U_{n+1}'\,)'$, where the $U_m = Y_m - \sum_{j=1}^{m-1} \Phi_{j,m-1}\,Y_{m-j}$, for $m = 1, \ldots, n+1$, with $U_1 \equiv Y_1$, are of the same form as the "error" vectors $U_{m,t}$ defined below, in (3.19) of Section 3.3.3, such that the covariance matrix of $\mathbf{u}$, $\text{Cov}(\,\mathbf{u}\,) = \Phi\,\Gamma\,\Phi' = D$, is block diagonal with $\text{Cov}(\,U_m\,) = \Sigma_{m-1}$.

As an alternative to the "autoregressive representation"

$$Y_m - \sum_{j=1}^{m-1} \Phi_{j,m-1}\,Y_{m-j} = U_m, \qquad m = 1, \ldots, n+1,$$

associated with the transformation $\Phi\,\mathbf{y} = \mathbf{u}$, the "inverse" relation $\mathbf{y} = \Phi^{-1}\mathbf{u} = \Theta\,\mathbf{u}$ may also be considered. In this relation, the matrix $\Theta = \Phi^{-1}$ is block upper triangular such that $\Gamma = \Theta\,D\,\Theta'$, so that Θ' is the "square-root-free block" Cholesky factor of Γ. This matrix Θ will be denoted as

$$\Theta = \begin{bmatrix} I_k & 0 & 0 & . & . & . & 0 \\ \Theta_{11} & I_k & 0 & . & . & . & 0 \\ \Theta_{22} & \Theta_{12} & I_k & . & . & . & 0 \\ . & . & . & . & . & . & . \\ . & . & . & . & . & . & . \\ \Theta_{nn} & \Theta_{n-1,n} & \Theta_{n-2,n} & . & . & . & I_k \end{bmatrix} .$$

The transformation $\mathbf{y} = \Theta \, \mathbf{u}$ implies the "innovations representation" for the $\boldsymbol{Y}_m$ as

$$\begin{aligned} \boldsymbol{Y}_m &= \boldsymbol{U}_m + \sum_{j=1}^{m-1} \Theta_{j,m-1} \, \boldsymbol{U}_{m-j} \\ &= (\, \boldsymbol{Y}_m - \tilde{\boldsymbol{Y}}_{m-1}(1) \,) + \sum_{j=1}^{m-1} \Theta_{j,m-1} \, (\, \boldsymbol{Y}_{m-j} - \tilde{\boldsymbol{Y}}_{m-j-1}(1) \,) , \end{aligned} \tag{3.18}$$

where $\tilde{\boldsymbol{Y}}_{m-1}(1) = \sum_{j=1}^{m-1} \Phi_{j,m-1} \, \boldsymbol{Y}_{m-j}$ denotes the minimum mean square error one-step linear predictor of $\boldsymbol{Y}_m$ based on $\boldsymbol{Y}_{m-1}, \ldots, \boldsymbol{Y}_1$. Also, $\boldsymbol{U}_m = \boldsymbol{Y}_m - \tilde{\boldsymbol{Y}}_{m-1}(1)$ represents the one-step ahead error of the best linear predictor, and the $\boldsymbol{U}_m$ are referred to as the (one-step) innovations of $\boldsymbol{Y}_1, \boldsymbol{Y}_2, \ldots, \boldsymbol{Y}_{n+1}$. Although the matrices Θ_{jm} in (3.18) could be determined from the AR coefficient matrices Φ_{jm} through the relation $\Phi \, \Theta = I$, it can readily be shown that these coefficients can be calculated more directly through the following innovations algorithm [e.g., see Brockwell and Davis (1987, Chap. 11)]. In this algorithm, the coefficient matrices Θ_{jm} and the error covariance matrices $\Sigma_m = \text{Cov}(\, \boldsymbol{U}_{m+1} \,)$ are determined directly in terms of the autocovariance matrices $\Gamma(j)$ of the process $\{ \boldsymbol{Y}_t \}$ recursively as

$$\Theta_{m-j,m} = (\, \Gamma(m-j)' - \sum_{i=0}^{j-1} \Theta_{m-i,m} \, \Sigma_i \, \Theta_{j-i,j}' \,) \, \Sigma_j^{-1}, \quad j = 0, 1, \ldots, m-1 ,$$

$$\Sigma_m = \Gamma(0) - \sum_{i=0}^{m-1} \Theta_{m-i,m} \, \Sigma_i \, \Theta_{m-i,m}' ,$$

starting with $\Sigma_0 = \Gamma(0)$ and, hence, $\Theta_{11} = \Gamma(1)' \, \Gamma(0)^{-1} = \Phi_{11}$. In the innovations representation (3.18), since the innovations $\boldsymbol{U}_m$ are mutually uncorrelated, the coefficient matrices Θ_{jm} have the interpretation as

$$\Theta_{jm} = \text{Cov}(\, \boldsymbol{Y}_{m+1}, \, \boldsymbol{U}_{m+1-j} \,) \, \{ \, \text{Cov}(\, \boldsymbol{U}_{m+1-j} \,) \, \}^{-1} = \text{Cov}(\, \boldsymbol{Y}_{m+1}, \, \boldsymbol{U}_{m+1-j} \,) \, \Sigma_{m-j}^{-1} .$$

Hence, the recursive relations for the $\Theta_{m-j,m}$ follow from the relation

$$\begin{aligned} \Theta_{m-j,m} \, \Sigma_j &= \text{Cov}(\, \boldsymbol{Y}_{m+1}, \, \boldsymbol{U}_{j+1} \,) = \text{Cov}(\, \boldsymbol{Y}_{m+1}, \, \boldsymbol{Y}_{j+1} - \textstyle\sum_{i=0}^{j-1} \Theta_{j-i,j} \, \boldsymbol{U}_{i+1} \,) \\ &= \Gamma(m-j)' - \sum_{i=0}^{j-1} \Theta_{m-i,m} \, \Sigma_i \, \Theta_{j-i,j}' . \end{aligned}$$

In addition, from (3.18) we have $\boldsymbol{Y}_{m+1} = \boldsymbol{U}_{m+1} + \sum_{i=0}^{m-1} \Theta_{m-i,m} \boldsymbol{U}_{i+1}$, so that

$$\Gamma(0) = \text{Cov}(\boldsymbol{Y}_{m+1}) = \text{Cov}(\boldsymbol{U}_{m+1} + \sum_{i=0}^{m-1} \Theta_{m-i,m} \boldsymbol{U}_{i+1})$$
$$= \Sigma_m + \sum_{i=0}^{m-1} \Theta_{m-i,m} \Sigma_i \Theta'_{m-i,m} ,$$

and, hence, the above recursive relations for the Σ_m follow directly. The innovations representation (3.18) may be viewed as a finite sample analogue of the Wold decomposition (1.8) of Section 1.2.1 that represents the stationary process $\{\boldsymbol{Y}_t\}$ as an infinite moving average in terms of the white noise process $\{\varepsilon_t\}$, with coefficient matrices $\Psi_j = \text{Cov}(\boldsymbol{Y}_t, \varepsilon_{t-j}) \Sigma^{-1}$.

3.3.3 Partial Cross-Correlation Matrices for a Stationary Vector Process

Now, to define the related partial cross-correlation matrices, we thus consider the "error" vector

$$\boldsymbol{U}_{m,t} = \boldsymbol{Y}_t - \sum_{j=1}^{m-1} \Phi_{j,m-1} \boldsymbol{Y}_{t-j} = \boldsymbol{Y}_t - \Phi'_{(m-1)} \mathbf{Y}_{m-1,t-1} \tag{3.19}$$

resulting from the approximation by an AR model of order $m-1$. Similarly, consider the backward "error" vector given by

$$\boldsymbol{U}^*_{m,t-m} = \boldsymbol{Y}_{t-m} - \sum_{j=1}^{m-1} \Phi^*_{j,m-1} \boldsymbol{Y}_{t-m+j} = \boldsymbol{Y}_{t-m} - \Phi^{*\prime}_{(m-1)} \mathbf{Y}_{m-1,t-1} , \tag{3.20}$$

where the matrices Φ^*_{jm} were introduced following (3.16) and are obtained in an analogous manner to the method by which the Φ_{jm} are defined. That is, the Φ^*_{jm} satisfy the equations

$$\Gamma(-l) = \sum_{j=1}^{m} \Gamma(j-l) \Phi^{*\prime}_{jm} , \qquad l = 1, \ldots, m ,$$

and, hence, $\Phi^*_{(m)} = (\Phi^*_{mm}, \Phi^*_{m-1,m}, \ldots, \Phi^*_{1m})' = \Gamma_m^{-1} \Gamma^*_{(m)}$, where

$$\Gamma^*_{(m)} = E(\mathbf{Y}_{m,t-1} \boldsymbol{Y}'_{t-m-1}) = (\Gamma(m), \Gamma(m-1), \ldots, \Gamma(1))'$$

with $\mathbf{Y}_{m,t-1} = (\boldsymbol{Y}'_{t-1}, \ldots, \boldsymbol{Y}'_{t-m})'$. Note that $\boldsymbol{U}_{m,t}$ and $\boldsymbol{U}^*_{m,t-m}$ are both "residual" vectors from the multivariate regressions of $\boldsymbol{Y}_t$ and $\boldsymbol{Y}_{t-m}$, respectively, on the same set of predictor variables $\boldsymbol{Y}_{t-1}, \ldots, \boldsymbol{Y}_{t-m+1}$. Also define

$$\Sigma_{m-1} = \text{Cov}(\boldsymbol{U}_{m,t}) = \Gamma(0) - \sum_{j=1}^{m-1} \Gamma(-j) \Phi'_{j,m-1}$$

and

$$\Sigma^*_{m-1} = \text{Cov}(\boldsymbol{U}^*_{m,t-m}) = \Gamma(0) - \sum_{j=1}^{m-1} \Gamma(m-j) \Phi^{*\prime}_{m-j,m-1} ,$$

and let $V(m)$ and $V_*(m)$ denote diagonal matrices whose diagonal elements are the diagonal elements of the matrices $\Sigma_{m-1} = \text{Cov}(\boldsymbol{U}_{m,t})$ and $\Sigma^*_{m-1} = \text{Cov}(\boldsymbol{U}^*_{m,t-m})$, respectively. Then we can define the *partial cross-correlation matrix* at lag m as

$$P_m = \text{Corr}(\boldsymbol{U}^*_{m,t-m}, \boldsymbol{U}_{m,t}) = V_*(m)^{-1/2}\, E(\boldsymbol{U}^*_{m,t-m}\, \boldsymbol{U}'_{m,t})\, V(m)^{-1/2}, \qquad (3.21)$$

with

$$E(\boldsymbol{U}^*_{m,t-m}\, \boldsymbol{U}'_{m,t}) = \text{Cov}(\boldsymbol{U}^*_{m,t-m}, \boldsymbol{U}_{m,t})$$
$$= \Gamma(m) - \sum_{j=1}^{m-1} \Gamma(m-j)\, \Phi'_{j,m-1} = \Gamma(m) - \sum_{j=1}^{m-1} \Phi^*_{m-j,m-1}\, \Gamma(j),$$

since

$$E(\boldsymbol{U}^*_{m,t-m}\, \boldsymbol{U}'_{m,t}) = E[\,(\boldsymbol{Y}_{t-m} - \Phi^{*'}_{(m-1)}\, \mathbf{Y}_{m-1,t-1})\,(\boldsymbol{Y}_t - \Phi'_{(m-1)}\, \mathbf{Y}_{m-1,t-1})'\,]$$
$$= \Gamma(m) - \Phi^{*'}_{(m-1)}\, \Gamma_{(m-1)} - \Gamma^{*'}_{(m-1)}\, \Phi_{(m-1)} + \Phi^{*'}_{(m-1)}\, \Gamma_{m-1}\, \Phi_{(m-1)}$$
$$= \Gamma(m) - \Gamma^{*'}_{(m-1)}\, \Phi_{(m-1)} = \Gamma(m) - \Phi^{*'}_{(m-1)}\, \Gamma_{(m-1)}. \qquad (3.22)$$

The matrix P_m in (3.21) represents the cross-correlation matrix between the elements of $\boldsymbol{Y}_{t-m}$ and $\boldsymbol{Y}_t$, after adjustment of both for their dependence on the intervening values $\boldsymbol{Y}_{t-m+1}, \ldots, \boldsymbol{Y}_{t-1}$. It follows by the definition that each element of the matrix P_m is a (partial) correlation, and, in fact, the (i, j)th element of P_m is the partial correlation between the variables $Y_{i,t-m}$ and Y_{jt}, adjusting for $\boldsymbol{Y}_{t-m+1}, \ldots, \boldsymbol{Y}_{t-1}$.

In terms of the above notation for partial correlation matrices, it can be seen directly from (3.16) that the partial autoregressive matrix Φ_{mm} defined previously is given by $\Phi_{mm} = E(\boldsymbol{U}_{m,t}\, \boldsymbol{U}^{*'}_{m,t-m})\,\{\text{Cov}(\boldsymbol{U}^*_{m,t-m})\}^{-1}$, since $E(\boldsymbol{U}_{m,t}\, \boldsymbol{U}^{*'}_{m,t-m}) = \{\Gamma(m) - \Gamma^{*'}_{(m-1)}\, \Phi_{(m-1)}\}'$ from (3.22). Thus, the expression to obtain Φ_{mm}, as in (3.16), has the traditional interpretation as an ordinary regression coefficient matrix for the regression of $\boldsymbol{U}_{m,t}$ on $\boldsymbol{U}^*_{m,t-m}$, where these are the residual vectors from the regressions of $\boldsymbol{Y}_t$ and $\boldsymbol{Y}_{t-m}$, respectively, on the remaining predictor variables $\mathbf{Y}_{m-1,t-1}$ in the AR(m) model. Hence, we see that both $P_m = 0$ and $\Phi_{mm} = 0$ if and only if $E(\boldsymbol{U}_{m,t}\, \boldsymbol{U}^{*'}_{m,t-m}) = 0$, which occurs for vector AR(p) processes for all $m > p$. Thus, both P_m and Φ_{mm} share the same "cutoff" property for AR processes. Finally, we note that an alternate form of partial correlation matrices has been studied by Ansley and Newbold (1979), which is given by $P(m) = \Sigma^{*-1/2}_{m-1}\, E(\boldsymbol{U}^*_{m,t-m}\, \boldsymbol{U}'_{m,t})\, \Sigma^{-1/2}_{m-1} = \Sigma^{*1/2}_{m-1}\, \Phi'_{mm}\, \Sigma^{-1/2}_{m-1}$, and represents the partial correlations in terms of the "standardized" variables $\Sigma^{*-1/2}_{m-1}\, \boldsymbol{U}^*_{m,t-m}$ and $\Sigma^{-1/2}_{m-1}\, \boldsymbol{U}_{m,t}$. Morf, Vieira, and Kailath (1978) also considered the matrices $P(m)$ in the context of introducing an equivalent "normalized" form of the Whittle recursive algorithm (3.16)–(3.17) to obtain the (forward and backward) autoregression and partial autoregression matrices.

3.3.4 Partial Canonical Correlations for a Stationary Vector Process

In addition to the partial cross-correlation matrices P_m and $P(m)$ at lag m, we also consider the *partial canonical correlations* of the process $\{ Y_t \}$ at lag m. These lag-m partial canonical correlations, denoted as $1 \ge \rho_1(m) \ge \rho_2(m) \ge \cdots \ge \rho_k(m) \ge 0$, are the canonical correlations between the vectors Y_t and Y_{t-m}, after adjustment for the dependence of these variables on the intervening values $Y_{t-1}, \ldots, Y_{t-m+1}$. Hence, these are the canonical correlations between the residual series $U_{m,t}$ and $U^*_{m,t-m}$. So, according to (3.9), the (squared) partial canonical correlations $\rho_i^2(m)$ are the eigenvalues of the matrix

$$\{ \operatorname{Cov}(U_{m,t}) \}^{-1} E(U_{m,t} U^{*\prime}_{m,t-m}) \{ \operatorname{Cov}(U^*_{m,t-m}) \}^{-1} E(U^*_{m,t-m} U'_{m,t})$$

$$= \Sigma_{m-1}^{-1} \Phi_{mm} \Sigma^*_{m-1} \Phi'_{mm} = \Phi^{*\prime}_{mm} \Phi'_{mm}, \tag{3.23}$$

where

$$\Phi^*_{mm} = E(U^*_{m,t-m} U'_{m,t}) \{ \operatorname{Cov}(U_{m,t}) \}^{-1} = \Sigma^*_{m-1} \Phi'_{mm} \Sigma_{m-1}^{-1}$$

is the coefficient matrix of Y_t in the "backward" autoregression $Y_{t-m} = \sum_{j=1}^{m} \Phi^*_{j,m} Y_{t-m+j} + U^*_{m+1,t-m}$ of Y_{t-m} on $Y_{t-m+1}, \ldots, Y_t$. We also note that the matrix in (3.23) is equal to $\Sigma_{m-1}^{-1/2} P(m)' P(m) \Sigma_{m-1}^{1/2}$, so that the eigenvalues of (3.23), the (squared) partial canonical correlations $\rho_i^2(m)$, are also equal to the eigenvalues of the matrix $P(m)' P(m)$, that is, the singular values of the "normalized" partial correlation matrix $P(m)'$. Hence, from the discussion which follows (3.9), the number s of zero partial canonical correlations between Y_t and Y_{t-m} is equal to $s = k - r_m$ where $r_m = \operatorname{rank}(\Phi'_{mm}) = \operatorname{rank}(\Phi_{mm})$. Thus, examination of the partial canonical correlations $\rho_i(m)$, for various lags $m = 1, 2, \ldots,$ can be useful in the determination of certain reduced-rank structures in the AR coefficient matrices Φ_j of an AR(p) process, and this topic will be explored in more detail in Section 6.1 within the context of nested reduced-rank vector AR models.

EXAMPLE 3.2. We illustrate the above discussion of partial autoregression matrices with a bivariate ($k = 2$) ARMA(1,1) process $Y_t - \Phi Y_{t-1} = \varepsilon_t - \Theta \varepsilon_{t-1}$, with

$$\Phi = \begin{bmatrix} 1.2 & -0.5 \\ 0.6 & 0.3 \end{bmatrix}, \qquad \Theta = \begin{bmatrix} -0.6 & 0.3 \\ 0.3 & 0.6 \end{bmatrix}, \qquad \Sigma = \begin{bmatrix} 1.00 & 0.50 \\ 0.50 & 1.25 \end{bmatrix}.$$

The covariance matrix $\Gamma(0)$ of the process $\{ Y_t \}$ for this ARMA(1,1) example is obtained by solving the relation

$$\Gamma(0) - \Phi \Gamma(0) \Phi' = \Sigma - \Theta \Sigma \Phi' - (\Phi - \Theta) \Sigma \Theta',$$

as discussed in Section 2.3.3, and is equal to

$$\Gamma(0) = E(\, \boldsymbol{Y}_t \, \boldsymbol{Y}_t' \,) = \begin{bmatrix} 12.40 & 8.47 \\ 8.47 & 9.04 \end{bmatrix} .$$

Note that $\det\{\, I - \Phi\, z \,\} = 0$ for $z = 1.136 \pm 0.473\, i$ (and hence $|\, z \,| = 1.23$), and $\det\{\, I - \Theta\, z \,\} = 0$ for $z = \pm\, (\, 0.45\,)^{-1/2}$ (and hence $|\, z \,| = 1.49$) for the above values of Φ and Θ. The "residual" error covariance matrices Σ_m and the autoregressive parameter matrices $\Phi_{1m}, \ldots, \Phi_{mm}$ for this example are given in Table 3.1 for AR orders $m = 1$ to 5. The coefficient matrices Π_j in the infinite AR form $\boldsymbol{Y}_t = \sum_{j=1}^{\infty} \Pi_j\, \boldsymbol{Y}_{t-j} + \varepsilon_t$, obtained as described in (2.13) of Section 2.3.1 from $\Pi_1 = \Phi - \Theta$ and $\Pi_j = \Theta\, \Pi_{j-1} = \Theta^{j-1}(\, \Phi - \Theta\,)$ for $j > 1$, are also given in the last row of Table 3.1 for comparison. Note that minimizing $\det\{\, [\, 1 + (km/T)\,]\, \Sigma_m \,\}$, a population version of the FPE criterion of Akaike (1971), with respect to m yields $m = 3$ for $T = 50$ and $m = 4$ for $T = 100$, where T denotes sample series length and

$$\Sigma_m = E[\, (\, \boldsymbol{Y}_t - \textstyle\sum_{j=1}^{m} \Phi_{jm}\, \boldsymbol{Y}_{t-j}\,)\, (\, \boldsymbol{Y}_t - \sum_{j=1}^{m} \Phi_{jm}\, \boldsymbol{Y}_{t-j}\,)'\,].$$

In addition, we note that the partial canonical correlations $\rho_i(m)$ between $\boldsymbol{Y}_t$ and $\boldsymbol{Y}_{t-m}$ for this process, determined in accordance with (3.23), are $\{\, 0.638, 0.196\, \}$, $\{\, 0.263, 0.174\, \}$, $\{\, 0.245, 0.076\, \}$, and $\{\, 0.110, 0.073\, \}$, respectively, for lags $m = 2$ to 5. So these partial canonical correlations become relatively negligible after lag $m = 4$.

Table 3.1 Covariance Matrices Σ_m and Autoregressive Parameters of mth Order Autoregressive Model Approximations for the ARMA(1,1) Model of Example 3.2

m	Σ_m	Autoregressive Parameters $(\Phi_{1m}, \ldots, \Phi_{mm}) = \Gamma'_{(m)}\Gamma_m^{-1}$						
1	$\begin{bmatrix} 1.23 & 0.37 \\ 0.37 & 1.87 \end{bmatrix}$	$\begin{bmatrix} 1.32 & -0.62 \\ 0.66 & 0.15 \end{bmatrix}$						
2	$\begin{bmatrix} 1.10 & 0.54 \\ 0.54 & 1.38 \end{bmatrix}$	$\begin{bmatrix} 1.60 & -0.62 \\ 0.24 & -0.11 \end{bmatrix}$	$\begin{bmatrix} -0.45 & 0.27 \\ 0.75 & -0.16 \end{bmatrix}$					
3	$\begin{bmatrix} 1.03 & 0.49 \\ 0.49 & 1.32 \end{bmatrix}$	$\begin{bmatrix} 1.74 & -0.77 \\ 0.32 & -0.24 \end{bmatrix}$	$\begin{bmatrix} -0.80 & 0.31 \\ 0.57 & -0.23 \end{bmatrix}$	$\begin{bmatrix} 0.44 & -0.19 \\ 0.28 & -0.01 \end{bmatrix}$				
4	$\begin{bmatrix} 1.02 & 0.51 \\ 0.51 & 1.27 \end{bmatrix}$	$\begin{bmatrix} 1.77 & -0.77 \\ 0.29 & -0.27 \end{bmatrix}$	$\begin{bmatrix} -0.90 & 0.37 \\ 0.72 & -0.37 \end{bmatrix}$	$\begin{bmatrix} 0.61 & -0.24 \\ 0.08 & -0.04 \end{bmatrix}$	$\begin{bmatrix} -0.17 & 0.10 \\ 0.30 & -0.07 \end{bmatrix}$			
5	$\begin{bmatrix} 1.01 & 0.50 \\ 0.50 & 1.26 \end{bmatrix}$	$\begin{bmatrix} 1.79 & -0.79 \\ 0.30 & -0.29 \end{bmatrix}$	$\begin{bmatrix} -0.95 & 0.38 \\ 0.69 & -0.39 \end{bmatrix}$	$\begin{bmatrix} 0.74 & -0.33 \\ 0.16 & -0.11 \end{bmatrix}$	$\begin{bmatrix} -0.34 & 0.13 \\ 0.23 & -0.09 \end{bmatrix}$	$\begin{bmatrix} 0.19 & -0.08 \\ 0.12 & -0.01 \end{bmatrix}$		
∞	$\begin{bmatrix} 1.00 & 0.50 \\ 0.50 & 1.25 \end{bmatrix}$	$\begin{bmatrix} 1.80 & -0.80 \\ 0.30 & -0.30 \end{bmatrix}$	$\begin{bmatrix} -0.99 & 0.39 \\ 0.72 & -0.42 \end{bmatrix}$	$\begin{bmatrix} 0.81 & -0.36 \\ 0.14 & -0.14 \end{bmatrix}$	$\begin{bmatrix} -0.45 & 0.18 \\ 0.32 & -0.19 \end{bmatrix}$	$\begin{bmatrix} 0.36 & -0.16 \\ 0.06 & -0.06 \end{bmatrix}$	$\begin{bmatrix} -0.20 & 0.08 \\ 0.15 & -0.09 \end{bmatrix}$	$\cdots$

CHAPTER 4

Initial Model Building and Least Squares Estimation for Vector AR Models

In this chapter, preliminary model specification techniques based on sample statistics such as correlations and partial correlations are first introduced and discussed. Least squares estimation for vector AR models and associated tests of hypothesis for the order of the AR model are emphasized. Properties of least squares estimates for vector AR models are discussed. Additional methods for initial specification and selection of an appropriate ARMA model, including the use of canonical correlation methods and information-theoretic model selection criteria such as AIC and BIC, are also explored.

4.1 Sample Cross-Covariance and Correlation Matrices and Their Properties

4.1.1 Sample Estimates of Mean Vector and of Covariance and Correlation Matrices

Given a sample vector time series $\boldsymbol{Y}_1, \boldsymbol{Y}_2, \ldots, \boldsymbol{Y}_T$ of length T (possibly) from a stationary multivariate process with mean vector $\boldsymbol{\mu} = E(\boldsymbol{Y}_t)$ and autocovariance matrices $\Gamma(l) = E\,[(\boldsymbol{Y}_t - \boldsymbol{\mu})(\boldsymbol{Y}_{t+l} - \boldsymbol{\mu})']$ and autocorrelation matrices $\rho(l)$, one of the fundamental tools for preliminary model building purposes is to obtain sample estimates of the mean vector $\boldsymbol{\mu}$ and especially of the $\Gamma(l)$ and the $\rho(l)$. The preliminary estimate of $\boldsymbol{\mu}$ is given by the sample mean vector

$$\hat{\boldsymbol{\mu}} = \bar{\boldsymbol{Y}} = (\bar{Y}_1, \ldots, \bar{Y}_k)' = \frac{1}{T} \sum_{t=1}^{T} \boldsymbol{Y}_t .$$

The sample mean vector $\bar{\boldsymbol{Y}}$ is an unbiased estimator of $\boldsymbol{\mu}$, with $E(\bar{\boldsymbol{Y}}) = \boldsymbol{\mu}$ easily seen to hold, and for a stationary process $\{\boldsymbol{Y}_t\}$, the sample mean vector $\bar{\boldsymbol{Y}}$

converges to $\boldsymbol{\mu}$ as $T \to \infty$ under mild conditions (e.g., Hannan, 1970, Chap. 4). A measure of the precision of $\bar{\boldsymbol{Y}}$ as an estimator of $\boldsymbol{\mu}$ is given by

$$\mathrm{Cov}(\bar{\boldsymbol{Y}}) = \frac{1}{T^2} \sum_{t=1}^{T} \sum_{s=1}^{T} \Gamma(t-s) = \frac{1}{T} \sum_{l=-(T-1)}^{T-1} \left(1 - \frac{l}{T}\right) \Gamma(l).$$

Assuming that $\sum_{l=-\infty}^{\infty} \| \Gamma(l) \| < \infty$, a "large-sample" approximation for the covariance matrix expression for $\bar{\boldsymbol{Y}}$ is, thus, $\mathrm{Cov}(\bar{\boldsymbol{Y}}) \approx T^{-1} \sum_{l=-\infty}^{\infty} \Gamma(l)$, in the sense that $T\,\mathrm{Cov}(\bar{\boldsymbol{Y}}) \to \sum_{l=-\infty}^{\infty} \Gamma(l)$ as $T \to \infty$, and the last expression is recognized as being equal to $2\pi f(0)$, where $f(\lambda)$ denotes the spectral density matrix of the stationary process $\{\boldsymbol{Y}_t\}$. Hence, if the process has the infinite MA representation as $\boldsymbol{Y}_t = \boldsymbol{\mu} + \sum_{j=0}^{\infty} \Psi_j\, \varepsilon_{t-j}$, then from results of Section 1.1.3 the spectral density matrix is $f(\lambda) = (1/2\pi)\, \Psi(e^{i\lambda})\, \Sigma\, \Psi(e^{-i\lambda})'$. So the "large-sample" covariance matrix of $\bar{\boldsymbol{Y}}$ is then given by

$$\mathrm{Cov}(\bar{\boldsymbol{Y}}) \approx T^{-1}\, \Psi(1)\, \Sigma\, \Psi(1)' = T^{-1} \left(\sum_{j=0}^{\infty} \Psi_j\right) \Sigma \left(\sum_{j=0}^{\infty} \Psi_j'\right).$$

For example, for the vector MA(1) process $\boldsymbol{Y}_t = \boldsymbol{\mu} + \varepsilon_t - \Theta\, \varepsilon_{t-1}$, the expression is $\mathrm{Cov}(\bar{\boldsymbol{Y}}) \approx T^{-1} (I - \Theta)\, \Sigma\, (I - \Theta')$.

The sample estimate of the cross-covariance matrix at lag l, $\Gamma(l)$, is given by the *sample covariance matrix* at lag l, defined as

$$\hat{\Gamma}(l) = C(l) = \frac{1}{T} \sum_{t=1}^{T-l} (\boldsymbol{Y}_t - \bar{\boldsymbol{Y}})(\boldsymbol{Y}_{t+l} - \bar{\boldsymbol{Y}})', \qquad l = 0, 1, 2, \ldots, \tag{4.1}$$

where $\bar{\boldsymbol{Y}} = (\bar{Y}_1, \ldots, \bar{Y}_k)' = T^{-1} \sum_{t=1}^{T} \boldsymbol{Y}_t$ is the sample mean vector. In particular, $\hat{\Gamma}(0) = C(0) = T^{-1} \sum_{t=1}^{T} (\boldsymbol{Y}_t - \bar{\boldsymbol{Y}})(\boldsymbol{Y}_t - \bar{\boldsymbol{Y}})'$ is the sample covariance matrix of the $\boldsymbol{Y}_t$. Thus, $\hat{\Gamma}(l)$ has (i, j)th element

$$\hat{\gamma}_{ij}(l) = c_{ij}(l) = T^{-1} \sum_{t=1}^{T-l} (Y_{it} - \bar{Y}_i)(Y_{j,t+l} - \bar{Y}_j).$$

If $\{\boldsymbol{Y}_t\}$ is a stationary linear process, $\boldsymbol{Y}_t = \boldsymbol{\mu} + \sum_{j=0}^{\infty} \Psi_j\, \varepsilon_{t-j}$, $\sum_{j=0}^{\infty} \| \Psi_j \| < \infty$, where the ε_t are independent and identically distributed with $E(\varepsilon_t) = 0$ and $\mathrm{Cov}(\varepsilon_t) = \Sigma$, then the sample covariance matrix $\hat{\Gamma}(l)$ at lag l converges to $\Gamma(l)$ almost surely as $T \to \infty$ [e.g., see Hannan (1970, Chap. 4) and Hannan and Deistler (1988, Sec. 4.1)]. The *sample cross-correlations* are defined as

$$\hat{\rho}_{ij}(l) = r_{ij}(l) = c_{ij}(l) \,/\, [\, c_{ii}(0)\, c_{jj}(0)\,]^{1/2}, \qquad i, j = 1, \ldots, k. \tag{4.2}$$

For a stationary series, the $\hat{\rho}_{ij}(l)$ are sample estimates of the theoretical $\rho_{ij}(l)$; they are particularly useful in the model specification for a low order pure vector moving average model, since the MA(q) model has the property that $\rho_{ij}(l) = 0$ for all $l > q$.

4.1.2 Asymptotic Properties of Sample Correlations

Unfortunately, in general, the collection of sample estimates $\hat{\rho}_{ij}(l)$ may be rather difficult to interpret as a whole in the multivariate case because of the large number of terms that may need to be examined and the possible complex patterns that they may exhibit. Also, the sampling properties of the $\hat{\rho}_{ij}(l)$ are complicated and depend on the (unknown) theoretical values of $\rho_{ij}(l)$. For a stationary vector process when T is large, it is known (e.g., Hannan, 1970, Chap. 4) that a collection of sample $\hat{\rho}_{ij}(l)$ are asymptotically normally distributed with corresponding means $\rho_{ij}(l)$, but the (approximate) variances of the estimates and the covariances among the estimates are complicated. For example, from results of Bartlett (1955) (also see Roy, 1989), and assuming that the process $\{\boldsymbol{Y}_t\}$ has zero fourth-order cumulants as in the case of a Gaussian process, we have

$$\begin{aligned}\mathrm{Cov}(\,\hat{\rho}_{ij}(l),\,\hat{\rho}_{ij}(n)\,) \approx T^{-1} \sum_{u=-\infty}^{\infty} [\,& \rho_{ii}(u)\,\rho_{jj}(u+n-l) + \rho_{ij}(u+n)\,\rho_{ji}(u-l) \\ & - \rho_{ij}(l)\,(\,\rho_{ii}(u)\,\rho_{ij}(u+n) + \rho_{jj}(u)\,\rho_{ji}(u-n)\,) \\ & - \rho_{ij}(n)\,(\,\rho_{ii}(u)\,\rho_{ij}(u+l) + \rho_{jj}(u)\,\rho_{ji}(u-l)\,) \\ & + \rho_{ij}(l)\,\rho_{ij}(n)\,(\,(1/2)\,\rho_{ii}^2(u) + \rho_{ij}^2(u) + (1/2)\,\rho_{jj}^2(u)\,)\,].\end{aligned} \tag{4.3}$$

Setting $l = n$ in (4.3) we can obtain an expression for the asymptotic variance of $\hat{\rho}_{ij}(l)$, for example. Generally, expression (4.3) may not be of much value, but there are a few special cases that deserve consideration.

Special Cases. (a) Suppose that $\boldsymbol{Y}_t$ is a vector white noise process, with covariance matrix Σ and correlation matrix $\rho(0)$, so that $\rho_{ij}(l) = 0$ for $l \neq 0$. Then (4.3) yields

$$\mathrm{Var}(\,\hat{\rho}_{ij}(l)\,) \approx 1/T$$

for $l \neq 0$, while $\mathrm{Var}(\,\hat{\rho}_{ij}(0)\,) \approx (\,1 - \rho_{ij}^2(0)\,)^2/T$, $i \neq j$. Also we have $\mathrm{Cov}(\,\hat{\rho}_{ij}(l),\,\hat{\rho}_{ij}(-l)\,) \approx \rho_{ij}^2(0)\,/T$, so that $\mathrm{Corr}(\,\hat{\rho}_{ij}(l),\,\hat{\rho}_{ji}(l)\,) \approx \rho_{ij}^2(0)$, and

$$\mathrm{Cov}(\,\hat{\rho}_{ij}(l),\,\hat{\rho}_{ij}(n)\,) \approx 0$$

otherwise. It can also be established in this case that $\mathrm{Cov}(\,\hat{\rho}_{ij}(l),\,\hat{\rho}_{km}(n)\,) \approx 0$ for $l \neq n$, so that the $\hat{\rho}_{ij}(l)$ are not correlated at different lags.

(b) Suppose $\boldsymbol{Y}_t$ is a MA(q) process. In particular, first suppose we have an MA(1) process ($q = 1$), so that $\rho_{ij}(l) = 0$ for $l > 1$. Hence, in this case, from (4.3) we have

$$\mathrm{Var}(\,\hat{\rho}_{ii}(1)\,) \approx (\,1 - 3\,\rho_{ii}^2(1) + 4\,\rho_{ii}^4(1)\,)/T\,,$$

and $\mathrm{Var}(\,\hat{\rho}_{ij}(l)\,) \approx (\,1 + 2\,\rho_{ii}(1)\,\rho_{jj}(1)\,)/T$, for $l = \pm 2, \pm 3, \ldots$. This last result can be used to check the significance of individual sample cross-correlations $\hat{\rho}_{ij}(l)$ for lags $l > 1$ in assessing the appropriateness of a vector MA(1) model for the series $\boldsymbol{Y}_t$. Similarly, if $\boldsymbol{Y}_t$ is an MA(q) process, so that $\rho_{ij}(l) = 0$ for $l > q$, then we have $E(\,\hat{\rho}_{ij}(l)\,) \approx 0$ and

$$\mathrm{Var}(\,\hat{\rho}_{ij}(l)\,) \approx \frac{1}{T}\,[\,1 + 2\sum_{u=1}^{q} \rho_{ii}(u)\,\rho_{jj}(u)\,], \qquad \text{for} \quad |\,l\,| > q\,. \tag{4.4}$$

(c) One further special case that we note occurs when two series Y_{it} and Y_{jt} are mutually independent, so that $\rho_{ij}(l) = 0$ for all l. Then it follows from (4.3) that $\mathrm{Var}(\,\hat{\rho}_{ij}(l)\,) \approx T^{-1} \sum_{u=-\infty}^{\infty} \rho_{ii}(u)\,\rho_{jj}(u)$. Furthermore, suppose it is assumed, in addition, that one of the series Y_{it} is a univariate white noise series so that $\rho_{ii}(l) = 0$ for $l \neq 0$. Then we have $\mathrm{Var}(\,\hat{\rho}_{ij}(l)\,) \approx 1/T$ and $\mathrm{Cov}(\,\hat{\rho}_{ij}(l),\,\hat{\rho}_{ij}(n)\,) \approx \rho_{jj}(n-l)/T$, so that $\mathrm{Corr}(\,\hat{\rho}_{ij}(l),\,\hat{\rho}_{ij}(n)\,) \approx \rho_{jj}(n-l)$.

The results from (4.4) can be used to check the significance of individual $\hat{\rho}_{ij}(l)$ for $l > q$ when considering a low order MA(q) model for $\boldsymbol{Y}_t$. In practice, expression (4.4) for $\mathrm{Var}(\,\hat{\rho}_{ij}(l)\,)$ is used with the unknown $\rho_{ii}(u)$ replaced by the estimates $\hat{\rho}_{ii}(u)$. Also, because there are a large number of sample correlations to examine with more than one series, for convenience one might employ a summary of the sample correlation results as suggested by Tiao and Box (1981). In their method, the symbols $+$, $-$, and $.$ are used in the $(\,i,\,j\,)$ positions of the sample correlation matrices $\hat{\rho}(l)$ as a summary to indicate correlation values $\hat{\rho}_{ij}(l)$ that are greater than 2 estimated standard errors, less than $-$ 2 estimated standard errors, and within $\pm$ 2 estimated standard errors, respectively. In fact, as rough guidelines which are convenient for summarization, they used the two standard error limits $\pm\, 2/\sqrt{T}$ which are appropriate for a vector white noise series as noted in case (a) above.

Another useful procedure may be to summarize the various correlations in the matrix $\hat{\rho}(l)$ at a given lag l into a single simple statistic. Under the assumption that $\boldsymbol{Y}_t$ is a white noise process with $\mathrm{Cov}(\,\boldsymbol{Y}_t\,) = \Sigma$, it follows from results presented later in Section 5.5.2 that $\mathbf{c}_l = \mathrm{vec}\{\,C(l)\,\}$, where $C(l)$ is given in (4.1), is asymptotically normally distributed with zero mean vector and covariance matrix $\mathrm{Cov}(\,\mathbf{c}_l\,) = T^{-2}\,(T - l)\,(\,\Sigma \otimes \Sigma\,)$, and the $\mathbf{c}_l$ are independent for different lags l. Consequently, if $\hat{V}$ denotes the diagonal matrix with the same diagonal elements as in $C(0) = \hat{\Sigma}$, then with $\hat{\rho}(l) = \hat{V}^{-1/2}\,C(l)\,\hat{V}^{-1/2}$, the vector of sample correlations at lag l is

$$\mathbf{r}_l = \mathrm{vec}\{\,\hat{\rho}(l)\,\} = \mathrm{vec}\{\,\hat{V}^{-1/2} C(l)\,\hat{V}^{-1/2}\,\} = (\,\hat{V}^{-1/2} \otimes \hat{V}^{-1/2}\,)\,\mathbf{c}_l\,.$$

It follows that $\mathbf{r}_l$ has an asymptotic normal distribution with zero mean vector and covariance matrix $T^{-2}\,(T - l)\,(\,\rho(0) \otimes \rho(0)\,) \approx T^{-1}\,(\,\rho(0) \otimes \rho(0)\,)$, since $V^{-1/2}\,\Sigma\,V^{-1/2} = \rho(0)$. This result is consistent with that mentioned in the special case (a) above. Hence, the statistic

$$T\, \mathbf{c}_l' \,(\hat{\Sigma}^{-1} \otimes \hat{\Sigma}^{-1})\, \mathbf{c}_l = T \operatorname{tr}\{\, C(l)\, \hat{\Sigma}^{-1} C(l)'\, \hat{\Sigma}^{-1} \}$$

$$= T \operatorname{tr}\{\, \hat{\rho}(l)\, \hat{\rho}(0)^{-1} \hat{\rho}(l)'\, \hat{\rho}(0)^{-1} \} \qquad (4.5)$$

or, more accurately, $T^2\,(T-l)^{-1} \operatorname{tr}\{\, \hat{\rho}(l)\, \hat{\rho}(0)^{-1} \hat{\rho}(l)'\, \hat{\rho}(0)^{-1} \}$, has an asymptotic chi-squared distribution with k^2 degrees of freedom under the white noise assumption, and this statistic can be used to check the "combined" significance of the elements of the correlation matrix $\hat{\rho}(l)$ at each lag l. In addition, the statistics for different lags l are approximately independent under the white noise assumptions, and these statistics (which are quite easy to construct from the sample correlations) for various lags l may provide an additional useful overall summary of significance of the sample correlation structure for the stationary vector series $\boldsymbol{Y}_t$. Similar statistics based on (4.5) will also be useful in assessment of the adequacy of a fitted model by examination of the residual series for characteristics of a vector white noise process, and this aspect will be discussed in Section 5.5. It also follows from results discussed in Appendix A4 that the above statistic (4.5) can be expressed, approximately, in the asymptotically equivalent form as $-T \sum_{i=1}^{k} \log(\, 1 - \hat{\rho}_i^2\,)$, where the $\hat{\rho}_i$, $i = 1, \ldots, k$, are the sample canonical correlations between the vectors $\boldsymbol{Y}_t$ and $\boldsymbol{Y}_{t+l}$.

4.2 Sample Partial AR and Partial Correlation Matrices and Their Properties

Given a sample vector time series $\boldsymbol{Y}_1, \ldots, \boldsymbol{Y}_T$ of length T, sample estimates of the partial autoregression and partial correlation matrices will be useful for specification of the order in the case of an AR model, since from Section 3.3 these matrices Φ_{mm} and P_m have the characteristic property that they equal zero for $m > p$ under an AR(p) process. So, for given lag m, we let $\hat{\Phi}_{mm}$ and $\hat{P}_m$ denote the natural sample estimates of Φ_{mm} and P_m defined in Section 3.3. These are obtained from the appropriate matrix calculations, as in (3.16) of Section 3.3.2 and (3.21) of Section 3.3.3, respectively, by replacing theoretical covariance matrices $\Gamma(l)$ by their sample covariance matrix counterparts, $\hat{\Gamma}(l)$, defined in (4.1). In particular, $\hat{\Phi}_{mm}$ is the solution to the sample Yule-Walker equations $\hat{\Gamma}(l) = \sum_{j=1}^{m} \hat{\Gamma}(l-j)\, \hat{\Phi}_{jm}'$, $l = 1, \ldots, m$. We also let

$$\hat{\Sigma} = \hat{\Sigma}_m = \hat{\Gamma}(0) - \sum_{j=1}^{m} \hat{\Gamma}(-j)\, \hat{\Phi}_{jm}'$$

be the corresponding sample estimate of the error covariance matrix Σ.

4.2.1 Test for Order of AR Model Based on Sample Partial Autoregression Matrices

Then, for example, under the assumption that the process $\boldsymbol{Y}_t$ is AR(p), it follows from results to be presented in the next two Sections 4.3 and 4.4 that for any $m > p$ the elements of $\hat{\Phi}_{mm}$ have an asymptotic normal distribution with mean vector zero, such that $\hat{\boldsymbol{\phi}}_{mm} = \text{vec}(\hat{\Phi}_{mm}')$ has approximate covariance matrix equal to $T^{-1}(\Sigma \otimes \Sigma_{m-1}^{*-1})$, where Σ_{m-1}^{*} is as defined in Section 3.3.2 preceding (3.16) [in fact, $\Sigma_{m-1}^{*} = \Sigma_{p}^{*}$ is constant for $m > p$ in an AR(p) process]. Hence, with $\hat{\Sigma}_{m-1}^{*} = \hat{\Gamma}(0) - \sum_{j=1}^{m-1} \hat{\Gamma}(m-j)\, \hat{\Phi}_{m-j,m-1}^{*\prime}$ denoting the sample version of Σ_{m-1}^{*}, the Wald statistic

$$T\, \hat{\boldsymbol{\phi}}_{mm}' \,(\hat{\Sigma}^{-1} \otimes \hat{\Sigma}_{m-1}^{*})\, \hat{\boldsymbol{\phi}}_{mm} = T \text{ tr}\{\, \hat{\Phi}_{mm}\, \hat{\Sigma}_{m-1}^{*}\, \hat{\Phi}_{mm}'\, \hat{\Sigma}^{-1} \}$$

$$\approx T \text{ tr}\{\, \hat{\Phi}_{mm}\, \hat{\Sigma}_{m-1}^{*}\, \hat{\Phi}_{mm}'\, \hat{\Sigma}_{m-1}^{-1} \} \tag{4.6}$$

will have an asymptotic chi-squared distribution with k^2 degrees of freedom for $m > p$ under an AR(p) model. So this statistic (4.6) can be used as a test statistic to test for an AR(p) model, that is, to test $\Phi_m = 0$ in an AR model of order $m > p$. It follows from discussions presented in Appendix A4 in regard to multivariate linear regression models, and from the presentation in the next two sections, that the Yule-Walker estimates $\hat{\Phi}_{jm}$, $j = 1, \ldots, m$, for an AR(m) model are asymptotically equivalent to the (conditional) least squares estimates $\hat{\Phi}_{jm}$ obtained from the multivariate linear least squares regression fitting of the $\boldsymbol{Y}_t$ on the lagged values $\boldsymbol{Y}_{t-1}, \ldots, \boldsymbol{Y}_{t-m}$. It also follows from discussion of Appendix A4 that an asymptotically equivalent test statistic to (4.6), the likelihood ratio (LR) test statistic for $H_0 : \Phi_m = 0$, can be formulated in terms of the ratio $U_m = \det(S_m) / \det(S_{m-1})$ as

$$M_m = -\,[\, N - mk - 1 - 1/2\,]\, \log(U_m), \qquad N = T - m,$$

where $S_m = \sum_{t=m+1}^{T} \hat{\varepsilon}_t\, \hat{\varepsilon}_t'$ denotes the residual sum of squares matrix, as presented in Section 4.3, from the multivariate least squares regression of the AR(m) model. Hence, the chi-squared test statistics for the order of the vector AR model can be equivalently formulated in the context of multivariate linear regression as will be discussed in the next sections.

4.2.2 Equivalent Test Statistics Based on Sample Partial Correlation Matrices

For the sample partial correlation matrix $\hat{P}_m$, it can be established (Wei, 1990, Chap. 14) that the individual elements of $\hat{P}_m$ are approximately normally distributed with zero means and variances $1/T$, for $m > p$ under an AR(p) model. Hence, an appropriate chi-squared test statistic can be formulated in terms of the elements of the matrix $\hat{P}_m$, and this statistic can be used to test for an AR model of order less than m, that is, $\Phi_m = 0$, similar to the above test based on $\hat{\Phi}_{mm}$. However, this statistic will essentially be equivalent to the chi-squared statistic given above in (4.6) in terms of $\hat{\Phi}_{mm}$. Similarly, consider the sample version $\hat{P}(m) = \hat{\Sigma}_{m-1}^{*\,1/2}\, \hat{\Phi}_{mm}'\, \hat{\Sigma}_{m-1}^{-1/2}$ of the partial correlation matrix studied by Ansley and Newbold (1979). Since $\text{vec}(\hat{P}(m)) = (\hat{\Sigma}_{m-1}^{-1/2} \otimes \hat{\Sigma}_{m-1}^{*\,1/2})\, \text{vec}(\hat{\Phi}_{mm}')$, it follows that $\text{vec}(\hat{P}(m))$ has an approximate distribution which is $(\Sigma^{-1/2} \otimes \Sigma_{m-1}^{*\,1/2})$ times the $N(0, T^{-1}(\Sigma \otimes \Sigma_{m-1}^{*-1}))$ distribution, that is, $\text{vec}(\hat{P}(m))$ is distributed approximately as $N(0, T^{-1}(I_k \otimes I_k))$. So the individual elements of $\hat{P}(m)$ are asymptotically independently and normally distributed with zero means and variances $1/T$. Hence, the test statistic

$$T \operatorname{tr}\{ \hat{P}(m)' \hat{P}(m) \} = T \operatorname{tr}\{ \hat{\Phi}_{mm} \hat{\Sigma}^*_{m-1} \hat{\Phi}'_{mm} \hat{\Sigma}^{-1}_{m-1} \}$$

also has an asymptotic chi-squared distribution with k^2 degrees of freedom for an AR(p) process when $m > p$. But it can be readily seen that this last statistic is the same as the form of the test statistic given earlier in (4.6) in terms of $\hat{\Phi}_{mm}$, and these are, thus, also essentially equivalent test statistics.

The test statistic (4.6) and the asymptotically equivalent likelihood ratio (LR) test statistic U_m can also be expressed in terms of *sample partial canonical correlations* between $\boldsymbol{Y}_t$ and $\boldsymbol{Y}_{t-m}$, given $\boldsymbol{Y}_{t-1}, \ldots, \boldsymbol{Y}_{t-m+1}$. Denote these sample partial canonical correlations as $\hat{\rho}_i(m)$, $i = 1, \ldots, k$, which are obtained using the sample analogues of the methods discussed in Section 3.2 related to equation (3.9). That is, the $\hat{\rho}_i^2(m)$ are the eigenvalues of the matrix

$$\Big(\sum_{t=m+1}^{T} \hat{\boldsymbol{U}}_{m,t} \hat{\boldsymbol{U}}'_{m,t} \Big)^{-1} \Big(\sum_{t=m+1}^{T} \hat{\boldsymbol{U}}_{m,t} \hat{\boldsymbol{U}}^{*\prime}_{m,t-m} \Big)$$
$$\times \Big(\sum_{t=m+1}^{T} \hat{\boldsymbol{U}}^*_{m,t-m} \hat{\boldsymbol{U}}^{*\prime}_{m,t-m} \Big)^{-1} \Big(\sum_{t=m+1}^{T} \hat{\boldsymbol{U}}^*_{m,t-m} \hat{\boldsymbol{U}}'_{m,t} \Big),$$

where $\hat{\boldsymbol{U}}_{m,t} = \boldsymbol{Y}_t - \sum_{j=1}^{m-1} \hat{\Phi}_{j,m-1} \boldsymbol{Y}_{t-j}$ and $\hat{\boldsymbol{U}}^*_{m,t-m} = \boldsymbol{Y}_{t-m} - \sum_{j=1}^{m-1} \hat{\Phi}^*_{j,m-1} \boldsymbol{Y}_{t-m+j}$ are the "residual" vectors (zero mean has been assumed for $\{\boldsymbol{Y}_t\}$), and the $\hat{\Phi}_{j,m-1}$ and $\hat{\Phi}^*_{j,m-1}$ are the appropriate least squares estimates in the "forward" and "backward" vector autoregressions, respectively, of order $m-1$. Then in view of the sample analogue of (3.23) and the related discussion in Section 3.3.4, the above matrix is nearly the same as $\hat{\Sigma}^{-1}_{m-1} \hat{\Phi}_{mm} \hat{\Sigma}^*_{m-1} \hat{\Phi}'_{mm}$ $= \hat{\Sigma}^{-1/2}_{m-1} \hat{P}(m)' \hat{P}(m) \hat{\Sigma}^{1/2}_{m-1}$. So it follows that the statistic (4.6) is approximately

$$T \operatorname{tr}\{ \hat{\Phi}_{mm} \hat{\Sigma}^*_{m-1} \hat{\Phi}'_{mm} \hat{\Sigma}^{-1}_{m-1} \} = T \operatorname{tr}\{ \hat{P}(m)' \hat{P}(m) \} \approx T \sum_{i=1}^{k} \hat{\rho}_i^2(m). \tag{4.7}$$

In addition, from results of Appendix A4 related to (A4.8), the LR statistic can be expressed as $-T \log(U_m) = -T \sum_{i=1}^{k} \log(1 - \hat{\rho}_i^2(m))$, which is approximately the same as the right side of (4.7) under the null hypothesis of an AR(p) model of order $p < m$. Hence, these asymptotically equivalent forms of the test statistic also have asymptotic $\chi^2_{k^2}$ distributions and can be used to test the order of a vector AR model.

4.3 Conditional Least Squares Estimation of Vector AR Models

Let $\boldsymbol{Y}_t = (Y_{1t}, \ldots, Y_{kt})'$ be a k-dimensional vector series that follows the AR(p) model $\boldsymbol{Y}_t - \boldsymbol{\mu} = \sum_{j=1}^{p} \Phi_j (\boldsymbol{Y}_{t-j} - \boldsymbol{\mu}) + \varepsilon_t$, or $\boldsymbol{Y}_t = \boldsymbol{\delta} + \sum_{j=1}^{p} \Phi_j \boldsymbol{Y}_{t-j} + \varepsilon_t$, where $\boldsymbol{\delta} = (I - \Phi_1 - \cdots - \Phi_p)\boldsymbol{\mu}$ in the stationary case, and where the ε_t are vector white noise with zero mean and covariance matrix $\Sigma = \mathrm{Cov}(\varepsilon_t)$. The vector AR($p$) model can be considered to be approximately in the form of a multivariate linear model as discussed in Appendix A4, since we have $\boldsymbol{Y}_t - \boldsymbol{\mu} = B_*' \tilde{\boldsymbol{X}}_t + \varepsilon_t$, where $\tilde{\boldsymbol{X}}_t = [\,(\boldsymbol{Y}_{t-1} - \boldsymbol{\mu})', \ldots, (\boldsymbol{Y}_{t-p} - \boldsymbol{\mu})'\,]'$ and $B_*' = (\Phi_1, \ldots, \Phi_p)$, or $\boldsymbol{Y}_t = B' \boldsymbol{X}_t + \varepsilon_t$, where $\boldsymbol{X}_t = [\,1, \boldsymbol{Y}_{t-1}', \ldots, \boldsymbol{Y}_{t-p}'\,]'$ and $B' = (\boldsymbol{\delta}, \Phi_1, \ldots, \Phi_p)$. In the estimation of vector AR(p) models, it is generally preferable to use least squares (LS) estimates of the AR parameters rather than Yule-Walker estimates as discussed in the previous section, although these two types of estimates have the same asymptotic properties in the stationary case, because the least squares estimates have better sampling behavior for situations near the nonstationary boundary. Generally, the Yule-Walker estimators have larger biases than the least squares estimators in such situations due to the impact from "end effects" in the use of the sample covariance matrices $\hat{\Gamma}(l)$ defined in (4.1), e.g., see studies by Tjostheim and Paulsen (1983), Yamamoto and Kunitomo (1984) and Shaman and Stine (1988). Given available sample observations $\boldsymbol{Y}_1, \ldots, \boldsymbol{Y}_T$, we thus want to obtain least squares estimates of the AR matrix parameters Φ_j and Σ.

4.3.1 Least Squares Estimation for the Vector AR(1) Model

Consider first the stationary AR(1) model $\boldsymbol{Y}_t - \boldsymbol{\mu} = \Phi_1 (\boldsymbol{Y}_{t-1} - \boldsymbol{\mu}) + \varepsilon_t$, or $\boldsymbol{Y}_t = \boldsymbol{\delta} + \Phi_1 \boldsymbol{Y}_{t-1} + \varepsilon_t$, for $t = 2, 3, \ldots, T$. Letting $\mathbf{Y} = [\,\boldsymbol{Y}_2, \boldsymbol{Y}_3, \ldots, \boldsymbol{Y}_T\,]'$, $\mathbf{X} = [\,\mathbf{1}_N, (\boldsymbol{Y}_1, \boldsymbol{Y}_2, \ldots, \boldsymbol{Y}_{T-1})'\,]$, where $\mathbf{1}_N$ denotes an $N \times 1$ column vector of ones with $N = T-1$, and $\boldsymbol{\varepsilon} = [\,\varepsilon_2, \varepsilon_3, \ldots, \varepsilon_T\,]'$, we have $\mathbf{Y} = \mathbf{X} B + \boldsymbol{\varepsilon}$ with $B = (\boldsymbol{\delta}, \Phi_1)'$. This has the form of the multivariate linear model with $N = T-1$ vector observations, as discussed in Appendix A4. The *least squares estimator* (LSE) of $B_* = \Phi_1'$ is given by

$$\hat{\Phi}_1' = (\tilde{\mathbf{X}}' \tilde{\mathbf{X}})^{-1} \tilde{\mathbf{X}}' \tilde{\mathbf{Y}}$$

$$= \left[\sum_{t=2}^{T} (\boldsymbol{Y}_{t-1} - \bar{\boldsymbol{Y}}_{(1)})(\boldsymbol{Y}_{t-1} - \bar{\boldsymbol{Y}}_{(1)})' \right]^{-1} \sum_{t=2}^{T} (\boldsymbol{Y}_{t-1} - \bar{\boldsymbol{Y}}_{(1)})(\boldsymbol{Y}_t - \bar{\boldsymbol{Y}}_{(0)})', \quad (4.8)$$

where

$$\tilde{\mathbf{Y}} = [\,(\boldsymbol{Y}_2 - \bar{\boldsymbol{Y}}_{(0)}), \ldots, (\boldsymbol{Y}_T - \bar{\boldsymbol{Y}}_{(0)})\,]', \quad \tilde{\mathbf{X}} = [\,(\boldsymbol{Y}_1 - \bar{\boldsymbol{Y}}_{(1)}), \ldots, (\boldsymbol{Y}_{T-1} - \bar{\boldsymbol{Y}}_{(1)})\,]',$$

with $\bar{\boldsymbol{Y}}_{(0)} = N^{-1} \sum_{t=2}^{T} \boldsymbol{Y}_t$ and $\bar{\boldsymbol{Y}}_{(1)} = N^{-1} \sum_{t=2}^{T} \boldsymbol{Y}_{t-1}$. The estimator of Σ is $\hat{\Sigma} = [\,1/(N-(k+1))\,]\, S_1$, where $S_1 = \sum_{t=2}^{T} \hat{\varepsilon}_t \hat{\varepsilon}_t'$ is the residual sum of squares matrix, with $\hat{\varepsilon}_t = (\boldsymbol{Y}_t - \bar{\boldsymbol{Y}}_{(0)}) - \hat{\Phi}_1 (\boldsymbol{Y}_{t-1} - \bar{\boldsymbol{Y}}_{(1)})$ the residual vectors.

Now it has been proven [e.g., see Hannan (1970, Chap. 6)] that, if we define $\hat{\boldsymbol{\phi}} = \mathrm{vec}(\hat{\Phi}_1')$ and $\boldsymbol{\phi} = \mathrm{vec}(\Phi_1')$, then as $N = T-1 \to \infty$ the asymptotic distribution of $N^{1/2}(\hat{\boldsymbol{\phi}} - \boldsymbol{\phi})$ is multivariate normal $N(0, \Sigma \otimes \Gamma(0)^{-1})$, where $\Gamma(0) = E[\,(\boldsymbol{Y}_{t-1} - \boldsymbol{\mu})(\boldsymbol{Y}_{t-1} - \boldsymbol{\mu})'\,]$. Thus, it follows that for large N, the distribution of $\hat{\boldsymbol{\phi}}$ is approximately $N(\boldsymbol{\phi}, N^{-1}(\Sigma \otimes \Gamma(0)^{-1}))$. This approximate distribution can be consistently estimated as $\hat{\boldsymbol{\phi}} \sim N(\boldsymbol{\phi}, \hat{\Sigma} \otimes (\tilde{\mathbf{X}}' \tilde{\mathbf{X}})^{-1})$, since

$$N\,(\tilde{\mathbf{X}}'\tilde{\mathbf{X}})^{-1} = (N^{-1}\,\tilde{\mathbf{X}}'\tilde{\mathbf{X}})^{-1}$$

$$= (N^{-1}\sum_{t=2}^{T}(\boldsymbol{Y}_{t-1}-\bar{\boldsymbol{Y}}_{(1)})(\boldsymbol{Y}_{t-1}-\bar{\boldsymbol{Y}}_{(1)})')^{-1} \approx \hat{\Gamma}(0)^{-1} \to \Gamma(0)^{-1}$$

as $N \to \infty$, so that $(\tilde{\mathbf{X}}'\tilde{\mathbf{X}})^{-1} \approx N^{-1}\,\Gamma(0)^{-1}$ for large N. Hence, we see that the LSE of Φ_1 in the stationary AR(1) model has approximately the same form of distributional properties as the LSE in the standard multivariate linear model. Also, note that $\hat{\text{Corr}}(\hat{\boldsymbol{\phi}}) = \rho(\hat{\Sigma}) \otimes \rho((\tilde{\mathbf{X}}'\tilde{\mathbf{X}})^{-1})$, where, for example, $\rho(\Sigma) = P\,\Sigma\,P$ with $P = \text{Diag}\{\sigma_{11}^{-1/2}, \ldots, \sigma_{kk}^{-1/2}\}$ denotes the correlation matrix associated with Σ. Note that if we write $\Phi_1' = (\phi_1, \phi_2, \ldots, \phi_k)$, then the model is $Y_{it} - \mu_i = (\boldsymbol{Y}_{t-1} - \boldsymbol{\mu})'\,\phi_i + \varepsilon_{it}$, that is, ϕ_i are the (auto)regressive coefficients in the (auto)regression for the ith variable Y_{it}. Thus, we have

$$\hat{\text{Cov}}(\hat{\phi}_i) = \hat{\sigma}_{ii}\,(\tilde{\mathbf{X}}'\tilde{\mathbf{X}})^{-1}, \qquad \hat{\text{Cov}}(\hat{\phi}_i, \hat{\phi}_j) = \hat{\sigma}_{ij}\,(\tilde{\mathbf{X}}'\tilde{\mathbf{X}})^{-1},$$

and $\hat{\text{Corr}}(\hat{\phi}_i, \hat{\phi}_j) = \hat{\rho}_{ij}\;\rho((\tilde{\mathbf{X}}'\tilde{\mathbf{X}})^{-1})$, $i, j = 1, \ldots, k$.

The estimates $\hat{\boldsymbol{\phi}}$ and the associated estimates of individual variances and correlations between parameter estimates can be used to make inferences about individual elements of the autoregression matrix Φ_1. One may also wish to make simultaneous inference about all the elements of Φ_1, in particular, to test $H_0 : \Phi_1 = 0$ against $\Phi_1 \neq 0$. For this, we use the likelihood ratio (LR) test statistic

$$M_1 = -(N - k - 1 - 1/2)\log[\,\det(S_1)/\det(S_0)\,],$$

where $S_0 = \sum_{t=2}^{T}(\boldsymbol{Y}_t - \bar{\boldsymbol{Y}}_{(0)})(\boldsymbol{Y}_t - \bar{\boldsymbol{Y}}_{(0)})'$ is the "residual" sum of squares matrix under $H_0 : \Phi_1 = 0$. For large N, M_1 is approximately distributed as $\chi^2_{k^2}$ under $\Phi_1 = 0$, and $\Phi_1 = 0$ is rejected for large values of M_1. We note that the least squares estimates of Φ_1 and Σ described above, under the assumption of normality for the white noise series ε_t, are essentially the (conditional) maximum likelihood estimators of Φ_1 and Σ, when we consider the joint distribution of the observations $\boldsymbol{Y}_2, \boldsymbol{Y}_3, \ldots, \boldsymbol{Y}_T$, conditional on the value of the initial observation $\boldsymbol{Y}_1$.

Comment (*Asymptotic distribution of the LSE*). We briefly discuss the large sample result for the LSE that $N^{1/2}(\hat{\boldsymbol{\phi}} - \boldsymbol{\phi}) \xrightarrow{D} N(0, \Sigma \otimes \Gamma(0)^{-1})$ as $N \to \infty$. First note that

$$\hat{\Phi}_1' = (\tilde{\mathbf{X}}'\tilde{\mathbf{X}})^{-1}\,\tilde{\mathbf{X}}'\tilde{\mathbf{Y}} = \Phi_1' + (\tilde{\mathbf{X}}'\tilde{\mathbf{X}})^{-1}\,\tilde{\mathbf{X}}'\boldsymbol{\varepsilon} = \Phi_1' + (N^{-1}\tilde{\mathbf{X}}'\tilde{\mathbf{X}})^{-1}\,N^{-1}\tilde{\mathbf{X}}'\boldsymbol{\varepsilon}, \quad (4.9)$$

hence

$$N^{1/2}(\hat{\boldsymbol{\phi}} - \boldsymbol{\phi}) = \text{vec}[\,(N^{-1}\tilde{\mathbf{X}}'\tilde{\mathbf{X}})^{-1}\,N^{-1/2}\,\tilde{\mathbf{X}}'\boldsymbol{\varepsilon}\,]$$

$$= [\,I_k \otimes (N^{-1}\tilde{\mathbf{X}}'\tilde{\mathbf{X}})^{-1}\,]\,N^{-1/2}\,\text{vec}(\tilde{\mathbf{X}}'\boldsymbol{\varepsilon}).$$

Then notice that $N^{-1/2}\,\tilde{\mathbf{X}}'\boldsymbol{\varepsilon} \approx N^{-1/2}\sum_{t=2}^{T}(\boldsymbol{Y}_{t-1} - \boldsymbol{\mu})\,\varepsilon_t'$, so that $N^{-1/2}\,\text{vec}(\tilde{\mathbf{X}}'\boldsymbol{\varepsilon}) \approx N^{-1/2}\sum_{t=2}^{T}(I_k \otimes (\boldsymbol{Y}_{t-1} - \boldsymbol{\mu}))\,\varepsilon_t$. Now since the ε_t are independent and ε_t is independent of the past value Y_{t-1}, we find that

$$E[\,N^{-1/2}\,\mathrm{vec}(\,\tilde{\mathbf{X}}'\boldsymbol{\varepsilon}\,)\,] \approx N^{-1/2}\sum_{t=2}^{T} E[\,(\,I_k \otimes (\,Y_{t-1}-\boldsymbol{\mu}\,)\,)\,]\,E(\,\varepsilon_t\,) = 0\,,$$

$$\begin{aligned}\mathrm{Cov}(\,N^{-1/2}\,\mathrm{vec}(\,\tilde{\mathbf{X}}'\boldsymbol{\varepsilon}\,)\,)\\ &\approx N^{-1}\sum_{t=2}^{T}\sum_{s=2}^{T} E[\,(\,I_k \otimes (\,Y_{t-1}-\boldsymbol{\mu}\,)\,)\,\varepsilon_t\,\varepsilon_s'\,\,(\,I_k \otimes (\,Y_{s-1}-\boldsymbol{\mu}\,)'\,)\,]\\ &= N^{-1}\sum_{t=2}^{T}(\,\Sigma \otimes E[\,(\,Y_{t-1}-\boldsymbol{\mu}\,)(\,Y_{t-1}-\boldsymbol{\mu}\,)'\,]\,) = \Sigma \otimes \Gamma(0)\,,\end{aligned}$$

where in the above we have used the property that expectations can be expressed in terms of conditional expectations. Now, in fact, we can recognize $S_T = \sum_{t=2}^{T}(\,I_k \otimes (\,Y_{t-1}-\boldsymbol{\mu}\,)\,)\,\varepsilon_t$ as a vector martingale process, satisfying the property that $E(\,S_T \mid F_{T-1}\,) = S_{T-1}$ where F_{T-1} denotes the σ-field generated by $\{\,\varepsilon_{T-1}, \varepsilon_{T-2}, \ldots\,\}$. Then by use of a "martingale central limit theorem" for stationary processes (e.g., Billingsley, 1968, p. 206, and Brown, 1971) it can be established that as $N = T-1 \to \infty$, $N^{-1/2}\,\mathrm{vec}(\,\tilde{\mathbf{X}}'\boldsymbol{\varepsilon}\,) \approx N^{-1/2}\,S_T$ converges in distribution to $N(\,0,\,\Sigma \otimes \Gamma(0)\,)$. Also, we know that $N^{-1}\,\tilde{\mathbf{X}}'\tilde{\mathbf{X}}$ converges in probability to $\Gamma(0)$ as $N \to \infty$. Hence, by a limit property,

$$\begin{aligned}N^{1/2}(\,\hat{\boldsymbol{\phi}}-\boldsymbol{\phi}\,) &= (\,I \otimes (\,N^{-1}\tilde{\mathbf{X}}'\tilde{\mathbf{X}}\,)^{-1})\,N^{-1/2}\,\mathrm{vec}(\,\tilde{\mathbf{X}}'\boldsymbol{\varepsilon}\,)\\ &\approx (\,I \otimes \Gamma(0)^{-1})\,N^{-1/2}\,\mathrm{vec}(\,\tilde{\mathbf{X}}'\boldsymbol{\varepsilon}\,) \xrightarrow{D} N(\,0,\,\Sigma \otimes \Gamma(0)^{-1}\,)\,. \qquad (4.10)\end{aligned}$$

4.3.2 Least Squares Estimation for the Vector AR Model of General Order

The general stationary AR(m) model can be expressed as

$$Y_t - \boldsymbol{\mu} = \sum_{j=1}^{m} \Phi_j\,(\,Y_{t-j}-\boldsymbol{\mu}\,) + \varepsilon_t = \Phi_{(m)}'\,\tilde{X}_t + \varepsilon_t\,,$$

where $\tilde{X}_t = [\,(\,Y_{t-1}-\boldsymbol{\mu}\,)', \ldots, (\,Y_{t-m}-\boldsymbol{\mu}\,)'\,]'$ and $\Phi_{(m)}' = (\,\Phi_1, \ldots, \Phi_m\,)$. Equivalently, we can express the model as

$$Y_t = \boldsymbol{\delta} + \sum_{j=1}^{m} \Phi_j\,Y_{t-j} + \varepsilon_t = B'\,X_t + \varepsilon_t\,,$$

where $X_t = [\,1, Y_{t-1}', \ldots, Y_{t-m}'\,]'$ and $B' = (\,\boldsymbol{\delta}, \Phi_1, \ldots, \Phi_m\,)$. As in the AR(1) model, we define, with $N = T - m$, the $N \times k$ data matrix $\mathbf{Y} = [\,Y_{m+1}, Y_{m+2}, \ldots, Y_T\,]'$, $\boldsymbol{\varepsilon} = [\,\varepsilon_{m+1}, \ldots, \varepsilon_T\,]'$, and the $N \times (mk+1)$ matrix $\mathbf{X}$ whose typical row is $Z_{t,m}' = [\,1, Y_{t-1}', \ldots, Y_{t-m}'\,]$, $t = m+1, \ldots, T$. Then we have $\mathbf{Y} = \mathbf{X}\,B + \boldsymbol{\varepsilon}$, which has the general form of the multivariate linear model with $N = T - m$ vector observations, as discussed in Appendix A4.

(We often will consider the situation where AR models of successively higher orders are fitted to the series, with M being the maximum order considered. Hence, our discussion relates to $m = 1, 2, \ldots, M$.)

It follows that the LS estimator of $B_* = \Phi_{(m)}$ when fitting an AR(m) model to the series is

$$\hat{\Phi}_{(m)} = (\tilde{\mathbf{X}}'\tilde{\mathbf{X}})^{-1}\tilde{\mathbf{X}}'\tilde{\mathbf{Y}}, \tag{4.11}$$

where

$$\tilde{\mathbf{X}}'\tilde{\mathbf{X}} = \sum_{t=m+1}^{T} \tilde{\boldsymbol{Z}}_{t,m}\tilde{\boldsymbol{Z}}'_{t,m}, \qquad \tilde{\mathbf{X}}'\tilde{\mathbf{Y}} = \sum_{t=m+1}^{T} \tilde{\boldsymbol{Z}}_{t,m}(\boldsymbol{Y}_t - \bar{\boldsymbol{Y}}_{(0)})', \tag{4.12}$$

and $\tilde{\mathbf{Y}} = [(\boldsymbol{Y}_{m+1} - \bar{\boldsymbol{Y}}_{(0)}), \ldots, (\boldsymbol{Y}_T - \bar{\boldsymbol{Y}}_{(0)})]'$, the $N \times mk$ matrix $\tilde{\mathbf{X}}$ has typical row $\tilde{\boldsymbol{Z}}'_{t,m} = [(\boldsymbol{Y}_{t-1} - \bar{\boldsymbol{Y}}_{(1)})', \ldots, (\boldsymbol{Y}_{t-m} - \bar{\boldsymbol{Y}}_{(m)})']$, $t = m+1, \ldots, T$, with $\bar{\boldsymbol{Y}}_{(i)} = N^{-1}\sum_{t=m+1}^{T}\boldsymbol{Y}_{t-i}$. The estimate of Σ is

$$\hat{\Sigma} = (N - (km+1))^{-1} S_m, \qquad \text{where} \qquad S_m = \sum_{t=m+1}^{T} \hat{\varepsilon}_t \hat{\varepsilon}'_t$$

is the residual sum of squares matrix, with $\hat{\varepsilon}_t = (\boldsymbol{Y}_t - \bar{\boldsymbol{Y}}_{(0)}) - \hat{\Phi}'_{(m)}\tilde{\boldsymbol{Z}}_{t,m}$ the residual vectors.

We define $\hat{\boldsymbol{\phi}} = \text{vec}(\hat{\Phi}_{(m)})$ and $\boldsymbol{\phi} = \text{vec}(\Phi_{(m)})$. Then it has been proven [e.g., see Hannan (1970, Chap. 6)] that, under stationarity, $N^{1/2}(\hat{\boldsymbol{\phi}} - \boldsymbol{\phi})$ converges in distribution to $N(0, \Sigma \otimes \Gamma_m^{-1})$ as $N \to \infty$, where the $km \times km$ matrix $\Gamma_m = E[\tilde{\boldsymbol{X}}_t \tilde{\boldsymbol{X}}'_t]$ has (i, j)th block matrix equal to $\Gamma(i-j)$. The justification for this asymptotic distribution result follows along the same lines as in (4.10) for the stationary AR(1) case, with $\hat{\boldsymbol{\phi}} = (I_k \otimes (\tilde{\mathbf{X}}'\tilde{\mathbf{X}})^{-1})\,\text{vec}(\tilde{\mathbf{X}}'\tilde{\mathbf{Y}})$ so that

$$N^{1/2}(\hat{\boldsymbol{\phi}} - \boldsymbol{\phi}) \approx (I_k \otimes (N^{-1}\tilde{\mathbf{X}}'\tilde{\mathbf{X}})^{-1})\,N^{-1/2}\,\text{vec}(\tilde{\mathbf{X}}'\boldsymbol{\varepsilon}) \xrightarrow{D} N(0, \Sigma \otimes \Gamma_m^{-1}), \tag{4.13}$$

since $N^{-1/2}\,\text{vec}(\tilde{\mathbf{X}}'\boldsymbol{\varepsilon}) \approx N^{-1/2}\sum_{t=m+1}^{T}(I_k \otimes \tilde{\boldsymbol{X}}_t)\,\varepsilon_t \xrightarrow{D} N(0, \Sigma \otimes \Gamma_m)$. In particular, it is easily seen that

$$N^{-1}\tilde{\mathbf{X}}'\tilde{\mathbf{X}} = N^{-1}\sum_{t=m+1}^{T} \tilde{\boldsymbol{Z}}_{t,m}\tilde{\boldsymbol{Z}}'_{t,m} \approx \hat{\Gamma}_m$$

converges in probability to Γ_m as $N \to \infty$, where $\hat{\Gamma}_m$ has (i, j)th block matrix equal to $\hat{\Gamma}(i-j)$. So we can consistently estimate the approximate large sample distribution of $\hat{\boldsymbol{\phi}}$ as $N(\boldsymbol{\phi}, \hat{\Sigma} \otimes (\tilde{\mathbf{X}}'\tilde{\mathbf{X}})^{-1})$, since $(\tilde{\mathbf{X}}'\tilde{\mathbf{X}})^{-1} \approx N^{-1}\hat{\Gamma}_m^{-1}$ consistently estimates $N^{-1}\Gamma_m^{-1}$ for large N. Thus, we see that in the general order AR model, the LSE $\hat{\boldsymbol{\phi}}$ has approximately the same distributional properties as the LSE in the standard multivariate linear model. Hence, inferences about individual elements of $\boldsymbol{\phi}$ may be based on the approximate distribution $\hat{\boldsymbol{\phi}} \sim N(\boldsymbol{\phi}, \hat{\Sigma} \otimes (\tilde{\mathbf{X}}'\tilde{\mathbf{X}})^{-1})$.

4.3.3 Likelihood Ratio Testing for the Order of the AR Model

In some cases, we may want to test the significance of the mth-order AR matrix, that is, test $H_0 : \Phi_m = 0$ against $\Phi_m \neq 0$, when an AR(m) model has been fitted to the series. For this, based on the results associated with (A4.5) in Appendix A4 for *likelihood ratio* (LR) testing in the multivariate linear model, we use the LR statistic

$$M_m = -\,(\,N - mk - 1 - 1/2\,)\log[\,\det(\,S_m\,) / \det(\,S_{m-1}\,)\,], \qquad N = T - m\,, \quad (4.14)$$

where S_{m-1} is the residual sum of squares matrix obtained from fitting an AR model of order $m-1$ to the *same* set of $N = T - m$ vector observations as used in fitting the AR(m) model (i.e., S_{m-1} is the residual sum of squares matrix under $\Phi_m = 0$). For large N, when $\Phi_m = 0$ is true, we have that M_m is approximately distributed as $\chi^2_{k^2}$, and we reject $H_0 : \Phi_m = 0$ for large values of M_m. [Recall that the LR statistic in (4.14) has other asymptotically equivalent forms, as noted in Section 4.2, which are expressible either as a Wald statistic in terms of the least squares estimate $\hat{\Phi}_m$ of Φ_m, e.g., see (4.6) and (A4.6) of Appendix A4, or in terms of the sample partial canonical correlations $\hat{\rho}_i(m)$, $i = 1, \ldots, k$, between $\boldsymbol{Y}_t$ and $\boldsymbol{Y}_{t-m}$, given $\boldsymbol{Y}_{t-1}, \ldots, \boldsymbol{Y}_{t-m+1}$, e.g., see (4.7) and (A4.8).] Hence, we see that in this approach we have in mind the procedure of fitting successively higher-order AR models to the series, and for each order testing the significance of the last coefficient matrix Φ_m included in the model. However, note that the testing procedure for $\Phi_m = 0$ based on M_m does not address the issue of whether the matrix Φ_m may be nonzero but possess special reduced-rank structure, and models with such special structure and associated estimation and testing procedures will be discussed in Section 6.1. As in the AR(1) case, we again note that the least squares estimates of the Φ_j and Σ described above, under normality of the white noise series ε_t, are essentially the (conditional) maximum likelihood estimators of the Φ_j and Σ, when we consider the joint distribution of the observations $\boldsymbol{Y}_{m+1}, \boldsymbol{Y}_{m+2}, \ldots, \boldsymbol{Y}_T$, conditional on the values of the initial observations $\boldsymbol{Y}_1, \ldots, \boldsymbol{Y}_m$.

4.3.4 Derivation of the Wald Statistic for Testing the Order of the AR Model

We comment that instead of arranging the LS parameter estimator $\hat{\Phi}_{(m)} = (\,\tilde{\mathbf{X}}'\tilde{\mathbf{X}}\,)^{-1}\,\tilde{\mathbf{X}}'\tilde{\mathbf{Y}}$ as $\hat{\boldsymbol{\phi}}_{(m)} = \operatorname{vec}(\,\hat{\Phi}_{(m)}\,)$, we could also consider the parameter estimates arranged as

$$\hat{\boldsymbol{\phi}}^*_{(m)} = \operatorname{vec}(\,\hat{\Phi}'_{(m)}\,) = \operatorname{vec}[\,\hat{\Phi}_1, \hat{\Phi}_2, \ldots, \hat{\Phi}_m\,]\,.$$

Similar to the previous developments in (4.13), it can be shown that for this arrangement we have that $N^{1/2}(\,\hat{\boldsymbol{\phi}}^*_{(m)} - \boldsymbol{\phi}^*_{(m)}\,)$ converges in distribution to $N(\,0, \Gamma_m^{-1} \otimes \Sigma\,)$ as $N \to \infty$. Now, in particular, suppose the process is truly an AR(p) process with $p < m$, so that $\Phi_m = 0$. Then consider the partition of Γ_m as

$$\Gamma_m = \begin{bmatrix} \Gamma_{m-1} & \Gamma^*_{(m-1)} \\ \Gamma^{*\prime}_{(m-1)} & \Gamma(0) \end{bmatrix},$$

where $\Gamma^{*\prime}_{(m-1)} = [\,\Gamma(m-1), \ldots, \Gamma(1)\,]$. It follows from a standard matrix inversion result (Rao, 1973, p. 33) that Γ_m^{-1} has the lower right $k \times k$ block element of the form D^{-1}, where $D = \Gamma(0) - \Gamma^{*\prime}_{(m-1)}\, \Gamma^{-1}_{m-1}\, \Gamma^*_{(m-1)}$ and we note that the matrix D is also interpretable as the matrix Σ^*_{m-1} as discussed in Section 3.3 in regard to partial correlation matrices. It then follows that $\text{vec}(\,\hat{\Phi}_m\,)$ is distributed approximately as $N(\,\text{vec}(\,\Phi_m\,), N^{-1}\,(\,D^{-1} \otimes \Sigma\,)\,)$ or, equivalently, $\text{vec}(\,\hat{\Phi}_m'\,)$ is distributed approximately as $N(\,\text{vec}(\,\Phi_m'\,), N^{-1}\,(\,\Sigma \otimes D^{-1}\,)\,)$. Hence, from previous arguments, as in (A4.6) of Appendix A4, an asymptotically equivalent test statistic to the LR procedure for testing $H_0 : \Phi_m = 0$ is the *Wald statistic*

$$N\,\text{vec}(\,\hat{\Phi}_m\,)'\,(\,\hat{D} \otimes \hat{\Sigma}^{-1}\,)\,\text{vec}(\,\hat{\Phi}_m\,) = N\,\text{vec}(\,\hat{\Phi}_m'\,)'\,(\,\hat{\Sigma}^{-1} \otimes \hat{D}\,)\,\text{vec}(\,\hat{\Phi}_m'\,)$$

which is directly seen to be distributed asymptotically as $\chi^2_{k^2}$ under $\Phi_m = 0$. Note that this is the same as the form of the test statistic mentioned in (4.6).

EXAMPLE 4.1. Let $\boldsymbol{Y}_t = (\,Y_{1t}, Y_{2t}\,)'$ denote the bivariate time series of logarithms of the annual sales of mink and muskrat furs, respectively, by the Hudson's Bay Company for the years 1850–1911, with $T = 62$ annual observations. These data have been analyzed previously by many authors, including Chan and Wallis (1978), Jenkins and Alavi (1981), Cooper and Wood (1982), and Terasvirta (1985). Time series plots of these two series are displayed in Figure 4.1. Sample cross-correlation matrices for lags 1 through 10, together

Table 4.1 Sample Correlation Matrices $\hat{\rho}(l)$ for the Log Mink and Muskrat Furs Data (with indicator symbols for $\pm 2\,T^{-1/2}$ limits).

l	1		2		3		4		5	
$\hat{\rho}(l)$	0.69	−0.13	0.29	−0.26	−0.03	−0.16	−0.24	0.05	−0.33	0.25
	0.44	0.74	0.35	0.40	0.24	0.24	0.07	0.15	−0.12	0.11
	+	.	+	−	.	.	.	.	−	.
	+	+	+	+	.	.	.	.	.	.
l	6		7		8		9		10	
$\hat{\rho}(l)$	−0.30	0.36	−0.13	0.42	0.09	0.41	0.33	0.31	0.39	0.08
	−0.34	0.09	−0.47	0.13	−0.48	0.27	−0.26	0.40	−0.01	0.39
	−	+	.	+	.	+	+	+	+	.
	−	+	−	.	−	+	−	+	.	+

with the summary indicator symbols in terms of $+$, $-$, and . are shown in Table 4.1. Fitting AR models of orders 1, 2, 3, 4, 5, and 6 by least squares yields the results that are also summarized in Table 4.2. Given in Table 4.2 are values of the likelihood ratio test statistic M_m of (4.14) as well as values of the (normalized) AIC model selection criterion (e.g., Akaike, 1976), $\text{AIC}_m = \log(\,|\tilde{\Sigma}_m|\,) + 2\,mk^2/N$, $N = T - m$. In particular, for orders 1, 2, and 3

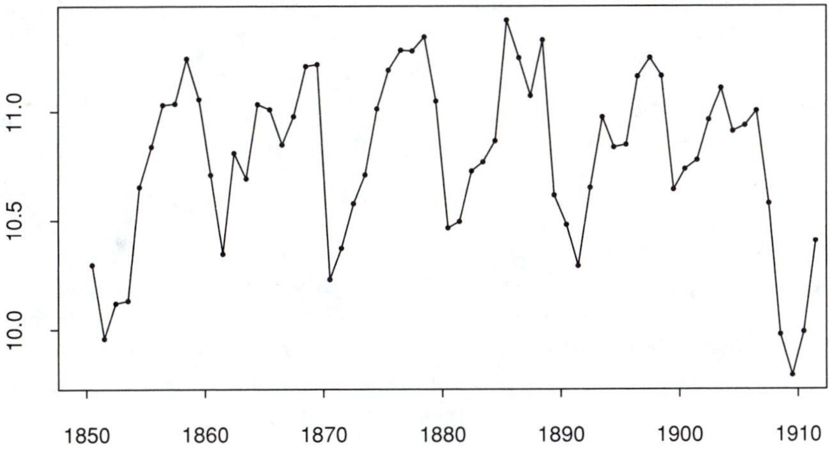

(a) Y_{1t} : Mink Fur Series

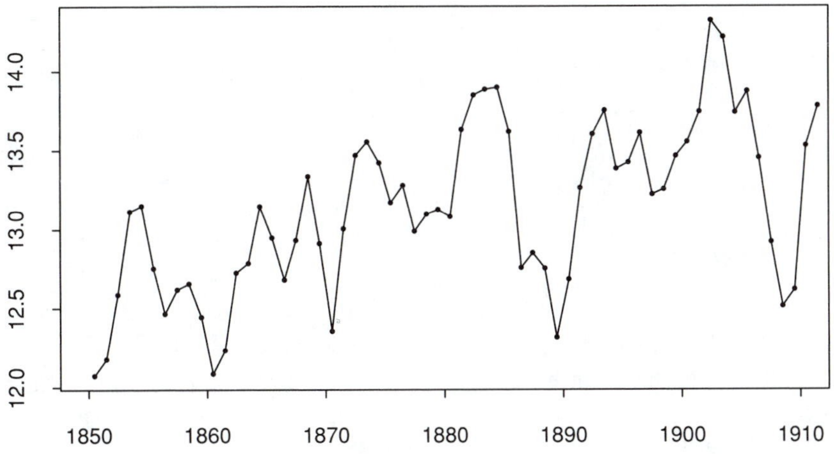

(b) Y_{2t} : Muskrat Fur Series

Figure 4.1. Logarithms of Annual Sales of Mink and Muskrat Furs by Hudson's Bay Company for the Years 1850 Through 1911

Table 4.2 Summary of Results from Fitting Autoregressive Models to the Log Mink and Muskrat Furs Data. [$\text{AIC}_m = \log(\,|\tilde{\Sigma}_m|\,) + 2m\,k^2/N$, $N = T - m$.]

m (AR order)	1	2	3	4	5	6
$\|\tilde{\Sigma}_m\|$ ($\times 10^{-2}$)	0.399835	0.308037	0.246736	0.199617	0.176035	0.146034
M_m statistic	128.85	13.70	12.59	9.37	6.29	8.98
AIC_m	−5.391	−5.516	−5.598	−5.665	−5.641	−5.672

we find that $\det(\tilde{\Sigma}_1) = 0.399835 \times 10^{-2}$, $\det(\tilde{\Sigma}_2) = 0.308037 \times 10^{-2}$, and $\det(\tilde{\Sigma}_3) = 0.246736 \times 10^{-2}$, where $\tilde{\Sigma}_m = S_m/N$ denotes an estimate of Σ not "corrected for degrees of freedom". To test $\Phi_2 = 0$ in an AR(2) model, we use

$$M_2 = -\,[(T-2) - (2)(2) - 1 - 1/2\,]\log(\,|S_2|\,/\,|S_1|\,) = 13.70,$$

with the statistic M_2 distributed as chi-squared with 4 degrees of freedom under H_0. So we reject $\Phi_2 = 0$ at $\alpha = 0.05$. Similarly, to test $\Phi_3 = 0$ in an AR(3) model, we find $M_3 = 12.59$, and we also reject $\Phi_3 = 0$. From subsequent fitting of higher-order AR models, we do not reject that higher-order terms are zero and, hence, we find that a vector AR(3) model might be acceptable. The LS estimates from the AR(3) model (with estimated standard errors in parenthesis) are given as

$$\hat{\Phi}_1 = \begin{bmatrix} 0.708 & 0.469 \\ (0.125) & (0.127) \\ & \\ -0.699 & 1.151 \\ (0.133) & (0.135) \end{bmatrix}, \qquad \hat{\Phi}_2 = \begin{bmatrix} 0.277 & -0.680 \\ (0.193) & (0.185) \\ & \\ 0.316 & -0.380 \\ (0.206) & (0.197) \end{bmatrix},$$

$$\hat{\Phi}_3 = \begin{bmatrix} -0.504 & 0.332 \\ (0.149) & (0.121) \\ & \\ 0.050 & 0.108 \\ (0.159) & (0.129) \end{bmatrix}, \quad \text{with} \quad \tilde{\Sigma} = \begin{bmatrix} 0.049915 & 0.018759 \\ 0.018759 & 0.056482 \end{bmatrix}.$$

Clearly both coefficient estimates in the second row of the matrix $\hat{\Phi}_3$ are insignificant and they can be omitted from the model. The possibility of a low order mixed ARMA model for these data may also be explored, particularly because the values of the LR statistics at lags 4, 5 and 6 are moderately large, and also because the AIC_m values do not show a clear minimum at lag $m = 3$ but are relatively constant beyond lag 3. In fact, we find later in Section 5.7 that an ARMA(2,1) model provides a preferable fit.

4.4 Relation of LSE to Yule-Walker Estimate for Vector AR Models

We briefly discuss the relation between the LSE and Yule-Walker (YW) estimates for the vector AR model, considering first the AR(1) model. Recall that for a stationary vector AR(1) model, the Yule-Walker equation for Φ_1 is $\Gamma(0)\,\Phi_1' = \Gamma(1)$ and, hence, $\Phi_1' = \Gamma(0)^{-1}\,\Gamma(1)$. The sample version yields the Yule-Walker (YW) estimator of Φ_1 as $\hat{\Phi}_1' = \hat{\Gamma}(0)^{-1}\,\hat{\Gamma}(1)$, where $\hat{\Gamma}(l) = C(l) = T^{-1}\sum_{t=1}^{T-l}(\boldsymbol{Y}_t - \bar{\boldsymbol{Y}})(\boldsymbol{Y}_{t+l} - \bar{\boldsymbol{Y}})'$, $l = 0, 1, \ldots$. For the LSE of Φ_1 we have

$$\hat{\Phi}_1' = (\tilde{\mathbf{X}}'\tilde{\mathbf{X}})^{-1}\,\tilde{\mathbf{X}}'\tilde{\mathbf{Y}} = (N^{-1}\tilde{\mathbf{X}}'\tilde{\mathbf{X}})^{-1}\,N^{-1}\tilde{\mathbf{X}}'\tilde{\mathbf{Y}},$$

with $N = T - 1$, where

$$N^{-1}\tilde{\mathbf{X}}'\tilde{\mathbf{X}} = N^{-1}\sum_{t=2}^{T}(\boldsymbol{Y}_{t-1} - \bar{\boldsymbol{Y}}_{(1)})(\boldsymbol{Y}_{t-1} - \bar{\boldsymbol{Y}}_{(1)})',$$

$$N^{-1}\tilde{\mathbf{X}}'\tilde{\mathbf{Y}} = N^{-1}\sum_{t=2}^{T}(\boldsymbol{Y}_{t-1} - \bar{\boldsymbol{Y}}_{(1)})(\boldsymbol{Y}_t - \bar{\boldsymbol{Y}}_{(0)})'.$$

Now it is clear that for large T, $N^{-1}\tilde{\mathbf{X}}'\tilde{\mathbf{X}}$ and $\hat{\Gamma}(0)$ are asymptotically equivalent, as are $N^{-1}\tilde{\mathbf{X}}'\tilde{\mathbf{Y}}$ and $\hat{\Gamma}(1)$, since these two sets of quantities differ only by negligible "end effects". Hence, we see that the LS and the YW estimators of Φ_1 are essentially equivalent for large T. Also, from the relation $\Sigma = \Gamma(0) - \Gamma(1)'\,\Gamma(0)^{-1}\,\Gamma(1) = \Gamma(0) - \Phi_1\,\Gamma(0)\,\Phi_1'$, we obtain the "YW estimator" of Σ as $\hat{\Sigma} = \hat{\Gamma}(0) - \hat{\Phi}_1\,\hat{\Gamma}(0)\,\hat{\Phi}_1'$. Note that the LS estimator of Σ can be expressed as

$$\hat{\Sigma} = (N - k)^{-1}\,(\tilde{\mathbf{Y}} - \tilde{\mathbf{X}}\hat{\Phi}_1')'(\tilde{\mathbf{Y}} - \tilde{\mathbf{X}}\hat{\Phi}_1') = (N - k)^{-1}\,[\,\tilde{\mathbf{Y}}'\tilde{\mathbf{Y}} - \hat{\Phi}_1\,\tilde{\mathbf{X}}'\tilde{\mathbf{X}}\,\hat{\Phi}_1'\,],$$

where for large T it is clear that $(N - k)^{-1}\,\tilde{\mathbf{Y}}'\tilde{\mathbf{Y}} \approx (N - k)^{-1}\,\tilde{\mathbf{X}}'\tilde{\mathbf{X}} \approx \hat{\Gamma}(0)$. Hence, we see that the LS estimate is $\hat{\Sigma} \approx \hat{\Gamma}(0) - \hat{\Phi}_1\,\hat{\Gamma}(0)\,\hat{\Phi}_1'$, and so the LS and YW estimators of Σ are also essentially equivalent (except that as we have presented it here, the YW estimator has not been corrected for degrees of freedom).

It also follows that whether the process $\{\boldsymbol{Y}_t\}$ is AR(1) or not, when we compute the sample version of the partial autoregression matrix of order one, that is, an estimate of $\Phi_{11}' = \Gamma(0)^{-1}\,\Gamma(1)$, asymptotically equivalent estimators are obtained from the sample Yule-Walker estimator or the least squares autoregression estimator. And finally, since $N^{-1}\,\tilde{\mathbf{X}}'\tilde{\mathbf{X}} \approx \hat{\Gamma}(0)$, an asymptotically equivalent estimate for the covariance matrix of the estimator $\hat{\boldsymbol{\phi}} = \text{vec}(\hat{\Phi}_1')$ is $\widehat{\text{Cov}}(\hat{\boldsymbol{\phi}}) \approx \hat{\Sigma} \otimes N^{-1}\,\hat{\Gamma}(0)^{-1} \approx \hat{\Sigma} \otimes (\tilde{\mathbf{X}}'\tilde{\mathbf{X}})^{-1}$.

For the general AR(m) model, the *Yule-Walker estimates* of the Φ_j are obtained from the sample Yule-Walker equations, $\sum_{j=1}^{m}\hat{\Gamma}(l-j)\,\hat{\Phi}_j' = \hat{\Gamma}(l)$, $l = 1, 2, \ldots, m$. In matrix form, the YW estimator is $\hat{\Phi}_{(m)} = \hat{\Gamma}_m^{-1}\,\hat{\Gamma}_{(m)}$, where

$\hat{\Gamma}_{(m)} = [\ \hat{\Gamma}(1)', \ldots, \hat{\Gamma}(m)'\ ]'$, and $\hat{\Gamma}_m$ is $km \times km$ with $(\,i,\, j\,)$th block matrix equal to $\hat{\Gamma}(i-j)$. Of course, the YW estimates $\hat{\Phi}_j$ for the AR(m) model can be computed recursively using the same Whittle recursion algorithm as presented in (3.16)–(3.17) of Section 3.3.2, with the sample covariance matrices $\hat{\Gamma}(j)$ used in place of the $\Gamma(j)$. Now, concerning the LS estimator given in (4.11) and with $\tilde{\boldsymbol{Z}}_{t,m} = [\ (\boldsymbol{Y}_{t-1} - \bar{\boldsymbol{Y}}_{(1)})', \ldots, (\boldsymbol{Y}_{t-m} - \bar{\boldsymbol{Y}}_{(m)})'\ ]'$, when T is large and m is relatively small it is rather clear that $N^{-1}\tilde{\mathbf{X}}'\tilde{\mathbf{X}} = N^{-1}\sum_{t=m+1}^{T} \tilde{\boldsymbol{Z}}_{t,m}\,\tilde{\boldsymbol{Z}}_{t,m}' \approx \hat{\Gamma}_m$ and $N^{-1}\tilde{\mathbf{X}}'\tilde{\mathbf{Y}} = N^{-1}\sum_{t=m+1}^{T} \tilde{\boldsymbol{Z}}_{t,m}\,(\boldsymbol{Y}_t - \bar{\boldsymbol{Y}}_{(0)})' \approx \hat{\Gamma}_{(m)}$, so that the LSE of $\Phi_{(m)}$,

$$\hat{\Phi}_{(m)} = (\,\tilde{\mathbf{X}}'\tilde{\mathbf{X}}\,)^{-1}\,\tilde{\mathbf{X}}'\tilde{\mathbf{Y}} = (\,N^{-1}\tilde{\mathbf{X}}'\tilde{\mathbf{X}}\,)^{-1}\,N^{-1}\tilde{\mathbf{X}}'\tilde{\mathbf{Y}} \approx \hat{\Gamma}_m^{-1}\,\hat{\Gamma}_{(m)}\,.$$

That is, the least squares estimator of $\Phi_{(m)}$ is approximately equivalent to the Yule-Walker estimator of $\Phi_{(m)}$, and both estimators have the same asymptotic normal distribution as given previously in (4.13). For comparison of the two estimators, it may be instructive to note that the YW estimates can be viewed as estimates that are obtained using usual LS estimation for an AR(m) model but where the observed time series $\boldsymbol{Y}_1, \boldsymbol{Y}_2, \ldots, \boldsymbol{Y}_T$ is artificially extended at both ends, by appending m vector values equal to the sample mean vector $\bar{\boldsymbol{Y}} = T^{-1}\sum_{t=1}^{T} \boldsymbol{Y}_t$ to both the beginning and the end of the observed series of length T. This fact might provide some insight into the poorer finite sample behavior of the YW estimator relative to the LS estimator, due to the impact from the undesirable "end effects".

It also follows that when fitting an AR model of order m to the series, the LSE of the mth coefficient matrix is also an estimator of the partial autoregression matrix of order m, Φ_{mm}, and is asymptotically equivalent to that obtained from the sample Yule-Walker equations. That is, if the LSE from fitting an AR model of order m is denoted by $\hat{\Phi}_{(m)} = [\ \hat{\Phi}_{1m}, \hat{\Phi}_{2m}, \ldots, \hat{\Phi}_{mm}\ ]'$, then $\hat{\Phi}_{mm}$ is the sample estimator of the partial autoregressive matrix of order m, and we may use the likelihood ratio test statistic M_m in (4.14) [or asymptotically equivalent forms of the test statistic, such as (4.6) or (4.7), as discussed in Section 4.2] to test the overall significance of $\hat{\Phi}_{mm}$.

However, it should be mentioned that the preceding asymptotic normal distribution results for the least squares and the Yule-Walker estimators, and the asymptotic equivalence of these two estimators, apply only for the case of stationary AR processes. When the vector AR process is nonstationary, or near nonstationary, i.e., roots of $\det\{\,\Phi(B)\,\} = 0$ are on, or close to, the unit circle, it is known that the LS estimator of the Φ_j still performs consistently, whereas the Yule-Walker estimator may behave much more poorly with considerable bias, and, hence, the LSE is generally to be preferred. The consistency properties and asymptotic distribution theory of least squares estimators for nonstationary vector AR models which have roots of $\det\{\,\Phi(B)\,\} = 0$ on the unit circle have been established by Tsay and Tiao (1990), Sims, Stock, and Watson (1990), and Ahn and Reinsel (1990), among others. Consideration of certain special but important cases of the general nonstationary vector AR model, in which the only "nonstationary roots" in the AR operator $\Phi(B)$ are roots equal to one (unit roots), will be given in Section 6.3 and some detailed results concerning estimation of such models will be presented there.

4.5 Additional Techniques for Specification of Vector ARMA Models

After fitting AR models of each order $m = 1, 2, \ldots$ by least squares as discussed in Section 4.3, we may assess the appropriateness of an AR(m) model by considering the sequence of LR statistics M_m, and also by considering the diagonal elements of the estimated white noise covariance matrix $\hat{\Sigma}_m$ (that is, the estimated one-step ahead forecast error variances), which give an indication of how the model fit improves as the order of the AR model is increased. However, in addition, after each AR(m) fit, $m = 1, 2, \ldots, M$, one may also want to examine the cross-correlation matrices of the sample residuals $\hat{\varepsilon}_t$. These will give additional information as to the appropriateness of a fitted AR(m) model. The fitted AR(m) model would be deemed appropriate if the residual correlation matrices behave similar to those of a vector white noise process $\{\varepsilon_t\}$; formal overall goodness-of-fit tests for an estimated AR (or ARMA) model, based on correlation matrices of the residuals $\hat{\varepsilon}_t$ taken over several lags, can be applied and these will be discussed in Section 5.5. An alternative in the examination of the residual series $\hat{\varepsilon}_t$ for model adequacy is the use of AR order determination criteria, which will be discussed in the next subsection, applied to the residual series.

We should note that for many vector mixed ARMA processes, it may be necessary to fit pure AR models of rather high order before we obtain an approximate model that seems appropriate. (This is because the mixed ARMA model may have an infinite AR representation in which the coefficient matrices Π_j do not decay very quickly with increasing lag j.) However, it may not be desirable to use such a high order AR model as a final fitted model, due to the large number of estimated parameters involved with such a model. In such cases, we may inspect the pattern of residual correlations after fitting low order AR models, with the possibility that at least in some situations these will suggest a low order mixed ARMA model that may be appropriate for the series, with a resulting reduction in the number of model parameters required relative to a high order pure AR model.

For example, consider the vector ARMA(1,1) model $(I - \Phi B)\boldsymbol{Y}_t = (I - \Theta B)\varepsilon_t$. If an AR(1) model is fit to the series $\boldsymbol{Y}_t$, then the YW estimate $\hat{\Phi}_{11} = \hat{\Gamma}(1)' \hat{\Gamma}(0)^{-1}$ (or the LS estimate) will be an asymptotically biased estimate of Φ, and, in fact, $\hat{\Phi}_{11}$ converges in probability to $\Gamma(1)' \Gamma(0)^{-1} \equiv \Phi_{11}$. Thus, approximately, the residuals after the AR(1) fit, $\hat{\varepsilon}_t = \boldsymbol{Y}_t - \hat{\Phi}_{11} \boldsymbol{Y}_{t-1} \approx \boldsymbol{Y}_t - \Phi_{11} \boldsymbol{Y}_{t-1}$, will follow the "model" $\hat{\varepsilon}_t = (I - \Phi_{11} B)(I - \Phi B)^{-1}(I - \Theta B)\varepsilon_t$, which, in general, could have a complicated correlation structure not sufficiently close to that of an MA(1) process to enable ready identification. However, in some cases, Φ_{11} may not be too different from Φ, and the sample correlations of the residuals $\hat{\varepsilon}_t$ will behave approximately like those of an MA(1) model. Hence, in such a case, examination of residual correlations after the AR(1) fit would lead to the correct identification of the ARMA(1,1) model. In other cases, a more complex model might be incorrectly suggested. For another example, in the seasonal model $(I - \Phi B) \boldsymbol{Y}_t = (I - \Theta_{12} B^{12}) \varepsilon_t$, when an AR(1) model is fit, we obtain approximately $\Phi_{11} = \Gamma(1)' \Gamma(0)^{-1}$. But the generalized YW equation [see (2.14) of Section 2.3.2] yields $\Gamma(1)' = \Phi \Gamma(0) - \Theta_{12} \Sigma \Psi'_{11} = \Phi \Gamma(0) - \Theta_{12} \Sigma (\Phi')^{11}$, which may be close to $\Gamma(1)' = \Phi \Gamma(0)$ when Φ is not "too large", since then we might have $\Phi^{11} \approx 0$. Hence, in this case, we might expect to have $\Phi_{11} = \Gamma(1)' \Gamma(0)^{-1} = \Phi - \Theta_{12} \Sigma (\Phi')^{11} \Gamma(0)^{-1} \approx \Phi$.

4.5.1 Use of Order Selection Criteria for Model Specification

The likelihood ratio testing procedure and associated partial correlation and partial canonical correlation analyses are useful to determine the overall AR order in cases when a low order vector AR model is appropriate for the data. In more complicated situations, when a low order AR model does not seem to provide an adequate representation for the series, this might be taken as an indication that low order mixed ARMA models should be considered and such models can be estimated by maximum likelihood. Then, however, more general model selection procedures are needed to determine an adequate low order mixed ARMA model for the series. Even in the case of model selection for AR models, alternate procedures to LR testing methods may be more suitable for certain purposes. In general, various *model selection criteria* such as AIC, BIC, and FPE could be used to aid in the most appropriate choice of model. The (normalized by T) AIC model selection criterion (e.g., Akaike, 1974b, 1976) is given by

$$\begin{aligned} \text{AIC}_r &= [- 2 \log(\text{ maximized likelihood }) + 2r] / T \\ &\approx \log(|\tilde{\Sigma}_r|) + 2r / T + \text{constant}, \end{aligned}$$

where r denotes the number of parameters estimated by maximum likelihood in the vector ARMA model and $\tilde{\Sigma}_r$ is the corresponding ML residual covariance matrix estimate of Σ. The BIC criterion from Schwarz (1978) takes the similar form $\text{BIC}_r = \log(|\tilde{\Sigma}_r|) + r \log(T) / T$, and, hence, BIC_r imposes a greater "penalty factor" for the number of estimated model parameters than does AIC_r. A similar criterion, which was proposed by Hannan and Quinn (1979) and Quinn (1980) for AR models, is intermediate between AIC and BIC and is given by $\text{HQ}_r = \log(|\tilde{\Sigma}_r|) + 2r \log(\log(T)) / T$. The FPE ("final prediction error") criterion, suggested by Akaike (1971) for selection of vector AR(m) models, is $\text{FPE}_m = \det\{ [1 + (mk/T)] \hat{\Sigma}_m \}$, where $\hat{\Sigma}_m = (T/(T - mk)) \tilde{\Sigma}_m$ is the estimate of Σ adjusted for degrees of freedom. The FPE criterion is based on the result, derived in Section 5.6 [e.g., see (5.43)], that an approximate covariance matrix of one-step ahead forecast errors when forecasting from a vector AR(m) model with parameters that have been estimated (by LS or ML estimation) is $[\,1 + (mk/T)\,]\,\Sigma$.

These model selection criteria are used to compare various models fitted by maximum likelihood to the series such that the fitted model that yields a minimum value of the given criterion is chosen. Properties of order selection criteria such as AIC, BIC, and their generalizations applied to selection of the order in vector AR models have been investigated by Quinn (1980) and Paulsen (1984) and empirically by Lutkepohl (1985), among others, and by Hannan (1981) and Hannan and Deistler (1988, Chap. 5) for the more general case of fitting vector ARMA models. For instance, under a true stationary vector AR(p) model, Quinn (1980) showed that the order m which minimizes a criterion of the form $\phi(\,m\,) = \log(\,|\tilde{\Sigma}_m|\,) + 2\,m\,C_T \log(\log(T))\,/\,T$, $C_T > 1$, where $\tilde{\Sigma}_m$ is the ML residual covariance matrix estimate from fitting an mth-order AR model, is strongly consistent for the true AR order p as $T \rightarrow \infty$ if and only if lim sup $C_T > 1$. For the nonstationary vector AR model where some roots of $\det\{\,\Phi(B)\,\} = 0$ are equal to one, Paulsen (1984) established weak consistency for the AR order m selected by criteria of the form $\log(\,|\tilde{\Sigma}_m|\,) + m\,K_T\,/\,T$ if and only if K_T increases to $+\infty$ and $K_T/T \rightarrow 0$ as $T \rightarrow \infty$.

The use of order selection criteria, based on fitting vector AR models of various orders $m = 0, 1, \ldots,$ to a series, has also been proposed and studied by Pukkila and Krishnaiah (1988) as a testing procedure for assessing whether a series is a white noise process. In their procedure, the white noise null hypothesis for the series is accepted if the AR order selected by the use of the given order determination criterion (e.g., AIC, BIC, HQ) is equal to zero. This proposal has also been extended by Koreisha and Pukkila (1993), in the following way, for selection of the order of a vector AR model for a series $\boldsymbol{Y}_t$ by use of the procedure applied to the residuals $\hat{\varepsilon}_t$ from the fitted AR models. For this, vector AR models of given orders $m = 0, 1, \ldots$ are fitted by least squares to the series $\boldsymbol{Y}_t$, and the residuals $\hat{\varepsilon}_t(m)$ from the fitted AR model of order m are obtained. Then the procedure of Pukkila and Krishnaiah (1988) is applied to this residual series, and if the order selection criterion leads to selection of an AR model of order greater than zero for the residual series $\hat{\varepsilon}_t(m)$, then the residuals are viewed as not having satisfied the white noise test and, hence, an AR model of order m for the original series $\boldsymbol{Y}_t$ is rejected. The smallest AR order m for which the corresponding residual series $\hat{\varepsilon}_t(m)$ is accepted, by the procedure of AR fitting and model selection criterion, to be a white noise process is then selected as the appropriate AR order model for the series $\boldsymbol{Y}_t$.

4.5.2 Sample Canonical Correlation Analysis Methods

As an alternative to the direct estimation by maximum likelihood of various models, in more complex modeling cases, initial model specification techniques that involve canonical correlation and other related techniques, such as those discussed by Tiao and Tsay (1989), Cooper and Wood (1982), Akaike (1976), and others, may be considered. Specifically, in the approach of Tiao and Tsay

(1989) the developments of Section 3.2 are explored for the sample data. Thus, following (3.9) one is lead to examine the sample canonical correlations $\hat{\rho}_i(j)$ related to the matrix

$$(\textstyle\sum_t \mathbf{Y}_{m,t}\mathbf{Y}'_{m,t})^{-1}(\sum_t \mathbf{Y}_{m,t}\mathbf{Y}'_{n,t-j-1})$$

$$\times(\textstyle\sum_t \mathbf{Y}_{n,t-j-1}\mathbf{Y}'_{n,t-j-1})^{-1}(\sum_t \mathbf{Y}_{n,t-j-1}\mathbf{Y}'_{m,t}), \qquad (4.15)$$

where $\mathbf{Y}_{m,t} = (\boldsymbol{Y}'_t, \boldsymbol{Y}'_{t-1}, \ldots, \boldsymbol{Y}'_{t-m})'$ (and zero mean has been assumed), for various values of lag $j = 0, 1, \ldots$ and $m = 0, 1, \ldots$, and $n \geq m$.

Tiao and Tsay (1989) use a chi-squared test statistics approach, based on the smallest eigenvalues (squared sample canonical correlations) $\hat{\rho}_i^2(j)$ of (4.15), to test the number of zero canonical correlations. From this, they obtain a preliminary assessment not only of the overall orders p, q of the model but also information on more detailed and possibly simplifying features, such as reduced-rank structure and lower-order scalar component models, of the vector ARMA(p,q) model through the concept of scalar components models within this framework. As described in Section 3.2.2, within the vector ARMA(p,q) model structure for $\boldsymbol{Y}_t$, a linear combination $z_t = \boldsymbol{a}'\boldsymbol{Y}_t$ is said to follow a scalar component model of order (p^*, q^*), $p^* \leq p$, $q^* \leq q$, if the vector $\boldsymbol{a}$ has the properties that $\boldsymbol{a}'\Phi_{p^*} \neq 0$, $\boldsymbol{a}'\Phi_j = 0$ for $j > p^*$, $\boldsymbol{a}'\Theta_{q^*} \neq 0$, and $\boldsymbol{a}'\Theta_j = 0$ for $j > q^*$, so that $z_t - \sum_{j=1}^{p^*} \boldsymbol{a}'\Phi_j \boldsymbol{Y}_{t-j} = \boldsymbol{a}'\varepsilon_t - \sum_{j=1}^{q^*} \boldsymbol{a}'\Theta_j \varepsilon_{t-j}$. The approach of Tiao and Tsay (1989) is to identify (construct), through the sample canonical correlation analyses associated with (4.15), a set of k such linearly independent linear combinations of $\boldsymbol{Y}_t$ that follow scalar component models of lowest possible orders, thereby uncovering the simplifying structure in the parameters of the vector ARMA(p,q) model. The ability to reveal possible simplifying structures in the parameterization of the vector ARMA model is important because of the need to substantially reduce the number of parameters that require estimation in the model, if possible. Procedures related to the concept of scalar component models and canonical correlation analysis will be illustrated in more detail in Section 6.1 in connection with modeling through the use of reduced-rank vector AR models.

Canonical correlation methods have also been proposed previously for vector ARMA modeling by Akaike (1976) and Cooper and Wood (1982). Their approach is to use canonical correlation analysis to construct a minimal dimension state-space representation for the vector ARMA model, where the state vector consists of a basis of the prediction space at time t, the collection of forecasts $\hat{\boldsymbol{Y}}_t(l)$ for all lead times $l > 0$. This basis is determined by performing a canonical correlation analysis between the vector of present and past values, $\mathbf{U}_t = (\boldsymbol{Y}'_t, \boldsymbol{Y}'_{t-1}, \ldots, \boldsymbol{Y}'_{t-m})'$, and the vector of future values, $\mathbf{V}_t = (\boldsymbol{Y}'_{t+1}, \boldsymbol{Y}'_{t+2}, \ldots)'$. In practice, the finite lag m used to construct the vector of present and past values $\mathbf{U}_t$ may be fixed by use of an order determination criterion such as AIC applied to fitting of vector AR models of various orders. The canonical correlation analysis is performed sequentially by adding elements to

$\mathbf{V}_t$ one at a time (starting with $\mathbf{V}_t^* = (Y_{1,t+1})$) until k zero canonical correlations between $\mathbf{U}_t$ and the $\mathbf{V}_t$ are determined. Akaike (1976) uses an AIC-type criterion, called "DIC", to judge whether the smallest sample canonical correlation can be taken to be zero, while Cooper and Wood (1982) use a traditional chi-squared statistic approach to assess the significance of the smallest canonical correlation, although as pointed out by Tsay (1989a), to be valid in the presence of a moving average component this statistic needs to be modified.

At any given stage in the procedure, if the smallest sample canonical correlation between $\mathbf{U}_t$ and $\mathbf{V}_t^*$ is judged to be 0 and $Y_{i,t+K_i+1}$ is the most recent variable to be included in $\mathbf{V}_t^*$, then $\hat{Y}_{it}(K_i+1)$ is identified as being linearly dependent on the forecasts of the preceding elements in the vector $\mathbf{V}_t^*$ of future values. This is because a linear combination of $Y_{i,t+K_i+1}$ in terms of the remaining elements of $\mathbf{V}_t^*$ is identified which is uncorrelated with the past. Specifically, using the notation of (3.6) of Section 3.1.2, the linear combination $\phi_0(i)' \boldsymbol{Y}_{t+K_i+1} - \sum_{j=1}^{K_i} \phi_j(i)' \boldsymbol{Y}_{t+K_i+1-j}$ of the future values $\mathbf{V}_t^*$ is (in theory) determined to be uncorrelated with the past $\mathbf{U}_t$, where $\phi_j(i)'$ denotes the ith row of the AR coefficient matrix $\mathbf{\Phi}_j^{\#}$ in the echelon canonical ARMA model form (3.4). (Also note that in the sample canonical correlation analysis, the coefficients of the sample canonical variate for $\mathbf{V}_t^*$ corresponding to the smallest sample canonical correlation provide (initial) estimates for the unknown AR parameters of the $\phi_j(i)'$ in the ith row of the echelon canonical ARMA model form.) To continue the procedure, in the preceding case where the smallest sample canonical correlation is judged to be zero, the most recently added variable $Y_{i,t+K_i+1}$ is removed from $\mathbf{V}_t^*$ and all future values $Y_{i,t+K_i+j}$, $j \geq 1$, of the ith variable are also excluded from consideration in $\mathbf{V}_t^*$ (because their forecasts would also necessarily be dependent on the preceding forecasts). The procedure is then completed when k sample canonical correlations which are judged to be zero have been determined.

This canonical correlation analysis procedure will identify a basis for the prediction space of future values. Hence, existence of the zero canonical correlations at various stages in the sequential procedure can determine the special structure for the parameters of the vector ARMA model in the echelon canonical ARMA form (3.4) of Section 3.1.2, through association with the prediction relation for model (3.4) similar to (2.22) of Section 2.5.3,

$$\mathbf{\Phi}_0^{\#} \hat{\boldsymbol{Y}}_t(l) - \sum_{j=1}^{p} \mathbf{\Phi}_j^{\#} \hat{\boldsymbol{Y}}_t(l-j) = \boldsymbol{\delta} - \sum_{j=l}^{q} \Theta_j^{\#} \varepsilon_{t+l-j} . \tag{4.16}$$

Specifically, for the ith variable it is determined in the canonical correlation analysis procedure that $\hat{Y}_{it}(K_i+1)$ is linearly dependent on its predecessor forecasts, and, hence, from (4.16) this indicates that the ith rows of the AR matrices $\mathbf{\Phi}_j^{\#}$ must be zero for $j > K_i$, and similarly for the MA matrices $\Theta_j^{\#}$. That is, in the notation of equation (3.6) of Section 3.1.2, it is determined that the forecasts satisfy

$$\phi_0(i)' \hat{Y}_t(K_i+1) - \sum_{j=1}^{K_i} \phi_j(i)' \hat{Y}_t(K_i+1-j) = \delta_i, \quad \text{for} \quad i = 1, \ldots, k. \qquad (4.17)$$

In addition there is indicated the presence of zero elements in certain positions in the ith rows of the lower lag AR matrices $\Phi_j^{\#}$, $j \le K_i$, corresponding to variables which had been excluded from $\mathbf{V}_t^*$ in the canonical correlation analysis procedure. It has been noted by Akaike (1974c, 1976) that this canonical correlation analysis is equivalent in principle to determination of the first M (equal to the McMillan degree) linearly independent rows of the Hankel matrix H as discussed in Section 3.1. The corresponding (minimum) lags or indices K_1, $K_2, \ldots, K_k$, for which $\hat{Y}_{it}(K_i+1)$ is linearly dependent on the preceding forecasts, in the theoretical structure of the vector ARMA model are known as the Kronecker indices of the model. These quantities have been discussed in some detail in Section 3.1, and the connection between the structure of the ARMA model implied by the predictive relations (4.16) and by the covariance matrix relations (3.5) for model (3.4), or between relations (4.17) and (3.6), is direct. Thus, the above canonical correlation analysis procedure leads to specification of the Kronecker indices K_1, $K_2, \ldots, K_k$ and, hence, to specification of the corresponding echelon canonical form (3.4) of the vector ARMA model for the process $\{Y_t\}$. The state-space representation of the vector ARMA model in terms of prediction variables, the Kronecker indices of the model, and related notions will be discussed in further detail in Section 7.2.

4.5.3 Order Determination Using Linear LSE Methods for the Vector ARMA Model

Before we proceed, in the next chapter, to the discussion of maximum likelihood estimation techniques for the general vector ARMA model, we will consider one other technique that may be useful at the preliminary model specification stage. This is a multivariate generalization of a technique explored by Hannan and Rissanen (1982) for univariate ARMA models, and related methods in the vector case were discussed by Spliid (1983). The technique is also discussed for the vector case by Granger and Newbold (1986, Chap. 8).

In the proposed procedure, one first obtains estimates of the innovations series ε_t in a potential vector ARMA(p,q) model by approximation of the model by a (sufficiently high order) AR model of order m^*. The order m^* of the approximating AR model might be chosen by use of a selection criterion such as AIC, for example, which yields the value of m for which $\log(\,|\tilde{\Sigma}_m|\,) + 2\,mk^2/T$ is minimized. From the selected AR(m^*) model, one obtains residuals $\tilde{\varepsilon}_t = Y_t - \sum_{j=1}^{m^*} \hat{\Phi}_{jm^*} Y_{t-j}$. In the second stage of the procedure, one regresses Y_t on $Y_{t-1}, \ldots, Y_{t-p}$ and $\tilde{\varepsilon}_{t-1}, \ldots, \tilde{\varepsilon}_{t-q}$, for various values of p and q. That is, one estimates (approximate) models of the form

$$\boldsymbol{Y}_t = \sum_{j=1}^{p} \Phi_j \, \boldsymbol{Y}_{t-j} - \sum_{j=1}^{q} \Theta_j \, \tilde{\varepsilon}_{t-j} + \varepsilon_t \,, \tag{4.18}$$

by linear least squares regression, and let $\tilde{\Sigma}_{p,q}$ denote the estimated error covariance matrix (uncorrected for degrees of freedom) based on the ordinary least squares residuals obtained from estimation of (4.18). Then by application of the BIC criterion (Schwarz, 1978), the order (p,q) of the ARMA model is chosen as the one which minimizes $\log(\, | \tilde{\Sigma}_{p,q} | \,) + (\, p + q \,) \, k^2 \log(T)/T$. Use of this procedure may lead to one or two ARMA models that seem highly promising, and these models can be finally estimated by maximum likelihood procedures to be discussed in the next chapter. The models can subsequently be checked by performing residual analysis checks for model adequacy. The appeal of the procedure is that computation of maximum likelihood estimates, which are much more computationally expensive to obtain, over a wide range of possible ARMA models is avoided. Furthermore, the parameter estimates obtained by this procedure are generally fairly efficient relative to maximum likelihood and provide excellent starting values for the maximum likelihood iterations.

Similar linear estimation procedures have been proposed by Hannan and Kavalieris (1984) and Poskitt (1992), among others, for specification of the echelon form (3.4) of the vector ARMA model. Briefly, in the second stage (approximate) models of the echelon form (3.4) (with $\Phi_0^{\#} = \Theta_0^{\#}$ lower triangular and having ones on the diagonal),

$$\boldsymbol{Y}_t = (\, I - \Phi_0^{\#} \,) \, (\, \boldsymbol{Y}_t - \tilde{\varepsilon}_t \,) + \sum_{j=1}^{p} \Phi_j^{\#} \, \boldsymbol{Y}_{t-j} - \sum_{j=1}^{q} \Theta_j^{\#} \, \tilde{\varepsilon}_{t-j} + \varepsilon_t \,, \tag{4.19}$$

where the $\tilde{\varepsilon}_t$ are residuals from the first stage AR model fitting, are estimated by linear least squares regression, for a variety of different specifications of the set of Kronecker indices $\{ K_1, \ldots, K_k \}$ and, hence, a variety of different echelon ARMA form models. A model selection criterion such as BIC is evaluated for each estimated model, and the Kronecker index structure and corresponding echelon form model is chosen as the one which minimizes the selection criterion. This procedure will suggest a few distinct ARMA model structures as most likely, and these models can be estimated by maximum likelihood methods, compared, and checked for adequacy by methods discussed in the next chapter.

More recently, for the univariate model case Kavalieris (1991) has proposed an improvement in the above type of model selection procedures that are based on linear least squares estimation in the second stage. In the procedures described above, such as the one associated with (4.18), the error covariance matrix estimate $\tilde{\Sigma}_{p,q}$ is based on the least squares residuals $\hat{\varepsilon}_t = \boldsymbol{Y}_t - \sum_{j=1}^{p} \hat{\Phi}_j \, \boldsymbol{Y}_{t-j} + \sum_{j=1}^{q} \hat{\Theta}_j \, \tilde{\varepsilon}_{t-j}$, where $\hat{\Phi}_j$ and $\hat{\Theta}_j$ are the least squares estimates obtained from (4.18) in the second stage. In the modification, the estimate $\tilde{\Sigma}_{p,q}$ is based on the innovations as computed recursively, in the usual way, from the ARMA(p,q) model as $\hat{\varepsilon}_t = \boldsymbol{Y}_t - \sum_{j=1}^{p} \hat{\Phi}_j \boldsymbol{Y}_{t-j} + \sum_{j=1}^{q} \hat{\Theta}_j \hat{\varepsilon}_{t-j}$, $t = 1, \ldots, T$, instead of on ordinary least squares residuals with the ε_{t-j} as regressor variables, and, hence, $\tilde{\Sigma}_{p,q} = T^{-1} \sum_{t=1}^{T} \hat{\varepsilon}_t \hat{\varepsilon}_t'$. (This recursive computation of the innovations $\hat{\varepsilon}_t$ will be discussed in Chapter 5 in relation to ML estimation of ARMA models.) In the modified procedure, this alternate residual covariance matrix estimate $\tilde{\Sigma}_{p,q}$ is then used in the evaluation of the model selection criterion such as BIC. For some simulated univariate ARMA(p,q) examples, Kavalieris (1991) has indicated that this modification improves on the performance of the model selection procedures in terms of increased percentage of correct selection of the model orders.

EXAMPLE 4.2. We consider the bivariate time series of U.S. fixed investment and change in business inventories. These data are quarterly, seasonally adjusted, and have been given in Lutkepohl (1991). The data for the time period 1947–1971 are plotted in Figure 4.2. Since the investment series is clearly nonstationary, the first differences of this series (also displayed in Figure 4.2) are considered for analysis as series Y_{1t} together with the change in business inventories as series Y_{2t}, resulting in $T = 99$ quarterly observations. Sample cross-correlation matrices of the series $\boldsymbol{Y}_t = (Y_{1t}, Y_{2t})'$ for lags 1 through 12 are shown in Table 4.3, and these sample autocorrelations and cross-correlations are also displayed up to 18 lags in Figure 4.3. These correlations show some exponential decaying and damped sinusoidal behavior as a function of lag l, and a low order pure MA model is not suggested by the patterns of the correlation matrices. Fitting AR models of orders $m = 1, \ldots, 6$ by least squares yields the results on the LR test statistic M_m of (4.14) for testing $H_0 : \Phi_m = 0$ in an AR(m) model, as well as values of the AIC and HQ model selection criteria, which are presented in Table 4.4. These results all indicate that, among pure AR models, a second-order AR(2) model is the most appropriate for these data.

Table 4.3 Sample Correlation Matrices $\hat{\rho}(l)$ for the Bivariate Quarterly Series of First Differences of U.S. Fixed Investment and U.S. Changes in Business Inventories.

l	1		2		3		4		5		6	
$\hat{\rho}(l)$	0.47	0.25	0.10	0.33	−0.12	0.28	−0.32	0.27	−0.30	0.20	−0.21	0.07
	−0.04	0.70	−0.32	0.50	−0.29	0.32	−0.21	0.11	−0.10	0.09	0.10	0.08
l	7		8		9		10		11		12	
$\hat{\rho}(l)$	−0.14	−0.01	−0.09	−0.10	0.13	−0.02	0.19	0.08	0.13	0.11	0.03	0.19
	0.20	0.07	0.19	0.06	0.11	0.06	0.06	0.12	0.02	0.12	−0.03	0.10

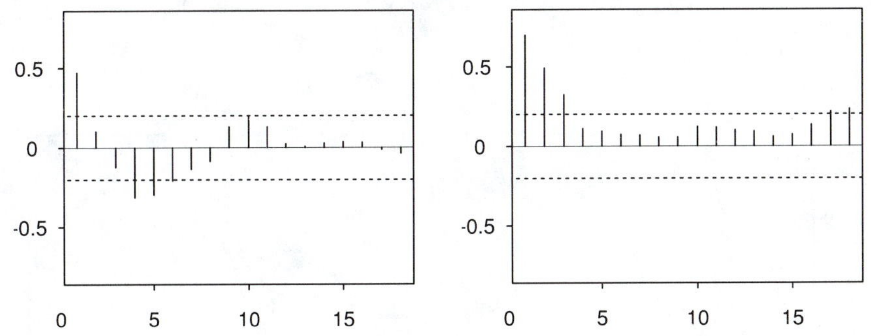

(a) Sample Autocorrelations $\hat{\rho}_{11}(l)$ and $\hat{\rho}_{22}(l)$

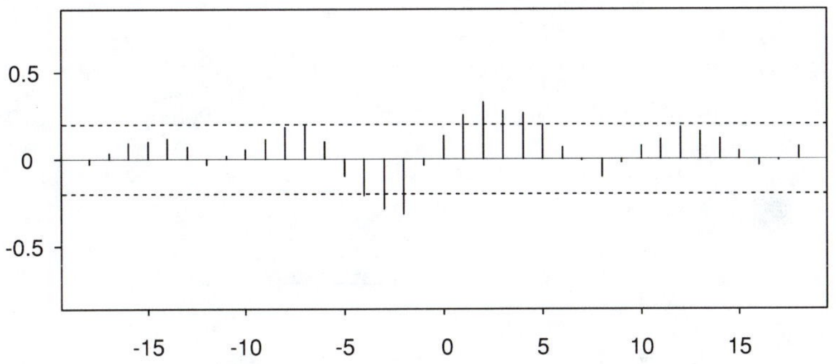

(b) Sample Cross-correlations $\hat{\rho}_{12}(l)$

Figure 4.3. Sample Auto- and Cross-correlations, $\hat{\rho}_{ij}(l)$, for the Bivariate Series of First Differences of U.S. Fixed Investment and U.S. Changes in Business Inventories

Table 4.4 Summary of Results from Fitting Autoregressive Models to the U.S. Business Investment and Inventories Data. [$\text{AIC}_m = \log(\,|\tilde{\Sigma}_m|\,) + 2\,m\,k^2/N$, and $\text{HQ}_m = \log(\,|\tilde{\Sigma}_m|\,) + 2\,m\,k^2 \log(\log(N))/N$, $N = T - m$.]

m (AR order)	1	2	3	4	5	6
$\lvert\tilde{\Sigma}_m\rvert$ ($\times 10^2$)	0.940432	0.756595	0.733917	0.676727	0.642977	0.645372
M_m statistic	97.04	20.32	2.57	7.55	5.42	0.88
AIC_m	4.625	4.491	4.546	4.552	4.589	4.683
HQ_m	4.668	4.577	4.675	4.725	4.808	4.947

The LS estimates from the AR(2) model (with estimated standard errors in parenthesis), as well as the (conditional) ML estimate of Σ, are given as

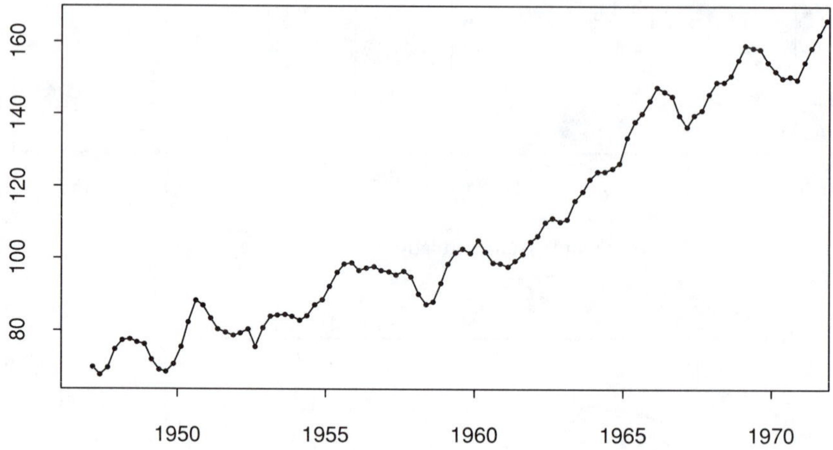

(a) Y^*_{1t} : Fixed Investment Series

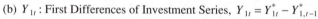

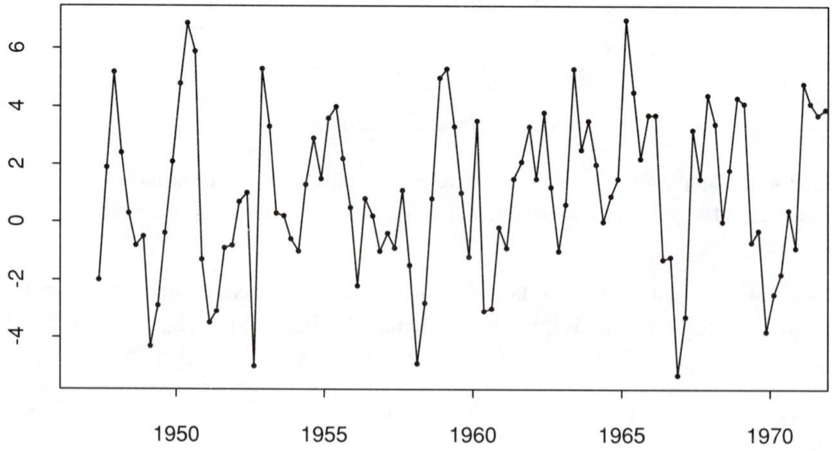

(b) Y_{1t} : First Differences of Investment Series, $Y_{1t} = Y^*_{1t} - Y^*_{1,t-1}$

Figure 4.2. Quarterly (Seasonally Adjusted) U.S. Fixed Investment and Business Inventories Data (in billions) for the Period 1947 Through 1971

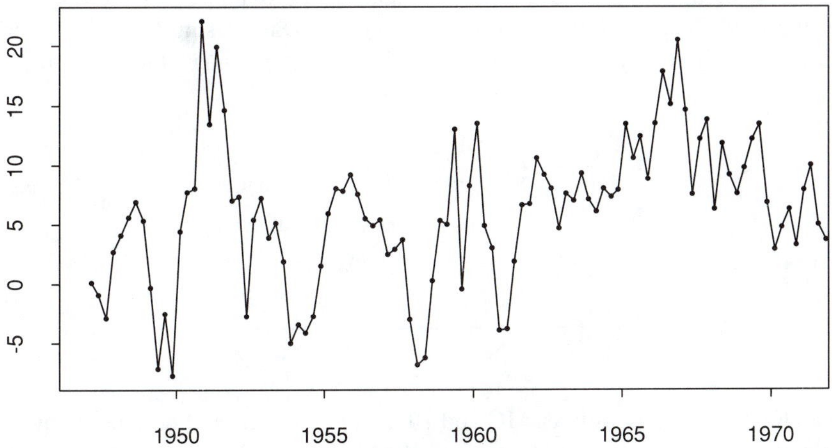

(c) Y_{2t} : Changes in Business Inventories Series

Figure 4.2. (continued)

$$\hat{\Phi}_1 = \begin{bmatrix} \underset{(0.093)}{0.502} & \underset{(0.057)}{0.131} \\ \\ \underset{(0.166)}{0.234} & \underset{(0.101)}{0.616} \end{bmatrix}, \qquad \hat{\Phi}_2 = \begin{bmatrix} \underset{(0.095)}{-0.138} & \underset{(0.055)}{-0.226} \\ \\ \underset{(0.168)}{0.240} & \underset{(0.098)}{0.050} \end{bmatrix},$$

$$\tilde{\Sigma} = \begin{bmatrix} 4.9690 & 1.6449 \\ 1.6449 & 15.7707 \end{bmatrix},$$

with $\det(\tilde{\Sigma}) = 75.6595$. The coefficient estimates in the second row of the matrix $\hat{\Phi}_2$, in particular, are not significant and could be omitted from the model, and possibly one or two other coefficients as well.

We also consider the possibility of a mixed ARMA model for these data by use of the linear least squares regression estimation methods, associated with the LS fitting of (4.18), for model specification, as discussed earlier in this section. The AR order m^* used at the first stage was chosen as $m^* = 5$, and linear least squares regression estimates were obtained for ARMA(p,q) models for each combination of $p = 0, 1, 2, 3$ and $q = 0, 1, 2$. The resulting values for the Quinn (1980) criterion are presented in Table 4.5, where the error covariance matrix estimate $\tilde{\Sigma}_{p,q}$ was obtained by the modification method suggested by Kavalieris (1991), as discussed earlier.

Table 4.5 Summary of Results for Preliminary ARMA(p,q) Order Determination Procedure, Based on Linear Least Squares Regression Estimation Method, for the U.S. Business Investment and Inventories Data, Using Quinn's Criteria with Values Given by $\mathrm{HQ}(p, q) = \log(|\tilde{\Sigma}_{p,q}|) + 2(p+q)k^2 \log(\log(T))/T$.

	q		
p	0	1	2
0	5.573	4.968	4.979
1	4.669	4.589	4.707
2	4.578	4.708	4.860
3	4.679	4.808	5.183

The HQ criterion, as well as AIC and BIC, applied to these linear least squares estimation results all suggest that an ARMA(1,1) model might be essentially equivalent to the AR(2) model in terms of fit, and that these two models are clearly superior to the other models considered. Hence, the ARMA(1,1) model was estimated by (conditional) maximum likelihood estimation methods, as will be discussed subsequently in Section 5.1, and the results are given as

$$\hat{\Phi}_1 = \begin{bmatrix} \underset{(0.157)}{0.421} & \underset{(0.067)}{-0.206} \\ \underset{(0.194)}{0.532} & \underset{(0.082)}{0.738} \end{bmatrix}, \qquad \hat{\Theta}_1 = \begin{bmatrix} \underset{(0.178)}{-0.072} & \underset{(0.083)}{-0.331} \\ \underset{(0.260)}{0.326} & \underset{(0.125)}{0.124} \end{bmatrix},$$

$$\tilde{\Sigma} = \begin{bmatrix} 4.9038 & 1.6532 \\ 1.6532 & 15.7847 \end{bmatrix},$$

with $\det(\tilde{\Sigma}) = 74.6724$. Again, the coefficient estimates in the second row of the matrix $\hat{\Theta}_1$ are not significant and could be omitted from the model. It is clear from these estimation results that the ARMA(1,1) model provides a nearly equivalent fit to the AR(2) model. For instance, we consider the coefficient matrices Ψ_j in the infinite MA representation for $\boldsymbol{Y}_t$ implied by the AR(2) and ARMA(1,1) models. For the AR(2) model, the Ψ_j are determined from $\Psi_1 = \Phi_1$, $\Psi_j = \Phi_1 \Psi_{j-1} + \Phi_2 \Psi_{j-2}$ for $j > 1$ ($\Psi_0 = I$), and, hence, the Ψ_j are given as

$$\Psi_1 = \begin{bmatrix} 0.50 & 0.13 \\ 0.23 & 0.62 \end{bmatrix}, \quad \Psi_2 = \begin{bmatrix} 0.14 & -0.08 \\ 0.50 & 0.46 \end{bmatrix}, \quad \Psi_3 = \begin{bmatrix} 0.02 & -0.14 \\ 0.47 & 0.33 \end{bmatrix},$$

$$\Psi_4 = \begin{bmatrix} -0.06 & -0.12 \\ 0.36 & 0.17 \end{bmatrix}, \quad \Psi_5 = \begin{bmatrix} -0.09 & -0.09 \\ 0.23 & 0.06 \end{bmatrix}, \quad \Psi_6 = \begin{bmatrix} -0.09 & -0.06 \\ 0.12 & -0.00 \end{bmatrix},$$

and so on, while those for the ARMA(1,1) model are determined from $\Psi_1 = \Phi_1 - \Theta_1$, $\Psi_j = \Phi_1 \Psi_{j-1}$, $j > 1$, and so are given as

$$\Psi_1 = \begin{bmatrix} 0.49 & 0.13 \\ 0.21 & 0.61 \end{bmatrix}, \quad \Psi_2 = \begin{bmatrix} 0.16 & -0.07 \\ 0.41 & 0.52 \end{bmatrix}, \quad \Psi_3 = \begin{bmatrix} -0.02 & -0.14 \\ 0.39 & 0.34 \end{bmatrix},$$

$$\Psi_4 = \begin{bmatrix} -0.09 & -0.13 \\ 0.28 & 0.18 \end{bmatrix}, \quad \Psi_5 = \begin{bmatrix} -0.09 & -0.09 \\ 0.16 & 0.06 \end{bmatrix}, \quad \Psi_6 = \begin{bmatrix} -0.07 & -0.05 \\ 0.07 & -0.00 \end{bmatrix}.$$

Thus, we see that the Ψ_j coefficient matrices are very similar for both models, implying, in particular, that forecasts $\hat{\boldsymbol{Y}}_t(l)$ and the covariance matrices $\Sigma(l) = \sum_{j=0}^{l-1} \Psi_j \, \Sigma \, \Psi_j'$ of the l-step ahead forecast errors $\boldsymbol{e}_t(l) = \boldsymbol{Y}_{t+l} - \hat{\boldsymbol{Y}}_t(l)$ obtained from the two models, AR(2) and ARMA(1,1), are nearly identical.

We now examine the residuals $\hat{\varepsilon}_t$ from the fitted AR(2) model, for example. In particular, the residual correlation matrices $\hat{\rho}_\varepsilon(l)$, which will be discussed in Section 5.5, were obtained for the AR(2) model and are presented in Table 4.6.

Table 4.6 Residual Correlation Matrices $\hat{\rho}_\varepsilon(l)$ from the AR(2) Model Fitted to the Series of First Differences of U.S. Fixed Investment and U.S. Changes in Business Inventories Data.

l	1		2		3		4		5		6	
$\hat{\rho}_\varepsilon(l)$	0.01	0.01	0.06	−0.02	0.03	0.02	−0.10	0.09	−0.09	0.13	−0.01	−0.02
	0.02	0.01	−0.02	−0.04	0.11	0.06	−0.04	−0.20	−0.03	0.08	0.07	0.09
l	7		8		9		10		11		12	
$\hat{\rho}_\varepsilon(l)$	0.03	0.07	−0.15	−0.13	0.21	0.07	0.07	0.12	0.14	0.02	0.01	0.12
	0.08	0.03	0.08	−0.09	0.05	−0.17	0.08	0.05	0.03	0.04	−0.16	−0.01

One notable feature of these residual correlations is the (marginally) significant correlation at lag 4 for the second residual series $\hat{\varepsilon}_{2t}$. This feature may be related to a weak seasonal structure that may still exist in the quarterly ("seasonally adjusted") series $\boldsymbol{Y}_t$, and so to accommodate this feature we consider a modification to the AR(2) model by inclusion of a MA coefficient matrix Θ_4 at the quarterly seasonal lag of 4 in the model, i.e., we consider the model of the form $\boldsymbol{Y}_t - \Phi_1 \boldsymbol{Y}_{t-1} - \Phi_2 \boldsymbol{Y}_{t-2} = \boldsymbol{\delta} + \varepsilon_t - \Theta_4 \varepsilon_{t-4}$. After eliminating the off-diagonal coefficient terms from the ML estimate $\hat{\Theta}_4$ and the coefficient term in the (2,1) position of $\hat{\Phi}_2$, which were found to be clearly nonsignificant, and reestimating the simplified model, we arrive at the estimated model given by

$$\hat{\Phi}_1 = \begin{bmatrix} 0.470 & 0.145 \\ (0.097) & (0.059) \\ & \\ 0.385 & 0.652 \\ (0.154) & (0.101) \end{bmatrix}, \qquad \hat{\Phi}_2 = \begin{bmatrix} -0.135 & -0.184 \\ (0.096) & (0.057) \\ & \\ 0.000 & 0.157 \\ (\text{---}) & (0.099) \end{bmatrix},$$

$$\hat{\Theta}_4 = \begin{bmatrix} 0.306 & 0.000 \\ (0.105) & (\text{---}) \\ & \\ 0.000 & 0.457 \\ (\text{---}) & (0.099) \end{bmatrix}, \quad \text{with} \quad \tilde{\Sigma} = \begin{bmatrix} 4.8443 & 1.6873 \\ 1.6873 & 14.3734 \end{bmatrix},$$

and $\det(\tilde{\Sigma}) = 66.7815$. Examination of the residuals from this fitted model gives no indication of inadequacy of the model, and so this model is accepted as an adequate representation for the bivariate series. The fitted model implies that the changes in business inventories series Y_{2t} has a significant influence on the (first differences of) investments, but there appears to be less influence in the feedback from investments to the changes in inventories series. In addition, there is only a small degree of contemporaneous correlation suggested, since the correlation between the residual series $\hat{\varepsilon}_{1t}$ and $\hat{\varepsilon}_{2t}$ can be estimated from $\tilde{\Sigma}$ as equal to 0.202. Similar overall model fitting results are also obtained when the ARMA$(1,1)$ model is modified to accommodate the correlation structure at the seasonal lag of 4 by considering the model of the form $\boldsymbol{Y}_t - \Phi_1\, \boldsymbol{Y}_{t-1} = \boldsymbol{\delta} + \varepsilon_t - \Theta_1\, \varepsilon_{t-1} - \Theta_4\, \varepsilon_{t-4}$.

APPENDIX A4

Review of the General Multivariate Linear Regression Model

Let $\boldsymbol{Y}_t = (Y_{1t}, \ldots, Y_{kt})'$ be a k-dimensional random vector of response variables and $\boldsymbol{X}_t = (x_{1t}, \ldots, x_{rt})'$ be r-dimensional input variables. We consider a multivariate linear model of the form $Y_{it} = \boldsymbol{X}_t' \beta_i + \varepsilon_{it}$, $i = 1, \ldots, k$, or

$$\boldsymbol{Y}_t' = \boldsymbol{X}_t' B + \varepsilon_t', \qquad t = 1, \ldots, T, \tag{A4.1}$$

where $B = (\beta_1, \ldots, \beta_k)$ and the $\varepsilon_t = (\varepsilon_{1t}, \ldots, \varepsilon_{kt})'$ are distributed independently as multivariate normal $N(0, \Sigma)$, so that $\Sigma = E(\varepsilon_t \varepsilon_t')$. Now given the T observations $\boldsymbol{Y}_1, \ldots, \boldsymbol{Y}_T$ and $\boldsymbol{X}_1, \ldots, \boldsymbol{X}_T$, we define the $T \times k$ data matrix $\mathbf{Y} = (\boldsymbol{Y}_1, \ldots, \boldsymbol{Y}_T)'$, the $T \times r$ matrix $\mathbf{X} = (\boldsymbol{X}_1, \ldots, \boldsymbol{X}_T)'$, and the $T \times k$ matrix $\boldsymbol{\varepsilon} = (\varepsilon_1, \ldots, \varepsilon_T)'$. Then we have the multivariate linear model $\mathbf{Y} = \mathbf{X} B + \boldsymbol{\varepsilon}$, with $B = (\beta_1, \ldots, \beta_k)$. The ith column of B, β_i, is the vector of regression coefficients for the ith response variable. Now it is well known that the *maximum likelihood estimator* (MLE) of B is the same as the least squares (LS) estimator and, hence, is given by

$$\hat{B} = (\mathbf{X}'\mathbf{X})^{-1} \mathbf{X}'\mathbf{Y}, \qquad \text{that is}, \qquad \hat{\beta}_i = (\mathbf{X}'\mathbf{X})^{-1} \mathbf{X}'\mathbf{y}_i, \qquad i = 1, \ldots, k, \tag{A4.2}$$

where $\mathbf{y}_i = (Y_{i1}, \ldots, Y_{iT})'$ is the ith column of $\mathbf{Y}$. Also, the usual unbiased estimator of the error covariance matrix Σ is

$$\begin{aligned} \hat{\Sigma} &= \frac{1}{T - r} (\mathbf{Y} - \mathbf{X} \hat{B})'(\mathbf{Y} - \mathbf{X} \hat{B}) \\ &= \frac{1}{T - r} \sum_{t=1}^{T} (\boldsymbol{Y}_t - \hat{B}' \boldsymbol{X}_t)(\boldsymbol{Y}_t - \hat{B}' \boldsymbol{X}_t)' \end{aligned} \tag{A4.3}$$

or $\hat{\Sigma} = (T - r)^{-1} \sum_{t=1}^{T} \hat{\varepsilon}_t \hat{\varepsilon}_t'$, where the $\hat{\varepsilon}_t = \boldsymbol{Y}_t - \hat{B}' \boldsymbol{X}_t$ are the residual vectors. Note that the *likelihood function* is

$$\begin{aligned} &L(B, \Sigma ; \mathbf{Y}) \\ &\quad = (2\pi)^{-kT/2} \, |\Sigma|^{-T/2} \exp[(-1/2) \textstyle\sum_{t=1}^{T} (\boldsymbol{Y}_t - B' \boldsymbol{X}_t)' \Sigma^{-1} (\boldsymbol{Y}_t - B' \boldsymbol{X}_t)], \end{aligned}$$

which can be shown to be maximized for B equal to $\hat{B}$ in (A4.2) and Σ equal to $\tilde{\Sigma} = [(T - r)/T] \hat{\Sigma}$.

A4.1 Properties of the Maximum Likelihood Estimator of the Regression Matrix

We now consider the distributional properties of the MLE $\hat{B}$. One convenient way to do this is to express the multivariate linear model in the form of a univariate model. We introduce the "*vec*" *operator* which transforms a matrix into a column vector by stacking the columns of the matrix below each other. Thus, with the $T \times k$ matrix $\mathbf{Y} = (\mathbf{y}_1, \ldots, \mathbf{y}_k)$ defined above, we have

$\mathbf{y} = \text{vec}(\mathbf{Y}) = (\mathbf{y}_1', \ldots, \mathbf{y}_k')'$, and $\mathbf{e} = \text{vec}(\boldsymbol{\varepsilon}) = (\mathbf{e}_1', \ldots, \mathbf{e}_k')'$, where $\mathbf{e}_i$ is the ith column of $\boldsymbol{\varepsilon}$. We also define the *Kronecker product* of two matrices as follows. Let $A = (a_{ij})$ be $m \times n$ and $C = (c_{ij})$ be $p \times q$. Then $A \otimes C$ is an $mp \times nq$ matrix of the form $A \otimes C = [(a_{ij} C)]$ called the Kronecker product of A and C. A useful property which relates the "vec" operation and the Kronecker product is that if $Z = A\,B\,C$, then

$$\text{vec}(Z) = \text{vec}(A\,B\,C) = (C' \otimes A)\,\text{vec}(B).$$

We also have the fundamental relations $(A \otimes B)(C \otimes D) = (A\,C \otimes B\,D)$, when the matrices are conformable, and $(A \otimes B)^{-1} = (A^{-1} \otimes B^{-1})$, when inverses exist. Also, $\text{tr}(A\,B'\,C\,B) = [\text{vec}(B)]'\,(A \otimes C')\,\text{vec}(B)$.

Now from $\mathbf{Y} = \mathbf{X}\,B + \boldsymbol{\varepsilon}$, we have

$$\mathbf{y} = \text{vec}(\mathbf{Y}) = \text{vec}(\mathbf{X}\,B) + \text{vec}(\boldsymbol{\varepsilon}) = (I_k \otimes \mathbf{X})\,\boldsymbol{\beta} + \mathbf{e},$$

where $\boldsymbol{\beta} = \text{vec}(B) = (\beta_1', \ldots, \beta_k')'$, and $\mathbf{e} = \text{vec}(\boldsymbol{\varepsilon}) = (\mathbf{e}_1', \ldots, \mathbf{e}_k')'$. Note that $\text{Cov}(\mathbf{e}) = \Omega = \Sigma \otimes I_T = [(\sigma_{ij}\,I_T)]$, that is, $\text{Cov}(\mathbf{e}_i, \mathbf{e}_j) = \sigma_{ij}\,I_T$, where σ_{ij} denotes the (i, j)th element of Σ. The above model has the form of a univariate general linear model, so it is known that the MLE of $\boldsymbol{\beta}$ is the same as the generalized least squares (GLS) estimator of $\boldsymbol{\beta}$,

$$\begin{aligned}\hat{\boldsymbol{\beta}} &= [(I_k \otimes \mathbf{X})'\,\Omega^{-1}\,(I_k \otimes \mathbf{X})]^{-1}\,(I_k \otimes \mathbf{X})'\,\Omega^{-1}\,\mathbf{y} \\ &= [(\Sigma^{-1} \otimes \mathbf{X}'\mathbf{X})]^{-1}\,(\Sigma^{-1} \otimes \mathbf{X}')\,\mathbf{y} \\ &= [(I_k \otimes \mathbf{X}'\mathbf{X})]^{-1}\,(I_k \otimes \mathbf{X}')\,\mathbf{y} = (I_k \otimes (\mathbf{X}'\mathbf{X})^{-1}\,\mathbf{X}')\,\mathbf{y},\end{aligned}$$

so it is seen that $\hat{\beta}_i = (\mathbf{X}'\mathbf{X})^{-1}\,\mathbf{X}'\,\mathbf{y}_i$, $i = 1, \ldots, k$, the usual LS estimator for each i. It is also immediately seen from the above relation that $\hat{\boldsymbol{\beta}}$ has a multivariate normal distribution with mean

$$E(\hat{\boldsymbol{\beta}}) = (I_k \otimes (\mathbf{X}'\mathbf{X})^{-1}\,\mathbf{X}')\,E(\mathbf{y}) = \boldsymbol{\beta}$$

and

$$\begin{aligned}\text{Cov}(\hat{\boldsymbol{\beta}}) &= (I_k \otimes (\mathbf{X}'\mathbf{X})^{-1}\,\mathbf{X}')\,(\Sigma \otimes I_T)\,(I_k \otimes \mathbf{X}\,(\mathbf{X}'\mathbf{X})^{-1}) \\ &= \Sigma \otimes (\mathbf{X}'\mathbf{X})^{-1},\end{aligned} \tag{A4.4}$$

that is, $E(\hat{\beta}_i) = \beta_i$, $\text{Cov}(\hat{\beta}_i, \hat{\beta}_j) = \sigma_{ij}\,(\mathbf{X}'\mathbf{X})^{-1}$. In practice, since Σ is unknown, the covariance matrix of $\hat{\boldsymbol{\beta}}$ is estimated by $\widehat{\text{Cov}}(\hat{\boldsymbol{\beta}}) = \hat{\Sigma} \otimes (\mathbf{X}'\mathbf{X})^{-1}$. Also, note that the correlation matrix of $\hat{\boldsymbol{\beta}}$ is given by

$$\text{Corr}(\hat{\boldsymbol{\beta}}) = \rho(\Sigma) \otimes \rho((\mathbf{X}'\mathbf{X})^{-1}),$$

where $\rho(\Sigma) = P\,\Sigma\,P$ with $P = \text{Diag}\{\sigma_{11}^{-1/2}, \ldots, \sigma_{kk}^{-1/2}\}$ and $\rho((\mathbf{X}'\mathbf{X})^{-1}) = Q\,(\mathbf{X}'\mathbf{X})^{-1}\,Q$ with $Q = \text{Diag}\{a_{11}^{-1/2}, \ldots, a_{rr}^{-1/2}\}$ and the a_{ii} are the diagonal elements of $(\mathbf{X}'\mathbf{X})^{-1}$.

A4.2 Likelihood Ratio Test of Linear Hypothesis About Regression Coefficients

Consider $\mathbf{X}$ partitioned as $\mathbf{X} = (\mathbf{X}_1, \mathbf{X}_2)$ and corresponding $B = (B_1', B_2')'$, so that $\mathbf{Y} = \mathbf{X}_1 B_1 + \mathbf{X}_2 B_2 + \boldsymbol{\varepsilon}$, where $\mathbf{X}_1$ is $T \times r_1$, $\mathbf{X}_2$ is $T \times r_2$. It may typically be of interest in the multivariate linear model context to test the null hypothesis $H_0 : B_2 = 0$ against the alternative $B_2 \neq 0$. Using the *likelihood ratio* (LR) testing approach, it is easy to show that the LR test statistic for testing $B_2 = 0$ is $\lambda = U^{T/2}$, where $U = \det(S)/\det(S_1)$,

$$S = (\mathbf{Y} - \mathbf{X}\hat{B})'(\mathbf{Y} - \mathbf{X}\hat{B}), \qquad S_1 = (\mathbf{Y} - \mathbf{X}_1\tilde{B}_1)'(\mathbf{Y} - \mathbf{X}_1\tilde{B}_1),$$

S is the residual sum of squares matrix from fitting the full regression model, while S_1 is the residual sum of squares matrix from fitting the reduced model with $B_2 = 0$ and $\tilde{B}_1 = (\mathbf{X}_1'\mathbf{X}_1)^{-1}\mathbf{X}_1'\mathbf{Y}$. This result follows directly by noting that the value of the multivariate normal likelihood function evaluated at the MLEs $\hat{B}$ and $\tilde{\Sigma} = S/T = [(T - r)/T]\hat{\Sigma}$ is equal to a constant times $|S|^{-T/2}$, since in the exponent we have

$$\sum_{t=1}^{T} \hat{\varepsilon}_t' \tilde{\Sigma}^{-1} \hat{\varepsilon}_t = \mathrm{tr}\{\tilde{\Sigma}^{-1} \sum_{t=1}^{T} \hat{\varepsilon}_t \hat{\varepsilon}_t'\} = \mathrm{tr}\{\tilde{\Sigma}^{-1} T \tilde{\Sigma}\} = k\,T.$$

Thus, the likelihood ratio for testing $H_0 : B_2 = 0$ is

$$\lambda = L(\tilde{B}, \tilde{\Sigma}_1 ; \mathbf{Y})/L(\hat{B}, \tilde{\Sigma} ; \mathbf{Y}) = |S_1|^{-T/2}/|S|^{-T/2} = U^{T/2},$$

where $\tilde{B} = (\tilde{B}_1', 0')'$ and $\tilde{\Sigma}_1 = S_1/T$. Now it can be shown that for moderate and large sample size T, the test statistic

$$M = -[T - r_1 - r_2 + (r_2 - k - 1)/2]\log(U) \tag{A4.5}$$

is approximately distributed as chi-squared with $r_2 k$ degrees of freedom, $\chi^2_{r_2 k}$, under the null hypothesis $B_2 = 0$, and the LR test of $H_0 : B_2 = 0$ rejects when $M >$ constant, where the constant is determined from the $\chi^2_{r_2 k}$-distribution.

The above result can be applied in general to fitting models of the form $\mathbf{Y} = \sum_{j=1}^{p} \mathbf{X}_j B_j + \boldsymbol{\varepsilon}$ sequentially, introducing first the variables $\mathbf{X}_1$ in the model, then $\mathbf{X}_2$, and so forth, and, at each stage, testing the significance of the latest included terms $\mathbf{X}_l B_l$, i.e., testing $H_0 : B_l = 0$ in the model $\mathbf{Y} = \sum_{j=1}^{l} \mathbf{X}_j B_j + \boldsymbol{\varepsilon}$, sequentially, for each $l = 1, 2, \ldots, p$. We use the test statistic

$$M_l = -[T - r_1 - \cdots - r_l + (r_l - k - 1)/2]\log(U_l),$$

where $U_l = \det(S_l)/\det(S_{l-1})$, and S_l is the residual sum of squares matrix obtained from the current "full" model $\mathbf{Y} = \sum_{j=1}^{l} \mathbf{X}_j B_j + \boldsymbol{\varepsilon}$, while S_{l-1} is the residual sum of squares matrix obtained when the regression term $\mathbf{X}_l B_l$ is omitted from the model. The test statistic M_l has approximately a chi-squared distribution with $r_l k$ degrees of freedom under the null hypothesis $B_l = 0$.

Note that the main application of these estimation and testing results in the multivariate time series context will be in terms of the sequential fitting of multivariate AR(p) models of the form $Y_t = \sum_{j=1}^{p} \Phi_j Y_{t-j} + \varepsilon_t$ to k-dimensional vector time series Y_t.

Remark. Many basic results on the multivariate linear model may be found in Chapter 8 of Anderson (1984); also see Chapters 5 and 6 of Srivastava and Khatri (1979).

A4.3 Asymptotically Equivalent Forms of the Test of Linear Hypothesis

In the model $\mathbf{Y} = \mathbf{X}_1 B_1 + \mathbf{X}_2 B_2 + \boldsymbol{\varepsilon}$, write $\mathbf{X}'\mathbf{X} = A = [\,(A_{ij})\,]$, where $A_{ij} = \mathbf{X}_i'\mathbf{X}_j$, $i, j = 1, 2$, and let $A_{22.1} = A_{22} - A_{21}A_{11}^{-1}A_{12}$. Then it can be shown that

$$\begin{aligned} S_1 &= (\mathbf{Y} - \mathbf{X}_1 \tilde{B}_1)'(\mathbf{Y} - \mathbf{X}_1 \tilde{B}_1) \\ &= (\mathbf{Y} - \mathbf{X}\hat{B})'(\mathbf{Y} - \mathbf{X}\hat{B}) + \hat{B}_2'(\mathbf{X}_2 - \mathbf{X}_1 A_{11}^{-1}A_{12})'(\mathbf{X}_2 - \mathbf{X}_1 A_{11}^{-1}A_{12})\hat{B}_2 \\ &= S + \hat{B}_2'(A_{22} - A_{21}A_{11}^{-1}A_{12})\hat{B}_2 = S + \hat{B}_2' A_{22.1} \hat{B}_2, \end{aligned}$$

since $\mathbf{Y} - \mathbf{X}_1 \tilde{B}_1 = (\mathbf{Y} - \mathbf{X}\hat{B}) + (\mathbf{X}_2 - \mathbf{X}_1 A_{11}^{-1}A_{12})\hat{B}_2$ and the two terms on the right are orthogonal. (The relation follows because the "normal" equations $\mathbf{X}_1'\mathbf{Y} = \mathbf{X}_1'\mathbf{X}_1\hat{B}_1 + \mathbf{X}_1'\mathbf{X}_2\hat{B}_2 = A_{11}\hat{B}_1 + A_{12}\hat{B}_2$ imply that $\tilde{B}_1 = \hat{B}_1 + A_{11}^{-1}A_{12}\hat{B}_2$.) Thus, we have the likelihood ratio

$$U = \det(S)/\det(S_1) = |S|\,/\,|S + \hat{B}_2' A_{22.1}\hat{B}_2| = 1/|I + S^{-1}\hat{B}_2' A_{22.1}\hat{B}_2|.$$

Now if $\hat{\lambda}_1 \geq \hat{\lambda}_2 \geq \cdots \geq \hat{\lambda}_k \geq 0$ denote the eigenvalues of $S^{-1}\hat{B}_2' A_{22.1}\hat{B}_2$, i.e., the roots of $|\hat{\lambda} I - S^{-1}\hat{B}_2' A_{22.1}\hat{B}_2| = 0$, then $(1 + \hat{\lambda}_i)$ are the eigenvalues of $I + S^{-1}\hat{B}_2' A_{22.1}\hat{B}_2$. Since $|I + S^{-1}\hat{B}_2' A_{22.1}\hat{B}_2| = \Pi_{i=1}^{k}(1 + \hat{\lambda}_i)$, it follows that

$$\begin{aligned} -\log(U) &= \sum_{i=1}^{k} \log(1 + \hat{\lambda}_i) \\ &\approx \sum_{i=1}^{k} \hat{\lambda}_i = \mathrm{tr}\{S^{-1}\hat{B}_2' A_{22.1}\hat{B}_2\} = \hat{\boldsymbol{\beta}}_2'(S^{-1} \otimes A_{22.1})\hat{\boldsymbol{\beta}}_2, \end{aligned} \tag{A4.6}$$

where $\hat{\boldsymbol{\beta}}_2 = \mathrm{vec}(\hat{B}_2)$, the approximation in (A4.6) holding under the null hypothesis since the $\hat{\lambda}_i$ are near zero under $H_0 : B_2 = 0$. Now $\hat{\boldsymbol{\beta}}_2 = \mathrm{vec}(\hat{B}_2)$ is distributed as $N(0, \Sigma \otimes A_{22.1}^{-1})$ under H_0, since $\hat{\boldsymbol{\beta}} = \mathrm{vec}(\hat{B})$ is distributed as $N(\boldsymbol{\beta}, \Sigma \otimes (\mathbf{X}'\mathbf{X})^{-1})$, from (A4.4), and $A_{22.1}^{-1}$ is the $r_2 \times r_2$ lower diagonal block of the matrix $(\mathbf{X}'\mathbf{X})^{-1}$ [e.g., see Rao (1973, p. 33)]. Hence, since $S/(T - r_1 - r_2) \xrightarrow{P} \Sigma$, by Slutsky theorem arguments applied in (A4.6) we find that, asymptotically,

$$- [\, T - r_1 - r_2 + (\, r_2 - k - 1\,)/2\,] \log(U)$$
$$\approx \hat{\boldsymbol{\beta}}_2' (\, \hat{\Sigma}^{-1} \otimes A_{22.1}) \, \hat{\boldsymbol{\beta}}_2 \approx \hat{\boldsymbol{\beta}}_2' (\, \Sigma^{-1} \otimes A_{22.1}) \, \hat{\boldsymbol{\beta}}_2 \, , \qquad \text{(A4.7)}$$

and the right-hand term is clearly seen to be distributed as $\chi^2_{r_2 k}$ under $\boldsymbol{\beta}_2 = 0$. The statistic $\hat{\boldsymbol{\beta}}_2' (\, \hat{\Sigma}^{-1} \otimes A_{22.1}) \, \hat{\boldsymbol{\beta}}_2 = \text{tr}\{\, \hat{\Sigma}^{-1} \hat{B}_2' A_{22.1} \hat{B}_2 \,\}$ is generally referred to as a *Wald statistic* and can be seen to be asymptotically equivalent to the LR testing procedure under H_0. (In multivariate analysis, this statistic is referred to as the "trace" statistic.) Tables of critical values of the *exact* distributions of multivariate statistics such as $-\log(\, U\,)$ and the Wald statistic, as well as more precise asymptotic approximations, are available for various values of k, r_1, r_2, and $T - (\, r_1 + r_2\,)$.

It is useful to note that because $S_1 = S + \hat{B}_2' A_{22.1} \hat{B}_2$, we have that $|\, \lambda I - S^{-1} \hat{B}_2' A_{22.1} \hat{B}_2 \,| = 0$ is equivalent to

$$|\, \{\, \lambda / (\, 1 + \lambda\,)\, \}\, I - S_1^{-1} \hat{B}_2' A_{22.1} \hat{B}_2 \,| = 0 ,$$

and this last equation has the form

$$|\, \{\, \lambda / (\, 1 + \lambda\,)\, \}\, I - \hat{\Sigma}_{YY.X_1}^{-1} \hat{\Sigma}_{YX_2.X_1} \hat{\Sigma}_{X_2X_2.X_1}^{-1} \hat{\Sigma}_{X_2Y.X_1} \,| = 0 .$$

The roots of this last equation, say $\hat{\rho}_i^2 = \hat{\lambda}_i / (\, 1 + \hat{\lambda}_i\,)$, are the (squared) sample *partial canonical correlations* between $\boldsymbol{Y}_t$ and $\boldsymbol{X}_{2t}$, given $\boldsymbol{X}_{1t}$ [compare with equation (3.9) of Section 3.2]. Hence, the roots of the first equation are given by $\hat{\lambda}_i = \hat{\rho}_i^2 / (\, 1 - \hat{\rho}_i^2\,)$, $i = 1, \ldots, k$, and the LR test statistic can also be expressed in terms of the $\hat{\rho}_i^2$ as

$$- \log(\, U\,) = \sum_{i=1}^{k} \log(\, 1 + \hat{\lambda}_i\,) = - \sum_{i=1}^{k} \log(\, 1 - \hat{\rho}_i^2\,) . \qquad \text{(A4.8)}$$

Finally, suppose that $B_2 \neq 0$ but that $\text{rank}(\, B_2\,) = s < \min(\, r_2, k\,) \equiv k$ (say). Then the regression coefficient matrix B_2 is said to have *reduced-rank* structure, and there are several reasons why it may be of interest to identify such a structure. From the above discussion, it will follow that to test the restriction $H_0 : \text{rank}(\, B_2\,) = s$, we can use the test statistic

$$- (\, T - r\,) \log(\, U^*\,) = - (\, T - r\,) \sum_{i=s+1}^{k} \log(\, 1 - \hat{\rho}_i^2\,) , \qquad \text{(A4.9)}$$

which is distributed asymptotically as $\chi^2_{(k-s)(r_2-s)}$ under H_0 (see Anderson, 1951). The MLE of B_2 under the reduced-rank restriction that $\text{rank}(\, B_2\,) = s$ has also been derived by Anderson (1951). The restricted MLE is given by $\hat{B}_2^* = \hat{B}_2 \hat{V} \hat{V}' \tilde{\Sigma}$, where $\hat{B}_2 = A_{22.1}^{-1} (\, \mathbf{X}_2 - \mathbf{X}_1 A_{11}^{-1} A_{12}\,)' \, \mathbf{Y}$ and $\tilde{\Sigma} = (1/T)\, S$, $S = (\, \mathbf{Y} - \mathbf{X} \hat{B}\,)' (\, \mathbf{Y} - \mathbf{X} \hat{B}\,)$, are the unrestricted MLEs of B_2 and Σ, respectively, and $\hat{V} = (\, \hat{V}_1, \ldots, \hat{V}_s\,)$, where the vectors $\hat{V}_i$ are solutions to

$$(\, \hat{\rho}_i^2 S_1 - \hat{B}_2' A_{22.1} \hat{B}_2\,)\, \hat{V}_i = 0 , \qquad i = 1, \ldots, s ,$$

corresponding to the s largest (squared) partial canonical correlations $\hat{\rho}_i^2$. The $\hat{V}_i$ are normalized by $\hat{V}_i' \tilde{\Sigma} \hat{V}_i = 1$ $(\hat{V}' \tilde{\Sigma} \hat{V} = I_s)$ or equivalently, $\hat{V}_i' \tilde{\Sigma}_1 \hat{V}_i = (1 - \hat{\rho}_i^2)^{-1}$, where $\tilde{\Sigma}_1 = (1/T)\, S_1$, $S_1 = (\mathbf{Y} - \mathbf{X}_1 \tilde{B}_1)'(\mathbf{Y} - \mathbf{X}_1 \tilde{B}_1)$. Thus, we see that the full rank unrestricted estimator $\hat{B}_2$ and the reduced-rank restricted estimator are related by $\hat{B}_2^* = \hat{B}_2 \hat{P}$, where $\hat{P} = \hat{V} \hat{V}' \tilde{\Sigma}$ is an idempotent matrix of rank s. The corresponding MLE of B_1 is then given by $\hat{B}_1^* = (\mathbf{X}_1' \mathbf{X}_1)^{-1} \mathbf{X}_1' (\mathbf{Y} - \mathbf{X}_2 \hat{B}_2^*)$. It also follows (Anderson, 1951) that the likelihood ratio test for the more refined test of H_0 : rank$(B_2) = s_1$ against the alternative that rank$(B_2) = s_2$ $(s_1 < s_2 \le k)$ is

$$-(T-r) \sum_{i=s_1+1}^{s_2} \log(1 - \hat{\rho}_i^2). \tag{A4.10}$$

CHAPTER 5

Maximum Likelihood Estimation and Model Checking for Vector ARMA Models

In this chapter, conditional and exact maximum likelihood (ML) estimation procedures for vector ARMA time series models are presented and their properties are examined. For conditional maximum likelihood, explicit iterative computation of the ML estimator in the form of generalized least squares estimation is presented, while for the exact likelihood method, two different approaches to computation of the exact likelihood function are developed. ML estimation of vector ARMA models under linear constraints on the parameters, and associated LR testing of the hypothesis of the linear constraints are examined. Model checking techniques for an estimated model, based on correlation matrix properties of model residuals, are also explored. The effect of parameter estimation errors on mean square error for prediction from an estimated model is also considered. Two numerical examples of fitting and checking vector ARMA models are also presented.

5.1 Conditional Maximum Likelihood Estimation for Vector ARMA Models

In this section we consider conditional maximum likelihood estimation for the vector ARMA model. The conditional likelihood approach for the vector model was examined first by Tunnicliffe Wilson (1973), and also by Reinsel (1979), Anderson (1980), Hannan and Kavalieris (1984), and Reinsel, Basu, and Yap (1992), among others. We examine the explicit computation of the conditional maximum likelihood estimates through explicit development of a modified Newton-Raphson (Gauss-Newton) procedure, and provide explicit evaluation for the gradient and (approximate) Hessian of the conditional log-likelihood function. The computations are shown to have the familiar form of generalized least squares estimation of a certain multivariate linear model, providing an appealing and easily interpretable framework for the maximum likelihood estimation in the vector ARMA(p,q) model.

5.1.1 *Conditional Likelihood Function for the Vector ARMA Model*

Thus, we consider conditional maximum likelihood estimation of parameters for the vector ARMA(p,q) model,

$$\boldsymbol{Y}_t - \sum_{i=1}^{p} \Phi_i \boldsymbol{Y}_{t-i} = \varepsilon_t - \sum_{i=1}^{q} \Theta_i \varepsilon_{t-i}, \tag{5.1}$$

based on a sample of T vector observations $\boldsymbol{Y}_t$, $t = 1, 2, \ldots, T$. The conditional likelihood approach is based on the assumption that initial observations $\boldsymbol{Y}_0, \boldsymbol{Y}_{-1}, \ldots, \boldsymbol{Y}_{1-p}$ are also available (for convenience of notation) and these are considered as fixed, and it uses an approximation involving the initial disturbances by setting $\varepsilon_0 = \varepsilon_{-1} = \cdots = \varepsilon_{1-q} = 0$. (So T is the "effective" number of observations.) The ε_t, $t = 1, \ldots, T$, are assumed to be independent and normally distributed with mean vector 0 and nonsingular covariance matrix Σ.

We define the $T \times k$ matrices $\mathbf{Y} = (\boldsymbol{Y}_1, \ldots, \boldsymbol{Y}_T)'$, $\boldsymbol{\varepsilon} = (\varepsilon_1, \ldots, \varepsilon_T)'$, with $B^i \mathbf{Y} = (\boldsymbol{Y}_{1-i}, \ldots, \boldsymbol{Y}_{T-i})'$, $B^i \boldsymbol{\varepsilon} = (\varepsilon_{1-i}, \ldots, \varepsilon_{T-i})'$. Then the model for $\mathbf{Y}$ can be expressed as

$$\mathbf{Y} - \sum_{i=1}^{p} B^i \mathbf{Y} \Phi_i' = \boldsymbol{\varepsilon} - \sum_{i=1}^{q} B^i \boldsymbol{\varepsilon} \Theta_i'.$$

The model can be expressed in vector form using the vec operator and the relation $\text{vec}(ABC) = (C' \otimes A)\,\text{vec}(B)$. Define the vectors

$$\mathbf{y} = \text{vec}(\mathbf{Y}') = (\boldsymbol{Y}_1', \ldots, \boldsymbol{Y}_T')', \qquad \mathbf{e} = \text{vec}(\boldsymbol{\varepsilon}') = (\varepsilon_1', \ldots, \varepsilon_T')',$$

as well as $B^i \mathbf{y} = \text{vec}[\,(B^i \mathbf{Y})'\,]$, $B^i \mathbf{e} = \text{vec}[\,(B^i \boldsymbol{\varepsilon})'\,]$, $\boldsymbol{\phi}_i = \text{vec}(\Phi_i)$, $i = 1, \ldots, p$, and $\boldsymbol{\theta}_i = \text{vec}(\Theta_i)$, $i = 1, \ldots, q$. Then the model can be expressed as

$$\mathbf{y} - \sum_{i=1}^{p} (I_T \otimes \Phi_i) B^i \mathbf{y} = \mathbf{e} - \sum_{i=1}^{q} (I_T \otimes \Theta_i) B^i \mathbf{e}, \tag{5.2}$$

and also in another useful form as

$$\mathbf{y} - \sum_{i=1}^{p} (B^i \mathbf{Y} \otimes I_k) \boldsymbol{\phi}_i = \mathbf{e} - \sum_{i=1}^{q} (B^i \boldsymbol{\varepsilon} \otimes I_k) \boldsymbol{\theta}_i. \tag{5.3}$$

We introduce the $T \times T$ lag matrix L which has ones on the (sub)diagonal directly below the main diagonal and zeros elsewhere. Under the approximation of zero initial values for the ε_t, $B^i \boldsymbol{\varepsilon}$ in (5.3) is replaced by $L^i \boldsymbol{\varepsilon} = (0, \ldots, 0, \varepsilon_1, \ldots, \varepsilon_{T-i})'$ and, hence, $B^i \mathbf{e}$ in (5.2) becomes $(L^i \otimes I_k)\,\mathbf{e} = (0', \ldots, 0', \varepsilon_1', \ldots, \varepsilon_{T-i}')'$. Thus, from (5.2) and (5.3) we obtain the relation

$$\mathbf{y} - \sum_{i=1}^{p} (B^i \mathbf{Y} \otimes I_k) \boldsymbol{\phi}_i = \mathbf{e} - \sum_{i=1}^{q} (L^i \boldsymbol{\varepsilon} \otimes I_k) \boldsymbol{\theta}_i$$

$$= \mathbf{e} - \sum_{i=1}^{q} (L^i \otimes \Theta_i) \mathbf{e} = \Theta \, \mathbf{e} , \tag{5.4}$$

where $\Theta = (I_T \otimes I_k) - \sum_{i=1}^{q} (L^i \otimes \Theta_i)$. Hence, on the assumption of normality of the ε_t, since $\mathbf{e}$ is $N(0, I_T \otimes \Sigma)$, the approximate (*conditional*) *log-likelihood function* can be written as

$$l = -\frac{T}{2} \log |\Sigma| - \frac{1}{2} \sum_{t=1}^{T} \varepsilon_t' \Sigma^{-1} \varepsilon_t$$

$$= -\frac{T}{2} \log |\Sigma| - \frac{1}{2} \mathbf{e}' (I_T \otimes \Sigma^{-1}) \mathbf{e}$$

$$= -\frac{T}{2} \log |\Sigma| - \frac{1}{2} \mathbf{w}' \Theta'^{-1} (I_T \otimes \Sigma^{-1}) \Theta^{-1} \mathbf{w} ,$$

where $\mathbf{w} = \mathbf{y} - \sum_{i=1}^{p} (B^i \mathbf{Y} \otimes I_k) \boldsymbol{\phi}_i$, with $\mathbf{w} = (\boldsymbol{W}_1', \ldots, \boldsymbol{W}_T')'$ and $\boldsymbol{W}_t = \boldsymbol{Y}_t - \sum_{i=1}^{p} \Phi_i \boldsymbol{Y}_{t-i}$.

5.1.2 Likelihood Equations for Conditional ML Estimation

Now we consider the maximization of l with respect to the parameters $\boldsymbol{\phi}_i$, $\boldsymbol{\theta}_i$, and Σ. For fixed $\boldsymbol{\phi}_i$, $\boldsymbol{\theta}_i$, it is clear that maximization with respect to Σ yields $\hat{\Sigma} = T^{-1} \sum_{t=1}^{T} \varepsilon_t \varepsilon_t' = \boldsymbol{\varepsilon}' \boldsymbol{\varepsilon} / T$, where $\text{vec}(\boldsymbol{\varepsilon}') = \mathbf{e} = \Theta^{-1} \mathbf{w}$. The partial derivatives of l with respect to the $\boldsymbol{\phi}_j$ and $\boldsymbol{\theta}_j$ are given by $\partial l / \partial \boldsymbol{\phi}_j = - (\partial \mathbf{e}' / \partial \boldsymbol{\phi}_j) (I_T \otimes \Sigma^{-1}) \mathbf{e}$ and $\partial l / \partial \boldsymbol{\theta}_j = - (\partial \mathbf{e}' / \partial \boldsymbol{\theta}_j) (I_T \otimes \Sigma^{-1}) \mathbf{e}$. Hence, from (5.4), these are

$$\frac{\partial l}{\partial \boldsymbol{\phi}_j} = (B^j \mathbf{Y} \otimes I_k)' \, \Theta'^{-1} (I_T \otimes \Sigma^{-1}) \, \Theta^{-1} (\mathbf{y} - \sum_{i=1}^{p} (B^i \mathbf{Y} \otimes I_k) \boldsymbol{\phi}_i) ,$$

for $j = 1, \ldots, p$, and

$$\frac{\partial l}{\partial \boldsymbol{\theta}_j} = - (L^j \boldsymbol{\varepsilon} \otimes I_k)' \, \Theta'^{-1} (I_T \otimes \Sigma^{-1}) \, \Theta^{-1} (\mathbf{y} - \sum_{i=1}^{p} (B^i \mathbf{Y} \otimes I_k) \boldsymbol{\phi}_i) ,$$

for $j = 1, \ldots, q$, where $\boldsymbol{\varepsilon}$ is expressible in terms of the observations through the relation $\text{vec}(\boldsymbol{\varepsilon}') = \mathbf{e} = \Theta^{-1} \mathbf{w}$. For the derivatives with respect to the $\boldsymbol{\theta}_j$, we have used the relation

$$\mathbf{e}' = \mathbf{w}' + \sum_{i=1}^{q} \mathbf{e}' (L'^i \otimes \Theta_i') = \mathbf{w}' + \sum_{i=1}^{q} \boldsymbol{\theta}_i' (\boldsymbol{\varepsilon}' L'^i \otimes I_k)$$

from (5.4) to obtain $\partial \mathbf{e}'/\partial \boldsymbol{\theta}_j = (\boldsymbol{\varepsilon}' L'^j \otimes I_k) + (\partial \mathbf{e}'/\partial \boldsymbol{\theta}_j) \sum_{i=1}^{q} (L'^i \otimes \Theta_i')$, so that $\partial \mathbf{e}'/\partial \boldsymbol{\theta}_j = (\boldsymbol{\varepsilon}' L'^j \otimes I_k)\, \Theta'^{-1}$. Defining the vector

$$\boldsymbol{\beta} = (\boldsymbol{\phi}_1', \ldots, \boldsymbol{\phi}_p', \boldsymbol{\theta}_1', \ldots, \boldsymbol{\theta}_q')'$$

and the matrix

$$\mathbf{Z} = [\, (B\,\mathbf{Y} \otimes I_k), \ldots, (B^p\,\mathbf{Y} \otimes I_k), -(L\,\boldsymbol{\varepsilon} \otimes I_k), \ldots, -(L^q\,\boldsymbol{\varepsilon} \otimes I_k)\,],$$

we can express these derivatives collectively in a convenient form as

$$\frac{\partial l}{\partial \boldsymbol{\beta}} = \mathbf{Z}'\, \Theta'^{-1} (I_T \otimes \Sigma^{-1})\, \Theta^{-1} (\mathbf{y} - \sum_{i=1}^{p} (B^i\,\mathbf{Y} \otimes I_k)\, \boldsymbol{\phi}_i)$$

$$= \mathbf{Z}'\, \Theta'^{-1} (I_T \otimes \Sigma^{-1})\, \mathbf{e}. \tag{5.5}$$

That is, since

$$l = -(T/2) \log |\Sigma| - (1/2)\, \mathbf{e}' (I_T \otimes \Sigma^{-1})\, \mathbf{e} = -(T/2) \log |\Sigma| - (1/2) \sum_{t=1}^{T} \varepsilon_t' \Sigma^{-1} \varepsilon_t,$$

we have $\partial l/\partial \boldsymbol{\beta} = -(\partial \mathbf{e}'/\partial \boldsymbol{\beta})(I_T \otimes \Sigma^{-1})\, \mathbf{e} = -\sum_{t=1}^{T} (\partial \varepsilon_t'/\partial \boldsymbol{\beta})\, \Sigma^{-1} \varepsilon_t$, with $\partial \mathbf{e}/\partial \boldsymbol{\beta}' = -\Theta^{-1}\mathbf{Z}$.

In the case with $q = 0$, that is, a pure AR(p) model with $\Theta = I_{kT}$, we have $\boldsymbol{\beta} = \text{vec}(\Phi_{(p)}')$ in the notation of Section 4.3 with $\Phi_{(p)}' = (\Phi_1, \ldots, \Phi_p)$ and $\mathbf{Z} = \mathbf{X} \otimes I_k$, where $\mathbf{X} = (B\,\mathbf{Y}, B^2\,\mathbf{Y}, \ldots, B^p\,\mathbf{Y})$. Then the likelihood equations simplify to

$$\partial l/\partial \boldsymbol{\beta} = \mathbf{Z}' (I_T \otimes \Sigma^{-1})(\mathbf{y} - \mathbf{Z}\,\boldsymbol{\beta}) = (I_{kp} \otimes \Sigma^{-1})\, \mathbf{Z}' (\mathbf{y} - \mathbf{Z}\,\boldsymbol{\beta}) = 0,$$

which implies that the conditional MLE in the AR(p) model is equal to

$$\hat{\boldsymbol{\beta}} = (\mathbf{Z}'\mathbf{Z})^{-1}\, \mathbf{Z}'\mathbf{y} = ((\mathbf{X}'\mathbf{X})^{-1}\, \mathbf{X}' \otimes I_k)\, \mathbf{y}$$

or, equivalently, $\hat{\Phi}_{(p)}' = \mathbf{Y}'\mathbf{X}\, (\mathbf{X}'\mathbf{X})^{-1}$, which is the same as the (conditional) least squares estimate presented in Section 4.3 [e.g., as given by (4.11)].

For $q > 0$, the likelihood equations (5.5) are highly nonlinear in the parameters $\boldsymbol{\beta}$, and so these equations need to be solved by iterative numerical procedures such as the Newton-Raphson method. Newton-Raphson equations for an approximate maximum likelihood estimator (MLE) $\hat{\boldsymbol{\beta}}$ are

$$-\left[\frac{\partial^2 l}{\partial \boldsymbol{\beta}\, \partial \boldsymbol{\beta}'} \right]_{\boldsymbol{\beta}_0} (\hat{\boldsymbol{\beta}} - \boldsymbol{\beta}_0) = \left[\frac{\partial l}{\partial \boldsymbol{\beta}} \right]_{\boldsymbol{\beta}_0}, \tag{5.6}$$

where $\boldsymbol{\beta}_0$ is an initial estimate of $\boldsymbol{\beta}$ and the estimate $\tilde{\Sigma} = \tilde{\boldsymbol{\varepsilon}}'\, \tilde{\boldsymbol{\varepsilon}}/T$ from a previous iteration is used for Σ. In general, if the initial estimate $\boldsymbol{\beta}_0$ is consistent of order $O_p(T^{-1/2})$, then the estimate $\hat{\boldsymbol{\beta}}$ obtained at the first iteration is consistent and asymptotically equivalent to the MLE.

5.1.3 Iterative Computation of the Conditional MLE by GLS Estimation

To carry out the iterations in (5.6) it is useful to have a convenient expression for the Hessian matrix of second partial derivatives. It can be shown that on neglecting terms which, when divided by T, converge to zero in probability as $T \to \infty$, we obtain the approximation

$$-\left[\frac{\partial^2 l}{\partial \boldsymbol{\beta}\, \partial \boldsymbol{\beta}'} \right] \approx \left(\frac{\partial \mathbf{e}'}{\partial \boldsymbol{\beta}} \right) (I_T \otimes \Sigma^{-1}) \left(\frac{\partial \mathbf{e}}{\partial \boldsymbol{\beta}'} \right) = \sum_{t=1}^{T} \left[\frac{\partial \varepsilon_t'}{\partial \boldsymbol{\beta}} \right] \Sigma^{-1} \left[\frac{\partial \varepsilon_t}{\partial \boldsymbol{\beta}'} \right]$$

$$= \mathbf{Z}' \Theta'^{-1} (I_T \otimes \Sigma^{-1}) \Theta^{-1} \mathbf{Z}. \tag{5.7}$$

From (5.5) it follows that the approximation involves neglecting terms that have the form of inner products of the vector $\mathbf{e}$ with the rows of the matrix of derivatives of $\mathbf{Z}' \Theta'^{-1} (I_T \otimes \Sigma^{-1})$ with respect to a parameter β_i, that is, terms of the form $-\{ \partial (\mathbf{Z}' \Theta'^{-1}) / \partial \beta_i \} (I_T \otimes \Sigma^{-1}) \mathbf{e}$. Thus, these terms have the form of cross-term summations over t of ε_t's times linear combinations of lagged $\boldsymbol{Y}_t$'s and lagged ε_t's, and these converge to zero in probability when divided by T because the lagged $\boldsymbol{Y}_t$'s and lagged ε_t's are independent of the current ε_t. Hence, using initial estimates

$$\boldsymbol{\beta}_0 = (\tilde{\boldsymbol{\phi}}_1', \ldots, \tilde{\boldsymbol{\phi}}_p', \tilde{\boldsymbol{\theta}}_1', \ldots, \tilde{\boldsymbol{\theta}}_q')' = \text{vec}(\tilde{\Phi}_1, \ldots, \tilde{\Phi}_p, \tilde{\Theta}_1, \ldots, \tilde{\Theta}_q),$$

we let

$$\tilde{\Theta} = (I_T \otimes I_k) - \textstyle\sum_{i=1}^{q} (L^i \otimes \tilde{\Theta}_i), \qquad \tilde{\mathbf{e}} = \tilde{\Theta}^{-1} (\mathbf{y} - \textstyle\sum_{i=1}^{p} (B^i \mathbf{Y} \otimes I_k) \tilde{\boldsymbol{\phi}}_i),$$

with $\text{vec}(\tilde{\boldsymbol{\varepsilon}}') = \tilde{\mathbf{e}}$, and $\tilde{\Sigma} = \tilde{\boldsymbol{\varepsilon}}' \tilde{\boldsymbol{\varepsilon}} / T$, and let $\tilde{\mathbf{Z}}$ denote the matrix $\mathbf{Z}$ with $L^j \tilde{\boldsymbol{\varepsilon}}$ in place of $L^j \boldsymbol{\varepsilon}$, $j = 1, \ldots, q$. Then using (5.5)–(5.7), the modified Newton-Raphson equations for $\hat{\boldsymbol{\beta}}$ have solution of the form

$$\hat{\boldsymbol{\beta}} = \boldsymbol{\beta}_0 + \left[\tilde{\mathbf{Z}}' \tilde{\Theta}'^{-1} (I_T \otimes \tilde{\Sigma}^{-1}) \tilde{\Theta}^{-1} \tilde{\mathbf{Z}} \right]^{-1} \tilde{\mathbf{Z}}' \tilde{\Theta}'^{-1} (I_T \otimes \tilde{\Sigma}^{-1}) \tilde{\mathbf{e}}$$

$$= \boldsymbol{\beta}_0 + \left[\bar{\mathbf{Z}}' (I_T \otimes \tilde{\Sigma}^{-1}) \bar{\mathbf{Z}} \right]^{-1} \bar{\mathbf{Z}}' (I_T \otimes \tilde{\Sigma}^{-1}) \tilde{\mathbf{e}}, \tag{5.8}$$

where $\bar{\mathbf{Z}} = \tilde{\Theta}^{-1} \tilde{\mathbf{Z}}$. We note that $\tilde{\mathbf{e}} = \text{vec}(\tilde{\boldsymbol{\varepsilon}}')$ is easily computed recursively from $\tilde{\Theta} \tilde{\mathbf{e}} = \mathbf{y} - \sum_{i=1}^{p} (B^i \mathbf{Y} \otimes I_k) \tilde{\boldsymbol{\phi}}_i = \mathbf{y} - \sum_{i=1}^{p} (I_T \otimes \tilde{\Phi}_i) B^i \mathbf{y}$ as

$$\tilde{\varepsilon}_t = \boldsymbol{Y}_t - \sum_{i=1}^{p} \tilde{\Phi}_i \boldsymbol{Y}_{t-i} + \sum_{i=1}^{q} \tilde{\Theta}_i \tilde{\varepsilon}_{t-i}, \qquad t = 1, \ldots, T, \tag{5.9}$$

with $\tilde{\varepsilon}_0 = \tilde{\varepsilon}_{-1} = \cdots = \tilde{\varepsilon}_{1-q} = 0$. Also, the columns of the matrix of independent variables $\bar{\mathbf{Z}} = \tilde{\Theta}^{-1} \tilde{\mathbf{Z}}$ in (5.8) may similarly be computed recursively from the relation $\tilde{\Theta} \bar{\mathbf{Z}} = \tilde{\mathbf{Z}}$. Specifically, let $\bar{\mathbf{Z}} = (\bar{\mathbf{U}}_1, \ldots, \bar{\mathbf{U}}_p, \bar{\mathbf{V}}_1, \ldots, \bar{\mathbf{V}}_q)$ with $\bar{\mathbf{U}}_j = \tilde{\Theta}^{-1} (B^j \mathbf{Y} \otimes I_k)$ and $\bar{\mathbf{V}}_j = - \tilde{\Theta}^{-1} (L^j \tilde{\boldsymbol{\varepsilon}} \otimes I_k)$. Then with $\bar{\mathbf{U}}_j$ partitioned as $(\overline{U}_{j1}', \overline{U}_{j2}', \ldots, \overline{U}_{jT}')'$, where the $\overline{U}_{jt}$ are $k \times k^2$ matrices, $\overline{\mathbf{U}}_j$ can be computed from $\tilde{\Theta}\, \overline{\mathbf{U}}_j = (B^j\, \mathbf{Y} \otimes I_k)$ recursively as

$$\overline{U}_{jt} = \sum_{i=1}^{q} \tilde{\Theta}_i\, \overline{U}_{j,t-i} + (Y_{t-j}' \otimes I_k), \qquad t = 1, 2, \ldots, T,$$

with $\overline{U}_{jt} \equiv 0$, $t \leq 0$. The calculation of the $\overline{\mathbf{V}}_j$ can be performed similarly, with

$$\overline{V}_{jt} = \sum_{i=1}^{q} \tilde{\Theta}_i\, \overline{V}_{j,t-i} - (\tilde{\varepsilon}_{t-j}' \otimes I_k), \qquad t = 1, 2, \ldots, T.$$

Hence, the only other major computation involved in the procedure is the solving of the "normal" equations represented by the right side of (5.8).

It is also interesting to note that equations (5.8) can be equivalently expressed in the form

$$\hat{\boldsymbol{\beta}} = \left[\overline{\mathbf{Z}}' (I_T \otimes \tilde{\Sigma}^{-1})\, \overline{\mathbf{Z}} \right]^{-1} \overline{\mathbf{Z}}' (I_T \otimes \tilde{\Sigma}^{-1})\, \tilde{\Theta}^{-1} (\mathbf{y} - \sum_{i=1}^{q} (L^i \tilde{\boldsymbol{\varepsilon}} \otimes I_k)\, \tilde{\boldsymbol{\theta}}_i). \tag{5.10}$$

This has the interpretation of being the generalized least squares estimator associated with the model represented by the following identity based on (5.3),

$$\mathbf{y} - \sum_{i=1}^{q} (L^i \tilde{\boldsymbol{\varepsilon}} \otimes I_k)\, \tilde{\boldsymbol{\theta}}_i = \sum_{i=1}^{p} (B^i\, \mathbf{Y} \otimes I_k)\, \boldsymbol{\phi}_i - \sum_{i=1}^{q} (L^i \tilde{\boldsymbol{\varepsilon}} \otimes I_k)\, \boldsymbol{\theta}_i + \tilde{\Theta}\, \mathbf{e}$$
$$- \sum_{i=1}^{q} [\, L^i (\tilde{\boldsymbol{\varepsilon}} - \boldsymbol{\varepsilon}) \otimes I_k \,] (\tilde{\boldsymbol{\theta}}_i - \boldsymbol{\theta}_i). \tag{5.11}$$

Equivalently, (5.11) is expressible as

$$Y_t - \sum_{i=1}^{q} \tilde{\Theta}_i\, \tilde{\varepsilon}_{t-i} = \sum_{i=1}^{p} \Phi_i\, Y_{t-i} - \sum_{i=1}^{q} \Theta_i\, \tilde{\varepsilon}_{t-i} + \varepsilon_t - \sum_{i=1}^{q} \tilde{\Theta}_i\, \varepsilon_{t-i}$$
$$- \sum_{i=1}^{q} (\tilde{\Theta}_i - \Theta_i) (\tilde{\varepsilon}_{t-i} - \varepsilon_{t-i}), \qquad t = 1, \ldots, T. \tag{5.12}$$

Thus, the generalized least squares (GLS) estimator in (5.10) results from (5.11) when the last term on the right side of (5.11) is neglected, and the disturbance term $\tilde{\Theta}\, \mathbf{e}$ is approximated as having covariance matrix $\tilde{\Theta} (I_T \otimes \tilde{\Sigma})\, \tilde{\Theta}'$.

In practice, the basic iterative procedure as in (5.8) may need to be modified in certain ways, for example, by use of a scale factor adjustment to the vector increment on the right side of (5.8) to avoid the problem of overshoot and to help ensure that the likelihood function actually increases at each iteration. On convergence, the covariance matrix of the maximum likelihood estimator $\hat{\boldsymbol{\beta}}$ may be estimated by $[\, \hat{\mathbf{Z}}'\, \hat{\Theta}'^{-1} (I_T \otimes \hat{\Sigma}^{-1})\, \hat{\Theta}^{-1}\, \hat{\mathbf{Z}} \,]^{-1}$. The initial estimates needed to start the iterations can be obtained by a first stage vector autoregression to obtain initial residuals $\tilde{\varepsilon}_t$ followed by a least squares regression of $\boldsymbol{Y}_t$ on $\boldsymbol{Y}_{t-1}, \ldots, \boldsymbol{Y}_{t-p}$ and $\tilde{\varepsilon}_{t-1}, \ldots, \tilde{\varepsilon}_{t-q}$ as in the second stage of the procedure by Hannan and Rissanen (1982) or one step of the procedure of Spliid (1983) [e.g., refer to (4.18) in Section 4.5.3]. An alternate method to obtain initial estimates [e.g., see Hannan (1975) and An, Chen, and Hannan (1983)] involves solving the sample moment equations analogous to (2.14) of Section 2.3.2, $\hat{\Gamma}(l) = \sum_{j=1}^{p} \hat{\Gamma}(l-j)\, \tilde{\Phi}_j'$, $l = q+1, \ldots, q+p$, for the AR estimates $\tilde{\Phi}_j$, from which estimates of the MA(q) process covariance matrices can be formed as the sample covariance matrices of $\tilde{\boldsymbol{W}}_t = \boldsymbol{Y}_t - \sum_{j=1}^{p} \tilde{\Phi}_j \boldsymbol{Y}_{t-j}$, and then estimates $\tilde{\Theta}_j$ of the MA coefficient matrices can be obtained by use of the factorization algorithm of Tunnicliffe Wilson (1972). We note that the above discussion on the ML estimation procedure for the vector ARMA model easily extends to the case in which exogenous input variables are also included in the model, and, in fact, this was considered in a more general dynamic simultaneous equations econometric model context by Reinsel (1979).

5.1.4 Asymptotic Distribution for the MLE in the Vector ARMA Model

Concerning the asymptotic distribution theory for the MLE $\hat{\boldsymbol{\beta}}$, stationarity and identifiability assumptions for the ARMA model are made, and for the white noise process $\{\varepsilon_t\}$, it is assumed that $E(\varepsilon_t \mid \boldsymbol{F}_{t-1}) = 0$ and $E(\varepsilon_t \varepsilon_t' \mid \boldsymbol{F}_{t-1}) = \Sigma$, where $\boldsymbol{F}_{t-1}$ denotes the σ-field generated by $\{\varepsilon_{t-1}, \varepsilon_{t-2}, \ldots\}$, and that the ε_t possess finite fourth moments. Then, under these assumptions, it has been proven (Dunsmuir and Hannan, 1976; Deistler, Dunsmuir and Hannan, 1978; Hannan and Deistler, 1988, Chap. 4; and Rissanen and Caines, 1979) that $\hat{\boldsymbol{\beta}}$ is strongly consistent for $\boldsymbol{\beta}$ (that is, $\hat{\boldsymbol{\beta}}$ converges almost surely to $\boldsymbol{\beta}$ as $T \to \infty$), and that $T^{1/2}(\hat{\boldsymbol{\beta}} - \boldsymbol{\beta})$ converges in distribution as $T \to \infty$ to multivariate normal $N(0, V^{-1})$, where V is the *asymptotic information matrix* of $\boldsymbol{\beta}$. To describe the form of this asymptotic information matrix V of $\boldsymbol{\beta}$ more explicitly, define the $k \times k^2$ (stationary) process $\bar{\boldsymbol{U}}_t$ by $\bar{\boldsymbol{U}}_t = \sum_{i=1}^{q} \Theta_i \bar{\boldsymbol{U}}_{t-i} + (\boldsymbol{Y}_t' \otimes I_k)$, and similarly define $\bar{\boldsymbol{V}}_t$ by $\bar{\boldsymbol{V}}_t = \sum_{i=1}^{q} \Theta_i \bar{\boldsymbol{V}}_{t-i} - (\varepsilon_t' \otimes I_k)$, and set $\bar{\boldsymbol{Z}}_t = (\bar{\boldsymbol{U}}_{t-1}, \ldots, \bar{\boldsymbol{U}}_{t-p}, \bar{\boldsymbol{V}}_{t-1}, \ldots, \bar{\boldsymbol{V}}_{t-q})'$, which is $k^2(p+q) \times k$. From the preceding developments, $\bar{\boldsymbol{Z}}_t$ is recognized as being equal to the process $\bar{\boldsymbol{Z}}_t = -\partial \varepsilon_t' / \partial \boldsymbol{\beta}$. Then V is given by

$$V = E(\bar{\boldsymbol{Z}}_t \Sigma^{-1} \bar{\boldsymbol{Z}}_t') = \lim_{T \to \infty} T^{-1} E(-\partial^2 l / \partial \boldsymbol{\beta}\, \partial \boldsymbol{\beta}'). \tag{5.13}$$

For example, in the AR(1) model we have $\bar{\boldsymbol{U}}_t = (\boldsymbol{Y}_t' \otimes I_k)$ and $\bar{\boldsymbol{Z}}_t' = \bar{\boldsymbol{U}}_{t-1}$, so that

$$V = E(\bar{\boldsymbol{Z}}_t \Sigma^{-1} \bar{\boldsymbol{Z}}_t') = E(\boldsymbol{Y}_{t-1} \boldsymbol{Y}_{t-1}' \otimes \Sigma^{-1}) = \Gamma(0) \otimes \Sigma^{-1},$$

with $V^{-1} = \Gamma(0)^{-1} \otimes \Sigma$ similar to results in Section 4.3.1. For the MA(1) model $\boldsymbol{Y}_t = \varepsilon_t - \Theta_1 \varepsilon_{t-1}$, we have

$$\overline{\boldsymbol{V}}_t = \Theta_1 \, \overline{\boldsymbol{V}}_{t-1} - (\, \varepsilon_t' \otimes I_k \,) = - \sum_{j=0}^{\infty} \Theta_1^j \, (\, \varepsilon_{t-j}' \otimes I_k \,) = - \sum_{j=0}^{\infty} (\, \varepsilon_{t-j}' \otimes \Theta_1^j \,) \, ,$$

and $\overline{\boldsymbol{Z}}_t' = \overline{\boldsymbol{V}}_{t-1}$, so that

$$V = E[\, \textstyle\sum_{j=0}^{\infty} (\, \varepsilon_{t-j} \, \varepsilon_{t-j}' \otimes \Theta_1'^j \, \Sigma^{-1} \, \Theta_1^j \,) \,] = \Sigma \otimes \displaystyle\sum_{j=0}^{\infty} \Theta_1'^j \, \Sigma^{-1} \, \Theta_1^j \, .$$

Closed form expressions, in terms of spectral relations, were given for the elements of the information matrix V of the general vector ARMA(p,q) model by Newton (1978).

The asymptotic distribution of the MLE $\hat{\boldsymbol{\beta}}$ can be established following fairly standard arguments, based on relations as in (5.6) and (5.8) obtained from Taylor expansion of the vector of partial derivatives (5.5) of the log-likelihood function. Hence, it can be derived that

$$\begin{aligned} T^{1/2} \, (\, \hat{\boldsymbol{\beta}} - \boldsymbol{\beta} \,) &= \left[\, T^{-1} \, \overline{\mathbf{Z}}' \, (\, I_T \otimes \Sigma^{-1}) \, \overline{\mathbf{Z}} \, \right]^{-1} T^{-1/2} \, \overline{\mathbf{Z}}' \, (\, I_T \otimes \Sigma^{-1}) \, \mathbf{e} \; + o_p(1) \\ &\approx \left[\, T^{-1} \sum_{t=1}^{T} \overline{Z}_t \, \Sigma^{-1} \, \overline{Z}_t' \, \right]^{-1} T^{-1/2} \sum_{t=1}^{T} \overline{Z}_t \, \Sigma^{-1} \, \varepsilon_t \\ &\xrightarrow{D} N(\, 0 \, , \, V^{-1} \,) \, , \end{aligned} \tag{5.14}$$

where $T^{-1} \sum_{t=1}^{T} \overline{Z}_t \, \Sigma^{-1} \, \overline{Z}_t' \xrightarrow{P} V$ as $T \to \infty$, and

$$T^{-1/2} \sum_{t=1}^{T} \overline{Z}_t \, \Sigma^{-1} \, \varepsilon_t \xrightarrow{D} N(\, 0, \, V \,)$$

as $T \to \infty$ by use of a martingale central limit theorem, noting that

$$\mathrm{Cov}(\, \overline{Z}_t \, \Sigma^{-1} \, \varepsilon_t \,) = E(\, \overline{Z}_t \, \Sigma^{-1} \, \varepsilon_t \, \varepsilon_t' \, \Sigma^{-1} \, \overline{Z}_t' \,) = E(\, \overline{Z}_t \, \Sigma^{-1} \, \overline{Z}_t' \,) = V$$

from the property that $E(\, \varepsilon_t \, \varepsilon_t' \mid \boldsymbol{F}_{t-1} \,) = \Sigma$.

5.2 ML Estimation and LR Testing of ARMA Models Under Linear Restrictions

5.2.1 ML Estimation of Vector ARMA Models with Linear Constraints on the Parameters

In practice, it will often be desired to perform ML estimation of the vector ARMA model with constraints imposed on the AR and MA parameter coefficient matrices Φ_j and Θ_j. In the simplest situations, these constraints will take the form of specifying that certain individual elements of the coefficient matrices are equal to zero. Estimation of the model with the zero constraints imposed is desirable because it will lead to improved precision in the estimates of the remaining unknown parameters. We mention that the estimation procedure as in (5.8) for $\hat{\boldsymbol{\beta}}$ is easily modified to include the case where some elements of the matrices $\Phi_1, \ldots, \Phi_p, \Theta_1, \ldots, \Theta_q$ are specified to equal zero. We merely modify the definitions of the matrix $\tilde{\Theta}$ and the vector $\tilde{\mathbf{e}}$ accordingly, and in the matrix $\tilde{\mathbf{Z}}$ we delete the columns of $(B^j \mathbf{Y} \otimes I_k)$ and $(L^j \tilde{\boldsymbol{\varepsilon}} \otimes I_k)$ corresponding to the zero elements of $\boldsymbol{\phi}_j = \text{vec}(\Phi_j)$ and $\boldsymbol{\theta}_j = \text{vec}(\Theta_j)$, respectively. Similar asymptotic theory as in (5.14) holds for the ML estimator $\hat{\boldsymbol{\beta}}$ when elements of the matrices Φ_j and Θ_j are specified to equal zero. In this case the corresponding columns of $\bar{Z}_t'$ (and hence rows of $\bar{Z}_t$) are deleted in the asymptotic information matrix $V = E(\bar{Z}_t \Sigma^{-1} \bar{Z}_t')$.

More generally, we consider estimation of the parameters $\boldsymbol{\beta} = (\boldsymbol{\phi}_1', \ldots, \boldsymbol{\phi}_p', \boldsymbol{\theta}_1', \ldots, \boldsymbol{\theta}_q')'$ in the vector ARMA model (5.1) under the imposition of linear constraints on $\boldsymbol{\beta}$ of the form $\boldsymbol{\beta} = R\boldsymbol{\gamma}$, where R is a $k^2(p+q) \times s$ known matrix and $\boldsymbol{\gamma}$ is an $s \times 1$ vector of unknown (unrestricted) parameters, $s < k^2(p+q)$. Linear constraints of interest can always be expressed in this form, and this form is equivalent to expressing the constraints in another common form as $H\boldsymbol{\beta} = 0$, where H is a certain $(k^2(p+q) - s) \times k^2(p+q)$ known matrix related to R. Under the linear constraints, ML estimation of the vector ARMA model proceeds under the reparameterization of the model in terms of the unknown parameters $\boldsymbol{\gamma}$. Then, similar to (5.5) and (5.7), the likelihood equations are obtained as

$$\frac{\partial l}{\partial \boldsymbol{\gamma}} = \frac{\partial \boldsymbol{\beta}'}{\partial \boldsymbol{\gamma}} \frac{\partial l}{\partial \boldsymbol{\beta}} = R' \frac{\partial l}{\partial \boldsymbol{\beta}} = R' \mathbf{Z}' \Theta'^{-1} (I_T \otimes \Sigma^{-1}) \mathbf{e}, \tag{5.15}$$

and the approximate Hessian matrix is

$$-\left[\frac{\partial^2 l}{\partial \boldsymbol{\gamma} \partial \boldsymbol{\gamma}'}\right] = -R' \left[\frac{\partial^2 l}{\partial \boldsymbol{\beta} \partial \boldsymbol{\beta}'}\right] R \approx R' \mathbf{Z}' \Theta'^{-1} (I_T \otimes \Sigma^{-1}) \Theta^{-1} \mathbf{Z} R. \tag{5.16}$$

Thus, ML estimation can be carried out through modified Newton-Raphson iterations similar to (5.8) as

$$\hat{\boldsymbol{\gamma}} = \boldsymbol{\gamma}_0 + \left[R' \bar{\mathbf{Z}}' (I_T \otimes \tilde{\Sigma}^{-1}) \bar{\mathbf{Z}} R\right]^{-1} R' \bar{\mathbf{Z}}' (I_T \otimes \tilde{\Sigma}^{-1}) \tilde{\mathbf{e}}, \tag{5.17}$$

where $\bar{\mathbf{Z}} = \tilde{\Theta}^{-1} \tilde{\mathbf{Z}}$ and $\boldsymbol{\gamma}_0$ is the estimate of $\boldsymbol{\gamma}$ at the previous iteration.

The asymptotic information matrix of $\boldsymbol{\gamma}$, V_γ, is also readily obtained as

$$V_\gamma = \lim_{T \to \infty} T^{-1} E(-\partial^2 l / \partial \boldsymbol{\gamma} \partial \boldsymbol{\gamma}') = E(R' \bar{Z}_t \Sigma^{-1} \bar{Z}_t' R) = R' V R,$$

and the asymptotic theory for the restricted MLE $\hat{\boldsymbol{\gamma}}$ of $\boldsymbol{\gamma}$ is readily obtained similar to (5.14) such that $T^{1/2}(\hat{\boldsymbol{\gamma}} - \boldsymbol{\gamma}) \xrightarrow{D} N(0, V_\gamma^{-1}) \equiv N(0, \{R' V R\}^{-1})$. Of course, the restricted ML estimator of the full parameter vector $\boldsymbol{\beta}$ under the linear constraints $\boldsymbol{\beta} = R\boldsymbol{\gamma}$ is given by $\hat{\boldsymbol{\beta}}_R = R\hat{\boldsymbol{\gamma}}$, and the corresponding ML residual covariance matrix estimate $\hat{\Sigma}_R$ of Σ is obtained from $\hat{\boldsymbol{\beta}}_R$ in the usual manner from calculation of the residual vectors as in (5.9). The asymptotic distribution of the restricted ML estimator $\hat{\boldsymbol\beta}_R$ is thus such that $T^{1/2}(\hat{\boldsymbol\beta}_R - \boldsymbol\beta) \xrightarrow{D} N(0, R\{R' V R\}^{-1} R')$, and it is readily established that $R\{R' V R\}^{-1} R' < V^{-1}$, so that the asymptotic covariance matrix of the ML estimator of $\boldsymbol\beta$ under the constraints is less than that of the unrestricted ML estimator.

In the simple case where only zero constraints on individual coefficients in the matrices Φ_j and Θ_j are specified, the $s \times k^2(p+q)$ matrix R' is a "selection matrix" such that R' has exactly one element in each row equal to one and the other elements are zero, and $\boldsymbol\gamma$ ($\boldsymbol\gamma = R' \boldsymbol\beta$ in this case) consists of the remaining parameters of $\boldsymbol\beta$ that are not specified to equal zero. The rows of R that are zero correspond to the elements of $\boldsymbol\beta$ that are specified to equal zero in $\boldsymbol\beta = R\,\boldsymbol\gamma$. Then, as noted previously, the use of $\bar{\mathbf{Z}} R = \Theta^{-1}\tilde{\mathbf{Z}} R$, in place of $\bar{\mathbf{Z}} = \Theta^{-1}\tilde{\mathbf{Z}}$, in the estimation procedure (5.17) merely consists of deleting the columns of $(B^j \mathbf{Y} \otimes I_k)$ and $(L^j \tilde{\boldsymbol\varepsilon} \otimes I_k)$ corresponding to the elements of $\boldsymbol\phi_j = \text{vec}(\Phi_j)$ and $\boldsymbol\theta_j = \text{vec}(\Theta_j)$, respectively, that are specified to equal zero. In addition, the asymptotic information matrix $V_\gamma = R' V R$ of $\boldsymbol\gamma$ in this case is obtained by deleting the rows and columns of V that correspond to the parameters of $\boldsymbol\beta$ which are specified to equal zero under the restrictions $\boldsymbol\beta = R\,\boldsymbol\gamma$.

5.2.2 *LR Testing of the Hypothesis of the Linear Constraints*

In addition to ML estimation of the vector ARMA model under the linear constraints, the appropriateness of the linear constraints can be tested using the LR test procedure for $H_0 : \boldsymbol\beta = R\,\boldsymbol\gamma$. It follows that this procedure leads to the LR test statistic $-T \log(\,|\tilde\Sigma| / |\tilde\Sigma_R|\,)$ where $\tilde\Sigma$ denotes the unrestricted ML estimate of Σ. Under H_0, the LR statistic is asymptotically distributed as chi-squared with $k^2(p+q) - s$ degrees of freedom (e.g., Kohn, 1979). The asymptotically equivalent Wald test statistic procedure could also be used to test H_0, where the *Wald statistic* for testing $H_0 : H\,\boldsymbol\beta = 0$ has the form

$$T\,(H\,\hat{\boldsymbol\beta})' \{ H\,\hat{V}^{-1} H' \}^{-1} (H\,\hat{\boldsymbol\beta}) \tag{5.18}$$

with $\hat{\boldsymbol\beta}$ denoting the unrestricted MLE of $\boldsymbol\beta$ and $\hat V$ denoting a consistent estimator of the asymptotic information matrix V of $\boldsymbol\beta$ in (5.13), such as V evaluated at the MLE $\hat{\boldsymbol\beta}$ or $-T^{-1}(\partial^2 l / \partial\boldsymbol\beta\,\partial\boldsymbol\beta')_{\hat{\boldsymbol\beta}}$. For a simple illustration, in a vector AR(p) model with $\boldsymbol Y_t$ partitioned as $\boldsymbol Y_t = (\boldsymbol Y_{1t}', \boldsymbol Y_{2t}')'$, where $\boldsymbol Y_{1t}$ and $\boldsymbol Y_{2t}$ are of dimensions k_1 and k_2 ($k_1 + k_2 = k$), respectively, we may be interested in testing the hypothesis that the variables $\boldsymbol Y_{2t}$ "do not cause" $\boldsymbol Y_{1t}$, that is, there is no feedback from $\boldsymbol Y_{2t}$ to $\boldsymbol Y_{1t}$ [e.g., see Granger and Newbold (1986, Sections 7.3 and 8.5) for discussion]. In the AR(p) model $\boldsymbol Y_t = \sum_{j=1}^{p} \Phi_j \boldsymbol Y_{t-j} + \varepsilon_t$, if the coefficient matrices Φ_j are partitioned accordingly as

$$\Phi_j = \begin{bmatrix} \Phi_{j,11} & \Phi_{j,12} \\ \Phi_{j,21} & \Phi_{j,22} \end{bmatrix}, \qquad j = 1, \ldots, p,$$

the test for noncausality is a test of the hypothesis $H_0 : \Phi_{j,12} = 0$, for $j = 1, \ldots, p$. The number of unknown parameters under H_0 is $s = p\,(k_1^2 + k_2 k)$, and the degrees of freedom in the asymptotic chi-squared distribution of the LR test statistic under H_0 is $k^2 p - s = k_1 k_2 p$.

5.2.3 ML Estimation of Vector ARMA Models in the Echelon Canonical Form

The estimation procedure as presented thus far for the vector ARMA model (5.1) is also easily extended to include ML estimation of parameters when the model is considered in the echelon canonical ARMA form given as in (3.4) of Section 3.1.2,

$$\Phi_0^{\#} Y_t - \sum_{i=1}^{p} \Phi_i^{\#} Y_{t-i} = \Theta_0^{\#} \varepsilon_t - \sum_{i=1}^{q} \Theta_i^{\#} \varepsilon_{t-i},$$

with $\Phi_0^{\#} = \Theta_0^{\#}$ lower triangular and having ones on the diagonal. Usually, some elements of the coefficient matrices in this model will be specified to equal zero, e.g., as implied by the specification of the Kronecker indices $K_1, \ldots, K_k$ of the process. As in the previous discussion, these zero constraints can be expressed as $\boldsymbol{\beta} = R\,\boldsymbol{\gamma}$, where now $\boldsymbol{\beta} = (\boldsymbol{\phi}_0^{\#\prime}, \boldsymbol{\phi}_1^{\#\prime}, \ldots, \boldsymbol{\phi}_p^{\#\prime}, \boldsymbol{\theta}_1^{\#\prime}, \ldots, \boldsymbol{\theta}_q^{\#\prime})'$ with $\boldsymbol{\phi}_i^{\#} = \mathrm{vec}(\Phi_i^{\#})$, $i = 0, 1, \ldots, p$, and $\boldsymbol{\theta}_i^{\#} = \mathrm{vec}(\Theta_i^{\#})$, $i = 1, \ldots, q$. As in the development for estimation of the vector ARMA model in the standard form (5.1), we express the model in a similar form to (5.4) as

$$\begin{aligned}(\mathbf{Y} \otimes I_k)\boldsymbol{\phi}_0^{\#} - \sum_{i=1}^{p} (B^i \mathbf{Y} \otimes I_k)\boldsymbol{\phi}_i^{\#} &= (\boldsymbol{\varepsilon} \otimes I_k)\boldsymbol{\theta}_0^{\#} - \sum_{i=1}^{q} (L^i \boldsymbol{\varepsilon} \otimes I_k)\boldsymbol{\theta}_i^{\#} \\ &= (I_T \otimes \Theta_0^{\#})\mathbf{e} - \sum_{i=1}^{q} (L^i \otimes \Theta_i^{\#})\mathbf{e} = \Theta^{\#}\mathbf{e}, \qquad (5.19)\end{aligned}$$

where $\Theta^{\#} = (I_T \otimes \Theta_0^{\#}) - \sum_{i=1}^{q} (L^i \otimes \Theta_i^{\#})$ and $\boldsymbol{\theta}_0^{\#} = \boldsymbol{\phi}_0^{\#}$. The log-likelihood for this model is of the same form as in the standard ARMA model, and the first partial derivatives are thus obtained similarly as

$$\frac{\partial l}{\partial \boldsymbol{\beta}} = \mathbf{Z}^{\#\prime}\, \Theta^{\#\prime -1}(I_T \otimes \Sigma^{-1})\,\mathbf{e}, \qquad (5.20)$$

where

$$\begin{aligned}\mathbf{Z}^{\#} = [\,-((\mathbf{Y} - \boldsymbol{\varepsilon}) \otimes I_k),\ (B\,\mathbf{Y} \otimes I_k), \ldots, (B^p\,\mathbf{Y} \otimes I_k), \\ -(L\boldsymbol{\varepsilon} \otimes I_k), \ldots, -(L^q \boldsymbol{\varepsilon} \otimes I_k)\,].\end{aligned}$$

The approximate Hessian matrix is given by

$$-\left[\frac{\partial^2 l}{\partial \boldsymbol{\beta}\, \partial \boldsymbol{\beta}'}\right] = \mathbf{Z}^{\#\prime}\, \Theta^{\#\prime -1}(I_T \otimes \Sigma^{-1})\, \Theta^{\#-1}\mathbf{Z}^{\#}. \qquad (5.21)$$

The ML estimator of the unrestricted parameters $\boldsymbol{\gamma}$ is thus obtained through the Newton-Raphson iterations as in (5.8) or (5.17), with $\tilde{\mathbf{Z}}^{\#}$ and $\tilde{\Theta}^{\#}$ in place of $\tilde{\mathbf{Z}}$ and $\tilde{\Theta}$, respectively. Also, as discussed previously, the appropriate rows and columns of the matrices in (5.20)–(5.21) are deleted, that is, the appropriate columns of the matrix $\mathbf{Z}^{\#}$ are deleted, that correspond to parameter elements in $\boldsymbol{\beta}$ specified to equal zero (or specified to equal one in the case of the diagonal elements of $\Phi_0^{\#}$). The asymptotic properties of the resulting ML estimator $\hat{\boldsymbol{\gamma}}$ for the vector ARMA model in this form can be developed similar to the results given in (5.14).

Although the above "conditional" ML procedures discussed in Sections 5.1 and 5.2 are asymptotically equivalent to more "exact" ML procedures, to improve the performance of the GLS estimation procedures such as the one associated with (5.8), for small or moderate series length T, it might be preferable to calculate the "exact" residuals $\hat{\varepsilon}_t$ in (5.9) using starting values $\hat{\varepsilon}_0, \ldots, \hat{\varepsilon}_{1-q}$ obtained by the method of backcasting (e.g., Nicholls and Hall, 1979, Hillmer and Tiao, 1979) as will be discussed in Section 5.3. Also, the exact residuals can be used in the formation of $\tilde{\mathbf{Z}}$ (i.e., we can replace $L^j \tilde{\boldsymbol{\varepsilon}}$ by $B^j \tilde{\boldsymbol{\varepsilon}}$), and the "backcast method" could also be employed in the computation of the columns of $\bar{\mathbf{Z}} = \tilde{\Theta}^{-1} \tilde{\mathbf{Z}}$ [i.e., in the decomposition and inversion of the exact covariance matrix for the MA(q) process $\boldsymbol{W}_t = \Theta(B)\, \varepsilon_t$]. We will next discuss the details concerning the exact likelihood function in Section 5.3.

5.3 Exact Likelihood Function for Vector ARMA Models

The exact likelihood function for the vector ARMA(p,q) model,

$$\boldsymbol{Y}_t - \sum_{j=1}^{p} \Phi_j\, \boldsymbol{Y}_{t-j} = \varepsilon_t - \sum_{j=1}^{q} \Theta_j\, \varepsilon_{t-j}\,, \tag{5.22}$$

was derived by Hillmer and Tiao (1979) and Nicholls and Hall (1979) [also see Hall and Nicholls (1980)], while Osborn (1977) and Phadke and Kedem (1978) previously obtained expressions for the vector MA(q) model case. In this section we provide a derivation using a similar approach to these authors, and we also give some explicit details concerning the computation of the exact likelihood function for the vector ARMA model (5.22). Given T vector observations $\boldsymbol{Y}_1, \ldots, \boldsymbol{Y}_T$, define the vectors $\mathbf{y} = (\boldsymbol{Y}_1', \boldsymbol{Y}_2', \ldots, \boldsymbol{Y}_T')'$, $\mathbf{e} = (\varepsilon_1', \varepsilon_2', \ldots, \varepsilon_T')'$, $\mathbf{y}_* = (\boldsymbol{Y}_{1-p}', \ldots, \boldsymbol{Y}_0')'$, $\mathbf{e}_* = (\varepsilon_{1-q}', \ldots, \varepsilon_0')'$, and $\mathbf{a}_* = (\mathbf{y}_*', \mathbf{e}_*')'$. The relations in model (5.22) for $\mathbf{y}$ expressed in vector form can be written as

$$\Phi\, \mathbf{y} = \Theta\, \mathbf{e} + F\, \mathbf{a}_*\,, \tag{5.23}$$

where $\Phi = (I_T \otimes I_k) - \sum_{i=1}^{p} (L^i \otimes \Phi_i)$, $\Theta = (I_T \otimes I_k) - \sum_{i=1}^{q} (L^i \otimes \Theta_i)$, and L denotes the $T \times T$ lag matrix that has ones on the (sub)diagonal directly below the main diagonal and zeros elsewhere. The matrix F in (5.23) has the form

$$F = \begin{bmatrix} A_p & B_q \\ 0 & 0 \end{bmatrix},$$

where

$$A_p = \begin{bmatrix} \Phi_p & \Phi_{p-1} & . & . & . & \Phi_1 \\ 0 & \Phi_p & . & . & . & \Phi_2 \\ . & . & . & . & . & . \\ . & . & . & . & . & . \\ 0 & 0 & . & . & . & \Phi_p \end{bmatrix} \quad \text{and} \quad B_q = -\begin{bmatrix} \Theta_q & \Theta_{q-1} & . & . & . & \Theta_1 \\ 0 & \Theta_q & . & . & . & \Theta_2 \\ . & . & . & . & . & . \\ . & . & . & . & . & . \\ 0 & 0 & . & . & . & \Theta_q \end{bmatrix}.$$

Note that $\mathbf{a}_*$ denotes the vector of presample values of the $\boldsymbol{Y}_t$, $t = 1-p, \ldots, 0$, and ε_t, $t = 1-q, \ldots, 0$, that are needed in (5.23).

Since the vector $\mathbf{e}$ in (5.23) is independent of the presample vector $\mathbf{a}_*$, with $\text{Cov}(\mathbf{e}) = I_T \otimes \Sigma$, the $kT \times kT$ covariance matrix of $\mathbf{y}$ can be expressed as

$$\Gamma_T = E(\mathbf{y}\,\mathbf{y}') = \Phi^{-1}\,[\,\Theta\,(I_T \otimes \Sigma)\,\Theta' + F\,\Omega\,F'\,]\,\Phi'^{-1}, \tag{5.24}$$

where $\Omega = E(\mathbf{a}_*\,\mathbf{a}_*')$ denotes the covariance matrix of $\mathbf{a}_*$. The matrix Ω has the form

$$\Omega = \begin{bmatrix} \Gamma_p & C' \\ C & I_q \otimes \Sigma \end{bmatrix},$$

where Γ_p is the $kp \times kp$ matrix with $\Gamma(j-i) = E(\boldsymbol{Y}_{t-j}\,\boldsymbol{Y}_{t-i}')$ in the (i, j)th block. The matrix C in Ω is $C = E(\mathbf{e}_*\,\mathbf{y}_*') = (I_q \otimes \Sigma)\,\Psi$, where

$$\Psi = \begin{bmatrix} . & . & . & . & \Psi_{q-2}' & \Psi_{q-1}' \\ . & . & . & . & \Psi_{q-3}' & \Psi_{q-2}' \\ . & . & . & . & . & . \\ . & . & . & . & . & . \\ 0 & . & . & . & I_k & \Psi_1' \\ 0 & . & . & . & 0 & I_k \end{bmatrix}$$

is a $kq \times kp$ matrix, with (i, j)th block element defined by $\Sigma^{-1}E(\varepsilon_{i-q}\,\boldsymbol{Y}_{j-p}')$ for $i = 1, \ldots, q$ and $j = 1, \ldots, p$, such that $E(\varepsilon_u\,\boldsymbol{Y}_v') = \Sigma\,\Psi_{v-u}'$ for $v \geq u$ and 0 otherwise. The matrices Ψ_j are the coefficients in the infinite MA operator $\Psi(B) = \Phi(B)^{-1}\,\Theta(B) = \sum_{j=0}^{\infty} \Psi_j\,B^j$, $\Psi_0 = I$. Recall that the Ψ_j are easily determined recursively through the equations (2.12) given in Section 2.3.1. The autocovariance matrices $\Gamma(j)$ in Γ_p for the vector ARMA(p,q) process $\{\boldsymbol{Y}_t\}$ defined by (5.22) can directly be determined in terms of the coefficient matrices Φ_j and Θ_j, and Σ [e.g., see Nicholls and Hall (1979)]. Specifically, the $\Gamma(j)$ are determined by solving the relations from (2.14) of Section 2.3.2,

$$\Gamma(l) - \sum_{j=1}^{p} \Gamma(l-j)\,\Phi_j' = \begin{cases} -\sum_{j=l}^{q} \Psi_{j-l}\,\Sigma\,\Theta_j' , & l = 0, 1, \ldots, q \\ 0, & l = q+1, \ldots, p, \end{cases}$$

noting that $\Gamma(-j) = \Gamma(j)'$.

5.3.1 Expressions for the Exact Likelihood Function and Exact Backcasts

Using a standard matrix inversion formula for Γ_T in (5.24) (Rao, 1973, p. 33), we find that

$$\Gamma_T^{-1} = \Phi' [\, \Gamma_0^{-1} - \Gamma_0^{-1} F\, D^{-1} F' \,\Gamma_0^{-1} \,]\, \Phi, \tag{5.25}$$

where $\Gamma_0 = \Theta\,(I_T \otimes \Sigma)\,\Theta'$, so that $\Gamma_0^{-1} = \Theta'^{-1}(I_T \otimes \Sigma^{-1})\,\Theta^{-1}$, and $D = \Omega^{-1} + F'\,\Gamma_0^{-1} F$. It is noted that Ω^{-1} may conveniently be expressed as

$$\Omega^{-1} = \begin{bmatrix} K^{-1} & -K^{-1}\Psi' \\ -\Psi K^{-1} & (I_q \otimes \Sigma^{-1}) + \Psi K^{-1} \Psi' \end{bmatrix},$$

where $K = \Gamma_p - \Psi'\,(I_q \otimes \Sigma)\,\Psi$. We define the vector $\mathbf{w} = \Phi\,\mathbf{y} = (\, W_1', \ldots, W_T' \,)'$, where $W_t = Y_t - \sum_{j=1}^{t-1} \Phi_j\, Y_{t-j}$, $t = 1, \ldots, p$, $W_t = Y_t - \sum_{j=1}^{p} \Phi_j\, Y_{t-j}$, $t > p$. From (5.24) we also have the relations involving determinants given by

$$|\Gamma_T| = |\Phi|^{-2}\; |\Gamma_0 + F\,\Omega\,F'| = |\Gamma_0|\; |\Omega|\; |\Omega^{-1} + F'\,\Gamma_0^{-1} F| = |\Sigma|^T\; |\Omega|\; |D|$$

and $|\Omega| = |I_q \otimes \Sigma|\; |K| = |\Sigma|^q\; |K|$. Thus, ignoring the normalizing factor involving 2π, the (*exact*) *likelihood function* can be written as

$$L(\,\eta; \mathbf{y}\,) = |\Sigma|^{-(T+q)/2}\, |D|^{-1/2}\, |K|^{-1/2}$$

$$\times \exp\{ -\frac{1}{2}\, \mathbf{w}' [\, \Gamma_0^{-1} - \Gamma_0^{-1} F\, D^{-1} F'\, \Gamma_0^{-1}] \mathbf{w} \}, \tag{5.26}$$

where η denotes the elements of the matrices Φ_j, Θ_j, and Σ. The quadratic form in the exponent in (5.26) can also be conveniently expressed as

$$\mathbf{y}'\, \Gamma_T^{-1}\, \mathbf{y} = \mathbf{w}'\, \Gamma_0^{-1}\, \mathbf{w} - \hat{\mathbf{a}}_*'\, D\, \hat{\mathbf{a}}_*, \tag{5.27}$$

where $\hat{\mathbf{a}}_* = D^{-1}\, F'\, \Gamma_0^{-1}\, \mathbf{w}$.

Now it is easily verified from (5.23) and (5.25) that the vector $\hat{\mathbf{a}}_*$ as defined actually equals $\hat{\mathbf{a}}_* = E(\,\mathbf{a}_*\, \mathbf{y}'\,)\, \{\, E(\,\mathbf{y}\,\mathbf{y}'\,)\,\}^{-1}\, \mathbf{y} = E(\,\mathbf{a}_* \mid \mathbf{y}\,)$, the vector of predictors of the presample values $\mathbf{a}_*$ given the data $\mathbf{y}$, or the so-called "*backcasted*" *values* of $\mathbf{a}_*$. In addition, we find that the MSE matrix of $\hat{\mathbf{a}}_*$ (i.e., the conditional covariance matrix of $\mathbf{a}_*$, given $\mathbf{y}$) is equal to $\text{Cov}(\,\mathbf{a}_* - \hat{\mathbf{a}}_*\,) = \text{Cov}(\,\mathbf{a}_* \mid \mathbf{y}\,) = \text{Cov}(\,\mathbf{a}_*\,) - \text{Cov}(\,\hat{\mathbf{a}}_*\,) = D^{-1}$ using (5.24) and (5.25). With the vector $\mathbf{e}$ as defined by (5.23), we also note the following identity:

$$\mathbf{e}'\,(\,I_T \otimes \Sigma^{-1}\,)\,\mathbf{e} + \mathbf{a}_*'\,\Omega^{-1}\,\mathbf{a}_* = (\,\mathbf{w} - F\,\mathbf{a}_*\,)'\,\Gamma_0^{-1}\,(\,\mathbf{w} - F\,\mathbf{a}_*\,) + \mathbf{a}_*'\,\Omega^{-1}\,\mathbf{a}_*$$

$$= \mathbf{w}'\,\Gamma_0^{-1}\,\mathbf{w} - \hat{\mathbf{a}}_*'\,D\,\hat{\mathbf{a}}_* + (\,\mathbf{a}_* - \hat{\mathbf{a}}_*\,)'\,D\,(\,\mathbf{a}_* - \hat{\mathbf{a}}_*\,). \qquad (5.28)$$

This identity provides another interpretation for the determination of the density function of the vector $\mathbf{y}$ as given in (5.26)–(5.27). The left side of (5.28) represents the terms (quadratic form) in the exponent of the joint density of $(\,\mathbf{e},\,\mathbf{a}_*\,)$, the middle expression in (5.28) corresponds to the joint density of $(\,\mathbf{y},\,\mathbf{a}_*\,)$ under the transformation $\Phi\,\mathbf{y} = \Theta\,\mathbf{e} + F\,\mathbf{a}_*$, and the right side of (5.28) corresponds to the same density when expressed as the product of the marginal density of $\mathbf{y}$ [with quadratic form as given in (5.27)] and the conditional density of $\mathbf{a}_*$ given $\mathbf{y}$ [with quadratic form given by the last term on the right side of (5.28)]. Hence, the marginal density of $\mathbf{y}$ is obtained from this joint density of $(\,\mathbf{y},\,\mathbf{a}_*\,)$ by the standard procedure of "integrating out" the variables $\mathbf{a}_*$.

For computations in (5.27), we also note that

$$\mathbf{w}'\,\Gamma_0^{-1}\,\mathbf{w} = \mathbf{y}'\,\Phi'\,\Theta'^{-1}(\,I_T \otimes \Sigma^{-1}\,)\,\Theta^{-1}\,\Phi\,\mathbf{y} = \mathbf{e}_0'\,(\,I_T \otimes \Sigma^{-1}\,)\,\mathbf{e}_0 = \sum_{t=1}^{T} \varepsilon_t^{0\prime}\,\Sigma^{-1}\,\varepsilon_t^0\,,$$

where the components of the vector $\mathbf{e}_0 = \Theta^{-1}\Phi\,\mathbf{y} = (\,\varepsilon_1^{0\prime},\,\ldots,\,\varepsilon_T^{0\prime}\,)'$ are easily computed recursively from $\Theta\,\mathbf{e}_0 = \Phi\,\mathbf{y}$ as the "conditional residuals",

$$\varepsilon_t^0 = \boldsymbol{Y}_t - \sum_{j=1}^{p} \Phi_j\,\boldsymbol{Y}_{t-j} + \sum_{j=1}^{q} \Theta_j\,\varepsilon_{t-j}^0\,, \qquad t = 1,\,\ldots,\,T\,,$$

with all presample values set equal to zero (i.e., $\mathbf{a}_*^0 \equiv 0$). We note that use of only the term $\mathbf{w}'\,\Gamma_0^{-1}\,\mathbf{w}$ in the exponent and ignoring the term $|\,\Omega\,|^{-1/2}\;|\,D\,|^{-1/2}$ in (5.26) corresponds to the approximate likelihood (conditional on $\mathbf{a}_* = 0$) considered by Tunnicliffe Wilson (1973), Anderson (1980), and others. Also note then that $\hat{\mathbf{a}}_* = D^{-1}F'\,\Theta'^{-1}(\,I_T \otimes \Sigma^{-1}\,)\,\mathbf{e}_0 = D^{-1}F'\,\mathbf{u}$, and the vector $\mathbf{u} = (\,U_1',\,\ldots,\,U_T'\,)' = \Theta'^{-1}(\,I_T \otimes \Sigma^{-1}\,)\,\mathbf{e}_0$ can easily be computed through a backward recursion as

$$\boldsymbol{U}_t = \Sigma^{-1}\varepsilon_t^0 + \sum_{j=1}^{q} \Theta_j'\,\boldsymbol{U}_{t+j}\,, \qquad t = 1,\,\ldots,\,T\,, \qquad \text{with } \boldsymbol{U}_{T+1} = \cdots = \boldsymbol{U}_{T+q} = 0\,.$$

This relation clearly illustrates the "back forecasting" nature of the predicted values $\hat{\mathbf{a}}_* = D^{-1}F'\,\mathbf{u}$, a term used by Box and Jenkins (1976) in the univariate case. Additionally, the computation of the matrix $M = \Theta^{-1}F$ required in $D = \Omega^{-1} + F'\Theta'^{-1}(\,I_T \otimes \Sigma^{-1}\,)\,\Theta^{-1}F$ is also conveniently performed from the relation $\Theta\,M = F$ similar to the computation of the ε_t^0.

In addition, as a consequence of (5.28), when $\mathbf{a}_*$ is set equal to $\hat{\mathbf{a}}_*$ and, thus, from (5.23) $\hat{\mathbf{e}} = \Theta^{-1}(\,\Phi\,\mathbf{y} - F\,\hat{\mathbf{a}}_*\,)$, one obtains the identity

$$\mathbf{w}'\,\Gamma_0^{-1}\,\mathbf{w} - \hat{\mathbf{a}}_*'\,D\,\hat{\mathbf{a}}_* = \hat{\mathbf{e}}'\,(\,I_T \otimes \Sigma^{-1})\,\hat{\mathbf{e}} + \hat{\mathbf{a}}_*'\,\Omega^{-1}\,\hat{\mathbf{a}}_* \tag{5.29}$$

which may also be obtained by direct substitution of $\mathbf{w} = \mathbf{\Phi}\,\mathbf{y} = \Theta\,\hat{\mathbf{e}} + F\,\hat{\mathbf{a}}_*$ into (5.27). We also note from the inversion formula for Ω mentioned below (5.25) that

$$\mathbf{a}_*'\,\Omega^{-1}\,\mathbf{a}_* = \mathbf{e}_*'\,(\,I_q \otimes \Sigma^{-1})\,\mathbf{e}_* + (\,\mathbf{y}_* - \Psi'\,\mathbf{e}_*\,)'\,K^{-1}\,(\,\mathbf{y}_* - \Psi'\,\mathbf{e}_*\,), \tag{5.30}$$

where the first term on the right side of (5.30) corresponds to the marginal density of $\mathbf{e}_*$, while the second term corresponds to the conditional density of $\mathbf{y}_*$ given $\mathbf{e}_*$. Using (5.30) with $\mathbf{a}_* = \hat{\mathbf{a}}_*$ in (5.29), we find that an alternate expression for the term (5.27) in the exponent of the exact likelihood function is

$$\begin{aligned}
&\mathbf{w}'\,\Gamma_0^{-1}\,\mathbf{w} - \hat{\mathbf{a}}_*'\,D\,\hat{\mathbf{a}}_* \\
&\qquad = \hat{\mathbf{e}}'\,(\,I_T \otimes \Sigma^{-1})\,\hat{\mathbf{e}} + \hat{\mathbf{e}}_*'\,(\,I_q \otimes \Sigma^{-1})\,\hat{\mathbf{e}}_* + (\,\hat{\mathbf{y}}_* - \Psi'\,\hat{\mathbf{e}}_*\,)'\,K^{-1}\,(\,\hat{\mathbf{y}}_* - \Psi'\,\hat{\mathbf{e}}_*\,) \\
&\qquad = \sum_{t=1-q}^{T} \hat{\varepsilon}_t'\,\Sigma^{-1}\,\hat{\varepsilon}_t + (\,\hat{\mathbf{y}}_* - \Psi'\,\hat{\mathbf{e}}_*\,)'\,K^{-1}\,(\,\hat{\mathbf{y}}_* - \Psi'\,\hat{\mathbf{e}}_*\,),
\end{aligned} \tag{5.31}$$

where the $\hat{\varepsilon}_t$ are the elements of the vectors $\hat{\mathbf{e}}_*$ and $\hat{\mathbf{e}}$. Hence, the $\hat{\varepsilon}_t$ are the "exact residuals" which satisfy

$$\hat{\varepsilon}_t = \boldsymbol{Y}_t - \sum_{j=1}^{p} \Phi_j\,\boldsymbol{Y}_{t-j} + \sum_{j=1}^{q} \Theta_j\,\hat{\varepsilon}_{t-j}, \qquad t = 1, \ldots, T, \tag{5.32}$$

where the presample values for the $\boldsymbol{Y}_t$ and ε_t used in (5.32) are the predicted (backcasted) values $\hat{\boldsymbol{Y}}_t$, $t = 1-p, \ldots, 0$, $\hat{\varepsilon}_t$, $t = 1-q, \ldots, 0$. Note that for the special case of a pure MA(q) model, expression (5.31) simplifies such that the last term on the right side is not present, whereas for the pure AR(p) model, the right side of (5.31) reduces to $\sum_{t=1}^{T} \hat{\varepsilon}_t'\,\Sigma^{-1}\,\hat{\varepsilon}_t + \hat{\mathbf{y}}_*'\,\Gamma_p^{-1}\,\hat{\mathbf{y}}_*$, where $\hat{\mathbf{y}}_* = D^{-1}A_p'\,(\,I_p \otimes \Sigma^{-1})\,\mathbf{w}_p$ with $\mathbf{w}_p = (\,\boldsymbol{W}_1', \ldots, \boldsymbol{W}_p'\,)'$ and $D = \Gamma_p^{-1} + A_p'\,(\,I_p \otimes \Sigma^{-1}\,)\,A_p$.

Thus, expressions (5.26) with (5.27) and (5.31) provide simple convenient expressions for the exact likelihood function of the vector ARMA model which may be used together with nonlinear maximization algorithms, such as modified Newton-Raphson methods, to obtain exact maximum likelihood estimates of the parameters. Alternate approaches to the construction of the exact likelihood function in vector ARMA models are through use of an innovations algorithm method which will be discussed in the next section, or the related method of use of the state-space formulation of the ARMA model and associated Kalman filtering techniques, which will be discussed later in Sections 7.1 and 7.2. We now examine the preceding results for two special cases in more explicit detail.

5.3.2 Special Cases of the Exact Likelihood Results

(a) AR(1) model. For the model $\boldsymbol{Y}_t = \Phi_1\,\boldsymbol{Y}_{t-1} + \varepsilon_t$, the form of the exact likelihood is

$$L(\Phi_1, \Sigma; \mathbf{y}) = |\Sigma|^{-T/2} \, |\Gamma(0)|^{-1/2} \, |D|^{-1/2}$$

$$\times \exp[-(1/2)(\textstyle\sum_{t=1}^{T} \hat{\varepsilon}_t' \Sigma^{-1} \hat{\varepsilon}_t + \hat{Y}_0' \Gamma(0)^{-1} \hat{Y}_0)]$$

$$= |\Gamma(0)|^{-1/2} \, |\Sigma|^{-(T-1)/2} \exp[-(1/2)(Y_1' \Gamma(0)^{-1} Y_1 + \textstyle\sum_{t=2}^{T} \varepsilon_t' \Sigma^{-1} \varepsilon_t)]$$

$$\equiv f_1(Y_1) \; L_*(\Phi_1, \Sigma; Y_2, Y_3, \ldots, Y_T \mid Y_1),$$

where $\hat{\varepsilon}_t = \varepsilon_t = Y_t - \Phi_1 Y_{t-1}$, $t = 2, \ldots, T$, $\hat{\varepsilon}_1 = Y_1 - \Phi_1 \hat{Y}_0$, and we have the correspondences $\mathbf{a}_* = \mathbf{y}_* = Y_0$, $\Omega = \Gamma(0) = K$, $\Theta = I_{kT}$, $\Gamma_0 = I_T \otimes \Sigma$, $F' = (\Phi_1', 0, 0, \ldots, 0)$, and $D = \Gamma(0)^{-1} + \Phi_1' \Sigma^{-1} \Phi_1$. The "backcasted" value of Y_0 is $\hat{Y}_0 = D^{-1} \Phi_1' \Sigma^{-1} Y_1 \equiv \Phi_{11}^* Y_1$, where $\Phi_{11}^* = \Gamma(1) \Gamma(0)^{-1} \equiv D^{-1} \Phi_1' \Sigma^{-1}$ is the "backward" AR(1) coefficient. In expressing the equivalent forms of L above, we have used the identities

$$\hat{\varepsilon}_1' \Sigma^{-1} \hat{\varepsilon}_1 + \hat{Y}_0' \Gamma(0)^{-1} \hat{Y}_0 = (Y_1 - \Phi_1 \hat{Y}_0)' \Sigma^{-1} (Y_1 - \Phi_1 \hat{Y}_0) + \hat{Y}_0' \Gamma(0)^{-1} \hat{Y}_0$$

$$= Y_1' \Sigma^{-1} Y_1 - \hat{Y}_0' D \hat{Y}_0 \equiv Y_1' \Gamma(0)^{-1} Y_1,$$

based on equation (5.29) and the relation $\Gamma(0)^{-1} = \Sigma^{-1} - \Sigma^{-1} \Phi_1 D^{-1} \Phi_1' \Sigma^{-1}$ (e.g., Rao, 1973, p. 33), and

$$|D| = |\Gamma(0)^{-1} + \Phi_1' \Sigma^{-1} \Phi_1| = |\Gamma(0)^{-1}| \; |\Sigma^{-1}| \; |\Sigma + \Phi_1 \Gamma(0) \Phi_1'| = |\Sigma|^{-1},$$

using the identity $|I + E F| = |I + F E|$. The factor

$$L_* = |\Sigma|^{-(T-1)/2} \exp[-(1/2) \textstyle\sum_{t=2}^{T} \varepsilon_t' \Sigma^{-1} \varepsilon_t]$$

represents the conditional p.d.f. of $(Y_2, Y_3, \ldots, Y_T)$, given Y_1, and is recognized as the conditional likelihood function. Clearly, as T increases, the likelihood function L will tend to be dominated by the conditional likelihood factor L_*. So for large T, the contribution of f_1 to the exact likelihood L is relatively negligible, and maximization of the conditional likelihood will yield asymptotically equivalent estimates to the maximization of the exact likelihood. It is also clear that the conditional likelihood L_* has the form of the standard multivariate linear model so that the conditional MLEs of Φ_1 and Σ are the same as the least squares estimates examined in Section 4.3 (assuming $\boldsymbol{\mu} = 0$ in this case). In the general AR(p) model we have similar results, with the exact likelihood

$$L(\Phi_{(p)}, \Sigma; \mathbf{y}) = |\Gamma_p|^{-1/2} \, |\Sigma|^{-(T-p)/2}$$

$$\times \exp[-(1/2)(\mathbf{y}_p' \Gamma_p^{-1} \mathbf{y}_p + \textstyle\sum_{t=p+1}^{T} \varepsilon_t' \Sigma^{-1} \varepsilon_t)]$$

$$\equiv f_1(\mathbf{y}_p) \; L_*(\Phi_{(p)}, \Sigma; Y_{p+1}, \ldots, Y_T \mid \mathbf{y}_p),$$

where $\mathbf{y}_p = (Y_1', Y_2', \ldots, Y_p')'$ and $\Gamma_p = \mathrm{Cov}(\mathbf{y}_p)$, and the conditional likelihood L_* corresponds to the conditional p.d.f. of $(\boldsymbol{Y}_{p+1}, \boldsymbol{Y}_{p+2}, \ldots, \boldsymbol{Y}_T)$, given $\mathbf{y}_p$. Again L_* has the form of the likelihood in the general multivariate linear model, and the conditional MLEs in the vector AR(p) model are the same as the LSEs.

(b) MA(1) model. For the model $\boldsymbol{Y}_t = \varepsilon_t - \Theta_1 \varepsilon_{t-1}$ we have the correspondences $\mathbf{a}_* = \mathbf{e}_* = \varepsilon_0$, $\Omega = \Sigma$, $F' = -(\Theta_1', 0, 0, \ldots, 0)$,

$$D = \Sigma^{-1} + F' \Theta'^{-1} (I_T \otimes \Sigma^{-1}) \Theta^{-1} F = \sum_{j=0}^{T} \Theta_1'^{j} \Sigma^{-1} \Theta_1^{j} .$$

Then equivalent forms of the exact likelihood are

$$\begin{aligned} L(\Theta_1, \Sigma ; \mathbf{y}) &= |\Sigma|^{-(T+1)/2} \, |D|^{-1/2} \exp[-(1/2) \textstyle\sum_{t=0}^{T} \hat{\varepsilon}_t' \Sigma^{-1} \hat{\varepsilon}_t] \\ &= |\Sigma|^{-1/2} \, |D|^{-1/2} \, |\Sigma|^{-T/2} \\ &\quad \times \exp[-(1/2) (\textstyle\sum_{t=1}^{T} \varepsilon_t^{0\prime} \Sigma^{-1} \varepsilon_t^0 - \hat{\varepsilon}_0' D \hat{\varepsilon}_0)] \\ &\equiv f_1(\mathbf{y}) \, L_*(\Theta_1, \Sigma ; \boldsymbol{Y}_1, \boldsymbol{Y}_2, \ldots, \boldsymbol{Y}_T \mid \varepsilon_0 = 0), \end{aligned}$$

where $\hat{\varepsilon}_t = \boldsymbol{Y}_t + \Theta_1 \hat{\varepsilon}_{t-1}$, $t = 1, 2, \ldots, T$, are the "exact" residuals using the initial value $\hat{\varepsilon}_0$, and $\hat{\varepsilon}_0 = D^{-1} F' \Gamma_0^{-1} \mathbf{y} = -D^{-1} \sum_{t=1}^{T} \Theta_1'^{t} \Sigma^{-1} (\sum_{j=0}^{t-1} \Theta_1^{j} \boldsymbol{Y}_{t-j})$ is the "backcasted" value of ε_0, whereas $\varepsilon_t^0 = \boldsymbol{Y}_t + \Theta_1 \varepsilon_{t-1}^0$, $t = 1, 2, \ldots, T$, with $\varepsilon_0^0 = 0$ are the "conditional" residuals. (Note that $\varepsilon_t^0 = \sum_{j=0}^{t-1} \Theta_1^{j} \boldsymbol{Y}_{t-j}$.) The factor $L_* = |\Sigma|^{-T/2} \exp[-(1/2) \sum_{t=1}^{T} \varepsilon_t^{0\prime} \Sigma^{-1} \varepsilon_t^0]$ is the conditional likelihood of $\boldsymbol{Y}_1, \ldots, \boldsymbol{Y}_T$ (conditional on $\varepsilon_0 = 0$). For large T, the likelihood L will tend to be dominated by the conditional likelihood factor L_*. However, empirically the difference in estimation results between the exact and approximate (conditional) likelihood function may be substantial for small or moderate sample sizes T in situations where the roots of $\det\{ \Theta(B) \} = 0$ lie close to the boundary of the unit circle (boundary of the invertible region). This is because in these situations the effects of the initial approximation $\varepsilon_0^0 = 0$ may not die out quickly and the factor $|D|^{-1/2}$ in L will not be negligible since Θ_1^{j} will not die out very quickly. A numerical examination of certain expected value properties of the conditional and exact likelihood functions for the univariate MA(1) model by Osborn (1982) gives some analytic support to these conclusions. For the conditional likelihood L_* we note that the conditional MLE of Σ is given by $\hat{\Sigma} = T^{-1} \sum_{t=1}^{T} \varepsilon_t^0 \varepsilon_t^{0\prime}$. But for the exact likelihood, the exact MLE of Σ is more complicated because the matrix D in L involves Σ in a nontrivial way, although it seems reasonable to use $\hat{\Sigma} = T^{-1} \sum_{t=0}^{T} \hat{\varepsilon}_t \hat{\varepsilon}_t'$ as the estimate of Σ. It might be noted that this estimate of Σ based on the "exact" residuals $\hat{\varepsilon}_t$ will always yield a "smaller" value than the MLE from the conditional likelihood, in the sense that from (5.29) we have $\sum_{t=0}^{T} \hat{\varepsilon}_t' \Sigma^{-1} \hat{\varepsilon}_t < \sum_{t=1}^{T} \varepsilon_t^{0\prime} \Sigma^{-1} \varepsilon_t^0$, or $\mathrm{tr}(\Sigma^{-1} \sum_{t=0}^{T} \hat{\varepsilon}_t \hat{\varepsilon}_t') < \mathrm{tr}(\Sigma^{-1} \sum_{t=1}^{T} \varepsilon_t^0 \varepsilon_t^{0\prime})$, for fixed Σ.

5.4 Innovations Form of the Exact Likelihood Function for ARMA Models

An alternate approach to the method presented in Section 5.3 for construction of the exact likelihood function of the vector ARMA(p,q) model (5.22) is through the formation of the one-step innovations corresponding to the observations $\boldsymbol{Y}_t$, $t = 1, \ldots, T$, and their covariance matrices. We discuss this approach in the present section. The method is similar to that proposed by Ansley (1979) for the univariate ARMA model and presented by Brockwell and Davis (1987, Chap. 11) [see also the closely related earlier work by Rissanen and Barbosa (1969), Rissanen (1973a, 1973b), and Aasnaes and Kailath (1973)], and it is also related to the Kalman filtering method to be discussed more generally in Sections 7.1 and 7.2.

5.4.1 Use of Innovations Algorithm Approach for the Exact Likelihood

Using the notation of Section 5.3, from (5.23) the vector ARMA model relations are expressed in vector form as

$$\mathbf{w} = \Phi\, \mathbf{y} = \Theta\, \mathbf{e} + F\, \mathbf{a}_* ,$$

where $\mathbf{w} = \Phi\, \mathbf{y} = (\boldsymbol{W}_1', \ldots, \boldsymbol{W}_T')'$, with $\boldsymbol{W}_t = \boldsymbol{Y}_t - \sum_{j=1}^{t-1} \Phi_j\, \boldsymbol{Y}_{t-j}$, $t = 1, \ldots, p$, $\boldsymbol{W}_t = \boldsymbol{Y}_t - \sum_{j=1}^{p} \Phi_j\, \boldsymbol{Y}_{t-j}$, $t > p$. The covariance matrix of the vector $\mathbf{w}$ is

$$\Gamma_w = E(\mathbf{w}\, \mathbf{w}') = \Theta\, (I_T \otimes \Sigma)\, \Theta' + F\, \Omega\, F' ,$$

and it can be seen that $\mathrm{Cov}(\boldsymbol{W}_t, \boldsymbol{W}_{t+s}) = 0$ for $|s| > m = \max(p, q)$, since $\boldsymbol{W}_t = \varepsilon_t - \sum_{j=1}^{q} \Theta_j\, \varepsilon_{t-j}$ for $t > p$ is an MA(q) process. Hence, the matrix Γ_w has nonzero blocks only in a band about the main diagonal block of maximum bandwidth m (and of bandwidth q after the first m block rows). The matrix thus has a block (Cholesky) decomposition of the form $\Gamma_w = G\, D_T\, G'$, where G is a lower triangular block-band matrix with bandwidth corresponding to that of Γ_w, and with diagonal blocks equal to I_k. The matrix D_T is block diagonal, $D_T = \mathrm{Diag}(\Sigma_{1|0}, \Sigma_{2|1}, \ldots, \Sigma_{T|T-1})$. Thus, the covariance matrix of $\mathbf{y} = \Phi^{-1} \mathbf{w}$ can be expressed as $\Gamma_T = \Phi^{-1} \Gamma_w\, \Phi'^{-1} = \Phi^{-1} G\, D_T\, G'\, \Phi'^{-1}$, with $|\Gamma_T| = |D_T| = \prod_{t=1}^{T} |\Sigma_{t|t-1}|$ since G and Φ are both lower triangular matrices with diagonal blocks equal to I_k.

We define the vector $\mathbf{e}_T = G^{-1}\, \Phi\, \mathbf{y} = (\varepsilon_{1|0}', \varepsilon_{2|1}', \ldots, \varepsilon_{T|T-1}')'$, and we see that $\mathrm{Cov}(\mathbf{e}_T) = D_T$ and also that

$$\mathbf{y}'\, \Gamma_T^{-1}\, \mathbf{y} = \mathbf{e}_T'\, D_T^{-1}\, \mathbf{e}_T = \sum_{t=1}^{T} \varepsilon_{t|t-1}'\, \Sigma_{t|t-1}^{-1}\, \varepsilon_{t|t-1} .$$

Since the transformation from $\mathbf{y}$ to $\mathbf{e}_T$ is lower triangular, each $\varepsilon_{t|t-1}$ has a form equal to the corresponding $\boldsymbol{Y}_t$ minus a linear combination of

$Y_{t-1}, \ldots, Y_1$, and $\varepsilon_{t|t-1}$ is uncorrelated with the preceding values $Y_{t-1}, \ldots, Y_1$ (since it is uncorrelated with $\varepsilon_{t-1|t-2}, \ldots, \varepsilon_{1|0}$). It thus follows from this property that the $\varepsilon_{t|t-1}$ are equal to the one-step prediction errors $\varepsilon_{t|t-1} = Y_t - \hat{Y}_{t|t-1}$ where $\hat{Y}_{t|t-1} = E(Y_t \mid Y_{t-1}, \ldots, Y_1)$ denotes the one-step linear prediction of Y_t based on $Y_{t-1}, \ldots, Y_1$, and $\Sigma_{t|t-1} = \text{Cov}(\varepsilon_{t|t-1})$ is the one-step prediction error covariance matrix. The $\varepsilon_{t|t-1}$ are referred to as the *one-step innovations* of the Y_t, $t = 1, 2, \ldots, T$. We thus find from the above results that the (exact) likelihood function can be expressed in the *innovations form* as

$$L(\boldsymbol{\eta}; \mathbf{y}) = \left(\prod_{t=1}^{T} |\Sigma_{t|t-1}|^{-1/2}\right) \exp\left\{-(1/2) \sum_{t=1}^{T} \varepsilon_{t|t-1}' \Sigma_{t|t-1}^{-1} \varepsilon_{t|t-1}\right\}, \quad (5.33)$$

where $\varepsilon_{t|t-1} = Y_t - \hat{Y}_{t|t-1}$ and $\Sigma_{t|t-1} = \text{Cov}(\varepsilon_{t|t-1})$.

Note that since the elements $\varepsilon_{t|t-1}$ of $\mathbf{e}_T$ are determined through the relations $G\, \mathbf{e}_T = \Phi\, \mathbf{y}$, they are obtained as

$$\varepsilon_{t|t-1} = Y_t - \sum_{i=1}^{t-1} \Phi_i Y_{t-i} + \sum_{i=1}^{t-1} \Theta_{i,t-1} \varepsilon_{t-i|t-i-1}$$

for $t \le m$, starting with $\varepsilon_{1|0} = Y_1$, and

$$\varepsilon_{t|t-1} = Y_t - \sum_{i=1}^{p} \Phi_i Y_{t-i} + \sum_{i=1}^{q} \Theta_{i,t-1} \varepsilon_{t-i|t-i-1}, \quad (5.34)$$

for $t > m$. In the above equations, the matrices $-\Theta_{i,t-1}$ denote the nonzero block elements to the left of the diagonal block I_k in the tth block row of the lower triangular block-band matrix G. That is, for $t > m$ the tth block row of the matrix G has the form

$$[\ 0 \ \cdots \ 0 \ -\Theta_{q,t-1} \ \cdots \ -\Theta_{1,t-1} \ I_k \ 0 \ 0 \ \cdots \ 0 \ 0\ ].$$

These block elements $-\Theta_{i,t-1}$, $i = 1, \ldots, t-1$ for $t = 2, \ldots, m$ and $i = 1, \ldots, q$ for $t > m$, and the diagonal blocks (one-step prediction error covariance matrices) $\Sigma_{t|t-1}$, are obtained recursively for $t = 1, \ldots, T$ from the block (Cholesky) decomposition algorithm for the covariance matrix Γ_w of the W_t. The starting values for the recursions are

$$\Sigma_{1|0} = \text{Cov}(W_1) = \text{Cov}(Y_1) = \Gamma(0), \quad \Theta_{1,1} = -\text{Cov}(W_2, W_1)\{\text{Cov}(W_1)\}^{-1}.$$

We note that $W_t = \varepsilon_t - \sum_{j=1}^{q} \Theta_j \varepsilon_{t-j}$ is an MA(q) process for $t > p$, so that the block matrices $\Gamma_w(j-i) = \text{Cov}(W_i, W_j)$ for $i, j > p$, in the covariance matrix $\Gamma_w = \text{Cov}(\mathbf{w})$ are easily determined as those of the MA(q) process. That is, for $i, j > p$, from equation (2.3) of Section 2.1.2 we have $\Gamma_w(j-i) = \Gamma_w(l) = \sum_{h=l}^{q} \Theta_{h-l} \Sigma \Theta_h'$ for $l = j-i = 0, 1, \ldots, q$ (with $\Theta_0 = -I$), and $\Gamma_w(j-i) = \Gamma_w(l) = 0$ for $l = j-i > q$. In addition, as presented in the discussion of the general innovations algorithm in the Remark of Section 3.3.2, the recursions for the matrices $-\Theta_{i,t-1}$ and $\Sigma_{t|t-1}$, for $t > m$, reduce to

$$\Theta_{i,t-1} = -\,[\,\Gamma_w(i)' - \textstyle\sum_{j=i+1}^{q} \Theta_{j,t-1}\,\Sigma_{t-j\,|\,t-j-1}\,\Theta'_{j-i,t-i-1}\,]\,\Sigma^{-1}_{t-i\,|\,t-i-1}\,,$$

for $i = q,\, q-1, \ldots,\, 1$, with $\Theta_{q,t-1} = -\,\Gamma_w(q)'\,\Sigma^{-1}_{t-q\,|\,t-q-1}$, and

$$\Sigma_{t\,|\,t-1} = \Gamma_w(0) - \textstyle\sum_{j=1}^{q} \Theta_{j,t-1}\,\Sigma_{t-j\,|\,t-j-1}\,\Theta'_{j,t-1}$$

for $t = m+1, \ldots, T$. In comparison to the general innovations algorithm results discussed in Section 3.3.2, the notable feature in the present ARMA(p,q) model situation is that now the coefficient matrices $\Theta_{i,t-1} \equiv 0$ for $i > q$ (with $t > m$), which results from the block-band structure (of bandwidth q after the first m block rows) of the covariance matrix $\Gamma_w = \text{Cov}(\,\mathbf{w}\,)$ since $\boldsymbol{W}_t$ is an MA(q) process for $t > m$ with zero covariances for all lags greater than q.

From the relations $\mathbf{w} = G\,\mathbf{e}_T$ we have the representation $\boldsymbol{W}_t = \varepsilon_{t\,|\,t-1} - \sum_{i=1}^{q} \Theta_{i,t-1}\,\varepsilon_{t-i\,|\,t-i-1}$. Since the $\varepsilon_{t\,|\,t-1}$ are mutually uncorrelated, we see that $\Theta_{i,t-1} = -\,\text{Cov}(\,\boldsymbol{W}_t,\, \varepsilon_{t-i\,|\,t-i-1}\,)\,\Sigma^{-1}_{t-i\,|\,t-i-1}$. Note that this is similar to the relation $\Theta_i = -\,\text{Cov}(\,\boldsymbol{W}_t,\, \varepsilon_{t-i}\,)\,\Sigma^{-1}$ which holds for the actual ARMA(p,q) model (for $t > p$). So, for example, the initial matrix in the algorithm is $\Theta_{1,1} = -\,\text{Cov}(\,\boldsymbol{W}_2,\, \varepsilon_{1\,|\,0}\,)\,\Sigma^{-1}_{1\,|\,0} = -\,\text{Cov}(\,\boldsymbol{Y}_2 - \Phi_1\,\boldsymbol{Y}_1,\, \boldsymbol{Y}_1\,)\,\Gamma(0)^{-1}$, and hence $\Theta_{1,1} = -\,\Gamma(1)'\,\Gamma(0)^{-1} + \Phi_1$. For the specific case of a vector ARMA(1,1) model, the general recursions to determine the $\Theta_{1,t-1}$ yield

$$\Theta_{1,t-1} = -\,\text{Cov}(\,\boldsymbol{W}_t,\, \varepsilon_{t-1\,|\,t-2})\,\Sigma^{-1}_{t-1\,|\,t-2}$$

$$= -\,\text{Cov}(\,\varepsilon_t - \Theta_1\,\varepsilon_{t-1},\, \varepsilon_{t-1\,|\,t-2})\,\Sigma^{-1}_{t-1\,|\,t-2} = \Theta_1 \Sigma\,\Sigma^{-1}_{t-1\,|\,t-2}\,,$$

since ε_t is uncorrelated with $\varepsilon_{t-1\,|\,t-2}$ and $\text{Cov}(\,\varepsilon_{t-1},\, \varepsilon_{t-1\,|\,t-2}\,) = \text{Cov}(\,\varepsilon_{t-1},\, \boldsymbol{Y}_{t-1}\,) = \Sigma$. In addition, the recursions for the error covariance matrix $\Sigma_{t\,|\,t-1} = \text{Cov}(\,\varepsilon_{t\,|\,t-1}\,)$, in the vector ARMA(1,1) model, yield

$$\Sigma_{t\,|\,t-1} = \text{Cov}(\,\boldsymbol{Y}_t - \Phi_1\,\boldsymbol{Y}_{t-1} + \Theta_{1,t-1}\,\varepsilon_{t-1\,|\,t-2}\,)$$

$$= \text{Cov}(\,\varepsilon_t - \Theta_1\,\varepsilon_{t-1} + \Theta_{1,t-1}\,\varepsilon_{t-1\,|\,t-2}\,)$$

$$= \Sigma + \Theta_1\,\Sigma\,\Theta'_1 - \Theta_1\,\Sigma\,\Theta'_{1,t-1} = \Sigma + \Theta_1\,\Sigma\,\Theta'_1 - \Theta_{1,t-1}\,\Sigma_{t-1\,|\,t-2}\,\Theta'_{1,t-1}\,.$$

More directly, in the ARMA(1,1) example since $\boldsymbol{W}_t$ is an MA(1) process for $t > 1$, from the above recursive equations given for the general ARMA(p,q) model, we see that $\Theta_{1,t-1} = -\,\Gamma_w(1)'\,\Sigma^{-1}_{t-1\,|\,t-2} = \Theta_1\,\Sigma\,\Sigma^{-1}_{t-1\,|\,t-2}$ and $\Sigma_{t\,|\,t-1} = \Gamma_w(0) - \Theta_{1,t-1}\,\Sigma_{t-1\,|\,t-2}\,\Theta'_{1,t-1} = \Sigma + \Theta_1\,\Sigma\,\Theta'_1 - \Theta_{1,t-1}\,\Sigma_{t-1\,|\,t-2}\,\Theta'_{1,t-1}$.

5.4.2 Prediction of Vector ARMA Processes Using the Innovations Approach

The innovations approach can also be used to obtain l-step ahead (finite sample) predictors $\hat{\boldsymbol{Y}}_{t+l\,|\,t} = E[\,\boldsymbol{Y}_{t+l} \mid \boldsymbol{Y}_t, \ldots, \boldsymbol{Y}_1\,]$ in a convenient computational manner. To establish the necessary recursive prediction relations, we note that $\hat{\boldsymbol{Y}}_{t+l\,|\,t} = E[\,\hat{\boldsymbol{Y}}_{t+l\,|\,t+l-1} \mid \boldsymbol{Y}_t, \ldots, \boldsymbol{Y}_1\,]$ and that the one-step innovations $\varepsilon_{t+l-i\,|\,t+l-i-1} = \boldsymbol{Y}_{t+l-i} - \hat{\boldsymbol{Y}}_{t+l-i\,|\,t+l-i-1}$ are uncorrelated with $\boldsymbol{Y}_t, \ldots, \boldsymbol{Y}_1$ for

$i < l$. Thus, applying conditional expectations to the relation $\hat{Y}_{t+l\,|\,t+l-1} = \sum_{i=1}^{p} \Phi_i \, Y_{t+l-i} - \sum_{i=1}^{q} \Theta_{i,t+l-1} \, \varepsilon_{t+l-i\,|\,t+l-i-1}$, we obtain

$$\hat{Y}_{t+l\,|\,t} = \sum_{i=1}^{p} \Phi_i \, \hat{Y}_{t+l-i\,|\,t} - \sum_{i=l}^{q} \Theta_{i,t+l-1} \, \varepsilon_{t+l-i\,|\,t+l-i-1}, \qquad l = 1, 2, \ldots, q, \tag{5.35}$$

and $\hat{Y}_{t+l\,|\,t} = \sum_{i=1}^{p} \Phi_i \, \hat{Y}_{t+l-i\,|\,t}$ for $l > q$, where $\hat{Y}_{t+l-i\,|\,t} = Y_{t+l-i}$ for $i \geq l$. Hence, the predictors $Y_{t+l\,|\,t}$ can be determined recursively through the above relations (5.35), by continuing the recursions to obtain the matrices $-\Theta_{i,t+h}$ (for $h = 0, \ldots, l-1$) in the block (Cholesky) factorization. For large t, assuming invertibility of the MA model the $\Theta_{i,t+h}$ will be very nearly identical to Θ_i [$\lim_{t \to \infty} \Theta_{i,t} = \Theta_i$, e.g., see Ansley and Kohn (1985c)], and the above prediction relations (5.35) will reduce essentially to the same form as the relations (2.22) in Section 2.5.3.

5.5 Overall Checking for Model Adequacy

5.5.1 Residual Correlation Matrices, and Overall Goodness-of-Fit Test

Having fitted a vector AR(p) model by least squares, as in Section 4.3, or, more generally, fitted a vector ARMA(p,q) model by maximum likelihood, diagnostic tests for overall model adequacy can be examined based on residual autocorrelation matrices. Suppose a vector ARMA(p,q) model is estimated for a k-dimensional series Y_t by maximum likelihood, and the residuals $\hat{\varepsilon}_t = Y_t - \sum_{j=1}^{p} \hat{\Phi}_j \, Y_{t-j} + \sum_{j=1}^{q} \hat{\Theta}_j \, \hat{\varepsilon}_{t-j}$, $t = 1, \ldots, T$, are obtained. To check adequacy of the model, we look at residual autocorrelation matrices. Let

$$C_\varepsilon(l) = \frac{1}{T} \sum_{t=1}^{T-l} \hat{\varepsilon}_t \, \hat{\varepsilon}_{t+l}', \qquad l = 0, 1, \ldots, s,$$

with $C_\varepsilon(0) = T^{-1} \sum_{t=1}^{T} \hat{\varepsilon}_t \, \hat{\varepsilon}_t' \approx \hat{\Sigma}$, denote the *residual covariance matrices.* Also, let the *residual autocorrelation matrices* be denoted as

$$\hat{\rho}_\varepsilon(l) = \hat{V}_\varepsilon^{-1/2} \, C_\varepsilon(l) \, \hat{V}_\varepsilon^{-1/2} = \{ \, \hat{\rho}_{ij}(l) \, \},$$

where $\hat{V}_\varepsilon = \text{Diag}(\, \hat{\sigma}_{11}, \ldots, \hat{\sigma}_{kk} \,)$ and $\hat{\sigma}_{ii}$ are the diagonal elements of $C_\varepsilon(0) \approx \hat{\Sigma}$. Individual elements of the residual correlation matrices $\hat{\rho}_\varepsilon(l)$ and possible patterns in these values would typically be examined to assess model adequacy, in the sense of whether the residuals $\hat{\varepsilon}_t$ have sample correlation patterns similar to those of white noise. The two standard error limits, $\pm 2/\sqrt{T}$, appropriate for a vector white noise process, can be used as guidelines in assessing significance of individual residual correlations [although there is some effect on the sampling properties of the $\hat{\rho}_\varepsilon(l)$, especially at low lags l, due to the use of estimated parameters in forming the residuals $\hat{\varepsilon}_t$, e.g., see Hosking (1980) or Li and McLeod (1981)].

Moreover, overall goodness-of-fit tests on a collection of residual correlation matrices can be employed. These multivariate "portmanteau" tests have been studied by Hosking (1980), Poskitt and Tremayne (1982), and Li and McLeod (1981). An overall "*goodness-of-fit" test statistic*, similar to those defined by the above authors, is given by

$$Q_s = T^2 \sum_{l=1}^{s} (T-l)^{-1} \sum_{i=1}^{k} \sum_{j=1}^{k} r_{ij}(l)\, r_{ji}(-l)$$

$$= T^2 \sum_{l=1}^{s} (T-l)^{-1} \operatorname{tr}\{ C_\varepsilon(l)\, \hat{\Sigma}^{-1} C_\varepsilon(-l)\, \hat{\Sigma}^{-1} \}, \tag{5.36}$$

where $R(l) = C_\varepsilon(l)\, C_\varepsilon(0)^{-1} \equiv C_\varepsilon(l)\, \hat{\Sigma}^{-1} = \{ r_{ij}(l) \}$, $l = 1, 2, \ldots, s$. Note that the $r_{ij}(l)$ are not the usual residual correlations, but the statistic Q_s can equivalently be expressed in terms of the residual correlations $\hat{\rho}_{ij}(l)$ (Hosking, 1981b) as

$$Q_s = T^2 \sum_{l=1}^{s} (T-l)^{-1} \operatorname{tr}\{ \hat{\rho}_\varepsilon(l)\, \hat{\rho}_\varepsilon(0)^{-1}\, \hat{\rho}_\varepsilon(-l)\, \hat{\rho}_\varepsilon(0)^{-1} \},$$

since $\operatorname{tr}\{ \hat{\rho}_\varepsilon(l)\, \hat{\rho}_\varepsilon(0)^{-1}\, \hat{\rho}_\varepsilon(-l)\, \hat{\rho}_\varepsilon(0)^{-1} \} = \operatorname{tr}\{ \hat{V}_\varepsilon^{-1/2}\, C_\varepsilon(l)\, \hat{\Sigma}^{-1} C_\varepsilon(-l)\, \hat{\Sigma}^{-1}\, \hat{V}_\varepsilon^{1/2} \}$. Under the null hypothesis that the process $\boldsymbol{Y}_t$ follows an ARMA(p,q) model, the test statistic Q_s is approximately distributed as chi-squared with $k^2(s-p-q)$ degrees of freedom. Hence, the model is rejected as inadequate for large values of Q_s. The approximate distribution is valid provided that s is chosen large enough so that the matrix weights Ψ_j in the infinite MA representation for the ARMA model are "small" for all $j > s$. Alternatively, the essentially equivalent Lagrange multiplier or score tests for the vector ARMA model could be used instead of (5.36). These tests are basically designed to test the ARMA(p,q) model specification against alternatives such as ARMA($p,q+s$) or ARMA($p+s,q$), after having fit the ARMA(p,q) model but without actually fitting the higher-order model by maximum likelihood estimation, and such tests are discussed by Hosking (1981a) and Poskitt and Tremayne (1982). The Lagrange multiplier tests will be discussed shortly in the subsequent Section 5.5.3.

5.5.2 Asymptotic Distribution of Residual Covariances and Goodness-of-Fit Statistic

The basis for the null distribution of the portmanteau statistic Q_s in (5.36) stems, initially, from consideration of the distribution of sample covariances based on the true white noise process of independent random vectors ε_t,

$$C(l) = T^{-1} \sum_{t=1}^{T-l} \varepsilon_t\, \varepsilon_{t+l}', \qquad l = 1, 2, \ldots, s.$$

Let $\mathbf{c}_l = \operatorname{vec}\{ C(l) \}$ and note that

$$\mathbf{c}_l = \frac{1}{T} \sum_{t=1}^{T-l} (I_k \otimes \varepsilon_t) \varepsilon_{t+l} = \frac{1}{T} \sum_{t=1}^{T-l} (\varepsilon_{t+l} \otimes I_k) \varepsilon_t .$$

Hence, it follows that $E(\mathbf{c}_l) = 0$ and

$$\mathrm{Cov}(\mathbf{c}_l) = E(\mathbf{c}_l \mathbf{c}_l') = T^{-2} (T - l) (\Sigma \otimes \Sigma),$$

using the property that

$$E[(I_k \otimes \varepsilon_t) \varepsilon_{t+l} \varepsilon_{t+l}' (I_k \otimes \varepsilon_t')] = E(\varepsilon_{t+l} \varepsilon_{t+l}') \otimes E(\varepsilon_t \varepsilon_t') = \Sigma \otimes \Sigma ,$$

and it also follows that $\mathrm{Cov}(\mathbf{c}_l, \mathbf{c}_m) = 0$ for $l \neq m$. In addition, the vectors $T^{1/2} \mathbf{c}_l$, $l = 1, \ldots, s$, are jointly asymptotically normal (Hannan, 1970, p. 228) by application of a martingale or an s-dependence central limit theorem. Hence, since $\mathbf{c}_l' (\Sigma^{-1} \otimes \Sigma^{-1}) \mathbf{c}_l = \mathrm{tr}\{ \Sigma^{-1} C(l)' \Sigma^{-1} C(l) \} = \mathrm{tr}\{ C(l) \Sigma^{-1} C(-l) \Sigma^{-1} \}$, the terms $T^2 (T - l)^{-1} \mathrm{tr}\{ C(l) \Sigma^{-1} C(-l) \Sigma^{-1} \}$ are asymptotically distributed as chi-squared with k^2 degrees of freedom and are asymptotically independent for different lags l. Thus, we see that

$$T^2 \sum_{l=1}^{s} (T - l)^{-1} \mathrm{tr}\{ C(l) \Sigma^{-1} C(-l) \Sigma^{-1} \}$$

is asymptotically distributed as chi-squared with $k^2 s$ degrees of freedom. However, when the parameters of the vector ARMA(p,q) model are estimated by maximum likelihood and the estimated residuals $\hat{\varepsilon}_t$ are used to form the residual covariance matrices $C_\varepsilon(l)$, the joint asymptotic normal distribution of the vectors $T^{1/2} \mathrm{vec}\{ C_\varepsilon(l) \}$ is affected because of the common dependence among the $C_\varepsilon(l)$ on the estimated parameters. Specifically, it is shown by Hosking (1980), Poskitt and Tremayne (1982), and others that the vectors $T^{1/2} \mathrm{vec}\{ \Sigma^{-1/2} C_\varepsilon(l) \Sigma^{-1/2} \} = T^{1/2} (\Sigma^{-1/2} \otimes \Sigma^{-1/2}) \mathrm{vec}\{ C_\varepsilon(l) \}$, $l = 1, \ldots, s$, are jointly asymptotically normal with approximate covariance matrix that is idempotent with rank $k^2(s - p - q)$. From this it follows that the statistic Q_s in (5.36) has approximately the chi-squared distribution with $k^2(s - p - q)$ degrees of freedom, since the asymptotic distribution of the quadratic form

$$T^2 \sum_{l=1}^{s} (T - l)^{-1} \mathrm{vec}\{ C_\varepsilon(l) \}' (\Sigma^{-1} \otimes \Sigma^{-1}) \mathrm{vec}\{ C_\varepsilon(l) \}$$

$$= T^2 \sum_{l=1}^{s} (T - l)^{-1} \mathrm{tr}\{ C_\varepsilon(l) \Sigma^{-1} C_\varepsilon(-l) \Sigma^{-1} \}$$

is not affected by the replacement of Σ^{-1} by the consistent estimator $\hat{\Sigma}^{-1}$.

5.5.3 Use of the Score Test Statistic for Model Diagnostic Checking

The Lagrange multiplier or score test procedure is an alternative to the direct use of overfitting a model by ML estimation and is closely related to the portmanteau test procedure for model checking. The general score test or Lagrange multiplier test procedure was presented by Silvey (1959), and its use for vector ARMA models has been discussed by Kohn (1979), Hosking (1981a), and Poskitt and Tremayne (1982). A computational advantage of the score test procedure is that it requires ML estimation of parameters only under the null model under test, but it yields tests asymptotically equivalent to the corresponding likelihood ratio tests obtained by directly overfitting the model. For model diagnostic purposes, it is assumed that a vector ARMA(p,q) model has been fit by ML estimation to the observations $\boldsymbol{Y}_t$, and we want to assess the adequacy of the model by testing this null model against the alternative of an ARMA($p+s$,q) model or of an ARMA(p,$q+s$) model. That is, for the vector ARMA($p+s$,q) alternative we test $\Phi_{p+1} = \cdots = \Phi_{p+s} = 0$, while for the vector ARMA(p,$q+s$) alternative we test $\Theta_{q+1} = \cdots = \Theta_{q+s} = 0$. The score test procedure is based on the vector of first partial derivatives, or score vector, of the log-likelihood function with respect to the model parameters of the general alternative model, but evaluated at the ML estimates obtained under the null model.

The (conditional) log-likelihood function of the vector ARMA model is given in Section 5.1.1 by $l = -(T/2)\log|\Sigma| - (1/2)\,\mathbf{e}'(I_T \otimes \Sigma^{-1})\,\mathbf{e}$, where $\mathbf{e} = (\varepsilon_1', \ldots, \varepsilon_T')'$. The partial derivatives of l with respect to the parameters $\boldsymbol{\phi}_j = \text{vec}(\Phi_j)$ and $\boldsymbol{\theta}_j = \text{vec}(\Theta_j)$ are also given, in (5.5) of Section 5.1.2, as $\partial l/\partial \boldsymbol{\beta} = \mathbf{Z}'\,\Theta'^{-1}(I_T \otimes \Sigma^{-1})\,\mathbf{e} = \overline{\mathbf{Z}}'(I_T \otimes \Sigma^{-1})\,\mathbf{e}$, where $\overline{\mathbf{Z}} = \Theta^{-1}\mathbf{Z} = -\,\partial \mathbf{e}/\partial \boldsymbol{\beta}'$ and $\mathbf{Z}$ is given in Section 5.1.2. Recall from Section 5.1.3 that $\overline{\mathbf{Z}}$ has components

$$-\frac{\partial \mathbf{e}}{\partial \boldsymbol{\phi}_j'} = \overline{\mathbf{U}}_j = (\overline{\boldsymbol{U}}_{j1}', \overline{\boldsymbol{U}}_{j2}', \ldots, \overline{\boldsymbol{U}}_{jT}')' \tag{5.37a}$$

and

$$-\frac{\partial \mathbf{e}}{\partial \boldsymbol{\theta}_j'} = \overline{\mathbf{V}}_j = (\overline{\boldsymbol{V}}_{j1}', \overline{\boldsymbol{V}}_{j2}', \ldots, \overline{\boldsymbol{V}}_{jT}')'. \tag{5.37b}$$

The $k \times k^2$ component matrices $\overline{\boldsymbol{U}}_{jt} = -\,\partial \varepsilon_t/\partial \boldsymbol{\phi}_j'$ and $\overline{\boldsymbol{V}}_{jt} = -\,\partial \varepsilon_t/\partial \boldsymbol{\theta}_j'$ in (5.37) are computed recursively as

$$\overline{\boldsymbol{U}}_{jt} = \sum_{i=1}^{q} \hat{\Theta}_i\, \overline{\boldsymbol{U}}_{j,t-i} + (\boldsymbol{Y}_{t-j}' \otimes I_k), \qquad \overline{\boldsymbol{V}}_{jt} = \sum_{i=1}^{q} \hat{\Theta}_i\, \overline{\boldsymbol{V}}_{j,t-i} - (\hat{\varepsilon}_{t-j}' \otimes I_k),$$

for $t = 1, 2, \ldots, T$, with $\overline{\boldsymbol{U}}_{jt} \equiv 0$ and $\overline{\boldsymbol{V}}_{jt} \equiv 0$ for $t \le 0$, where $\hat{\varepsilon}_t = \boldsymbol{Y}_t - \sum_{i=1}^{p} \hat{\Phi}_i \boldsymbol{Y}_{t-i} + \sum_{i=1}^{q} \hat{\Theta}_i \hat{\varepsilon}_{t-i}$ are the residual vectors from the ML fitting of the ARMA(p,q) model and $\hat{\Phi}_i$ and $\hat{\Theta}_i$ are the corresponding ML estimates. As in (5.7), the large sample information matrix for the parameters $\boldsymbol{\beta}$ can be consistently estimated by $\{-\,\partial^2 l/\partial \boldsymbol{\beta}\,\partial \boldsymbol{\beta}'\}_{\hat{\beta}} \approx \overline{\mathbf{Z}}'(I_T \otimes \hat{\Sigma}^{-1})\,\overline{\mathbf{Z}}$, where $\hat{\Sigma} = T^{-1}\sum_{t=1}^{T} \hat{\varepsilon}_t \hat{\varepsilon}_t'$. It follows that the score test statistic for testing that the additional s matrix coefficient parameters are equal to zero is

$$\Lambda = \{ \partial l / \partial \boldsymbol{\beta}' \}_{\hat{\beta}} \; \{ -\partial^2 l / \partial \boldsymbol{\beta} \, \partial \boldsymbol{\beta}' \}^{-1}_{\hat{\beta}} \; \{ \partial l / \partial \boldsymbol{\beta} \}_{\hat{\beta}}$$
$$= \hat{\mathbf{e}}' \, (I_T \otimes \hat{\Sigma}^{-1}) \, \overline{\mathbf{Z}} \, \{ \overline{\mathbf{Z}}' \, (I_T \otimes \hat{\Sigma}^{-1}) \, \overline{\mathbf{Z}} \}^{-1} \, \overline{\mathbf{Z}}' \, (I_T \otimes \hat{\Sigma}^{-1}) \, \hat{\mathbf{e}} , \qquad (5.38)$$

where $\hat{\mathbf{e}} = (\hat{\varepsilon}_1', \ldots, \hat{\varepsilon}_T')'$, $\overline{\mathbf{Z}} = (\overline{\mathbf{U}}_1, \ldots, \overline{\mathbf{U}}_{p+s}, \overline{\mathbf{V}}_1, \ldots, \overline{\mathbf{V}}_q)$ in the case of the vector ARMA(p+s,q) model alternative, and $\overline{\mathbf{Z}} = (\overline{\mathbf{U}}_1, \ldots, \overline{\mathbf{U}}_p, \overline{\mathbf{V}}_1, \ldots, \overline{\mathbf{V}}_{q+s})$ in the case of the vector ARMA(p,q+s) model alternative.

It has been noted by Godfrey (1979) in the univariate model, and by Poskitt and Tremayne (1982) in the vector ARMA case, that the computation of the test statistic in (5.38) can be given the interpretation as being equal to the "regression sum of squares" in an auxiliary generalized least squares estimation. That is, if the alternative model is ARMA(p+s,q), then we consider the auxiliary regression equation

$$\hat{\varepsilon}_t = \overline{\boldsymbol{U}}_{1t} \, \boldsymbol{\alpha}_1 + \cdots + \overline{\boldsymbol{U}}_{p+s,t} \, \boldsymbol{\alpha}_{p+s} + \overline{\boldsymbol{V}}_{1t} \, \boldsymbol{\beta}_1 + \cdots + \overline{\boldsymbol{V}}_{qt} \, \boldsymbol{\beta}_q + \boldsymbol{a}_t ,$$

whereas if the alternative model is ARMA(p,q+s), then we consider the auxiliary regression equation

$$\hat{\varepsilon}_t = \overline{\boldsymbol{U}}_{1t} \, \boldsymbol{\alpha}_1 + \cdots + \overline{\boldsymbol{U}}_{pt} \, \boldsymbol{\alpha}_p + \overline{\boldsymbol{V}}_{1t} \, \boldsymbol{\beta}_1 + \cdots + \overline{\boldsymbol{V}}_{q+s,t} \, \boldsymbol{\beta}_{q+s} + \boldsymbol{a}_t .$$

Let $\hat{\boldsymbol{a}}_t$ denote the residuals from the generalized least squares estimation of this regression equation, using $(I_T \otimes \hat{\Sigma})$ as the covariance matrix of the error vector $(\boldsymbol{a}_1', \ldots, \boldsymbol{a}_T')'$ in the GLS estimation. Then from (5.38) it is seen that Λ can be expressed, essentially, as

$$\Lambda = \sum_{t=1}^{T} \hat{\varepsilon}_t' \, \hat{\Sigma}^{-1} \, \hat{\varepsilon}_t - \sum_{t=1}^{T} \hat{\boldsymbol{a}}_t' \, \hat{\Sigma}^{-1} \, \hat{\boldsymbol{a}}_t = T \, [\, k - \mathrm{tr}\{ \hat{\Sigma}^{-1} \, T^{-1} \textstyle\sum_{t=1}^{T} \hat{\boldsymbol{a}}_t \, \hat{\boldsymbol{a}}_t' \} \,]$$

which is the "regression sum of squares" in the generalized least squares regression of the $\hat{\varepsilon}_t$'s on the $\overline{\boldsymbol{U}}_{jt}$'s and the $\overline{\boldsymbol{V}}_{jt}$'s. Under the null hypothesis that the fitted vector ARMA(p,q) model is correct, the score statistic Λ has an asymptotic χ^2-distribution with $s\,k^2$ degrees of freedom, and the null model is rejected as inadequate for large values of Λ. As emphasized by Godfrey (1979), Poskitt and Tremayne (1982), and others, rejection of the null model by the score test procedure should not be interpreted as evidence to adopt the specific alternative model involved, but simply as evidence against the adequacy of the (null) fitted model. Correspondingly, it is felt that a score test procedure of the above form will have reasonable power to detect inadequacies in the fitted model, even when the correct model is not that which is specified under the alternative. In particular, Poskitt and Tremayne (1982) indicate that the score test against a vector ARMA(p+s,q) model alternative is asymptotically identical to a test against a vector ARMA(p,q+s) alternative. Hence, the score test procedure may not be sensitive to the particular form of model specified under the alternative, but its performance will, of course, depend on the choice of the number s of additional matrix coefficient parameters specified.

For comparison with the portmanteau goodness-of-fit test statistic in (5.36), it is useful to note an additional feature of the score statistic Λ in (5.38). By the ML estimation procedure, it follows that the first partial derivatives, $\partial l / \partial \boldsymbol{\phi}_j$, $j = 1, \ldots, p$, and $\partial l / \partial \boldsymbol{\theta}_j$, $j = 1, \ldots, q$, will be identically equal to zero when evaluated at the ML estimates of the null model. Hence, the score vector, $\partial l / \partial \boldsymbol{\beta}$, will contain only $s\,k^2$ nonzero elements when evaluated at the ML estimates from the null model, these being the partial derivatives of l with respect to the additional $s\,k^2$ parameters of the alternative model. Thus, the score test in (5.38) can also be viewed as a quadratic form in these $s\,k^2$ nonzero values, whose matrix in the quadratic form is a consistent estimate of the inverse of the covariance matrix of these $s\,k^2$ score values when evaluated at the ML estimates obtained under the null model. Since these values are asymptotically normally distributed with zero means under the null model, the validity of the asymptotic $\chi^2_{sk^2}$-distribution under the null hypothesis is easily seen. It is also noted by Poskitt and Tremayne (1982) that the score test statistic procedure against the alternative of s additional coefficient matrix parameters is closely related to an appropriate test statistic based on the first s residual covariance matrices $C_\varepsilon(l)$ from the fitted model. The test statistic Λ in (5.38) is essentially a quadratic form involving these first s residual covariance matrices, but of a more complex form than the portmanteau test statistic in (5.36). [However, Hosking (1981a, pp. 225–226) has, in fact, established a close connection, asymptotically, between the portmanteau tests and the Lagrange multiplier tests when the alternative model is parameterized as a vector ARMA($p,q+s$) model with a multiplicative MA operator as in $\Phi(B)\,Y_t = \Theta(B)\,\Lambda(B)\,\varepsilon_t$ where $\Lambda(B)$ is an operator of order s.] As a specific illustration, suppose the fitted or null model is a pure vector AR(p) model, and the alternative model in the score test procedure is ARMA(p,s). Then it follows from the above developments that the variables $\overline{V}_{jt}$ are identical to $\overline{V}_{jt} = -(\hat{\varepsilon}_{t-j}' \otimes I_k)$, $j = 1, \ldots, s$, $t = 1, \ldots, T$, since $\Theta(B) \equiv I$ under the null model. Hence, the nonzero elements of the score vector are equal to

$$\overline{\mathbf{V}}_j'\,(I_T \otimes \hat{\Sigma}^{-1})\,\hat{\mathbf{e}} = \sum_{t=1}^{T} \overline{V}_{jt}'\,\hat{\Sigma}^{-1}\,\hat{\varepsilon}_t = -\sum_{t=1}^{T} (\hat{\varepsilon}_{t-j} \otimes I_k)\,\hat{\Sigma}^{-1}\,\hat{\varepsilon}_t$$

$$= -\operatorname{vec}\{\textstyle\sum_{t=1}^{T} \hat{\Sigma}^{-1}\,\hat{\varepsilon}_t\,\hat{\varepsilon}_{t-j}'\,\} = -T\operatorname{vec}\{\hat{\Sigma}^{-1}\,C_\varepsilon(j)'\,\},$$

for $j = 1, \ldots, s$, and the score test is thus directly seen to be a quadratic form involving these first s residual covariance matrices $C_\varepsilon(j)$. However, as noted by Godfrey (1979) in the univariate model, the number s of coefficient matrices specified to equal zero under the null hypothesis in the Lagrange multiplier test procedure is not really to be thought of in the same way as the number of residual covariance matrices s used in the portmanteau test statistic Q_s in (5.36), since, in particular, the validity of the Lagrange multiplier procedure does not require that s be large. These overall "goodness-of-fit" type of tests, either of the portmanteau test form (5.36) or the score test form (5.38), can be useful to identify some fundamental inadequacy in a fitted model, but they should be used in combination with (not as a substitute for) examination of the ARMA model residuals $\hat{\varepsilon}_t$ themselves and a more detailed study of the residual cross-correlation structure in particular.

5.6 Effects of Parameter Estimation Errors on Prediction Properties

In this section we will be concerned with the effect on prediction errors from using estimated parameters in the multivariate ARMA model, with particular emphasis on the vector autoregressive model. So we consider the vector AR(p) model, $Y_t = \sum_{i=1}^{p} \Phi_i Y_{t-i} + \varepsilon_t$, or $\Phi(B) Y_t = \varepsilon_t$. For the purposes of this section, it will be convenient to express the model in a simple state-space form, similar to that which will be discussed more generally in Section 7.1, as a vector first-order autoregressive model. With $\mathbf{Y}_t = (Y_t', Y_{t-1}', \ldots, Y_{t-p+1}')'$ and $\mathbf{e}_t = (\varepsilon_t', 0', \ldots, 0')'$ defined as kp-dimensional vectors, the AR(p) model can be expressed as the AR(1) relation

$$\mathbf{Y}_t = \Phi \, \mathbf{Y}_{t-1} + \mathbf{e}_t \, , \tag{5.39}$$

where

$$\Phi = \begin{bmatrix} \Phi_1 & \Phi_2 & . & . & . & \Phi_p \\ I & 0 & . & . & . & 0 \\ 0 & I & . & . & . & 0 \\ . & . & . & . & . & . \\ . & . & . & . & . & . \\ 0 & 0 & . & . & I & 0 \end{bmatrix} .$$

Note that $Y_t = E_1' \mathbf{Y}_t$, $\varepsilon_t = E_1' \mathbf{e}_t$, and $\mathbf{e}_t = E_1 \varepsilon_t$, where $E_1' = (I, 0, \ldots, 0)$ is a $(k \times kp)$ matrix. Also, from (5.39) $\mathbf{Y}_t$ has the representation $\mathbf{Y}_t = \sum_{i=0}^{\infty} \Phi^i \, \mathbf{e}_{t-i}$, so that $Y_t = E_1' \mathbf{Y}_t = \sum_{i=0}^{\infty} E_1' \Phi^i \, \mathbf{e}_{t-i} = \sum_{i=0}^{\infty} E_1' \Phi^i E_1 \, \varepsilon_{t-i}$. Thus, from the unique infinite moving average representation of Y_t from (2.6) of Section 2.2.1, we have that $\Psi_i = E_1' \Phi^i E_1$, that is, Ψ_i is the upper $(k \times k)$ block of the matrix Φ^i, with $\Psi(B) = \sum_{i=0}^{\infty} \Psi_i B^i = \Phi(B)^{-1}$.

Note from (5.39) that for $l = 1, 2, \ldots,$ we have $\mathbf{Y}_{t+l} = \Phi^l \, \mathbf{Y}_t + \sum_{i=0}^{l-1} \Phi^i \, \mathbf{e}_{t+l-i}$, so that Y_{t+l} can be expressed as

$$Y_{t+l} = E_1' \mathbf{Y}_{t+l} = E_1' \Phi^l \, \mathbf{Y}_t + \sum_{i=0}^{l-1} \Psi_i \, \varepsilon_{t+l-i} \, . \tag{5.40}$$

From (5.40) it follows that the minimum mean squared error (MSE) predictor of the future value Y_{t+l} based on $Y_t, Y_{t-1}, \ldots$ can be conveniently expressed as

$\hat{Y}_t(l) = E_1' \Phi^l \mathbf{Y}_t$ [which is, equivalently, equal to $\hat{Y}_t(l) = \sum_{j=1}^{p} \Phi_j \hat{Y}_t(l-j)$ as given in Section 2.5.3], with l-step ahead forecast error equal to $e_t(l) = Y_{t+l} - \hat{Y}_t(l) = \sum_{i=0}^{l-1} \Psi_i \varepsilon_{t+l-i}$ and l-step ahead mean squared forecast error matrix equal to $\Sigma(l) = \mathrm{Cov}(e_t(l)) = \sum_{i=0}^{l-1} \Psi_i \Sigma \Psi_i'$. Also note that we can write $\hat{Y}_t(l) = \mathrm{vec}(\mathbf{Y}_t' \Phi'^l E_1) = (I_k \otimes \mathbf{Y}_t') \boldsymbol{\phi}_l$, where $\boldsymbol{\phi}_l = \mathrm{vec}(\Phi'^l E_1)$ is formed from the first block column of the matrix Φ'^l.

5.6.1 Effects of Parameter Estimation Errors on Forecasting in the Vector AR(p) Model

When the parameters Φ_i in the vector AR(p) model are unknown, we suppose that they are estimated by least squares (conditional maximum likelihood) from a sample series of length T. As in Section 4.3, define $\Phi_{(p)}' = (\Phi_1, \ldots, \Phi_p)$ and $\boldsymbol{\phi} = \mathrm{vec}(\Phi_{(p)})$, and let $\hat{\boldsymbol{\phi}}$ denote the least squares or maximum likelihood estimator of $\boldsymbol{\phi}$ based on a sample of length T. Then we have the asymptotic distribution result that $T^{1/2}(\hat{\boldsymbol{\phi}} - \boldsymbol{\phi}) \sim N(0, \Sigma \otimes \Gamma_p^{-1})$, where $\Gamma_p = E[\mathbf{Y}_t \mathbf{Y}_t']$ is the $(kp \times kp)$ matrix with (i, j)th block matrix equal to $\Gamma(i-j) = E[Y_{t-i} Y_{t-j}']$. The lead l forecast corresponding to $\hat{Y}_t(l)$, but using estimated coefficients $\hat{\Phi}_i$, is $\tilde{Y}_t(l) = E_1' \hat{\Phi}^l \mathbf{Y}_t = (I_k \otimes \mathbf{Y}_t') \hat{\boldsymbol{\phi}}_l$, where $\hat{\boldsymbol{\phi}}_l = \mathrm{vec}(\hat{\Phi}'^l E_1)$ and $\hat{\Phi}$ is a matrix similar to Φ but with estimates $\hat{\Phi}_i$ in place of Φ_i. So the additional error in using estimated coefficients is the negative of $\tilde{Y}_t(l) - \hat{Y}_t(l) = E_1'(\hat{\Phi}^l - \Phi^l)\mathbf{Y}_t = (I_k \otimes \mathbf{Y}_t')(\hat{\boldsymbol{\phi}}_l - \boldsymbol{\phi}_l)$. Note that

$$\hat{\Phi}^l - \Phi^l = \sum_{i=0}^{l-1} \hat{\Phi}^i (\hat{\Phi} - \Phi) \Phi^{l-1-i} = \sum_{i=0}^{l-1} \hat{\Phi}^{l-1-i} (\hat{\Phi} - \Phi) \Phi^i .$$

Hence, it follows that $E_1'(\hat{\Phi}^l - \Phi^l) = \sum_{i=0}^{l-1} E_1' \hat{\Phi}^i E_1 \, E_1' (\hat{\Phi} - \Phi) \Phi^{l-1-i}$, so that

$$\hat{\boldsymbol{\phi}}_l - \boldsymbol{\phi}_l = \sum_{i=0}^{l-1} (E_1' \hat{\Phi}^i E_1 \otimes \Phi'^{l-1-i}) (\hat{\boldsymbol{\phi}} - \boldsymbol{\phi})$$

and, thus, we have $\tilde{Y}_t(l) - \hat{Y}_t(l) = (I_k \otimes \mathbf{Y}_t') \sum_{i=0}^{l-1} (E_1' \hat{\Phi}^i E_1 \otimes \Phi'^{l-1-i})(\hat{\boldsymbol{\phi}} - \boldsymbol{\phi})$. Then, on neglecting the terms of order $o_p(T^{-1/2})$, we obtain

$$\tilde{Y}_t(l) - \hat{Y}_t(l) = (I_k \otimes \mathbf{Y}_t') M_l' (\hat{\boldsymbol{\phi}} - \boldsymbol{\phi}), \tag{5.41}$$

where $M_l' = \sum_{i=0}^{l-1} (\Psi_i \otimes \Phi'^{l-1-i})$. By consistency of the $\hat{\Phi}_i$, $\hat{\Phi}^i - \Phi^i$ is $o_p(1)$ and $(\hat{\boldsymbol{\phi}} - \boldsymbol{\phi})$ is $O_p(T^{-1/2})$ as $T \to \infty$. Hence, the remainder of the approximation (5.41) is $o_p(T^{-1/2})$. The approximation (5.41) can also be obtained as a first degree Taylor expansion of $\tilde{Y}_t(l) = (I_k \otimes \mathbf{Y}_t') \hat{\boldsymbol{\phi}}_l$ about $\hat{\boldsymbol{\phi}} = \boldsymbol{\phi}$, by noting that $\partial \boldsymbol{\phi}_l / \partial \boldsymbol{\phi}' = M_l'$.

Now an assumption that has often been made in time series prediction problems of this nature [e.g., see Akaike (1971) and Yamamoto (1981)] is that the sample data $Y_1, \ldots, Y_T$ used in the parameter estimation are independent of the observations $\mathbf{Y}_t$ being used to initiate the forecasts, and, hence, $\hat{\boldsymbol{\phi}}$ and $\mathbf{Y}_t$

are independent. However, even in the more realistic situation where the forecast origin corresponds to the time period of the last observation used for parameter estimation, $\hat{\boldsymbol{\phi}}$ and $\mathbf{Y}_t$ with $t = T$ will still be asymptotically independent, in the sense that, omitting terms of $o_p(T^{-1/2})$, the covariance between $\hat{\boldsymbol{\phi}}$ and $\mathbf{Y}_t$ is zero. Thus, using the asymptotic distribution of $\hat{\boldsymbol{\phi}}$ and the (asymptotic) independence of $\hat{\boldsymbol{\phi}}$ and $\mathbf{Y}_t$, from (5.41) we have

$$E[\,(\tilde{Y}_t(l) - \hat{Y}_t(l))\,(\tilde{Y}_t(l) - \hat{Y}_t(l))'\,]$$

$$= E[\,(I_k \otimes \mathbf{Y}_t')\,M_l'(\hat{\boldsymbol{\phi}} - \boldsymbol{\phi})(\hat{\boldsymbol{\phi}} - \boldsymbol{\phi})'M_l\,(I_k \otimes \mathbf{Y}_t')\,]$$

$$= T^{-1}E[\,(I_k \otimes \mathbf{Y}_t')\,M_l'(\,\Sigma \otimes \Gamma_p^{-1})\,M_l\,(I_k \otimes \mathbf{Y}_t)\,]$$

$$= T^{-1}\sum_{i,j=0}^{l-1} \mathrm{tr}\{\Phi'^{\,l-1-i}\Gamma_p^{-1}\Phi^{l-1-j}\Gamma_p\}\,(\Psi_i\Sigma\Psi_j') \tag{5.42}$$

Finally, since the ε_{t+i}, $i > 0$, are independent of $\hat{\boldsymbol{\phi}}$ and $\mathbf{Y}_t$, and $Y_{t+l} - \hat{Y}_t(l) = [\,Y_{t+l} - \tilde{Y}_t(l)\,] + [\,\tilde{Y}_t(l) - \hat{Y}_t(l)\,]$, combining (5.42) with the MSE matrix result for $\tilde{Y}_t(l)$ we have the asymptotic covariance matrix of the prediction error l steps ahead using estimated coefficients is

$$\hat{\Sigma}(l) = E[\,(Y_{t+l} - \hat{Y}_t(l))\,(Y_{t+l} - \hat{Y}_t(l))'\,]$$

$$= \sum_{i=0}^{l-1} \Psi_i\Sigma\Psi_i' + T^{-1}\sum_{i=0}^{l-1}\sum_{j=0}^{l-1} \mathrm{tr}\{\,\Phi'^{\,l-1-i}\Gamma_p^{-1}\Phi^{l-1-j}\Gamma_p\,\}\,(\Psi_i\Sigma\Psi_j'). \tag{5.43}$$

This result has been obtained for the stationary vector AR model by Baillie (1979) and by Reinsel (1980). In the special case of one-step ahead prediction, i.e., $l = 1$, (5.43) reduces to $\hat{\Sigma}(1) = \Sigma + T^{-1}\mathrm{tr}\{\,I_{kp}\}\,\Sigma = [\,1 + (kp/T)\,]\,\Sigma$, a result which was obtained by Akaike (1971) and which forms the basis of his minimum MFPE criterion for determining the order of a multivariate autoregressive process. The result (5.43) readily extends to the nonstationary case of a vector AR model of the form $\Phi_1(B)\,D(B)\,Y_t = \varepsilon_t$, as in (2.17) of Section 2.4.1, where $D(B)$ is a nonstationary (differencing) operator which is assumed known, while $\Phi_1(B)$ is a stationary operator whose parameters are estimated by least squares. Then, with $W_t = D(B)\,Y_t$ denoting the stationary AR process and using fundamental relations between the predictions $\hat{Y}_t(l)$ and $\hat{W}_t(l)$, a result as in (5.43) can be shown to hold (e.g., Reinsel and Lewis, 1987). For the result in this nonstationary case, the Ψ_i are the coefficients in $\Psi(B) = \Phi(B)^{-1}$ with $\Phi(B) = \Phi_1(B)\,D(B)$, while the matrices Γ_p and Φ that occur in the second term in (5.43) correspond to quantities associated with the stationary AR model for W_t, $\Phi_1(B)\,W_t = \varepsilon_t$, e.g., $\Gamma_p = E[\,\mathbf{W}_t\,\mathbf{W}_t'\,]$ with $\mathbf{W}_t = (\,W_t', W_{t-1}', \ldots, W_{t-p+1}'\,)'$. In related work, Samaranayake and Hasza (1988) established a certain order of convergence to zero (as $T \to \infty$) for the error $\tilde{Y}_t(l) - \hat{Y}_t(l)$ in (5.41) for the special nonstationary model situation where Y_t is a Gaussian vector AR(p) process such that the autoregressive operator $\Phi(B)$ contains exactly one root equal to one with all other roots stationary, and the parameters of the model in (5.39) are estimated by least squares.

For the more general vector ARMA(p,q) model, a similar approach to obtain asymptotic prediction mean square error matrices when using estimated parameters can be taken, and results have been derived by Yamamoto (1981). However, the forms of the results are much more complicated and, hence, are of less practical usefulness. In particular, it is noted that for the general vector ARMA(p,q) model, the one-step ahead prediction MSE matrix with estimated parameters cannot be reduced to a simple form as in the vector AR model. For example, in the simple case of a vector MA(1) model, $\boldsymbol{Y}_t = \varepsilon_t - \Theta_1 \varepsilon_{t-1}$, it can be shown that the approximate one-step ahead prediction MSE matrix using the ML estimator $\hat{\Theta}_1$ is $\hat{\Sigma}(1) = \Sigma + (k/T) \sum_{j=0}^{\infty} \Theta_1^j V_1^{-1} \Theta_1^{\prime j}$, where $V_1 = \sum_{i=0}^{\infty} \Theta_1^{\prime i} \Sigma^{-1} \Theta_1^i$, and $T^{-1} V^{-1} = T^{-1} (\Sigma^{-1} \otimes V_1^{-1})$ represents the asymptotic covariance matrix of the MLE $\text{vec}(\hat{\Theta}_1 - \Theta_1)$ (e.g., see Section 5.1.4).

5.6.2 *Prediction Through Approximation by Autoregressive Model Fitting*

We now consider a somewhat different situation where the precise form of a parametric model for the series $\boldsymbol{Y}_t$ is not assumed to be known, and the approach used for prediction involves the fitting of vector autoregressive models of finite order. In this situation, however, it is not assumed that $\boldsymbol{Y}_t$ actually follows a finite order vector AR model. Instead, it is assumed only that $\boldsymbol{Y}_t$ has an infinite order vector AR representation as $\boldsymbol{Y}_t = \sum_{i=1}^{\infty} \Pi_i \boldsymbol{Y}_{t-i} + \varepsilon_t$, where $\sum_{i=1}^{\infty} \| \Pi_i \| < \infty$ and $\Pi(z) = I - \sum_{i=1}^{\infty} \Pi_i z^i$ satisfies $\det\{ \Pi(z) \} \neq 0$ for $|z| \leq 1$. A vector autoregressive model of finite order m is assumed to be fitted to a sample series of length T of the process, by least squares, and estimates $\hat{\Phi}_{1m}, \ldots, \hat{\Phi}_{mm}$ are obtained. These can be viewed as estimates of the matrices $\Phi_{1m}, \ldots, \Phi_{mm}$ which represent the solution to the theoretical Yule-Walker equations of order m, as in Sections 2.2.2 and 3.3.1, that is, $\sum_{i=1}^{m} \Gamma(j-i) \Phi_{im}^{\prime} = \Gamma(j)$, for $j = 1, 2, \ldots, m$. Lewis and Reinsel (1985) considered this situation under the assumption that the fitted AR order m is chosen as a function of T such that $m^3/T \to 0$ and $T^{1/2} \sum_{i=m+1}^{\infty} \| \Pi_i \| \to 0$ as $m, T \to \infty$.

For this case, let $\hat{\boldsymbol{\phi}}(m) = \text{vec}\{ (\hat{\Phi}_{1m}, \ldots, \hat{\Phi}_{mm})^{\prime} \}$ and let $\boldsymbol{\phi}(m) = \text{vec}\{ (\Pi_1, \ldots, \Pi_m)^{\prime} \}$. Then Lewis and Reinsel (1985) examined that asymptotic distribution behavior of the estimator $\hat{\boldsymbol{\phi}}(m)$ as $m, T \to \infty$. They established that

$$(T-m)^{1/2}\, l(m)^{\prime} [\, \hat{\boldsymbol{\phi}}(m) - \boldsymbol{\phi}(m) \,] / v_T \xrightarrow{D} N(0, 1),$$

where $v_T^2 = l(m)^{\prime} (\Sigma \otimes \Gamma_m^{-1})\, l(m)$ and $l(m)$ is any sequence of $(mk^2 \times 1)$ vectors with $\| l(m) \|^2$ bounded. In addition, consider the l-step ahead predictor based on the fitted AR model of order m, $\tilde{\boldsymbol{Y}}_{t,m}(l) = E_1^{\prime} \hat{\Phi}_{[m]}^{l} \mathbf{Y}_{t,m}$, where $\mathbf{Y}_{t,m} = (\boldsymbol{Y}_t^{\prime}, \boldsymbol{Y}_{t-1}^{\prime}, \ldots, \boldsymbol{Y}_{t-m+1}^{\prime})^{\prime}$ and $\hat{\Phi}_{[m]}$ is a $(km \times km)$ matrix similar to Φ in (5.39), but with estimates $\hat{\Phi}_{im}$ in place of the Φ_i in the previous case of a true AR model. Also, let $\hat{Y}_t(l)$ denote the optimal l-step ahead predictor of Y_{t+l} based on the true infinite order AR model representation, so that $\hat{Y}_t(l) = \sum_{i=1}^{\infty} \Pi_i \hat{Y}_t(l-i)$. This optimal predictor has prediction MSE matrix equal to $\Sigma(l) = E[(Y_{t+l} - \hat{Y}_t(l)) (Y_{t+l} - \hat{Y}_t(l))'] = \sum_{i=0}^{l-1} \Psi_i \Sigma \Psi_i'$, where $\Psi(B) = I + \sum_{i=1}^{\infty} \Psi_i B^i = \Pi(B)^{-1}$. Then, using the asymptotic properties of $\hat{\phi}(m) - \phi(m)$, it has been established that

$$(T/m)^{1/2} [\tilde{Y}_{t,m}(l) - \hat{Y}_t(l)] \xrightarrow{D} N(0, k\, \Sigma(l))$$

as $m, T \to \infty$. Hence, using this result we obtain an asymptotic approximation, as $m, T \to \infty$, that

$$E[(\tilde{Y}_{t,m}(l) - \hat{Y}_t(l)) (\tilde{Y}_{t,m}(l) - \hat{Y}_t(l))'] \approx (mk/T)\, \Sigma(l).$$

Therefore, writing the prediction error based on the fitted AR model of order m as $Y_{t+l} - \tilde{Y}_{t,m}(l) = [Y_{t+l} - \hat{Y}_t(l)] + [\hat{Y}_t(l) - \tilde{Y}_{t,m}(l)]$, and using similar reasoning as in the finite order AR model case, we obtain the asymptotic approximation for the prediction MSE matrix of the predictor $\tilde{Y}_{t,m}(l)$ as

$$\begin{aligned} \hat{\Sigma}_m(l) &= E[(Y_{t+l} - \tilde{Y}_{t,m}(l)) (Y_{t+l} - \tilde{Y}_{t,m}(l))'] \\ &\approx \Sigma(l) + (mk/T)\, \Sigma(l) = [1 + (mk/T)]\, \Sigma(l). \end{aligned} \tag{5.44}$$

A useful feature of the approximation (5.44) is its simplicity, which allows for easy computation and interpretation. The result implies that the asymptotic effect of parameter estimation in the vector autoregressive model fitting approach is to inflate the mean squared prediction error matrix $\Sigma(l)$ by a factor of $[1 + (mk/T)]$. One might also consider using the "finite sample" approximation for (5.44), as given by (5.43) in the case of a finite order AR model, but it is not clear which approximation would provide more accurate results. It is also noted that, in practical use, estimates $\hat{\Psi}_{im}$ and $\hat{\Sigma}_m$, obtained from the $\hat{\Phi}_{im}$, would need to be substituted in place of the Ψ_i and Σ in forming (5.44). Finally, in practice, one must choose the value of the AR order m to use for any given series length T, and it may be reasonable to use Akaike's (1971) FPE criterion, which was originally suggested to determine a finite order approximation to a true infinite order autoregressive process. Under this criterion, the order of a finite autoregressive process is selected by choosing the value of m which minimizes the determinant of an estimate, $[1 + (mk/T)]\, \hat{\Sigma}_m$, of the "final" one-step ahead mean squared prediction error matrix $\hat{\Sigma}_m(1)$.

5.7 Numerical Examples

In this section we consider two numerical examples to illustrate the model building, estimation, and model checking techniques for vector time series presented in the previous sections of the last two chapters. The first example involves a continuation of the analysis of the log mink and muskrat furs data that were previously discussed in Example 4.1 of Section 4.3. In the second example, data on billing and production schedule figures of a company are examined to investigate the bivariate modeling approach in a situation which a priori is expected to be of the form of a unidirectional transfer function model.

EXAMPLE 5.1 (Example 4.1, Cont.). Considering further the log mink and muskrat furs data from Example 4.1, it was noted that although an AR(3) model was indicated on the basis of the chi-squared M_m test statistics for $H_0 : \Phi_m = 0$, the values of M_m were still moderately large for the AR orders 4, 5, and 6. So we now consider the possibility of a mixed ARMA model for these data. For this, we apply the multivariate version of the procedure similar to that suggested by Hannan and Rissanen (1982), as discussed in Section 4.5.3. Since the AIC values given in Table 4.2 for pure AR models give a near minimum for the order $m^* = 4$, residuals $\tilde{\varepsilon}_t$ to be used in the regression procedure of model (4.18) were obtained from the vector AR(4) model. Then ARMA(p,q) models for each combination of $p = 1, 2, 3$ and $q = 0, 1, 2$ were estimated by least squares using the regression model (4.18), and the residual covariance matrices $\tilde{\Sigma}_{p,q}$ were obtained for each model fit. Instead of the BIC criterion, the criterion suggested by Quinn (1980) given as $\text{HQ}(p, q) = \log(|\tilde{\Sigma}_{p,q}|) + 2(p+q)k^2 \log(\log(T))/T$ was considered, and these values are presented in Table 5.1. It is seen that a clear minimum in these values is attained for the model with $p = 2$ and $q = 1$. [This was also true when the BIC criterion was employed, but then the minimum was not as pronounced and the AR(2) model gave nearly as small a value.]

Table 5.1 Summary of Results for Preliminary ARMA(p,q) Order Determination Procedure for the Log Mink and Muskrat Furs Data, Using Quinn's Criteria with Values Given by $\text{HQ}(p, q) = \log(|\tilde{\Sigma}_{p,q}|) + 2(p+q)k^2 \log(\log(T))/T$.

	q		
p	0	1	2
1	−5.339	−5.297	−5.232
2	−5.417	−5.514	−5.432
3	−5.456	−5.434	−5.471

Hence, an ARMA(2,1) model is fit to these data by an exact maximum likelihood procedure using the SCA computer system (Liu and Hudak, 1986). [Actually, the "exact" estimation procedure employed in SCA is only exact for the MA part of the ARMA model and uses the exact likelihood function of the $T-p$ vector observations of the transformed process $\boldsymbol{W}_t = \boldsymbol{Y}_t - \sum_{j=1}^{p} \Phi_j \boldsymbol{Y}_{t-j}$, $t = p+1, \ldots, T$, treated as an MA(q) process, as described in Section 4.4 of

Hillmer and Tiao (1979).] The diagonal elements of the matrix $\hat{\Theta}_1$ obtained from this fit were very small and insignificant, and so these coefficients were omitted and the model was reestimated. The resulting ML estimates were obtained as

$$\hat{\Phi}_1 = \begin{bmatrix} \underset{(0.084)}{0.816} & \underset{(0.253)}{-0.623} \\ \underset{(0.154)}{-1.116} & \underset{(0.092)}{1.074} \end{bmatrix}, \qquad \hat{\Phi}_2 = \begin{bmatrix} \underset{(0.175)}{-0.643} & \underset{(0.231)}{0.592} \\ \underset{(0.134)}{0.615} & \underset{(0.081)}{-0.133} \end{bmatrix},$$

$$\hat{\Theta}_1 = \begin{bmatrix} \underset{(---)}{0.000} & \underset{(0.260)}{-1.248} \\ \underset{(0.169)}{-0.801} & \underset{(---)}{0.000} \end{bmatrix}, \qquad \text{with} \qquad \tilde{\Sigma} = \begin{bmatrix} 0.035328 & 0.016937 \\ 0.016937 & 0.053272 \end{bmatrix}$$

and $\det(\tilde{\Sigma}) = 0.1595 \times 10^{-2}$. Residuals $\hat{\varepsilon}_t$ from this fitted model were examined, and, in particular, the residual correlation matrices $\hat{\rho}_\varepsilon(l)$ discussed in Section 5.5 were obtained. These residual correlation matrices for lags 1 through 10 are displayed in Table 5.2. When compared to the limits of $\pm 2\,T^{-1/2} \approx \pm 0.254$, there is no indication of inadequacy in the fitted ARMA(2,1) model (except for one significant autocorrelation at lag 10 in the residual mink data series).

Table 5.2 Residual Correlation Matrices $\hat{\rho}_\varepsilon(l)$ from the ARMA(2,1) Model Fitted to the Log Mink and Muskrat Furs Data (with indicator symbols for $\pm 2\,T^{-1/2}$ limits).

l	1		2		3		4		5	
$\hat{\rho}_\varepsilon(l)$	−0.07	−0.09	0.00	0.03	0.08	−0.06	0.01	0.04	0.18	0.15
	−0.03	−0.03	−0.09	0.02	−0.02	0.04	−0.03	0.08	0.10	−0.09
	.	.	.	.	.	.	.	.	.	.
	.	.	.	.	.	.	.	.	.	.

l	6		7		8		9		10	
$\hat{\rho}_\varepsilon(l)$	−0.24	0.05	0.14	0.06	−0.19	−0.14	0.13	0.22	0.27	0.06
	−0.19	−0.07	−0.02	−0.15	−0.18	−0.05	0.01	−0.02	0.16	−0.12
	.	.	.	.	.	.	.	.	+	.
	.	.	.	.	.	.	.	.	.	.

Also calculated were the portmanteau statistics Q_s in (5.36) for $s = 5$ and $s = 10$, and these yield the values of $Q_5 = 9.01$ and $Q_{10} = 35.72$ with 10 and 30 degrees of freedom, respectively. Comparison of these values with the appropriate chi-squared reference distributions does not give cause for concern about the adequacy of the ARMA(2, 1) model specification.

For illustration, from the above estimated ARMA(2, 1) model we computed forecasts for 10 steps (years) ahead, using equation (2.22) of Section 2.5.3, as $\hat{\boldsymbol{Y}}_t(1) = \hat{\Phi}_1 \boldsymbol{Y}_t + \hat{\Phi}_2 \boldsymbol{Y}_{t-1} + \hat{\boldsymbol{\delta}} - \hat{\Theta}_1 \hat{\varepsilon}_t$, and $\hat{\boldsymbol{Y}}_t(l) = \hat{\Phi}_1 \hat{\boldsymbol{Y}}_t(l-1) + \hat{\Phi}_2 \hat{\boldsymbol{Y}}_t(l-2) + \hat{\boldsymbol{\delta}}$, $l = 2, \ldots, 10$, with the estimated constant vector $\hat{\boldsymbol{\delta}} = (9.360, 6.209)'$. The estimated infinite MA coefficient matrices $\hat{\Psi}_j$ corresponding to the fitted ARMA(2, 1) model were obtained, based on (2.12) of Section 2.3.1, as $\hat{\Psi}_1 = \hat{\Phi}_1 - \hat{\Theta}_1$, $\hat{\Psi}_j = \hat{\Phi}_1 \hat{\Psi}_{j-1} + \hat{\Phi}_2 \hat{\Psi}_{j-2}$, $j = 2, \ldots, 10$, and, hence, the estimated forecast MSE matrices $\hat{\Sigma}(l)$ were also obtained, using (2.21) of Section 2.5.2, as $\hat{\Sigma}(l) = \sum_{j=0}^{l-1} \hat{\Psi}_j \hat{\Sigma} \hat{\Psi}_j'$. The matrices $\hat{\Psi}_j$ and $\hat{\Sigma}(l)$ from the ARMA(2, 1) model are displayed in Table 5.3. Examination of the individual coefficients (impulse response weights) in the matrices $\hat{\Psi}_j$ indicates that the coefficients each die out with an increase in j according to a damped sinusoidal pattern. The coefficients associated with the log-muskrat series seem to die out more slowly and have larger cyclic amplitude, so that from the infinite MA representation $\boldsymbol{Y}_t = \boldsymbol{\mu} + \Psi(B)\, \varepsilon_t$ of the process $\{\boldsymbol{Y}_t\}$, shocks ε_t have more of a longer-term influence on the muskrat series $\{Y_{2t}\}$ than on the mink series $\{Y_{1t}\}$. In particular, shocks ε_{1t} in the log-mink series seem to have much more (longer-term) influence on the muskrat series $\{Y_{2t}\}$ than shocks ε_{2t} in the log-muskrat series have on the mink series $\{Y_{1t}\}$.

Table 5.3 Infinite MA Coefficient Matrices $\hat{\Psi}_l$ and Prediction MSE Matrices $\hat{\Sigma}(l)$ ($\times 10^{-2}$) from Estimated ARMA(2, 1) Model Fitted to the Log Mink and Muskrat Furs Data.

l	1		2		3		4		5	
$\hat{\Psi}_l$	0.82	0.63	0.22	0.43	−0.14	0.39	−0.39	0.16	−0.43	0.01
	−0.32	1.07	−0.63	0.32	−0.38	0.10	−0.04	−0.10	0.36	−0.06
$\hat{\Sigma}(l)$	3.53	1.69	9.69	5.51	11.18	5.42	11.87	5.55	12.33	5.57
	1.69	5.33	5.51	10.68	5.42	11.96	5.55	12.40	5.57	12.46
l	6		7		8		9		10	
$\hat{\Psi}_l$	−0.35	−0.12	−0.19	−0.16	−0.03	−0.15	0.10	−0.10	0.16	−0.03
	0.64	0.04	0.76	0.19	0.73	0.30	0.59	0.37	0.42	0.37
$\hat{\Sigma}(l)$	12.99	5.06	13.64	4.09	14.01	3.15	14.14	2.64	14.20	2.63
	5.06	12.87	4.09	14.40	3.15	17.12	2.64	20.24	2.63	22.94

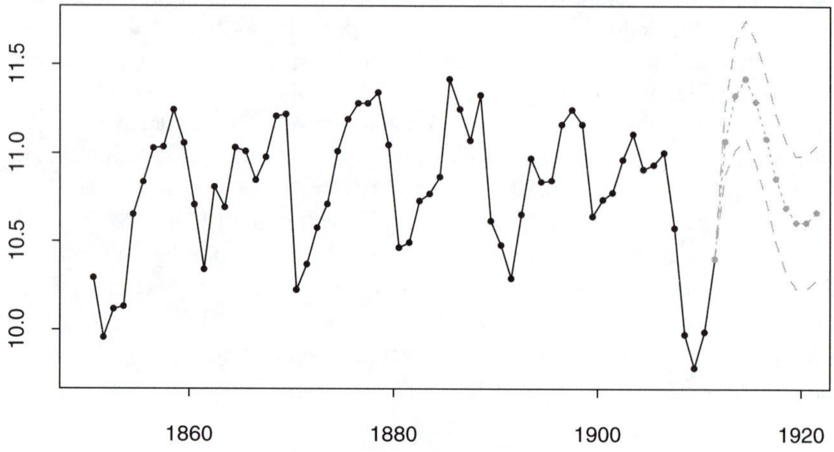

(a) Y_{1t} : Mink Fur Series

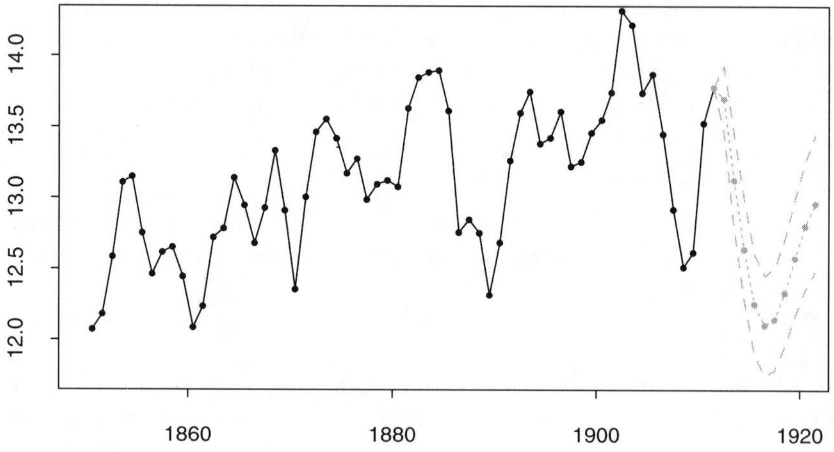

(b) Y_{2t} : Muskrat Fur Series

Figure 5.1. Logarithms of Annual Sales of Mink and Muskrat Furs by Hudson's Bay Company for the Years 1850 Through 1911, with Forecasts from ARMA(2,1) Model

The behavior of the diagonal elements of the matrices $\hat{\Sigma}(l)$ as l increases indicates that forecast errors associated with the log-muskrat series have larger variances at longer lead times than do those for the log-mink series, which is a reflection of the more nonstationary behavior of the log-muskrat series. The square roots $\{ \hat{\sigma}_{ii}(l) \}^{1/2}$ of the diagonal elements of the matrices $\hat{\Sigma}(l)$ provide the estimates of the standard deviations of the individual lead-l forecast errors, and approximate 95% prediction intervals for future values can be formed as $\hat{Y}_{it}(l) \pm 1.96 \{ \hat{\sigma}_{ii}(l) \}^{1/2}$. In Figure 5.1, the forecasts from the ARMA(2,1) model for the log-mink and log-muskrat series for 10 steps (years) ahead are depicted together with one standard deviation prediction bounds.

Finally, a few comments about the final fitted ARMA(2,1) model. First it is noted that the estimated AR(2) operator is

$$\hat{\Phi}(B) = \begin{bmatrix} 1 - 0.816\,B + 0.643\,B^2 & 0.623\,B - 0.592\,B^2 \\ 1.116\,B - 0.615\,B^2 & 1 - 1.074\,B + 0.133\,B^2 \end{bmatrix}$$

$$\approx \begin{bmatrix} 1 - 0.816\,B + 0.643\,B^2 & 0.608\,B\,(\,1 - B\,) \\ 1.116\,B - 0.615\,B^2 & (\,1 - 0.104\,B\,)(\,1 - B\,) \end{bmatrix} .$$

This implies that the model can be reformulated (approximately) as a particular ARMA(2,1) model in terms of the series Y_{1t} and $(\,1 - B\,)\,Y_{2t}$, with the second series Y_{2t} (the muskrat series) being nonstationary. This appears to be in agreement with the time series plots for the mink and muskrat series in Figures 4.1 and 5.1. Secondly, it is noted that the MA(1) operator is nearly noninvertible, with eigenvalues of $\hat{\Theta}_1$ equal to ± 0.999. This seems to be a rather unusual feature whose cause is not clear, but this model result deserves special care and attention.

As an alternative model specification approach, we consider the determination of the echelon canonical ARMA model form for these data. For this, we use the canonical correlation analyses suggested by Akaike (1976) and Cooper and Wood (1982), as discussed in Section 4.5.2, to determine the Kronecker indices for the process. For the vector of present and past values, we use a maximum of six time-lagged variables and set $\mathbf{U}_t = (\, \boldsymbol{Y}_t', \boldsymbol{Y}_{t-1}', \ldots, \boldsymbol{Y}_{t-5}' \,)'$. Then, for various vectors $\mathbf{V}_t^*$ of future variables, the squared sample canonical correlations between $\mathbf{V}_t^*$ and $\mathbf{U}_t$ are determined as the eigenvalues of the matrix similar to the matrix in (4.15) of Section 4.5.2. The resulting squared sample canonical correlations were obtained and are presented in Table 5.4. From these results, we interpret that the first occurrence of a small sample (squared) canonical correlation value (0.171), which is an indication of a zero canonical correlation between the future and the present and past, is obtained when $\mathbf{V}_t^* = (\, Y_{1,t+1}, Y_{2,t+1}, Y_{1,t+2}, Y_{2,t+2} \,)'$. This implies that the Kronecker index $K_2 = 1$, since it implies that a linear combination involving $Y_{2,t+2}$ in terms of the remaining variables in $\mathbf{V}_t^*$ is uncorrelated with the present and past vector $\mathbf{U}_t$. An additional small sample (squared) canonical correlation value (0.069 in addition to 0.158) occurs when $\mathbf{V}_t^* = (\, Y_{1,t+1}, Y_{2,t+1}, Y_{1,t+2}, Y_{2,t+2}, Y_{1,t+3} \,)'$, and this implies that $K_1 = 2$. Hence, this analysis leads to the specification of an ARMA(2,2) model in the echelon form of equation (3.4) of Section 3.1.2, with Kronecker indices $K_1 = 2$ and $K_2 = 1$.

Table 5.4 Results of the Canonical Correlation Analysis, Based on the Kronecker Indices Approach of Akaike (1976) and Cooper and Wood (1982), for the Log Mink and Muskrat Furs Data. Values are the squared sample canonical correlations between the present and past vector $\mathbf{U}_t = (\boldsymbol{Y}_t', \ldots, \boldsymbol{Y}_{t-5}')'$ and various future vectors $\mathbf{V}_t^*$.

Future Vector $\mathbf{V}_t^*$	Squared Canonical Correlations
$Y_{1,t+1}$	0.728
$Y_{1,t+1},\ Y_{2,t+1}$	0.852, 0.711
$Y_{1,t+1},\ Y_{2,t+1},\ Y_{1,t+2}$	0.874, 0.717, 0.332
$Y_{1,t+1},\ Y_{2,t+1},\ Y_{1,t+2},\ Y_{2,t+2}$	0.875, 0.720, 0.378, 0.171
$Y_{1,t+1},\ Y_{2,t+1},\ Y_{1,t+2},\ Y_{2,t+2},\ Y_{1,t+3}$	0.896, 0.740, 0.553, 0.158, 0.069
$Y_{1,t+1},\ Y_{2,t+1},\ Y_{1,t+2},\ Y_{2,t+2},\ Y_{1,t+3},\ Y_{2,t+3}$	0.896, 0.742, 0.611, 0.172, 0.126, 0.056

Fitting the echelon form model $\Phi_0^\# \boldsymbol{Y}_t - \Phi_1^\# \boldsymbol{Y}_{t-1} - \Phi_2^\# \boldsymbol{Y}_{t-2} = \boldsymbol{\delta} + \Theta_0^\# \varepsilon_t - \Theta_1^\# \varepsilon_{t-1} - \Theta_2^\# \varepsilon_{t-2}$, $\Phi_0^\# = \Theta_0^\#$, by conditional maximum likelihood, and eliminating a few parameters whose estimates were nonsignificant, we obtained the ML estimates

$$\hat{\Phi}_1^\# = \begin{bmatrix} \underset{(0.148)}{1.307} & 0 \\ \\ \underset{(\text{---})}{0.000} & \underset{(0.140)}{0.984} \end{bmatrix}, \qquad \hat{\Phi}_2^\# = \begin{bmatrix} \underset{(0.086)}{-0.704} & \underset{(0.038)}{0.046} \\ \\ 0 & 0 \end{bmatrix},$$

$$\hat{\Theta}_1^\# = \begin{bmatrix} \underset{(0.127)}{0.789} & \underset{(0.113)}{-0.666} \\ \\ \underset{(\text{---})}{0.000} & \underset{(0.196)}{-0.913} \end{bmatrix}, \qquad \hat{\Theta}_2^\# = \begin{bmatrix} \underset{(0.091)}{-0.289} & \underset{(0.102)}{0.323} \\ \\ 0 & 0 \end{bmatrix},$$

$$\hat{\Phi}_0^\# = \hat{\Theta}_0^\# = \begin{bmatrix} 1 & 0 \\ \\ \underset{(0.260)}{0.955} & 1 \end{bmatrix}, \qquad \text{with} \qquad \tilde{\Sigma} = \begin{bmatrix} 0.042317 & 0.019948 \\ 0.019948 & 0.057989 \end{bmatrix},$$

$\det(\tilde{\Sigma}) = 0.2056 \times 10^{-2}$, and $\text{AIC} = \log(\,|\tilde{\Sigma}|\,) + 2\,r\,/\,(T-2) = -5.854$, and the 0's denote values that are specified as equal to zero as a consequence of the echelon form structure implied by the Kronecker indices of $K_1 = 2$ and $K_2 = 1$.

These estimation results yield an ARMA(2,2) model [with reduced rank, rank(Φ_2, Θ_2) = 1, implied for the corresponding model expressed in "standard form"], and, on the basis of the estimate $\tilde{\Sigma}$ and the corresponding AIC value, this model is clearly preferable in terms of fit to the pure AR models estimated in Chapter 4. In addition, the MA(2) operator in this model is not close to the noninvertibility boundary as in the case of the ARMA(2,1) model presented above. This ARMA(2,2) model, with Kronecker indices $K_1 = 2$ and $K_2 = 1$, is also similar to the form of model identified by Cooper and Wood (1982) and by Hannan and Kavalieris (1984) for these data, using different methods.

EXAMPLE 5.2. As a second example, we consider data from Makridakis and Wheelwright (1978) consisting of weekly production schedule figures (Y_{1t}) and billing figures (Y_{2t}) of a company for $T = 100$ weeks. These data are plotted in Figure 5.2. From the context of the data, it is anticipated that these data will follow a unidirectional transfer function model structure with Y_{1t} as input in the model for Y_{2t}. Using rather traditional transfer function model building techniques such as detailed in Box and Jenkins (1976, Chap. 11), including the prewhitening of the input series Y_{1t}, the following structures were estimated and found to be adequate models: either model

$$(1 - 0.469\,B)\,Y_{1t} = 26.471 + a_{1t}, \qquad \hat{\sigma}^2_{a_1} = 2.426,$$

or

$$Y_{1t} = 49.808 + (1 + 0.497\,B)\,a_{1t}, \qquad \hat{\sigma}^2_{a_1} = 2.415,$$

for the series Y_{1t}, and

$$Y_{2t} = 71.764 + \frac{2.106\,B^3 - 1.965\,B^4}{1 - 1.541\,B + 0.793\,B^2}\,Y_{1t} + \frac{1 - 0.544\,B}{1 - 1.423\,B + 0.763\,B^2}\,a_{2t},$$

with $\hat{\sigma}^2_{a_2} = 3.370$. Since the second-order operators in the denominators on the right side of the transfer function equation are very similar, this last model equation can be simplified and reestimated to yield

Table 5.5 Sample Correlation Matrices $\hat{\rho}(l)$ for the Production Schedule and Billing Figures Data.

l	1		2		3		4		5	
$\hat{\rho}(l)$	0.47	0.06	0.17	0.16	0.20	0.51	0.21	0.41	0.17	0.13
	0.02	0.74	0.00	0.30	0.04	−0.13	0.08	−0.45	0.12	−0.61
l	6		7		8		9		10	
$\hat{\rho}(l)$	0.12	−0.10	0.16	−0.24	0.24	−0.27	0.23	−0.28	0.22	−0.17
	0.07	−0.55	0.02	−0.34	−0.01	−0.06	−0.01	0.18	0.03	0.32

$$(1 - 1.545\,B + 0.789\,B^2)\,Y_{2t}$$

$$= 18.403 + (2.069\,B^3 - 1.952\,B^4)\,Y_{1t} + (1 - 0.773\,B)\,a_{2t}\,,$$

with $\hat{\sigma}^2_{a_2} = 3.437$.

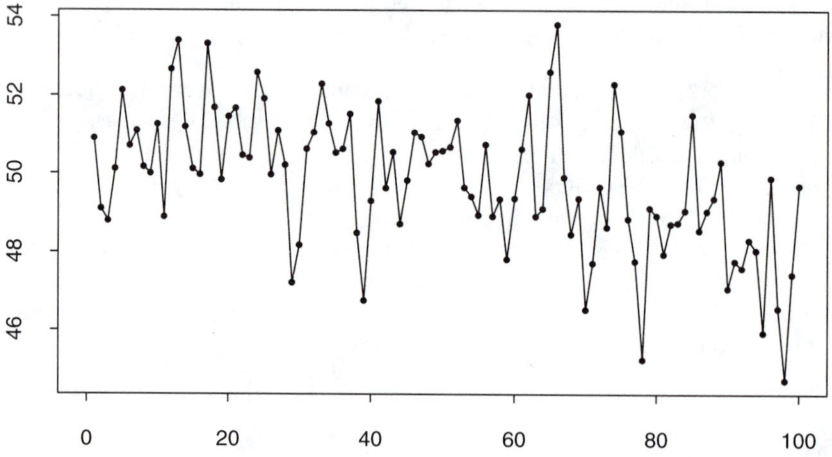

(a) Y_{1t} : Production Figures Series

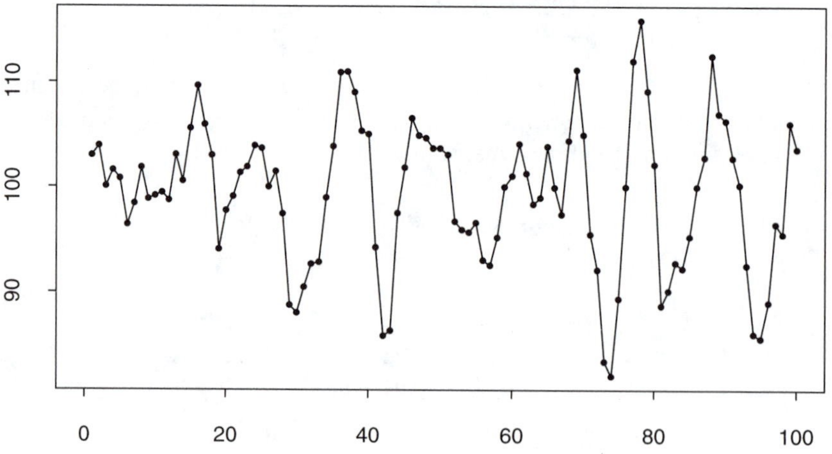

(b) Y_{2t} : Billing Figures Series

Figure 5.2. Weekly Data on Production Schedule and Billing Figures (in thousands) of a Company (from Makridakis and Wheelwright, 1978)

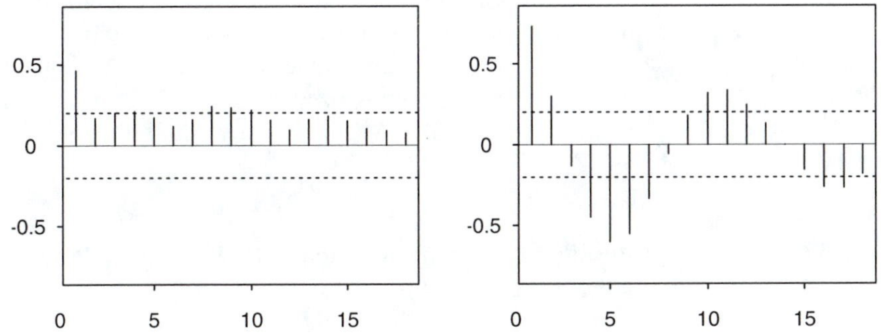

(a) Sample Autocorrelations $\hat{\rho}_{11}(l)$ and $\hat{\rho}_{22}(l)$

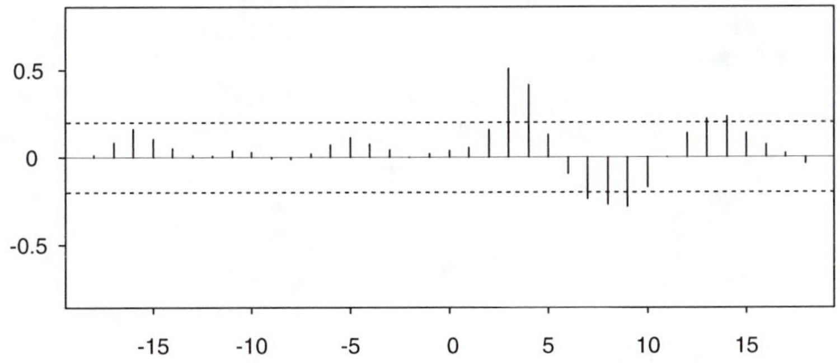

(b) Sample Cross-correlations $\hat{\rho}_{12}(l)$

Figure 5.3. Sample Autocorrelations and Cross-correlations, $\hat{\rho}_{ij}(l)$, for the Bivariate Series of the Production Schedule and Billing Figures Data of Example 5.2

We now consider the general bivariate modeling approach for these data, for comparison. The sample cross-correlation matrices for these data for lags 1 through 10 are given in Table 5.5, and they are also displayed graphically in Figure 5.3. Noting that the $\pm 2\, T^{-1/2}$ limits are ± 0.2, we see that the values $\hat{\rho}_{11}(l)$ tend to be relatively small after lag one and the values $\hat{\rho}_{21}(l) \equiv \hat{\rho}_{12}(-l)$, $l > 0$, are all quite insignificant, giving support to the unidirectional transfer function model structure. By contrast, the values $\hat{\rho}_{22}(l)$ do not die out quickly but according to a damped sinusoidal pattern. Because the cross-correlations $\hat{\rho}_{21}(l)$ between the two series are, in fact, all very small for $l > -2$ (the lag zero cross-correlation is 0.04), and, admittedly, in support of the time delay in the transfer function model above between Y_{2t} and Y_{1t}, we will use a realignment of the bivariate series and consider the shifted series $Y_{1t}^{*} = Y_{1,t-2}$ modeled together with Y_{2t}. It is hoped that this realignment will simplify the bivariate analysis of the two series. [For a brief discussion of the notion of alignment of series for multivariate time series analysis purposes, see Jenkins (1979, p. 116).] Fitting bivariate AR models of orders 1 through 6 by least squares to the aligned series $\boldsymbol{Y}_t = (Y^*_{1t}, Y_{2t})'$ gives the results summarized in Table 5.6. Based on both the LR test statistic values M_m and the AIC_m values, we find that the preferred AR model is of order 3.

Table 5.6 Summary of Results from Fitting AR Models to the Aligned Production Schedule and Billing Figures Data. [$\text{AIC}_m = \log(|\tilde{\Sigma}_m|) + 2mk^2/N$.]

m (AR order)	1	2	3	4	5	6
$\lvert\tilde{\Sigma}_m\rvert$ ($\times 10^2$)	0.320920	0.108382	0.083479	0.082621	0.075018	0.074860
M_m statistic	138.29	99.55	23.66	2.01	6.74	1.24
AIC_m	3.551	2.550	2.375	2.452	2.445	2.535

The LS estimates of the AR(3) model are given by

$$\hat{\Phi}_1 = \begin{bmatrix} 0.541 & 0.105 \\ (0.104) & (0.073) \\ & \\ 2.004 & 0.998 \\ (0.134) & (0.094) \end{bmatrix}, \quad \hat{\Phi}_2 = \begin{bmatrix} -0.386 & -0.195 \\ (0.199) & (0.109) \\ & \\ -0.832 & -0.052 \\ (0.257) & (0.140) \end{bmatrix},$$

$$\hat{\Phi}_3 = \begin{bmatrix} 0.492 & 0.096 \\ (0.180) & (0.061) \\ & \\ -0.794 & -0.372 \\ (0.232) & (0.079) \end{bmatrix}, \quad \text{with} \quad \tilde{\Sigma} = \begin{bmatrix} 2.27004 & -0.46795 \\ -0.46795 & 3.77389 \end{bmatrix},$$

and $\det(\tilde{\Sigma}) = 8.348$. Notice that the coefficients in the (1,2) positions of the $\hat{\Phi}_i$ matrices are rather insignificant. These coefficients could be omitted and then the model would have the simplifying structure of a transfer function form. Residual correlation matrices from the AR(3) model were examined, and all individual residual correlation values for lags 1 through 12 were found to be within the limits $\pm 2\,T^{-1/2} = \pm 0.2$. Hence, there is no indication of inadequacy in the model.

Because the chi-squared tests on the coefficient matrices $\hat{\Phi}_i$ were so highly significant for lags 1 and 2, but considerably less so for lag 3, an alternate model that we next consider for these data is a bivariate ARMA(2,1) model.

Estimates for this model are obtained by conditional maximum likelihood and yield the residual covariance matrix

$$\tilde{\Sigma} = \begin{bmatrix} 2.15639 & -0.07594 \\ -0.07594 & 3.37160 \end{bmatrix},$$

with $\det(\tilde{\Sigma}) = 7.265$. Since the AR(3) and the ARMA(2,1) models involve the same number of parameters, this result indicates a preference for the latter model. When insignificant coefficients are omitted from the ARMA(2,1) model, the final parameter estimates obtained are

$$\hat{\Phi}_1 = \begin{bmatrix} 0.000 & 0.000 \\ (\text{---}) & (\text{---}) \\ & \\ 2.024 & 1.531 \\ (0.125) & (0.033) \end{bmatrix}, \qquad \hat{\Phi}_2 = \begin{bmatrix} 0.000 & 0.000 \\ (\text{---}) & (\text{---}) \\ & \\ -1.862 & -0.785 \\ (0.138) & (0.029) \end{bmatrix},$$

$$\hat{\Theta}_1 = \begin{bmatrix} -0.498 & 0.000 \\ (0.088) & (\text{---}) \\ & \\ 0.000 & 0.702 \\ (\text{---}) & (0.082) \end{bmatrix}, \qquad \text{with} \qquad \tilde{\Sigma} = \begin{bmatrix} 2.44297 & -0.38194 \\ -0.38194 & 3.49827 \end{bmatrix}.$$

Residual correlations from this model are all within ± 0.2 for lags 1 through 12, and the portmanteau statistics Q_s in (5.36) for $s = 5$ and $s = 10$ gave values of $Q_5 = 16.86$ and $Q_{10} = 32.67$, with 14 and 34 degrees of freedom, respectively. Thus, this model gives no appearance of inadequacy. By comparison of this final bivariate model result with the transfer function structure indicated at the start of this example, we find that these two models are essentially identical.

CHAPTER 6

Reduced-Rank and Nonstationary Co-Integrated Models

In this chapter we present some additional topics concerning the modeling of vector time series. These include the examination of models which incorporate special structure in their parameterization, in particular, the nested reduced-rank models, which attempt to cope with the problem of the high dimensionality of the parameters in the vector models. Model specification methods, based on partial canonical correlation analysis, and parameter estimation will be presented for the nested reduced-rank AR models. We also consider estimation and testing issues relating to multivariate nonstationary models that contain unit roots in their AR operator, and the associated concept of cointegration among the components of a nonstationary vector process. Multiplicative seasonal vector ARMA models will be discussed as an additional special topic in this chapter.

6.1 Nested Reduced-Rank AR Models and Partial Canonical Correlation Analysis

We consider analysis of multivariate AR(p) models which may have certain simplifying reduced-rank structures in their coefficient matrices Φ_j. Specifically, we consider the vector AR(p) model

$$\boldsymbol{Y}_t = \sum_{j=1}^{p} \Phi_j \, \boldsymbol{Y}_{t-j} + \varepsilon_t , \tag{6.1}$$

where the ε_t are independent with mean vector $E(\varepsilon_t) = 0$ and nonsingular covariance matrix $\Sigma = \text{Cov}(\varepsilon_t)$. In the model (6.1), it is assumed that the matrices Φ_j have a particular reduced-rank structure, such that $\text{rank}(\Phi_j) = r_j \geq \text{rank}(\Phi_{j+1}) = r_{j+1}$, $j = 1, 2, \ldots, p-1$. So the Φ_j can be represented in the form $\Phi_j = A_j B_j$, where A_j is $k \times r_j$ and B_j is $r_j \times k$, and we further assume that range(A_j) $\supset$ range(A_{j+1}). Thus, we can write (6.1) as

$$\boldsymbol{Y}_t = \sum_{j=1}^{p} A_j B_j \boldsymbol{Y}_{t-j} + \varepsilon_t .$$

We refer to such models as *nested reduced-rank* AR models, and these models, which generalize earlier work by Velu, Reinsel, and Wichern (1986), have been studied by Ahn and Reinsel (1988). These models can result in more parsimonious parameterization, more detailed and possibly simplifying structures, and possibly more interesting and useful interpretations concerning the interrelations among the k time series.

6.1.1 Specification of Ranks Through Partial Canonical Correlation Analysis

To obtain information on the ranks of the matrices Φ_j in the nested reduced-rank model, a partial canonical correlation analysis approach can be used. One fundamental consequence of the above model is that there will exist at least $k - r_j$ *zero* partial canonical correlations between $\boldsymbol{Y}_t$ and $\boldsymbol{Y}_{t-j}$, given $\boldsymbol{Y}_{t-1}, \ldots, \boldsymbol{Y}_{t-j+1}$. This follows because we can find a $(k - r_j) \times k$ matrix F_j', whose rows are linearly independent, such that $F_j' A_j = 0$ and, hence, $F_j' A_i = 0$ for all $i \geq j$ because of the nested structure of the A_j. Therefore,

$$F_j' (\boldsymbol{Y}_t - \textstyle\sum_{i=1}^{j-1} \Phi_i \boldsymbol{Y}_{t-i}) = F_j' (\boldsymbol{Y}_t - \sum_{i=1}^{p} \Phi_i \boldsymbol{Y}_{t-i}) \equiv F_j' \varepsilon_t \tag{6.2}$$

is independent of $\mathbf{Y}_{j-1,t-1} = (\boldsymbol{Y}_{t-1}', \ldots, \boldsymbol{Y}_{t-j}')'$ and consists of $k - r_j$ linear combinations of $\mathbf{Y}_{j-1,t} = (\boldsymbol{Y}_t', \ldots, \boldsymbol{Y}_{t-j+1}')'$. Thus, $k - r_j$ zero partial canonical correlations between $\boldsymbol{Y}_t$ and $\boldsymbol{Y}_{t-j}$ occur as a result of the discussion following (3.9) in Section 3.2. Hence, performing a (partial) canonical correlation analysis for various values of $j = 1, 2, \ldots$ will identify the rank structure of the nested reduced-rank model, as well as the overall order p of the AR model.

The sample test statistic that can be used to (tentatively) specify the ranks is

$$C(j, s) = - (T - j - 1) \sum_{i=(k-s)+1}^{k} \log(1 - \hat{\rho}_i^2(j)) \tag{6.3}$$

for $s = 1, 2, \ldots, k$, where $1 \geq \hat{\rho}_1(j) \geq \hat{\rho}_2(j) \geq \cdots \geq \hat{\rho}_k(j) \geq 0$ are the *sample partial canonical correlations* between $\boldsymbol{Y}_t$ and $\boldsymbol{Y}_{t-j}$, given $\boldsymbol{Y}_{t-1}, \ldots, \boldsymbol{Y}_{t-j+1}$. [See Section 4.2.2, preceding equation (4.7), for further description of the sample partial canonical correlations $\hat{\rho}_i(j)$.] Under the null hypothesis that rank(Φ_j) $\leq k - s$ within the nested reduced-rank model framework, the statistic $C(j, s)$ is asymptotically distributed as chi-squared with s^2 degrees of freedom. Hence, if the value of the test statistic is not "unusually large", we would not reject the null hypothesis and might conclude that Φ_j has reduced rank equal to the smallest value r_j ($\equiv k - s_j$) for which the test does not reject the null hypothesis. Note, in particular, that when $s = k$ the statistic in (6.3) is (essentially) the same as the LR test statistic given in (4.14) for testing

$H_0 : \Phi_j = 0$ in an AR(j) model, as discussed previously in Sections 4.2 and 4.3, since it has been noted previously in Section 4.2.2 that $\log(U_j) = \log[\det(S_j)/\det(S_{j-1})] = \sum_{i=1}^{k} \log(1 - \hat{\rho}_i^2(j))$.

Example 6.1 (Example 4.2, Cont.). As a first brief illustration of the above testing procedure for investigation of the rank structure (and order) of an AR model, we consider again the bivariate quarterly time series from Example 4.2 on the first differences of U.S. fixed investment and U.S. changes in business inventories. The results of a partial canonical correlation analysis for these data, in terms of the (squared) partial canonical correlations $\hat{\rho}_i^2(j)$ for lags $j = 1, \ldots, 6$, and the associated test statistic values from (6.3) are displayed in Table 6.1. [The statistics $C(j, s)$ as presented in Table 6.1 have been adjusted slightly from the form given in (6.3), because the multiplier $(T - j - 1)$ in (6.3) has been replaced by $(T - j - jk - 1/2)$ to be more consistent with the form of the LR test statistic presented in (4.14) of Section 4.3.3.] Then note that for $s = k = 2$ the values of the test statistics in Table 6.1 are (essentially) the same as those for the LR test statistic M_j for testing $H_0 : \Phi_j = 0$ as given in Table 4.4. We see from the statistics in Table 6.1 that $\Phi_j = 0$ is not rejected for each $j > 2$, while $\Phi_2 = 0$ is clearly rejected, but we also find that the hypothesis that $\text{rank}(\Phi_2) = 1$ is clearly not rejected.

Table 6.1. Summary of Partial Canonical Correlation Analysis Between $\boldsymbol{Y}_t$ and $\boldsymbol{Y}_{t-j}$ and Associated Chi-Squared Test Statistics $C(j, s)$ from (6.3) for the Quarterly U.S. Business Investment and Inventories Data Example 6.1. (Test statistics are used to determine the appropriate AR order and ranks $k - s$ of coefficient matrices Φ_j.)

j	Squared Correlations	$C(j, s)$	
		$s = 1$	$s = 2$
1	0.240, 0.529	26.17	98.07
2	0.012, 0.189	1.12	20.54
3	0.001, 0.028	0.07	2.60
4	0.007, 0.078	0.62	7.64
5	0.011, 0.053	0.90	5.49
6	0.000, 0.011	0.00	0.90

Hence, these results suggest an AR(2) model with a reduced-rank structure for the coefficient matrix Φ_2 in the model. Subsequent estimation of this model by (conditional) maximum likelihood methods, which will be discussed shortly, resulted in the fitted model with

$$\hat{\Phi}_1 = \begin{bmatrix} 0.521 & 0.127 \\ 0.319 & 0.598 \end{bmatrix}, \quad \hat{\Phi}_2 = \hat{A}_2 \hat{B}_2 = \begin{bmatrix} 1 \\ \\ -0.450 \\ (0.443) \end{bmatrix} \begin{bmatrix} -0.176 & -0.215 \\ (0.088) & (0.054) \end{bmatrix},$$

and

$$\tilde{\Sigma} = \begin{bmatrix} 4.9793 & 1.6871 \\ 1.6871 & 15.9537 \end{bmatrix},$$

with $\det(\tilde{\Sigma}) = 76.5926$. (In the above results, the element in the first row of the matrix $\hat{A}_2$ is fixed at the value of 1 for normalization purposes.) Thus, we see that these reduced-rank estimates are similar to those from LS estimation of the AR(2) model presented in Example 4.2 of Section 4.5. In fact, since the estimate in the second row of $\hat{A}_2$ above is not significant, this coefficient could be set to zero as an alternative, which is equivalent to setting both the coefficients in the second row of Φ_2 to zero.

6.1.2 *Canonical Form for the Reduced-Rank Model*

It follows from the structure of the matrices $\Phi_j = A_j B_j$ that one can construct a nonsingular $k \times k$ matrix P such that the transformed series $Z_t = P\, Y_t$ will have the following simplified model structure. The model for Z_t is AR(p),

$$Z_t = P\, Y_t = \sum_{i=1}^{p} P\,(A_i B_i)\, P^{-1}\, P\, Y_{t-i} + P\, \varepsilon_t$$

$$= \sum_{i=1}^{p} A_i^* B_i^* Z_{t-i} + e_t \equiv \sum_{i=1}^{p} \Phi_i^* Z_{t-i} + e_t, \tag{6.4}$$

where $A_i^* = P A_i$, $B_i^* = B_i P^{-1}$, $\Phi_i^* = A_i^* B_i^*$, and $e_t = P\,\varepsilon_t$. The rows of P will consist of a basis of the rows of the various matrices F_j' indicated in (6.2). Thus, the matrix P can be chosen so that its last $k - r_j$ rows are orthogonal to the columns of A_j, and, hence, $A_i^* = [\, A_{i1}^{*\prime},\, 0' \,]'$ has its last $k - r_i$ rows which are zero. This implies that $\Phi_i^* = A_i^* B_i^*$ also has its last $k - r_i$ rows being zero. Hence, the "canonical form" of the model has

$$Z_t = \sum_{i=1}^{p} \begin{bmatrix} \Phi_{i1}^* \\ 0 \end{bmatrix} Z_{t-i} + e_t,$$

with the number $k - r_i$ of zero rows for $\Phi_i^* = [\, \Phi_{i1}^{*\prime},\, 0' \,]'$ increasing as i increases. Therefore, the components Z_{jt} of $Z_t = (\, Z_{1t}, Z_{2t}, \ldots, Z_{kt} \,)'$ are represented in an AR(p) model by fewer and fewer past lagged variables Z_{t-i} as j increases from 1 to k.

For the nested reduced-rank AR model with parameters $\Phi_j = A_j B_j$, for any $r_j \times r_j$ nonsingular matrix Q we can write $\Phi_j = A_j\, Q^{-1} Q\, B_j = \underline{A}_j\, \underline{B}_j$ where

$\bar{A}_j = A_j\, Q^{-1}$ and $\bar{B}_j = Q\, B_j$. Therefore, some normalization conditions on the A_j and B_j are required to ensure a unique set of parameters. Assuming the components of $\boldsymbol{Y}_t$ have been arranged so that the upper $r_j \times r_j$ block matrix of each A_j is full rank, this parameterization can be obtained by expressing the $\boldsymbol{\Phi}_j$ as $\boldsymbol{\Phi}_j = A_j\, B_j = A_1\, D_j\, B_j$, where A_1 is $k \times r_1$ and has certain elements "normalized" to fixed values of ones and zeros, and $D_j = [\, I_{r_j},\, 0\,]'$ is $r_1 \times r_j$. More specifically, the matrix A_1 can always be formed to be lower triangular with ones on the main diagonal, and, in fact, when augmented with the last $k - r_1$ columns of the identity matrix, the inverse of the resulting matrix can form the necessary matrix P of the canonical transformation discussed in (6.4). [See Ahn and Reinsel (1988) for further details concerning the normalization.] Thus, with these conventions we can write the model as $\boldsymbol{Y}_t = A_1 \sum_{i=1}^{p} D_i\, B_i\, \boldsymbol{Y}_{t-i} + \varepsilon_t$.

6.1.3 *Maximum Likelihood Estimation of Parameters in the Model*

We let $\boldsymbol{\beta}_0$ denote the $a \times 1$ vector of unknown parameters in the matrix A_1, where it follows that $a = \sum_{j=1}^{p} (\, k - r_j\,)(\, r_j - r_{j+1}\,)$. Also let $\boldsymbol{\beta}_j = \text{vec}(\, B_j'\,)$, a $kr_j \times 1$ vector, for $j = 1, \ldots, p$, and set $\boldsymbol{\gamma} = (\, \boldsymbol{\beta}_0', \boldsymbol{\beta}_1', \ldots, \boldsymbol{\beta}_p'\,)'$. The unknown parameters $\boldsymbol{\gamma}$ are estimated by (conditional) maximum likelihood, using a Newton-Raphson iterative procedure. The (conditional) log-likelihood for T observations $\boldsymbol{Y}_1, \boldsymbol{Y}_2, \ldots, \boldsymbol{Y}_T$ (given $\boldsymbol{Y}_0, \ldots, \boldsymbol{Y}_{1-p}$) is $l(\,\boldsymbol{\gamma}, \Sigma\,) = -\,(T/2) \log |\Sigma| - (1/2) \sum_{t=1}^{T} \varepsilon_t'\, \Sigma^{-1}\, \varepsilon_t$. It can be shown that

$$\frac{\partial l}{\partial \boldsymbol{\gamma}} = -\sum_{t=1}^{T} \frac{\partial \varepsilon_t'}{\partial \boldsymbol{\gamma}}\, \Sigma^{-1}\, \varepsilon_t = \sum_{t=1}^{T} H'\, \boldsymbol{U}_t\, \Sigma^{-1}\, \varepsilon_t\,, \tag{6.5}$$

with $\partial \varepsilon_t' / \partial \boldsymbol{\gamma} = H'\, \boldsymbol{U}_t$, where $\boldsymbol{U}_t = [\, I_k \otimes \boldsymbol{Y}_{t-1}', \ldots, I_k \otimes \boldsymbol{Y}_{t-p}'\,]'$, and H' is a specific matrix of dimension $(\, a + k \sum_{j=1}^{p} r_j\,) \times kp$ whose elements are given functions of the parameters $\boldsymbol{\gamma}$ (but not the observations). Also,

$$\frac{\partial^2 l}{\partial \boldsymbol{\gamma}\, \partial \boldsymbol{\gamma}'} \approx -\sum_{t=1}^{T} H'\, \boldsymbol{U}_t\, \Sigma^{-1} \boldsymbol{U}_t'\, H = -\, H'\, (\, \textstyle\sum_{t=1}^{T} \boldsymbol{U}_t\, \Sigma^{-1} \boldsymbol{U}_t'\,)\, H\,,$$

and an approximate Newton-Raphson iterative procedure to obtain the MLE of $\boldsymbol{\gamma}$ is given by

$$\begin{aligned}\hat{\boldsymbol{\gamma}}^{(i+1)} &= \hat{\boldsymbol{\gamma}}^{(i)} - \{\, \partial^2 l \,/\, \partial \boldsymbol{\gamma}\, \partial \boldsymbol{\gamma}'\,\}^{-1}_{\hat{\boldsymbol{\gamma}}^{(i)}}\, \{\, \partial l \,/\, \partial \boldsymbol{\gamma}\,\}_{\hat{\boldsymbol{\gamma}}^{(i)}} \\ &= \hat{\boldsymbol{\gamma}}^{(i)} + \{\, H'(\, \textstyle\sum_{t=1}^{T} \boldsymbol{U}_t\, \Sigma^{-1} \boldsymbol{U}_t'\,)\, H\,\}^{-1}_{\hat{\boldsymbol{\gamma}}^{(i)}}\, \{\, H' \textstyle\sum_{t=1}^{T} \boldsymbol{U}_t\, \Sigma^{-1}\, \varepsilon_t\,\}_{\hat{\boldsymbol{\gamma}}^{(i)}}\,,\end{aligned} \tag{6.6}$$

where $\hat{\boldsymbol{\gamma}}^{(i)}$ denotes the estimate at the ith iteration.

For large T, it can be shown that the MLE $\hat{\boldsymbol{\gamma}}$ is approximately distributed as $N(\,\boldsymbol{\gamma}, T^{-1} \hat{V}_T^{-1}\,)$, where $V_T = H'(\, T^{-1} \sum_{t=1}^{T} \boldsymbol{U}_t\, \Sigma^{-1} \boldsymbol{U}_t'\,)\, H$. That is, $T^{1/2}\,(\,\hat{\boldsymbol{\gamma}} - \boldsymbol{\gamma}\,)$ converges in distribution to $N(\,0, V^{-1}\,)$ as $T \to \infty$, where $V = H'\, E(\, \boldsymbol{U}_t\, \Sigma^{-1} \boldsymbol{U}_t'\,)\, H$. The result follows easily because

$$T^{1/2}(\hat{\boldsymbol{\gamma}}-\boldsymbol{\gamma}) \approx \{ T^{-1} \sum_{t=1}^{T} H' \boldsymbol{U}_t \Sigma^{-1} \boldsymbol{U}_t' H \}^{-1} H' T^{-1/2} \sum_{t=1}^{T} \boldsymbol{U}_t \Sigma^{-1} \varepsilon_t ,$$

and the factor $T^{-1/2} \sum_{t=1}^{T} \boldsymbol{U}_t \Sigma^{-1} \varepsilon_t$ is the same as the term which occurs in the asymptotic theory for the usual full rank least squares estimator in the vector AR(p) model [e.g., it is similar to the term involved in the asymptotic normal distribution of the LSE in (4.13) of Section 4.3.2]. Hence,

$$T^{-1/2} \partial l / \partial \boldsymbol{\gamma} = H' T^{-1/2} \sum_{t=1}^{T} \boldsymbol{U}_t \Sigma^{-1} \varepsilon_t \xrightarrow{D} H' N(0, G)$$

as $T \to \infty$, where G has (i, j)th block element equal to $\Sigma^{-1} \otimes \Gamma(i-j)$, and $H' T^{-1} \sum_{t=1}^{T} \boldsymbol{U}_t \Sigma^{-1} \boldsymbol{U}_t' H \xrightarrow{P} H' G H \equiv V$. Therefore, we see that

$$T^{1/2}(\hat{\boldsymbol{\gamma}}-\boldsymbol{\gamma}) \xrightarrow{D} N(0, V^{-1}) \qquad \text{where} \qquad V = H' G H .$$

In addition, the asymptotic distribution of the (reduced-rank) ML estimates $\hat{\Phi}_j = \hat{A}_j \hat{B}_j$ of the AR parameters Φ_j follows directly from the above result. It is determined from the relations

$$\begin{aligned}
\hat{\Phi}_j - \Phi_j &= \hat{A}_1 D_j \hat{B}_j - A_1 D_j B_j \\
&= (\hat{A}_1 - A_1) D_j B_j + A_1 D_j (\hat{B}_j - B_j) + (\hat{A}_1 - A_1) D_j (\hat{B}_j - B_j) \\
&= (\hat{A}_1 - A_1) D_j B_j + A_1 D_j (\hat{B}_j - B_j) + O_p(T^{-1}) .
\end{aligned}$$

From this, we find that $\hat{\boldsymbol{\phi}} - \boldsymbol{\phi} = H(\hat{\boldsymbol{\gamma}}-\boldsymbol{\gamma}) + O_p(T^{-1})$, where $\hat{\boldsymbol{\phi}} = \text{vec}\{ \hat{\Phi}_1', \ldots, \hat{\Phi}_p' \}$, $\boldsymbol{\phi} = \text{vec}\{ \Phi_1', \ldots, \Phi_p' \}$, and $H = \partial \boldsymbol{\phi} / \partial \boldsymbol{\gamma}'$, in fact, so that $T^{1/2}(\hat{\boldsymbol{\phi}} - \boldsymbol{\phi}) \xrightarrow{D} N(0, H V^{-1} H')$ as $T \to \infty$. In particular, it is noted that the collection of resulting reduced-rank estimates $\hat{\Phi}_j = \hat{A}_j \hat{B}_j$ of the Φ_j has smaller asymptotic covariance matrix than the corresponding full rank least squares estimates (Ahn and Reinsel, 1988), since $H V^{-1} H' = H \{ H' G H \}^{-1} H' < G^{-1}$. Note that the preceding developments and estimation results are similar to the results presented in Section 5.2 concerning the estimation of vector ARMA models with linear constraints imposed on the parameters, except that the constraints on the AR parameter vector $\boldsymbol{\phi}$ are now nonlinear in the nested reduced-rank AR model. Also, it follows that LR tests of various hypotheses concerning the ranks and other restrictions on the matrices Φ_j can be performed in the usual manner based on the ratio of determinants of the residual covariance matrix estimators, $\hat{\Sigma} = (1/T) \sum_{t=1}^{T} \hat{\varepsilon}_t \hat{\varepsilon}_t'$, in the "full" and "restricted" models, respectively. More details concerning the nested reduced-rank AR models are given by Ahn and Reinsel (1988).

6.1.4 Relation of Reduced-Rank AR Model with Scalar Component Models and Kronecker Indices

Tiao and Tsay (1989) have examined somewhat related structures for the general vector ARMA(p,q) models using the notion of scalar component ARMA models, which were discussed in Section 3.2.2. The concepts related to scalar component models have close connections with reduced-rank structures. More specifically, in the nested reduced-rank AR model, let u $(\leq k)$ be the number of distinct ranks of the Φ_j, that is, the distinct r_j's, and let l_i denote the smallest lag such that rank(Φ_j) $< r_{l_i}$ for all $j > l_i$, $i = 0, 1, 2, \ldots, u$, with $l_u = p$ and the conventions $l_0 = 0$ and $r_{l_0} = k$. So $\Phi_1, \ldots, \Phi_{l_1}$ have rank r_{l_1}, $\Phi_{l_1+1}, \ldots, \Phi_{l_2}$ have rank r_{l_2}, and so on. Thus, under the transformation $Z_t = P\, Y_t$ of the canonical form in (6.4), we can identify the components $Z_{1t}, \ldots, Z_{kt}$ as possessing, in the framework of Tiao and Tsay (1989), autoregressive scalar component models of nonincreasing orders $l_u > l_{u-1} > \cdots > l_1 > l_0$, with the number of scalar components of AR order l_i being equal to ($r_{l_i} - r_{l_{i+1}}$), with the convention $r_{l_i} = 0$ for $i > u$. Also, the structure of the reduced-rank autoregressive model in relation to the minimal dimension state-space representation in terms of a basis for the prediction space of all present and future values will be explored in some detail later in Section 7.2, through the use of the canonical form (6.4). In particular, it is noted from the concepts of Kronecker indices and McMillan degree that for the structure of the reduced-rank AR model, from (6.4) the Kronecker indices $K_1, \ldots, K_k$ can be seen to be equal to the lags $l_u, l_{u-1}, \ldots, l_1, l_0$, with multiplicities as indicated above, and the McMillan degree is $M = \sum_{i=0}^{u} (r_{l_i} - r_{l_{i+1}})\, l_i = \sum_{j=1}^{k} K_j$. It is easy to verify that the McMillan degree is, equivalently, also equal to $M = \sum_{l=1}^{p} r_l$, the sum of the ranks of the AR coefficient matrices Φ_l. In addition, from (6.4) we see that the reduced-rank AR model can also be represented as

$$\Phi_0^{\#}\, Y_t - \sum_{i=1}^{p} \Phi_i^{\#}\, Y_{t-i} = \Phi_0^{\#}\, \varepsilon_t ,$$

with $\Phi_0^{\#} = P = A^{-1}$, where A is the $k \times k$ matrix formed by augmenting the $k \times r_1$ matrix A_1 with the last ($k - r_1$) columns of the identity matrix (with $A = A_1$ when $r_1 = k$), and $\Phi_i^{\#} = A_i^{*}\, B_i = P\, A_i\, B_i = P\, A_1\, D_i\, B_i = [\, B_i', 0'\,]'$ having its last $k - r_i$ rows equal to zero. This relation can be viewed as an echelon form representation, as in (3.4) of Section 3.1.2, for the reduced-rank vector AR(p) model.

For a specific illustration, we note that the nested reduced-rank AR(2) model that was estimated for the quarterly U.S. business investment and inventories data in Example 6.1 can also equivalently be estimated in the "echelon form" $\Phi_0^{\#}\, Y_t - \Phi_1^{\#}\, Y_{t-1} - \Phi_2^{\#}\, Y_{t-2} = \Phi_0^{\#}\, \varepsilon_t$, with conditional ML estimates given by

$$\hat{\Phi}_0^{\#} = \begin{bmatrix} 1 & 0 \\ 0.450 & 1 \end{bmatrix}, \quad \hat{\Phi}_1^{\#} = \begin{bmatrix} 0.521 & 0.127 \\ 0.553 & 0.655 \end{bmatrix}, \quad \hat{\Phi}_2^{\#} = \begin{bmatrix} -0.176 & -0.215 \\ 0 & 0 \end{bmatrix},$$

such that $\hat{\Phi}_0^{\#-1}\, \hat{\Phi}_1^{\#} = \hat{\Phi}_1$ and $\hat{\Phi}_0^{\#-1}\, \hat{\Phi}_2^{\#} = \hat{\Phi}_2 = \hat{A}_2\, \hat{B}_2$ in Example 6.1, and with $\tilde{\Sigma}$ equal to the same value as given in Example 6.1. As was discussed in Section 3.1, the notion of nested reduced-rank model and its association with the echelon form representation can be directly extended to the vector ARMA model, leading to the specification of reduced-rank ARMA models of the form given in (3.8) of Section 3.1.3.

However, it is noted that the overall approach of Tiao and Tsay differs somewhat from the approach presented above. In their approach, within the vector AR framework, certain eigenvector information obtained from their "preliminary" sample canonical correlation analyses is used to obtain an estimate $\tilde{P}$ of the transformation matrix involved in the canonical form of (6.4). The series $\tilde{\boldsymbol{Z}}_t = \tilde{P}\,\boldsymbol{Y}_t$ is then constructed and the model for $\tilde{\boldsymbol{Z}}_t$ is estimated with the implied zero row structure from (6.4) incorporated in the parameter matrices Φ_i^*. In their approach, the matrix $\tilde{P}$ is not considered as parameters of the model for the original series $\boldsymbol{Y}_t$, but as a means of transformation to a series $\tilde{\boldsymbol{Z}}_t$ with simpler model structure, whereas, in the approach presented above, the matrix P (through the normalized coefficient matrix A_1) is included as a part of the estimation of the overall model for $\boldsymbol{Y}_t$ with the simplifying structure imposed through the nested reduced-rank constraints.

EXAMPLE 6.2. In Ahn and Reinsel (1988), monthly U.S. grain price data for the period January 1961 – October 1972 were considered. These data consist of a four-variate time series of prices on wheat flour, corn, wheat, and rye. The results of the partial canonical correlation analysis and the associated test statistic values from (6.3) are summarized in Table 6.2.

Table 6.2. Summary of Partial Canonical Correlation Analysis Between $\boldsymbol{Y}_t$ and $\boldsymbol{Y}_{t-j}$ and Associated Chi-Squared Test Statistics $C(j, s)$ from (6.3) for the Monthly Grain Price Data Example 6.2. (Test statistics are used to determine the appropriate AR order and ranks $k - s$ of coefficient matrices Φ_j.)

j	Squared Correlations	$C(j, s)$			
		$s=1$	$s=2$	$s=3$	$s=4$
1	0.580, 0.596, 0.702, 0.948	122.48	250.31	420.87	836.85
2	0.001, 0.002, 0.073, 0.188	0.11	0.38	11.07	40.24
3	0.002, 0.015, 0.020, 0.070	0.22	2.38	5.24	15.27
4	0.001, 0.007, 0.043, 0.092	0.13	1.05	7.17	20.54

From this, the tentatively identified model is a reduced-rank AR(2) model of the form

$$\boldsymbol{Y}_t = \Phi_1\,\boldsymbol{Y}_{t-1} + \Phi_2\,\boldsymbol{Y}_{t-2} + \varepsilon_t \equiv A_1\,(\,B_1\,\boldsymbol{Y}_{t-1} + D_2\,B_2\,\boldsymbol{Y}_{t-2}\,) + \varepsilon_t\,,$$

with $\text{rank}(\Phi_1) = 4$, $\text{rank}(\Phi_2) = 1$, and $D_2 = [\,1 \quad 0 \quad 0 \quad 0\,]'$. The maximum likelihood estimates of the parameters are obtained as

$$\hat{A}_1 = \begin{bmatrix} 1 & 0 & 0 & 0 \\ -0.015 & 1 & 0 & 0 \\ 0.323 & 0 & 1 & 0 \\ 0.193 & 0 & 0 & 1 \end{bmatrix}, \qquad \hat{B}_1 = \begin{bmatrix} 0.985 & -0.412 & 0.577 & -0.460 \\ 0.025 & 0.795 & 0.056 & -0.016 \\ -0.285 & -0.117 & 0.824 & 0.045 \\ -0.202 & -0.100 & 0.045 & 0.809 \end{bmatrix},$$

and $\hat{B}_2 = [\ -0.079 \quad 0.925 \quad -0.622 \quad 0.242\ ]$. It is found that the resulting reduced-rank estimates $\hat{\Phi}_j = \hat{A}_1 D_j \hat{B}_j$ are in close agreement with the usual full rank least squares estimates with estimated standard errors that are about 25% smaller than those of the corresponding full rank estimates. The canonical transformation can be obtained by use of the matrix $P = \hat{A}_1^{-1}$, and details concerning the model features for the resulting canonical series $\boldsymbol{Z}_t = P\,\boldsymbol{Y}_t$ are discussed by Ahn and Reinsel (1988). Here it is simply noted that the AR(2) model for $\boldsymbol{Z}_t$, $\boldsymbol{Z}_t = \Phi_1^* \boldsymbol{Z}_{t-1} + \Phi_2^* \boldsymbol{Z}_{t-2} + \boldsymbol{e}_t$, will be such that the last three rows of the matrix Φ_2^* are equal to zero, and, hence, the structure of the model for $\boldsymbol{Z}_t$ is much simplified.

6.2 Review of Estimation and Testing for Nonstationarity (Unit Roots) in Univariate ARIMA Models

In univariate time series modeling, it is rather common practice to difference a time series Y_t when the series exhibits some nonstationary features. The decision concerning the need for differencing is sometimes based, informally, on characteristics of the time series plot of Y_t and of its sample autocorrelation function (e.g., failure to dampen out sufficiently quickly). This has led more recently to an interest in more formal inference procedures concerning the appropriateness of a differencing operator (or a unit root in the AR operator) in the model. This, in turn, leads to an interest in estimation of parameters in unit-root nonstationary time series and the associated estimation and testing theory. Some of the developments in the estimation and testing for the univariate model with a unit root will now be briefly reviewed.

6.2.1 Limiting Distribution Results in the AR(1) Model with a Unit Root

We first examine the simple AR(1) model $Y_t = \phi\, Y_{t-1} + \varepsilon_t$, $t = 1, 2, \ldots, T$, $Y_0 = 0$, and consider testing for a random walk model, $H_0 : \phi = 1$. The least squares (LS) estimator of ϕ is given by

$$\hat{\phi} = \sum_{t=2}^{T} Y_{t-1} Y_t \,/\, \sum_{t=2}^{T} Y_{t-1}^2 = \phi + \sum_{t=2}^{T} Y_{t-1}\, \varepsilon_t \,/\, \sum_{t=2}^{T} Y_{t-1}^2 .$$

When $\phi = 1$, so that $Y_t = \sum_{j=0}^{t-1} \varepsilon_{t-j} + Y_0$, it can be shown that $T(\hat{\phi} - 1) = T^{-1} \sum_{t=2}^{T} Y_{t-1}\, \varepsilon_t \,/\, T^{-2} \sum_{t=2}^{T} Y_{t-1}^2 = O_p(1)$, bounded in probability as $T \to \infty$, with both the numerator and denominator possessing nondegenerate and non-normal limiting distributions. A representation for the limiting distribution of $T(\hat{\phi} - 1)$ has been given by Dickey and Fuller (1979), such that

$$T(\hat{\phi}-1) \xrightarrow{D} \frac{1}{2}(\Lambda^2 - 1)/\Gamma, \tag{6.7}$$

where $(\Gamma, \Lambda) = (\sum_{i=1}^{\infty} \gamma_i^2 Z_i^2, \sum_{i=1}^{\infty} 2^{1/2} \gamma_i Z_i)$, the Z_i are i.i.d. $N(0, 1)$ distributed r.v.'s, and $\gamma_i = 2(-1)^{i+1}/[(2i-1)\pi]$. An equivalent representation for the distribution (Chan and Wei, 1988) is given by

$$T(\hat{\phi}-1) \xrightarrow{D} \int_0^1 B(u)\, dB(u) \Big/ \int_0^1 B(u)^2\, du$$

$$= (1/2)(B(1)^2 - 1) \Big/ \int_0^1 B(u)^2\, du, \tag{6.8}$$

where $B(u)$ is a (continuous-parameter) standard Brownian motion process on $[0, 1]$. Tables for the percentiles of the limiting distribution of $T(\hat{\phi}-1)$ have been given by Fuller (1976). Also, the "Studentized" statistic

$$\hat{\tau} = (\hat{\phi}-1) / \{ S_\varepsilon (\sum_{t=2}^{T} Y_{t-1}^2)^{-1/2} \}, \tag{6.9}$$

where $S_\varepsilon^2 = (T-2)^{-1} \{ \sum_{t=2}^{T} Y_t^2 - \hat{\phi} \sum_{t=2}^{T} Y_{t-1} Y_t \}$ is the residual mean square, has been considered. The limiting distribution of the statistic $\hat{\tau}$ has been derived, with the representation given by

$$\hat{\tau} \xrightarrow{D} \int_0^1 B(u)\, dB(u) \Big/ \{ \int_0^1 B(u)^2\, du \}^{1/2},$$

and tables of the percentiles of this distribution under $H_0 : \phi = 1$ available in Fuller (1976) can be used to test H_0. The test rejects H_0 when $\hat{\tau}$ is "too negative".

6.2.2 *Unit-Root Distribution Results for General Order AR Models*

For higher-order AR(p) processes, $\phi(B) Y_t = \varepsilon_t$, we consider the case where $\phi(B) = \phi^*(B)(1-B)$ and $\phi^*(B)$ is a $(p-1)$th-order *stationary* AR operator. Hence, $\phi(B) Y_t = \phi^*(B)(1-B) Y_t = Y_t - Y_{t-1} - \sum_{j=1}^{p-1} \phi_j^* (Y_{t-j} - Y_{t-j-1})$, and testing for a unit root in $\phi(B)$ is equivalent to testing $\rho = 1$ in the model $Y_t = \rho Y_{t-1} + \sum_{j=1}^{p-1} \phi_j^* (Y_{t-j} - Y_{t-j-1}) + \varepsilon_t$, or testing $\rho - 1 = 0$ in the model

$$(Y_t - Y_{t-1}) = (\rho - 1) Y_{t-1} + \sum_{j=1}^{p-1} \phi_j^* (Y_{t-j} - Y_{t-j-1}) + \varepsilon_t .$$

In fact, for any AR(p) model $Y_t = \sum_{j=1}^{p} \phi_j Y_{t-j} + \varepsilon_t$, it is seen that the model can be expressed in an equivalent form as

$$W_t = (\rho - 1) Y_{t-1} + \sum_{j=1}^{p-1} \phi_j^* W_{t-j} + \varepsilon_t ,$$

where $W_t = Y_t - Y_{t-1}$, $\rho - 1 = -\phi(1) = \sum_{j=1}^{p} \phi_j - 1$, and $\phi_j^* = -\sum_{i=j+1}^{p} \phi_i$. [Hence, the existence of a unit root in the AR operator $\phi(B)$ is equivalent to $\rho = \sum_{j=1}^{p} \phi_j = 1$.] So let $(\hat{\rho}, \hat{\phi}_1^*, \ldots, \hat{\phi}_{p-1}^*)$ denote the usual least squares regression estimates for this model, obtained by regressing Y_t on $Y_{t-1}, W_{t-1}, \ldots, W_{t-p+1}$, where $W_t = Y_t - Y_{t-1}$, or, equivalently, let $(\hat{\rho} - 1, \hat{\phi}_1^*, \ldots, \hat{\phi}_{p-1}^*)$ denote the least squares estimates obtained by regressing $W_t = Y_t - Y_{t-1}$ on $Y_{t-1}, W_{t-1}, \ldots, W_{t-p+1}$. Then, under the unit-root model where $\rho = 1$ and $\phi^*(B) = 1 - \sum_{j=1}^{p-1} \phi_j^* B^j$ is stationary, it has been shown by Fuller (1976, Theorem 8.5.1 and Corollary 8.5.1) that $(\hat{\rho} - 1)/\{ S_\varepsilon (\sum_{t=p+1}^T Y_{t-1}^2)^{-1/2} \}$ has the same limiting distribution as the statistic $\hat{\tau}$ in (6.9) for the AR(1) case, while $(T - p)(\hat{\rho} - 1)\psi_*$, where $\psi_* = \sum_{j=0}^\infty \psi_j$, with $\psi(B) = \phi^*(B)^{-1}$, has approximately the same distribution as the statistic $T(\hat{\phi} - 1)$ for the AR(1) case. Also, it follows that the statistic, denoted as $\hat{\tau}$, formed by dividing $(\hat{\rho} - 1)$ by its "usual estimated standard error" from the least squares regression will be asymptotically equivalent to the statistic $(\hat{\rho} - 1)/\{ S_\varepsilon (\sum_{t=p+1}^T Y_{t-1}^2)^{-1/2} \}$ and, hence, will have the same limiting distribution as the statistic $\hat{\tau}$ for the AR(1) case.

The test statistic $\hat{\tau}$ formed in the above manner can be used to test $H_0 : \rho = 1$ in the AR(p) model, i.e., to test for a unit root in the AR(p) model $\phi(B) Y_t = \varepsilon_t$. Furthermore, it has been shown (Fuller, 1976, Theorem 8.5.1) that the limiting distribution of the LSE $(\hat{\phi}_1^*, \ldots, \hat{\phi}_{p-1}^*)$ for the parameters of the "stationary operator" $\phi^*(B)$ of the model is the same as the standard asymptotic distribution for LSE $\tilde{\phi}_1^*, \ldots, \tilde{\phi}_{p-1}^*$ obtained by regressing the stationary differenced series $W_t = Y_t - Y_{t-1}$ on $W_{t-1}, \ldots, W_{t-p+1}$. It is also noted that the above results extend to the case where a constant term is included in the least squares regression estimation, with the statistic analogous to $\hat{\tau}$ denoted as $\hat{\tau}_\mu$, although the limiting distribution for $\hat{\tau}_\mu$ is derived when the "true" value of the constant term in the model is equal to zero under the hypothesis that $\rho = 1$. Thus, for example, in the AR(1) model $Y_t = \phi Y_{t-1} + \delta + \varepsilon_t$, one obtains the LSE

$$\hat{\phi}_\mu = \sum_{t=2}^T (Y_{t-1} - \bar{Y}_{(1)})(Y_t - \bar{Y}_{(0)}) / \sum_{t=2}^T (Y_{t-1} - \bar{Y}_{(1)})^2,$$

as in Section 4.3.1. Then, when $\phi = 1$, the representation similar to (6.8) for the limiting distribution of $\hat{\phi}_\mu$ is given by

$$T(\hat{\phi}_\mu - 1) \xrightarrow{D} [\int_0^1 B(u)\, dB(u) - \xi B(1)] / [\int_0^1 B(u)^2\, du - \xi^2],$$

where $\xi = \int_0^1 B(u)\, du$, and it is assumed that $\delta = (1 - \phi)\mu = 0$ when $\phi = 1$. The corresponding test statistic for $H_0 : \phi = 1$ in the AR(1) case is

$$\hat{\tau}_\mu = (\hat{\phi}_\mu - 1) / \{ S_\varepsilon [\sum_{t=2}^T (Y_{t-1} - \bar{Y}_{(1)})^2]^{-1/2} \}, \qquad (6.10)$$

and tables of percentiles of the distribution of $\hat{\tau}_\mu$ when $\phi = 1$ are also available in Fuller (1976). We comment that these test procedures and other similar ones have also been extended for use in testing for a unit root in mixed ARIMA(p, 1,q) models, e.g., see Said and Dickey (1985) and Solo (1984a). The general result is that an appropriate "unit-root" test statistic for these models constructed from a standardized form of $\hat{\rho}$, based on Gaussian estimation, possesses the same limiting distribution as the statistic $\hat{\tau}$ (or $\hat{\tau}_\mu$) in the AR model, under the null model which contains a unit root $\rho = 1$. Results under somewhat more general conditions have been considered by Said and Dickey (1984) and Phillips and Perron (1988), among others. Also, a review of various results concerning testing for unit roots in univariate ARIMA models has been given by Dickey, Bell, and Miller (1986).

Example 6.3. We consider "deseasonalized" U.S. monthly housing starts data for the period January 1965 through December 1974. The deseasonalized data $Y_t = Y_t^* - \hat{S}_t$ are formed by subtracting the monthly average seasonal values $\hat{S}_t$ from the original series Y_t^*, where $\hat{S}_t = \hat{S}_{t,m} = (1/N) \sum_{i=0}^{N-1} Y_{12i+m}^*$ for month m and N denotes the number of years. The deseasonalized series Y_t is plotted in Figure 6.1(a), which is presented later in Section 6.3.6 in connection with Example 6.4. The resulting series Y_t is identified as an AR(2) model and estimated by least squares. The fitted model is $Y_t = 0.6875\, Y_{t-1} + 0.2620\, Y_{t-2} + \varepsilon_t$, or equivalently,

$$W_t = -0.0504\, Y_{t-1} - 0.2620\, W_{t-1} + \varepsilon_t ,$$

where $W_t = Y_t - Y_{t-1}$. Thus, $\hat{\rho} - 1 = -0.0504$, and the "Studentized" statistic is $\hat{\tau}_\mu = (\hat{\rho} - 1) / (\text{est. st. dev.} (\hat{\rho})) = -0.0504 / 0.0343 = -1.47$, which is not less than the 10% lower critical value -2.58 from Table 8.5.2 of Fuller (1976), so we cannot reject $H_0 : \rho - 1 = 0$. Thus, the model is approximately equivalent to $(1 + 0.262\, B)(1 - B)\, Y_t = \varepsilon_t$.

6.3 Nonstationary (Unit-Root) Multivariate AR Models, Estimation, and Testing

6.3.1 Unit-Root Nonstationary Vector AR Model, and the Error-Correction Form

We now consider multivariate AR models for processes $\{\boldsymbol{Y}_t\}$ which are nonstationary. We concentrate on situations where it is assumed that the only "nonstationary roots" in the AR operator $\Phi(B)$ are roots equal to one (unit roots), and we assume there are $d \le k$ such unit roots, with all other roots of $\det\{\Phi(B)\} = 0$ outside the unit circle. We note immediately that this implies that $\det\{\Phi(1)\} = 0$ so that the matrix $\Phi(1) = I - \sum_{j=1}^{p} \Phi_j$ does not have full rank. It is also assumed that $\text{rank}\{\Phi(1)\} = r$, with $r = k - d$, and it is further noted that this condition implies that each component of the first differences $\boldsymbol{W}_t = \boldsymbol{Y}_t - \boldsymbol{Y}_{t-1}$ will be stationary. The AR(p) model $\Phi(B)\, \boldsymbol{Y}_t = \boldsymbol{Y}_t - \sum_{j=1}^{p} \Phi_j\, \boldsymbol{Y}_{t-j} = \varepsilon_t$ can also be represented in the *error-correction form* (Engle and Granger, 1987) as $\Phi^*(B)(1 - B)\, \boldsymbol{Y}_t = -\Phi(1)\, \boldsymbol{Y}_{t-1} + \varepsilon_t$, that is,

$$\boldsymbol{W}_t = C\, \boldsymbol{Y}_{t-1} + \sum_{j=1}^{p-1} \Phi_j^* \, \boldsymbol{W}_{t-j} + \varepsilon_t \,, \tag{6.11}$$

where $\boldsymbol{W}_t = (1 - B)\, \boldsymbol{Y}_t = \boldsymbol{Y}_t - \boldsymbol{Y}_{t-1}$, $\Phi^*(B) = I - \sum_{j=1}^{p-1} \Phi_j^* B^j$, with $\Phi_j^* = -\sum_{i=j+1}^{p} \Phi_i$, and $C = -\Phi(1) = -(I - \sum_{j=1}^{p} \Phi_j)$.

From the assumptions, the matrix $I - \Phi(1) = \sum_{j=1}^{p} \Phi_j$ has d linearly independent eigenvectors associated with its d unit eigenvalues, while its remaining eigenvalues are less than one in absolute value. Let P and $Q = P^{-1}$ be $k \times k$ matrices such that $Q\, (\sum_{j=1}^{p} \Phi_j)\, P = \mathrm{Diag}[\, I_d, \Lambda_r \,] = J$, where J is the Jordan canonical form of $\sum_{j=1}^{p} \Phi_j$, so that $Q\, C\, P = J - I$ $= \mathrm{Diag}[\, 0, \Lambda_r - I_r \,]$. Hence, $C = P\, (J - I)\, Q = P_2\, (\Lambda_r - I_r)\, Q_2'$, where $P = [\, P_1, P_2 \,]$, $Q' = [\, Q_1, Q_2 \,]$, with P_1 and Q_1 being $k \times d$ matrices, so that C is of *reduced rank* $r < k$. Therefore, the error-correction form (6.11) can be written as

$$\boldsymbol{W}_t = A\, Q_2'\, \boldsymbol{Y}_{t-1} + \sum_{j=1}^{p-1} \Phi_j^* \, \boldsymbol{W}_{t-j} + \varepsilon_t \equiv A\, \boldsymbol{Z}_{2t-1} + \sum_{j=1}^{p-1} \Phi_j^* \, \boldsymbol{W}_{t-j} + \varepsilon_t \,, \tag{6.11$'$}$$

where $A = P_2\, (\Lambda_r - I_r)$ is $k \times r$ of rank r and $\boldsymbol{Z}_{2t} = Q_2'\, \boldsymbol{Y}_t$. It follows that although $\boldsymbol{Y}_t$ is nonstationary, the r linear combinations $\boldsymbol{Z}_{2t} = Q_2'\, \boldsymbol{Y}_t$ are stationary [since through (6.11′) the variables $\boldsymbol{Z}_{2t}$ are such that $\boldsymbol{Z}_{2t} - \Lambda_r\, \boldsymbol{Z}_{2t-1}$ is linearly expressible in terms of the stationary series $\{\boldsymbol{W}_t\}$ and $\{\varepsilon_t\}$ only and Λ_r is stable]. In this situation, $\boldsymbol{Y}_t$ is said to be *co-integrated* of rank r, and the rows of Q_2' are called *co-integrating vectors*.

Conversely, the d-dimensional series $\boldsymbol{Z}_{1t} = Q_1'\, \boldsymbol{Y}_t$ is purely nonstationary, with the unit-root nonstationary behavior of the series $\boldsymbol{Y}_t$ generated by $\boldsymbol{Z}_{1t}$. That is, we let $\boldsymbol{Z}_t = (\boldsymbol{Z}_{1t}', \boldsymbol{Z}_{2t}')' = Q\, \boldsymbol{Y}_t$, so that we obtain the representation $\boldsymbol{Y}_t = P\, \boldsymbol{Z}_t = P_1\, \boldsymbol{Z}_{1t} + P_2\, \boldsymbol{Z}_{2t}$. Then we see that $\boldsymbol{Y}_t$ is a linear combination of the d-dimensional purely nonstationary component $\boldsymbol{Z}_{1t}$ and the r-dimensional stationary component $\boldsymbol{Z}_{2t}$. The purely nonstationary component $\boldsymbol{Z}_{1t}$ may be viewed as the common nonstationary component or the d "common trends" among the $\boldsymbol{Y}_t$, with the interpretation that the nonstationary behavior in each of the k component series Y_{it} of $\boldsymbol{Y}_t$ is actually driven by d $(< k)$ common underlying stochastic trends $(\boldsymbol{Z}_{1t})$.

6.3.2 Asymptotic Properties of the Least Squares Estimator

Least squares estimates $\hat{C}$, $\hat{\Phi}_1^*, \ldots, \hat{\Phi}_{p-1}^*$ for the model (6.11) in error-correction form can be obtained, and the limiting distribution theory for these estimators has been derived by Ahn and Reinsel (1990). To describe the asymptotic results, note that the model in (6.11) can be expressed as

$$\boldsymbol{W}_t = CP\, \boldsymbol{Z}_{t-1} + \sum_{j=1}^{p-1} \Phi_j^* \, \boldsymbol{W}_{t-j} + \varepsilon_t = CP_1\, \boldsymbol{Z}_{1t-1} + CP_2\, \boldsymbol{Z}_{2t-1} + \sum_{j=1}^{p-1} \Phi_j^* \, \boldsymbol{W}_{t-j} + \varepsilon_t \,,$$

where $\boldsymbol{Z}_t = Q\, \boldsymbol{Y}_t = (\boldsymbol{Z}_{1t}', \boldsymbol{Z}_{2t}')'$, with $\boldsymbol{Z}_{1t} = Q_1'\, \boldsymbol{Y}_t$ purely nonstationary, whereas

$Z_{2t} = Q_2' \, Y_t$ is stationary, and note also that $CP_1 = 0$. In addition, from (6.11), the process

$$u_t = Q\,(W_t - C\, Y_{t-1}) = (Z_t - Z_{t-1}) - (J - I)\, Z_{t-1} = Z_t - J\, Z_{t-1}\,, \qquad (6.12)$$

where $J = \text{Diag}[\, I_d,\, \Lambda_r\,]$, is equal to $u_t = Q\,(\sum_{j=1}^{p-1} \Phi_j^*\, W_{t-j} + \varepsilon_t\,)$ and, hence, u_t is stationary. Therefore, u_t has the (stationary) infinite MA representation of the form $u_t = \Psi(B)\, a_t = \sum_{j=0}^{\infty} \Psi_j\, a_{t-j}$, with $a_t = Q\, \varepsilon_t$, and we let $\Psi = \Psi(1) = \sum_{j=0}^{\infty} \Psi_j$.

Now let $F = (C, \Phi_1^*, \ldots, \Phi_{p-1}^*)$, $X_{t-1} = (Y_{t-1}', W_{t-1}', \ldots, W_{t-p+1}')'$, and $X_{t-1}^* = (Z_{t-1}', W_{t-1}', \ldots, W_{t-p+1}')' \equiv Q^*\, X_{t-1}$, where $Q^* = \text{Diag}(Q, I_{k(p-1)})$, and assume that observations $Y_{1-p}, \ldots, Y_0, Y_1, \ldots, Y_T$ are available (for convenience of notation), with $W_t = Y_t - Y_{t-1}$. So the least squares estimator of F in the error-correction model (6.11) can be represented by

$$\begin{aligned}\hat{F} &= (\textstyle\sum_{t=1}^{T} W_t\, X_{t-1}' \,)\,(\sum_{t=1}^{T} X_{t-1} X_{t-1}' \,)^{-1} \\ &= (\textstyle\sum_{t=1}^{T} W_t\, X_{t-1}^{*\prime} \,)\,(\sum_{t=1}^{T} X_{t-1}^{*} X_{t-1}^{*\prime} \,)^{-1}\, Q^*.\end{aligned}$$

Then since the model is $W_t = F\, X_{t-1} + \varepsilon_t = F\, P^*\, X_{t-1}^* + \varepsilon_t$, where $P^* = \text{Diag}(P, I_{k(p-1)}) = Q^{*-1}$, we have

$$(\hat{F} - F)\, P^* = (\textstyle\sum_{t=1}^{T} \varepsilon_t\, X_{t-1}^{*\prime} \,)\,(\sum_{t=1}^{T} X_{t-1}^{*} X_{t-1}^{*\prime} \,)^{-1}.$$

With $U_{t-1} = (Z_{2t-1}', W_{t-1}', \ldots, W_{t-p+1}')'$, which is stationary, it is known that $T^{-3/2} \sum_{t=1}^{T} U_{t-1}\, Z_{1t-1}' \xrightarrow{P} 0$ as $T \to \infty$, and so it follows that the least squares estimator for the model (6.11) has the representation

$$\begin{aligned}&[\, T\, \hat{C}\, P_1,\; T^{1/2}\,(\hat{C}\, P_2 - C\, P_2,\, \hat{\Phi}_1^* - \Phi_1^*, \ldots, \hat{\Phi}_{p-1}^* - \Phi_{p-1}^* \,)\,] \\ &\qquad = \Big[\, P(T^{-1} \textstyle\sum a_t Z_{1t-1}')(T^{-2} \sum Z_{1t-1} Z_{1t-1}')^{-1}, \\ &\qquad\qquad (T^{-1/2} \textstyle\sum \varepsilon_t U_{t-1}')(T^{-1} \sum U_{t-1} U_{t-1}')^{-1} \Big] + o_p(1). \qquad (6.13)\end{aligned}$$

Then, the distribution theory [see Lemma 1 and Theorem 1 in Ahn and Reinsel (1990)] for the least squares estimator is such that $T\, \hat{C}\, P_1 \xrightarrow{D} P\, M$, where

$$M = \Sigma_a^{1/2}\, \{ \int_0^1 B_d(u)\, dB_k(u)' \,\}'\, \{ \int_0^1 B_d(u)\, B_d(u)'\, du \,\}^{-1}\, \Sigma_{a_1}^{-1/2}\, \Psi_{11}^{-1}\,, \qquad (6.14)$$

where $\Sigma_a = Q\, \Sigma\, Q' = \text{Cov}(a_t)$ with $\Sigma = \text{Cov}(\varepsilon_t)$, $\Sigma_{a_1} = [\, I_d, 0\,]\, \Sigma_a\, [\, I_d, 0\,]'$ is the upper left $d \times d$ block of the matrix Σ_a, and Ψ_{11} is the upper left $d \times d$ block of the matrix Ψ, $B_k(u)$ is a k-dimensional standard Brownian motion process, and $B_d(u) = \Sigma_{a_1}^{-1/2}\, [\, I_d, 0\,]\, \Sigma_a^{1/2}\, B_k(u)$ is a d-dimensional standard Brownian motion. It is also shown that $T^{1/2}\,(\hat{C}\, P_2 - C\, P_2)$ and $T^{1/2}\,(\hat{\Phi}_j^* - \Phi_j^*)$, $j = 1, \ldots, p-1$, have a joint limiting multivariate normal distribution as in stationary model situations, such that the "vec" of these terms has limiting distribution $N(0, \Gamma_u^{-1} \otimes \Sigma)$, where $\Gamma_u = \text{Cov}(U_t)$. Also, the LSEs $\hat{\Phi}_1^*, \ldots, \hat{\Phi}_{p-1}^*$ have the same asymptotic distribution as in the "stationary case", where Q_2 is known and one regresses W_t on the stationary variables $Z_{2t-1} = Q_2' Y_{t-1}$ and $W_{t-1}, \ldots, W_{t-p+1}$.

The above asymptotic distribution results are based, in part, on the behavior of the purely nonstationary d-dimensional process Z_{1t} for which $Z_{1t} = Z_{1t-1} + u_{1t}$ as in (6.12), where $u_t = (u_{1t}', u_{2t}')' = \sum_{j=0}^{\infty} \Psi_j a_{t-j}$. To briefly outline the developments of the asymptotic theory, we first also define the process V_{2t} such that $V_{2t} = V_{2t-1} + u_{2t}$, with $V_{20} = 0$, and let $V_t = (Z_{1t}', V_{2t}')'$, so that V_t satisfies

$$V_t = V_{t-1} + u_t = \sum_{j=0}^{t-1} u_{t-j} = \sum_{j=0}^{t-1} \Psi(B)\, a_{t-j}.$$

Then it can be established that, for $0 \le u \le 1$,

$$T^{-1/2} V_{[Tu]} = T^{-1/2} \sum_{t=1}^{[Tu]} u_t = T^{-1/2} \sum_{t=1}^{[Tu]} \Psi\, a_t + o_p(1) = T^{-1/2} M_{[Tu]} + o_p(1),$$

where $M_t = M_{t-1} + \Psi\, a_t$ is a k-dimensional vector random walk process, $\Psi = \Psi(1) = \sum_{j=0}^{\infty} \Psi_j$, and $[x]$ denotes the largest integer less than or equal to x. By an extension of the functional central limit theorem (e.g., Billingsley, 1968, p. 68) to the vector case, we, thus, have that $T^{-1/2} V_{[Tu]} = T^{-1/2} \sum_{t=1}^{[Tu]} \Psi\, a_t + o_p(1) \xrightarrow{D} \Psi\, \Sigma_a^{1/2} B_k(u)$, where $B_k(u)$ is a k-dimensional standard Brownian motion process. Then, by the continuous mapping theorem (Billingsley, 1968, Sec. 5),

$$T^{-2} \sum_{t=1}^{T} V_{t-1} V_{t-1}' = \int_0^1 (T^{-1/2} V_{[Tu]})(T^{-1/2} V_{[Tu]})'\, du + o_p(1)$$

$$\xrightarrow{D} \Psi\, \Sigma_a^{1/2} \left\{ \int_0^1 B_k(u)\, B_k(u)'\, du \right\} \Sigma_a^{1/2}\, \Psi'.$$

Similarly, it can be established that

$$T^{-1} \sum_{t=1}^{T} a_t V_{t-1}' = T^{-1} \sum_{t=1}^{T} a_t M_{t-1}' + o_p(1) \xrightarrow{D} \Sigma_a^{1/2} \left\{ \int_0^1 B_k(u)\, dB_k(u)' \right\}' \Sigma_a^{1/2}\, \Psi',$$

using Lemma 3.1 of Phillips and Durlauf (1986). From these results, with $V_t = (Z_{1t}', V_{2t}')'$ so that $Z_{1t} = [\, I_d, 0\,] V_t$, the form of the limiting distribution for

$$P\, (T^{-1} \textstyle\sum_{t=1}^{T} a_t Z_{1t-1}')\, (T^{-2} \sum_{t=1}^{T} Z_{1t-1} Z_{1t-1}')^{-1} \xrightarrow{D} P\, M,$$

where the representation for M is given by (6.14), can be readily established using the additional fact that $[\, I_d, 0\,]\, \Psi = \Psi_{11}\, [\, I_d, 0\,]$.

6.3.3 Reduced-Rank Estimation of the Error-Correction Form of the Model

When maximum likelihood estimation of the parameter matrix C in (6.11) is considered subject to the reduced-rank restriction that $\text{rank}(C) = r$, it is convenient to express C as $C = A\,B$ where A and B are $k \times r$ and $r \times k$ full rank matrices, respectively, with B normalized so that $B = [\, I_r,\ B_0 \,]$ where B_0 is an $r \times (k-r)$ matrix of unknown parameters. It is emphasized that the estimation of the model with the reduced-rank constraint imposed on C is equivalent to the estimation of the AR model with d unit roots explicitly imposed in the model. Hence, this is an alternative to (arbitrarily) differencing each component of the series $\boldsymbol{Y}_t$ prior to fitting a model in situations where the individual components tend to exhibit nonstationary behavior. This explicit form of modeling the nonstationarity may lead to a better understanding of the nature of nonstationarity among the different component series and may also improve longer-term forecasting over unconstrained model fits that do not explicitly incorporate unit roots in the model. Hence, there may be many desirable reasons to formulate and estimate the AR model with an appropriate number of unit roots explicitly incorporated in the model, and it is found that a particularly convenient way to do this is through the use of model (6.11) with the constraint $\text{rank}(C) = r$ imposed.

Maximum likelihood estimation of $C = A\,B = P_2\,(\Lambda_r - I_r)\,Q_2'$ and the Φ_j^* under the constraint that $\text{rank}(C) = r$ is presented by Ahn and Reinsel (1990), and the limiting distribution theory for these estimators is derived. Specifically, the model

$$\boldsymbol{W}_t = A\,B\,\boldsymbol{Y}_{t-1} + \sum_{j=1}^{p-1} \Phi_j^*\,\boldsymbol{W}_{t-j} + \varepsilon_t$$

is considered, where B is normalized as $B = [\, I_r,\ B_0 \,]$, and we define the vector of unknown parameters as $\boldsymbol{\beta} = (\boldsymbol{\beta}_0',\ \boldsymbol{\alpha}')'$, where $\boldsymbol{\beta}_0 = \text{vec}(B_0')$, and $\boldsymbol{\alpha} = \text{vec}\{\, (A,\ \Phi_1^*, \ldots, \Phi_{p-1}^*)' \,\}$. With $b = r(k-r) + rk + k^2(p-1)$ representing the number of unknown parameters in $\boldsymbol{\beta}$, define the $b \times k$ matrices $\boldsymbol{U}_t^* = [\, (A' \otimes H'\boldsymbol{Y}_{t-1})',\ I_k \otimes \tilde{\boldsymbol{U}}_{t-1}' \,]'$, where $\tilde{\boldsymbol{U}}_{t-1} = [\, (B\boldsymbol{Y}_{t-1})',\ \boldsymbol{W}_{t-1}', \ldots, \boldsymbol{W}_{t-p+1}' \,]'$ is stationary and $H' = [\, 0,\ I_{k-r} \,]$ is $(k-r) \times k$ such that $\boldsymbol{Y}_{2t} = H'\,\boldsymbol{Y}_t$ is taken to be purely nonstationary by assumption. Then based on T observations $\boldsymbol{Y}_1, \ldots, \boldsymbol{Y}_T$, the Gaussian estimator of $\boldsymbol{\beta}$ is obtained by the iterative approximate Newton-Raphson relations

$$\hat{\boldsymbol{\beta}}^{(i+1)} = \hat{\boldsymbol{\beta}}^{(i)} + \{\, \textstyle\sum_{t=1}^{T} \boldsymbol{U}_t^*\, \Sigma^{-1}\, \boldsymbol{U}_t^{*\prime} \,\}^{-1}_{\hat{\boldsymbol{\beta}}^{(i)}}\ \{\, \textstyle\sum_{t=1}^{T} \boldsymbol{U}_t^*\, \Sigma^{-1}\, \varepsilon_t \,\}_{\hat{\boldsymbol{\beta}}^{(i)}}, \qquad (6.15)$$

where $\hat{\boldsymbol{\beta}}^{(i)}$ denotes the estimate at the ith iteration.

Concerning the asymptotic distribution theory of the Gaussian estimators under the model where the unit roots are imposed, it is assumed that the iterations in (6.15) are started from an initial consistent estimator $\hat{\boldsymbol{\beta}}^{(0)}$ such that $D^*(\hat{\boldsymbol{\beta}}^{(0)} - \boldsymbol{\beta}) = \{\, T(\hat{\boldsymbol{\beta}}_0^{(0)} - \boldsymbol{\beta}_0)',\ T^{1/2}(\hat{\boldsymbol{\alpha}}^{(0)} - \boldsymbol{\alpha})' \,\}'$ is $O_p(1)$, where $D^* = \text{Diag}(\, T\,I_{rd},\ T^{1/2}\,I_{(b-rd)} \,)$. Then, using techniques and results similar to those for the unrestricted least squares estimator, it has been established that the Gaussian estimator $\hat{\beta} = (\hat{\beta}_0', \hat{\alpha}')'$ has the asymptotic representation

$$D^* (\hat{\beta} - \beta) = (D^{*-1} \sum_{t=1}^{T} U_t^* \Sigma^{-1} U_t^{*'} D^{*-1})^{-1} (D^{*-1} \sum_{t=1}^{T} U_t^* \Sigma^{-1} \varepsilon_t) + o_p(1)$$

and also that

$$D^{*-1} \sum_{t=1}^{T} U_t^* \Sigma^{-1} U_t^{*'} D^{*-1}$$

$$= \mathrm{Diag}[\, T^{-2} \sum_{t=1}^{T} (A' \otimes Y_{2t-1}) \Sigma^{-1} (A \otimes Y_{2t-1}'),$$

$$T^{-1} \sum_{t=1}^{T} (I_k \otimes \tilde{U}_{t-1}) \Sigma^{-1} (I_k \otimes \tilde{U}_{t-1}')\,] + o_p(1)$$

$$= \mathrm{Diag}[\, (A' \Sigma^{-1} A) \otimes (T^{-2} \sum_{t=1}^{T} Y_{2t-1} Y_{2t-1}'),$$

$$\Sigma^{-1} \otimes (T^{-1} \sum_{t=1}^{T} \tilde{U}_{t-1} \tilde{U}_{t-1}')\,] + o_p(1).$$

Consequently, it follows that the estimator $\hat{\beta}_0$ has the asymptotic representation

$$T(\hat{\beta}_0 - \beta_0) = \{ (A' \Sigma^{-1} A)^{-1} \otimes P_{21}'^{-1} (T^{-2} \sum_{t=1}^{T} Z_{1t-1} Z_{1t-1}')^{-1} \}$$

$$\times \{ T^{-1} \sum_{t=1}^{T} (A' \Sigma^{-1} \otimes Z_{1t-1}) \varepsilon_t \} + o_p(1),$$

where P_{21} is the $d \times d$ lower submatrix of P_1. To obtain the above representation, use is also made of the fact that the behavior of $Y_{2t} = H' Y_t = H' P Z_t = P_{21} Z_{1t} + P_{22} Z_{2t}$ is "dominated" by the behavior of the nonstationary component $P_{21} Z_{1t}$. From the above representation for $\hat{\beta}_0 = \mathrm{vec}(\hat{B}_0')$, it therefore follows directly that the estimator $\hat{B}_0$ has the asymptotic representation

$$T(\hat{B}_0 - B_0) = (A' \Sigma^{-1} A)^{-1} A' \Sigma^{-1} P (T^{-1} \sum_{t=1}^{T} a_t Z_{1t-1}')$$

$$\times (T^{-2} \sum_{t=1}^{T} Z_{1t-1} Z_{1t-1}')^{-1} P_{21}^{-1} + o_p(1).$$

Hence, it is established from previous results [see (6.13) and (6.14)] that

$$T(\hat{B}_0 - B_0) \xrightarrow{D} (A' \Sigma^{-1} A)^{-1} A' \Sigma^{-1} P M P_{21}^{-1}, \tag{6.16}$$

where the distribution of M is specified in (6.14). For the remaining parameters, α, in the model, it follows that

$$T^{1/2} (\hat{\alpha} - \alpha) = \left[I_k \otimes (T^{-1} \sum_{t=1}^{T} \tilde{U}_{t-1} \tilde{U}_{t-1}')^{-1} \right] T^{-1/2} \sum_{t=1}^{T} (I_k \otimes \tilde{U}_{t-1}) \varepsilon_t + o_p(1).$$

Hence, it is also shown that $T^{1/2}(\hat{\alpha} - \alpha) \xrightarrow{D} N(0, \Sigma \otimes \Gamma_{\tilde{u}}^{-1})$ where $\Gamma_{\tilde{u}} = \mathrm{Cov}(\tilde{U}_t)$, similar to results in stationary situations.

It is also established that a certain initial two-step estimator of B_0 and α, where estimates in the first step are obtained from the unrestricted least squares estimates of the parameters in model (6.11), is asymptotically equivalent to the Gaussian estimator as obtained through (6.15). Specifically, if the full rank least squares estimator $\hat{C}$ is partitioned as $\hat{C} = [\,\hat{C}_1, \hat{C}_2\,]$, where $\hat{C}_1$ is $k \times r$, then $\tilde{B}_0 = (\,\hat{C}_1' \hat{\Sigma}^{-1} \hat{C}_1\,)^{-1} \hat{C}_1' \hat{\Sigma}^{-1} \hat{C}_2$ is the estimator of B_0, where $\hat{\Sigma} = T^{-1} \sum_{t=1}^{T} \hat{\varepsilon}_t \hat{\varepsilon}_t'$ is the usual residual covariance matrix estimate of Σ. Then the estimates $\tilde{A}$, $\tilde{\Phi}_1^*, \ldots, \tilde{\Phi}_{p-1}^*$ are obtained in the second step by least squares regression of W_t on $\tilde{B}\, Y_{t-1}$, $W_{t-1}, \ldots, W_{t-p+1}$, where $\tilde{B} = [\, I_r, \tilde{B}_0\,]$. To motivate this initial estimator $\tilde{B}_0$, consider the Gaussian estimation of B_0 in the model when all other parameters A, $\Phi_1^*, \ldots, \Phi_{p-1}^*$, and Σ are known. Since $A\,B = [\,A, A\,B_0\,]$, we can express the model as

$$W_t - A Y_{1t-1} - \sum_{j=1}^{p-1} \Phi_j^*\, W_{t-j} = A\, B_0\, Y_{2t-1} + \varepsilon_t = (\,A \otimes Y_{2t-1}'\,)\, \boldsymbol{\beta}_0 + \varepsilon_t\,,$$

where $\boldsymbol{\beta}_0 = \mathrm{vec}(\,B_0'\,)$. Then it follows, from standard results on GLS estimation, that the Gaussian estimator of the unknown parameter $\boldsymbol{\beta}_0$ is

$$\begin{aligned}\tilde{\boldsymbol{\beta}}_0 &= \{\, \textstyle\sum_{t=1}^{T} (A' \otimes Y_{2t-1})\, \Sigma^{-1} (A \otimes Y_{2t-1}')\,\}^{-1} \\ &\qquad \times \textstyle\sum_{t=1}^{T} (A' \otimes Y_{2t-1})\, \Sigma^{-1} (W_t - A Y_{1t-1} - \sum_{j=1}^{p-1} \Phi_j^*\, W_{t-j}) \\ &= \{\, (A' \Sigma^{-1} A)^{-1} \otimes (\,\textstyle\sum_{t=1}^{T} Y_{2t-1} Y_{2t-1}'\,)^{-1}\,\} \\ &\qquad \times \{\, \textstyle\sum_{t=1}^{T} (A' \Sigma^{-1} \otimes Y_{2t-1})\, (W_t - A Y_{1t-1} - \sum_{j=1}^{p-1} \Phi_j^*\, W_{t-j})\,\}\,.\end{aligned}$$

If the unrestricted LS estimators $\hat{A} \equiv \hat{C}_1$, $\hat{\Phi}_1^*, \ldots, \hat{\Phi}_{p-1}^*$, and $\hat{\Sigma}$ are used as (initial) estimators for the actually unknown parameters on the right-hand side of the above expression, then we obtain the (initial) estimator for B_0 as

$$\begin{aligned}\tilde{B}_0 &= (\,\hat{A}' \hat{\Sigma}^{-1} \hat{A}\,)^{-1} \hat{A}' \hat{\Sigma}^{-1} \\ &\qquad \times \{\, \textstyle\sum_{t=1}^{T} (W_t - \hat{A} Y_{1t-1} - \sum_{j=1}^{p-1} \hat{\Phi}_j^*\, W_{t-j})\, Y_{2t-1}'\,\}\, \{\, \sum_{t=1}^{T} Y_{2t-1} Y_{2t-1}'\,\}^{-1} \\ &\equiv (\,\hat{C}_1' \hat{\Sigma}^{-1} \hat{C}_1\,)^{-1} \hat{C}_1' \hat{\Sigma}^{-1} \hat{C}_2\,,\end{aligned}$$

which is the initial two-step estimator of B_0 given previously. The asymptotic equivalence of the Gaussian estimator and this initial two-step estimator can also be explained, from this point of view, by the observation that both estimators, in fact, are also asymptotically equivalent to the Gaussian estimator of B_0 that is obtained when all other parameters in the model are known (as given above), with both $T(\,\hat{B}_0 - B_0\,)$ and $T(\,\tilde{B}_0 - B_0\,)$ having the same asymptotic representation

$$(A' \Sigma^{-1} A)^{-1} A' \Sigma^{-1}\, (\,T^{-1} \textstyle\sum_{t=1}^{T} \varepsilon_t Y_{2t-1}'\,)\, (\,T^{-2} \sum_{t=1}^{T} Y_{2t-1} Y_{2t-1}'\,)^{-1} + o_p(1)\,.$$

In addition, it is also useful to express the asymptotic distribution of $\hat{B}_0$ in an alternate form. From (6.14), define

$$B_d^*(u) = P_{21}\Psi_{11}\,\Sigma_{a_1}^{1/2}\,B_d(u) = P_{21}\Psi_{11}\,[\,I_d,\,0\,]\,\Sigma_a^{1/2}\,B_k(u)$$

and

$$B_r^*(u) = (A'\Sigma^{-1}A)^{-1}A'\Sigma^{-1}\,P\,\Sigma_a^{1/2}\,B_k(u)\,,$$

so that $B_d^*(u)$ and $B_r^*(u)$ are *independent* (since their cross-covariance matrix equals 0) Brownian motion processes of dimensions d and r, respectively, and with covariance matrices $P_{21}\Psi_{11}\,\Sigma_{a_1}\Psi_{11}'P_{21}'$ and $(A'\Sigma^{-1}A)^{-1}$, respectively. From (6.16) and (6.14), it is then established [also see Ahn and Reinsel (1990)] that the asymptotic distribution of the estimator $\hat{B}_0$ has the representation

$$T\,(\,\hat{B}_0 - B_0\,) \xrightarrow{D} \{\,\int_0^1 B_d^*(u)\,dB_r^*(u)'\,\}'\,\{\,\int_0^1 B_d^*(u)\,B_d^*(u)'\,du\,\}^{-1}\,.$$

Then, by use of a conditioning argument and the independence of the processes $\{B_d^*(u)\}$ and $\{B_r^*(u)\}$, the conditional distribution of $\mathrm{vec}\{\,\int_0^1 B_d^*(u)\,dB_r^*(u)'\,\}$, given $\{B_d^*(u)\}$, is $N(\,0,\,(A'\Sigma^{-1}A)^{-1}\otimes\int_0^1 B_d^*(u)\,B_d^*(u)'\,du\,)$, and by application of the continuous mapping theorem (Billingsley, 1968, Sec. 5), it is established that

$$\begin{aligned}
&\mathrm{vec}\{\,(\,\textstyle\sum_{t=1}^{T} H'\boldsymbol{Y}_{t-1}\boldsymbol{Y}_{t-1}'H\,)^{1/2}\,(\,\hat{B}_0 - B_0\,)'\,\} \\
&\qquad \xrightarrow{D} \mathrm{vec}\{\,(\,\int_0^1 B_d^*(u)\,B_d^*(u)'\,du\,)^{-1/2}\,(\,\int_0^1 B_d^*(u)\,dB_r^*(u)'\,)\,\} \\
&\qquad \equiv N(\,0,\,(A'\Sigma^{-1}A)^{-1}\otimes I_d\,).
\end{aligned}$$

Thus, although $\hat{B}_0 - B_0$ is not asymptotically normal in the usual sense, the last result indicates that a properly standardized form of $\hat{B}_0 - B_0$ does follow the usual normal theory asymptotics and this can be used for making inferences involving the estimator $\hat{B}_0$. [See Phillips (1991) for discussion of related concepts and results; there it is noted that the above asymptotic results for $\hat{B}_0$ fall within the "locally asymptotically mixed normal" theory, e.g., see LeCam (1986).] Thus, for example, under the assumption of cointegration with the specified rank r, LR or Wald tests of various hypotheses concerning the elements of B_0 can be performed in the usual way and the asymptotic null distribution of these test statistics is the usual chi-squared distribution with the appropriate degrees of freedom.

6.3.4 Likelihood Ratio Test for the Number of Unit Roots

Within the context of model (6.11), it is necessary to specify or determine the rank r of cointegration or the number d of unit roots in the model. Thus, it is also of interest to test the hypothesis H_0 : rank(C) $\le r$, which is equivalent to the hypothesis that the number of unit roots in the AR model is greater than or equal to d $(d = k - r)$. The *likelihood ratio test* for this hypothesis is considered by Johansen (1988) and by Reinsel and Ahn (1992). The likelihood ratio test statistic is given by $-T\log(\lambda) = -T\log(\,|S|/|S_0|\,)$, where $S = \sum_{t=1}^{T}\hat{\varepsilon}_t\,\hat{\varepsilon}_t'$ denotes the residual sum of squares matrix in the full or unconstrained model, while S_0 is the residual sum of squares matrix obtained under the reduced-rank restriction on C that rank$(\,C\,) = r$. It follows from work of Anderson (1951) in the multivariate linear model that the LR statistic can be expressed equivalently as

$$-T\log(\lambda) = -T\sum_{i=r+1}^{k}\log(\,1 - \hat{\rho}_i^2(p)\,), \tag{6.17}$$

where the $\hat{\rho}_i(p)$ are the $d = k - r$ smallest *sample partial canonical correlations* between $\boldsymbol{W}_t = \boldsymbol{Y}_t - \boldsymbol{Y}_{t-1}$ and $\boldsymbol{Y}_{t-1}$, given $\boldsymbol{W}_{t-1},\ldots,\boldsymbol{W}_{t-p+1}$. The limiting distribution for the likelihood ratio statistic has been derived and is given by

$$-T\log(\lambda) \xrightarrow{D} \mathrm{tr}\{\,(\int_0^1 B_d(u)\,dB_d(u)'\,)' \times (\int_0^1 B_d(u)\,B_d(u)'\,)^{-1}\,(\int_0^1 B_d(u)\,dB_d(u)'\,)\,\}, \tag{6.18}$$

where $B_d(u)$ is a d-dimensional standard Brownian motion process. Note that the asymptotic distribution of the likelihood ratio test statistic under H_0 depends only on d and not on any parameters or the order p of the AR model. Critical values of the asymptotic distribution of (6.18) have been obtained by simulation by Johansen (1988) and Reinsel and Ahn (1992) and can be used in the test of H_0. Some other approaches to testing for unit roots in nonstationary multivariate systems have been examined by Engle and Granger (1987), Stock and Watson (1988), Fountis and Dickey (1989), and Phillips and Ouliaris (1986).

The asymptotic distribution result (6.18) for the LR test statistic has been established by Reinsel and Ahn (1992) using the distribution theory for the full rank least squares and the Gaussian reduced-rank estimators of C. Specifically, it is shown that

$$-T\log(\,|S|/|S_0|\,)$$

$$= \mathrm{tr}\{\,\Sigma^{-1}(\,\hat{F} - \tilde{F}\,)\,P^*\sum_{t=1}^{T}\boldsymbol{X}_{t-1}^*\,\boldsymbol{X}_{t-1}^{*\prime}\,P^{*\prime}\,(\,\hat{F} - \tilde{F}\,)'\,\} + o_p(1)$$

$$= \mathrm{tr}\{\,\Sigma^{-1}\,T(\,\hat{C} - \hat{A}\hat{B}\,)\,P_1\,(\,T^{-2}\sum_{t=1}^{T}\boldsymbol{Z}_{1t-1}\,\boldsymbol{Z}_{1t-1}'\,)\,P_1'\,T(\,\hat{C} - \hat{A}\hat{B}\,)'\,\} + o_p(1),$$

where $\tilde{F} = (\,\hat{A}\hat{B},\,\tilde{\Phi}_1^*,\ldots,\tilde{\Phi}_{p-1}^*\,)$ denotes the Gaussian reduced-rank estimator obtained through procedure (6.15) under the restriction that rank$(\,C\,) = r$. This last expression indicates that the estimators of the "stationary" parameters of model (6.11) (that is, the parameter coefficients of the stationary series $\boldsymbol{Z}_{2t-1}$

and $\boldsymbol{W}_{t-1}, \ldots, \boldsymbol{W}_{t-p+1}$) are not affected asymptotically by the imposition of the reduced-rank structure for C and, consequently, the asymptotic distribution of the LR statistic for $H_0 : \text{rank}(C) \le r$ is not dependent on the distribution of the estimators for the "stationary" parameters, only on the distribution of the estimators for the "nonstationary" parameters $C P_1$ (that is, the parameter coefficients of the purely nonstationary series $\boldsymbol{Z}_{1t-1}$). Then the asymptotic representations for $T \hat{C} P_1$ and $T \hat{A} \hat{B} P_1$, such as in (6.13), can be used to obtain that

$$- T \log(\lambda) = \text{tr}\{ \Sigma_{a_1}^{-1} [I_d, 0] T^{-1} \sum_{t=1}^{T} \boldsymbol{a}_t \boldsymbol{Z}_{1t-1}' \times (T^{-2} \sum_{t=1}^{T} \boldsymbol{Z}_{1t-1} \boldsymbol{Z}_{1t-1}')^{-1} T^{-1} \sum_{t=1}^{T} \boldsymbol{Z}_{1t-1} \boldsymbol{a}_t' [I_d, 0]' \} + o_p(1),$$

and the asymptotic distribution (6.18) then follows directly from previous results.

Likelihood ratio testing of the more refined hypotheses $H_0 : \text{rank}(C) = r$ against the alternative that $\text{rank}(C) = r + 1$ (that is, $d = k - r$ unit roots versus $d - 1$ unit roots) can also be considered. It follows from the work of Anderson (1951) on reduced-rank regression models, and has been derived explicitly by Johansen (1991), that the likelihood ratio statistic for this test is given by $- T \log(1 - \hat{\rho}_{r+1}^2(p))$ where $\hat{\rho}_{r+1}(p)$ is the $(r + 1)$th sample partial canonical correlation between $\boldsymbol{W}_t = \boldsymbol{Y}_t - \boldsymbol{Y}_{t-1}$ and $\boldsymbol{Y}_{t-1}$, given $\boldsymbol{W}_{t-1}, \ldots, \boldsymbol{W}_{t-p+1}$. Johansen (1991) has derived the asymptotic distribution of the LR statistic for this situation under H_0 and showed that the limiting distribution of the LR statistic is that of the largest eigenvalue of the $d \times d$ matrix in (6.18). Generalization of these LR testing results to allow for a constant term in the vector AR model estimation, as will be discussed briefly in a subsequent subsection, was also considered by Johansen (1991). Percentage points of the limiting distribution for the LR test statistic for these more general situations have been tabulated through simulation by Johansen and Juselius (1990).

Remark. Many of the preceding estimation and LR testing results for the vector AR model (6.11) with unit roots have been extended to more general situations, such as to the vector ARMA model with unit roots by Yap and Reinsel (1992) and to the use of a finite order vector AR model as an approximation to an infinite AR order co-integrated process by Saikkonen (1992).

6.3.5 Reduced-Rank Estimation Through Partial Canonical Correlation Analysis

The iterative procedure (6.15) to obtain the Gaussian estimator $\hat{\boldsymbol{\beta}}$ can readily be modified to incorporate additional constraints, such as zero constraints or nested reduced-rank constraints on the stationary parameters Φ_j^* in (6.11), beyond the co-integrating constraint that $\text{rank}(C) = r$. However, when there are no constraints imposed other than $\text{rank}(C) = r$, the Gaussian reduced-rank estimator in model (6.11) can also be obtained explicitly through the *partial canonical* *correlation analysis*, based on the previous work of Anderson (1951) (e.g., see Appendix A4). This approach has been presented by Johansen (1988). To describe the results, let $\tilde{\boldsymbol{W}}_t$ and $\tilde{\boldsymbol{Y}}_{t-1}$ denote the residual vectors from the least squares regressions of $\boldsymbol{W}_t$ and $\boldsymbol{Y}_{t-1}$, respectively, on the lagged values $\boldsymbol{W}_{t-1}, \ldots, \boldsymbol{W}_{t-p+1}$, and let

$$S_{\tilde{w}\tilde{w}} = \frac{1}{T} \sum_{t=1}^{T} \tilde{\boldsymbol{W}}_t \tilde{\boldsymbol{W}}_t', \qquad S_{\tilde{w}\tilde{y}} = \frac{1}{T} \sum_{t=1}^{T} \tilde{\boldsymbol{W}}_t \tilde{\boldsymbol{Y}}_{t-1}', \qquad S_{\tilde{y}\tilde{y}} = \frac{1}{T} \sum_{t=1}^{T} \tilde{\boldsymbol{Y}}_{t-1} \tilde{\boldsymbol{Y}}_{t-1}'.$$

Then the sample partial canonical correlations $\hat{\rho}_i(p)$ between $\boldsymbol{W}_t$ and $\boldsymbol{Y}_{t-1}$, given $\boldsymbol{W}_{t-1}, \ldots, \boldsymbol{W}_{t-p+1}$, and the corresponding vectors $\hat{V}_i$ are the solutions to

$$(\hat{\rho}_i^2 \, S_{\tilde{w}\tilde{w}} - S_{\tilde{w}\tilde{y}} \, S_{\tilde{y}\tilde{y}}^{-1} \, S_{\tilde{y}\tilde{w}}) \, \hat{V}_i = 0 , \qquad i = 1, \ldots, k .$$

The *reduced-rank Gaussian estimator* of C can be expressed explicitly as $\tilde{C} = \hat{\Sigma} \hat{V} \hat{V}' \hat{C}$, where $\hat{C} = S_{\tilde{w}\tilde{y}} S_{\tilde{y}\tilde{y}}^{-1}$ is the full rank least squares estimator, $\hat{\Sigma} = T^{-1} \sum_{t=1}^{T} \hat{\varepsilon}_t \hat{\varepsilon}_t' = S_{\tilde{w}\tilde{w}} - S_{\tilde{w}\tilde{y}} S_{\tilde{y}\tilde{y}}^{-1} S_{\tilde{y}\tilde{w}}$ is the corresponding residual covariance matrix estimate of Σ from the full rank estimation, and $\hat{V} = (\hat{V}_1, \ldots, \hat{V}_r)$ are the vectors corresponding to the r largest partial canonical correlations $\hat{\rho}_i(p)$, $i = 1, \ldots, r$. The vectors $\hat{V}_i$ in $\hat{V}$ are normalized by $\hat{V}_i' \hat{\Sigma} \hat{V}_i = 1$ (so that $\hat{V}' \hat{\Sigma} \hat{V} = I_r$). Note that this form of the estimator provides the reduced-rank factorization as $\tilde{C} = (\hat{\Sigma} \hat{V}) (\hat{V}' \hat{C}) \equiv \hat{A} \hat{B}$, with $\hat{A} = \hat{\Sigma} \hat{V}$ satisfying the normalization $\hat{A}' \hat{\Sigma}^{-1} \hat{A} = I_r$. The Gaussian estimator for the other parameters $\Phi_1^*, \ldots, \Phi_{p-1}^*$ can be obtained by ordinary least squares regression of $\boldsymbol{W}_t - \tilde{C} \boldsymbol{Y}_{t-1}$ on the lagged variables $\boldsymbol{W}_{t-1}, \ldots, \boldsymbol{W}_{t-p+1}$.

An alternate equivalent representation for the reduced-rank estimator of C is $\tilde{C} = S_{\tilde{w}\tilde{y}} \hat{\boldsymbol{\delta}} \hat{\boldsymbol{\delta}}'$, where $\hat{\boldsymbol{\delta}} = (\hat{\delta}_1, \ldots, \hat{\delta}_r)$ and the $\hat{\delta}_i$ are the vector solutions to $(\hat{\rho}_i^2 \, S_{\tilde{y}\tilde{y}} - S_{\tilde{y}\tilde{w}} \, S_{\tilde{w}\tilde{w}}^{-1} \, S_{\tilde{w}\tilde{y}}) \, \hat{\delta}_i = 0$, $i = 1, \ldots, r$, normalized by $\hat{\boldsymbol{\delta}}' S_{\tilde{y}\tilde{y}} \hat{\boldsymbol{\delta}} = I_r$. This can be verified by use of the relations (e.g., see Anderson, 1984, pp. 489–490) $S_{\tilde{y}\tilde{y}}^{-1} S_{\tilde{y}\tilde{w}} \hat{V}_i (1 - \hat{\rho}_i^2)^{1/2} = \hat{\rho}_i \hat{\delta}_i$ and $\hat{\rho}_i \hat{V}_i (1 - \hat{\rho}_i^2)^{1/2} = S_{\tilde{w}\tilde{w}}^{-1} S_{\tilde{w}\tilde{y}} \hat{\delta}_i$. In this latter form of the reduced-rank estimator, we can write $\tilde{C} = (S_{\tilde{w}\tilde{y}} \hat{\boldsymbol{\delta}}) \hat{\boldsymbol{\delta}}' \equiv \hat{A} \hat{B}$ and take $\hat{B} = \hat{\boldsymbol{\delta}}'$, which shows that the estimated co-integrating vectors (rows of $\hat{B}$) are obtained from the vectors that determine the first r canonical variates, $\hat{\delta}_i' \boldsymbol{Y}_{t-1}$, of $\boldsymbol{Y}_{t-1}$ in the partial canonical correlation analysis. In addition, it is seen from the first form of the reduced-rank estimator $\tilde{C} = \hat{\Sigma} \hat{V} \hat{V}' \hat{C}$ given above that the Gaussian estimates of the "common trends" components $\boldsymbol{Z}_{1t} = Q_1' \boldsymbol{Y}_t$ can be obtained with $\hat{Q}_1 = (\hat{V}_{r+1}, \ldots, \hat{V}_k)$, so that the rows of $\hat{Q}_1'$ are the vectors $\hat{V}_i'$, $i = r + 1, \ldots, k$, corresponding to the d smallest partial canonical correlations, with the property that $\hat{Q}_1' \tilde{C} = 0$ since $\hat{Q}_1' \hat{A} = 0$.

6.3.6 Extension to Account for a Constant Term in the Estimation

It is well known that inclusion of a constant term in the estimation of the nonstationary AR model affects the limiting distribution of the estimator and test statistic. The previous estimation and LR testing results are readily extended to estimation of models with a constant term included. Corresponding to (6.11), the model considered is $W_t = \theta + C\, Y_{t-1} + \sum_{j=1}^{p-1} \Phi_j^*\, W_{t-j} + \varepsilon_t$, where θ is a $k \times 1$ vector of unknown constants such that $Q_1'\, \theta = 0$. This last assumption implies that the nonstationary components $Z_{1t} = Q_1'\, Y_t$ have zero drift. Similar to previous sections, we can express the (unconstrained) least squares estimator of $F = (\, C, \Phi_1^*, \ldots, \Phi_{p-1}^* \,)$ as

$$\hat{F}_\theta = \{ \textstyle\sum_{t=1}^T (W_t - \overline{W})(X_{t-1}^* - \overline{X}^*)' \} \{ \textstyle\sum_{t=1}^T (X_{t-1}^* - \overline{X}^*)(X_{t-1}^* - \overline{X}^*)' \}^{-1} Q^*,$$

where $\overline{W} = T^{-1} \sum_{t=1}^T W_t$ and $\overline{X}^* = T^{-1} \sum_{t=1}^T X_{t-1}^*$. The relevant limiting distribution results for estimation with a constant term included are then readily obtained, such as

$$T^{-1} \sum_{t=1}^T a_t\, (\, Z_{1t-1} - \overline{Z}_1)' \xrightarrow{D} \Sigma_a^{1/2} \{ \int_0^1 B_d(u)\, dB_k(u)' \,-\, \xi\, B_k(1)' \}' \, \Sigma_{a_1}^{1/2}\, \Psi_{11}'$$

and

$$T^{-2} \sum_{t=1}^T (\, Z_{1t-1} - \overline{Z}_1)(\, Z_{1t-1} - \overline{Z}_1)'$$

$$\xrightarrow{D} \Psi_{11}\, \Sigma_{a_1}^{1/2} \{ \int_0^1 B_d(u)\, B_d(u)'\, du \,-\, \xi\, \xi' \} \, \Sigma_{a_1}^{1/2}\, \Psi_{11}',$$

where $\overline{Z}_1 = T^{-1} \sum_{t=1}^T Z_{1t-1}$ and $\xi = \int_0^1 B_d(u)\, du$. Hence, limiting distribution results similar to those of (6.14), (6.16), and (6.18) can be obtained for the least squares and the Gaussian reduced-rank estimator and for the LR test statistic for $H_0 : \text{rank}(\, C\,) \le r$. The resulting limiting distributions have a similar form with $\{ \int_0^1 B_d(u)\, B_d(u)'\, du \,-\, \xi\, \xi' \}^{-1}$ and $\{ \int_0^1 B_d(u)\, dB_k(u)' \,-\, \xi\, B_k(1)' \}'$ in place of the corresponding terms that define M in (6.14) and in place of terms in (6.18). In particular, approximate percentage points of the limiting distribution similar to (6.18) for the case where a constant term is included in the estimation have been obtained by simulation and are given by Reinsel and Ahn (1992). The approximate critical values of the limiting distribution in (6.18) for the LR test of $H_0 : \text{rank}(\, C\,) \le r$ against a general alternative, for the cases without and with a constant term included in the estimation procedures, are reproduced from Reinsel and Ahn (1992) in Table 6.3 for values of $d = k - r = 1, 2, 3, 4, 5$.

Finally we note that for the nonstationary AR model $Y_t = \sum_{i=1}^p \Phi_i\, Y_{t-i} + \varepsilon_t$, the matrices Φ_i may also have simplifying structure such as nested reduced-rank structure discussed previously in Section 6.1. When they do have nested reduced-rank structure, it is noted that a similar structure is maintained for the coefficient matrices Φ_j^* in the error-correction form (6.11) of the model, because $\Phi_j^* = -\sum_{k=j+1}^p \Phi_k$. Thus, the estimation procedures under the "unit-root" constraint $\text{rank}(\, C\,) = r$ can be combined with those under the nested reduced-rank structure for the Φ_j^* as discussed previously. Some details concerning the estimation procedures and asymptotic distribution theory for the associated estimators have been presented by Reinsel and Ahn (1992). We comment here that asymptotic distribution theory for nonstationary vector AR models under more general model assumptions concerning the nature of roots of $\det\{\Phi(B)\} = 0$ on the unit circle has also been developed by Tsay and Tiao (1990) and Sims, Stock, and Watson (1990). However, these authors did not explicitly consider the estimation of models such as (6.11) with d unit roots [that is, $\text{rank}(C) = r$] imposed or the associated asymptotic distribution theory for estimators such as those presented in (6.15) and (6.16).

Table 6.3. Approximate Percentiles of the Limiting Distribution (6.18) for the Likelihood Ratio Test Statistic Under $H_0 : \text{rank}(C) = r = k - d$. The Table Values Were Constructed from 30,000 Simulations of the Quantity $\text{tr}\{(\sum_{t=1}^{T} \boldsymbol{a}_{1t}\boldsymbol{Z}_{1t-1}')(\sum_{t=1}^{T} \boldsymbol{Z}_{1t-1}\boldsymbol{Z}_{1t-1}')^{-1}(\sum_{t=1}^{T} \boldsymbol{a}_{1t}\boldsymbol{Z}_{1t-1}')'\}$ from a d-dimensional Random Walk Process $\boldsymbol{Z}_{1t} = \boldsymbol{Z}_{1t-1} + \boldsymbol{a}_{1t}$, with Series Length of $T = 2000$. The Second Portion of the Table Corresponds to the Case Where a Constant Term Is Included in the Estimation.

	Probability of a Smaller Value								
d	0.01	0.025	0.05	0.10	0.50	0.90	0.95	0.975	0.99
1	0.00	0.00	0.01	0.02	0.59	2.94	4.10	5.30	6.97
2	1.23	1.62	2.02	2.56	5.49	10.45	12.30	14.15	16.41
3	6.19	7.08	8.03	9.22	14.44	21.70	24.24	26.63	29.83
4	15.13	16.64	18.08	19.87	27.47	37.00	40.18	43.11	46.30
5	28.08	30.22	32.15	34.64	44.37	56.12	59.84	63.15	67.62
1	0.00	0.02	0.07	0.26	2.45	6.59	8.16	9.69	11.65
2	2.77	3.47	4.13	5.01	9.41	15.80	17.97	20.17	22.79
3	9.81	11.19	12.41	13.85	20.34	28.87	31.73	34.36	37.38
4	20.98	22.76	24.52	26.66	35.45	45.82	49.35	52.56	56.25
5	36.08	38.54	40.68	43.39	54.19	66.98	71.14	74.94	79.23

EXAMPLE 6.4. We briefly consider a simple numerical example involving the bivariate time series of seasonally adjusted monthly U.S. housing data consisting of housing-starts and housing-sold over the period 1965 through 1974. The deseasonalized data have been formed by subtracting monthly sample means from the data for each series. Time series plots of the resulting data are displayed in Figures 6.1(a) and (b). Using the partial canonical correlation methods discussed in Section 6.1.1, an AR(1) model was rather clearly specified for these data, since the values of the test statistics $C(j, s)$ from (6.3) were clearly not significant for lags $j > 1$. So we consider the AR(1) model for these

data, which in error-correction form is $\boldsymbol{W}_t = C\, \boldsymbol{Y}_{t-1} + \varepsilon_t$, where $\boldsymbol{W}_t = \boldsymbol{Y}_t - \boldsymbol{Y}_{t-1}$ and $C = \Phi_1 - I$.

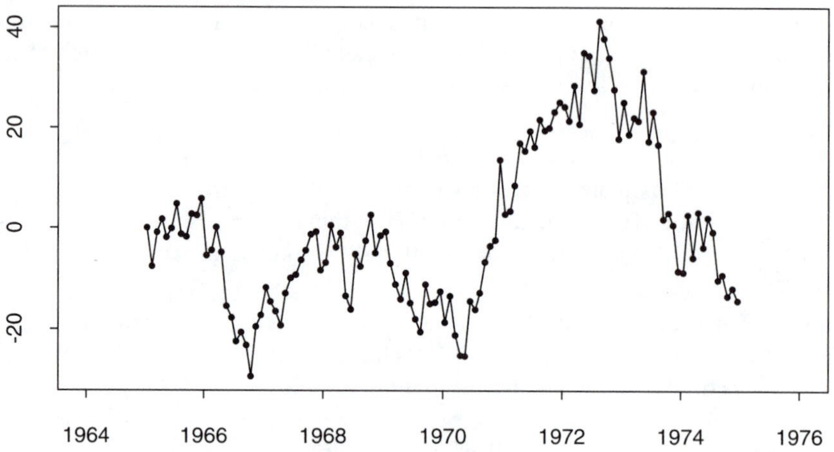

(a) Y_{1t} : Housing Starts Series

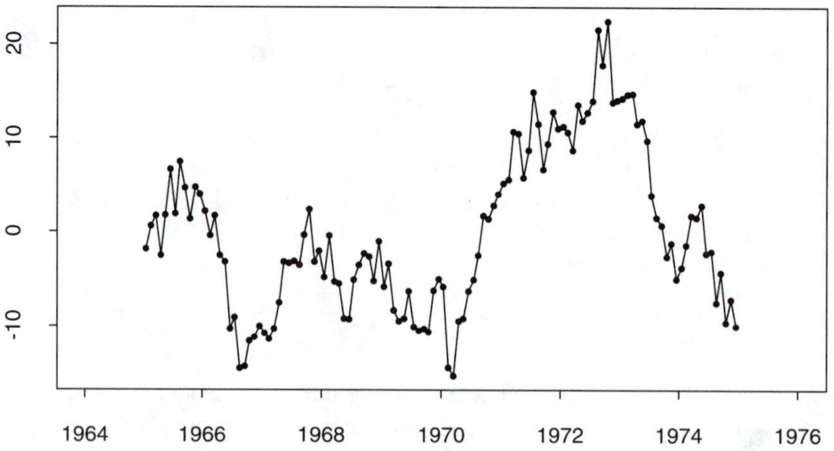

(b) Y_{2t} : Houses Sold Series

Figure 6.1. Seasonally Adjusted U.S. Monthly Housing Data (in thousands) for the Period January 1965 Through December 1974

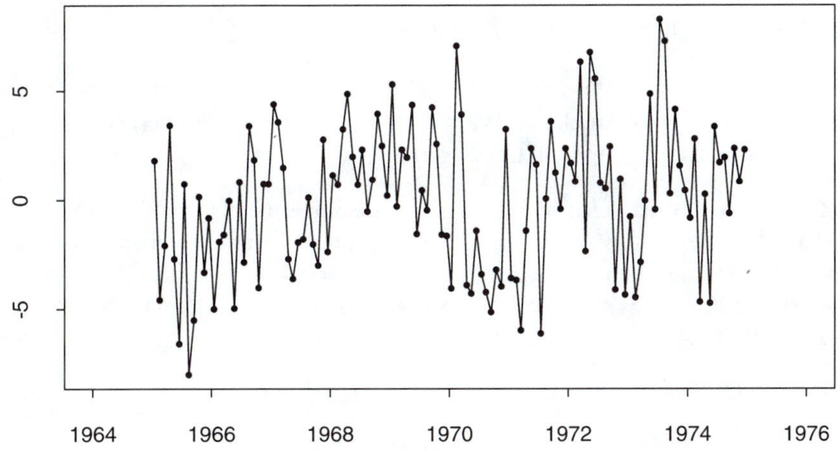

(c) Z_{2t} : Transformed Series, $Z_{2t} = 0.523\, Y_{1t} - 0.979\, Y_{2t}$

Figure 6.1. (continued)

The unrestricted least squares estimator of C and the error covariance matrix are obtained as

$$\hat{C} = \begin{bmatrix} -0.537 & 0.951 \\ 0.129 & -0.289 \end{bmatrix}, \qquad \hat{\Sigma} = \begin{bmatrix} 26.371 & 5.776 \\ 5.776 & 9.709 \end{bmatrix}.$$

Note that the eigenvalues of $\hat{C}$ are -0.785 and -0.041, and the latter value may be considered to be close to zero. So considering reduced-rank estimation of C with $\text{rank}(\ C\ ) = 1$ imposed, the final Gaussian estimator of C, which incorporates the unit root, and the residual covariance matrix are

$$\tilde{C} = \hat{A}\,\hat{B} = \begin{bmatrix} -0.523 \\ 0.141 \end{bmatrix} \begin{bmatrix} 1 & -1.872 \end{bmatrix} = \begin{bmatrix} -0.523 & 0.979 \\ 0.141 & -0.265 \end{bmatrix},$$

and

$$\tilde{\Sigma} = \begin{bmatrix} 26.589 & 5.966 \\ 5.966 & 9.875 \end{bmatrix}.$$

The likelihood ratio (LR) test statistic value for testing $H_0 : \text{rank}(\ C\ ) = 1$ is $-(\ T-1\ )\log(\ |\hat{\Sigma}|\ /\ |\tilde{\Sigma}|\ ) = 2.275$, which is not significant by comparison to the second portion of Table 6.3 with $d = 1$. Note the close agreement between the full rank estimates and the reduced-rank estimates for the nonstationary AR model with the unit root imposed. From these results, the linear combination $0.523\, Y_{1t} - 0.979\, Y_{2t} = Z_{2t}$ will be stationary and this series is exhibited in

Figure 6.1(c). Conversely, the series $Z_{1t} = 0.141\, Y_{1t} + 0.523\, Y_{2t}$ will follow a random walk model, that is, from previous discussions we have that $\boldsymbol{Z}_t = Q\, \boldsymbol{Y}_t$ follows the model $(1 - B)\, \boldsymbol{Z}_t = \text{Diag}(\, 0, -0.7883\,)\, \boldsymbol{Z}_{t-1} + \boldsymbol{a}_t$, or simply $\boldsymbol{Z}_t = \text{Diag}(\, 1, 0.2117\,)\, \boldsymbol{Z}_{t-1} + \boldsymbol{a}_t$. So Z_{1t} is a random walk and Z_{2t} is a stationary AR(1) process, and Y_{1t} and Y_{2t} are linear combinations of these two series. In the terminology of Stock and Watson (1988) and others, the series Z_{1t} might be referred to as the "common trend" series.

Comment. It may also be instructive to examine the nature of the model for the first differences $\boldsymbol{W}_t = (1 - B)\, \boldsymbol{Y}_t$ in the above example when only traditional ARMA models are considered for $\boldsymbol{W}_t$. For a k-dimensional AR(1) model $(I - \Phi B)\, \boldsymbol{Y}_t = \varepsilon_t$ with $d < k$ unit roots, from the relation $\Phi = P J Q = P_1 Q_1' + P_2 \Lambda_r Q_2'$ where $J = \text{Diag}[\, I_d, \Lambda_r\,]$, we find that

$$(I - P_2 Q_2' B)(I - \Phi B) = I - (I + P_2 \Lambda_r Q_2')\, B + P_2 \Lambda_r Q_2'\, B^2$$

$$= (I - P_2 \Lambda_r Q_2' B)(1 - B).$$

Hence, it follows that $(I - \Phi_1 B)(1 - B)\, \boldsymbol{Y}_t = (I - \Theta_1 B)\, \varepsilon_t$, where $\Phi_1 = P_2 \Lambda_r Q_2'$ and $\Theta_1 = P_2 Q_2'$, and so $\boldsymbol{W}_t = (1 - B)\, \boldsymbol{Y}_t$ satisfies an ARMA(1,1) model. However, the matrix Θ_1 has r eigenvalues equal to one (and the remaining d eigenvalues are equal to zero) and so the MA operator $(I - \Theta_1 B)$ is noninvertible. Furthermore, $\text{rank}(\, \Phi_1, \Theta_1\,) = r < k$, so that there is also a parameter nonidentifiability problem in the ARMA(1,1) model for $\boldsymbol{W}_t$. Thus, this illustrates the difficulties in modeling and parameter estimation that may result from "multivariate overdifferencing" when one considers the first differences $\boldsymbol{W}_t = (1 - B)\, \boldsymbol{Y}_t$ and attempts to model $\boldsymbol{W}_t$ using ARMA models. This example also illustrates one characteristic of error-correction models such as (6.11) that possess $d < k$ unit roots which holds generally (e.g., Engle and Granger, 1987). Namely, for such processes $\{\boldsymbol{Y}_t\}$ the infinite MA form for their first differences $\boldsymbol{W}_t$, $\boldsymbol{W}_t = (1 - B)\, \boldsymbol{Y}_t = \Psi^*(B)\, \varepsilon_t$, will be noninvertible with $\Psi^*(1)$ being of reduced rank d.

6.3.7 Forecast Properties for the Co-integrated Model

One additional important feature for the co-integrated model such as (6.11) concerns the properties of forecasts obtained from the model. Forecasts from the model (6.11) in which the d unit roots are explicitly imposed can be potentially more accurate for longer lead times than forecasts obtained from a model which uses unconstrained least squares estimates of AR parameters, and the forecasts from the two procedures may differ substantially. This is because the use of full rank unconstrained least squares estimates would typically lead to a stationary model representation in which forecasts are driven to the estimated series mean values for larger lead times while explicitly incorporating the unit roots yields a (partially) nonstationary model. In addition, the co-integrated model in the error-correction form (6.11) will produce more accurate forecasts in the short term than the less appropriate models that may result from "overdifferencing" and using a traditional stationary AR model for the vector of first differences $W_t = (1 - B)\, Y_t$. Engle and Yoo (1987) investigated the improvement in forecasting by use of the error-correction model with unit root imposed for a bivariate situation and found substantial improvements for longer-run forecasts over the use of full rank estimates for the AR model, and Reinsel and Ahn (1992) obtained similar findings in a more extensive investigation.

A notable feature of the forecasts $\hat{Y}_t(l)$ based on the error-correction model (6.11) with unit roots explicitly imposed is that the forecasts will tend to satisfy the equilibrium or co-integrating relations $Q_2' \hat{Y}_t(l) \approx 0$ for large l. As a simple illustration, consider the special case of an AR(1) model $Y_t = \Phi_1 Y_{t-1} + \varepsilon_t$, or $W_t = C\, Y_{t-1} + \varepsilon_t$, with $C = \Phi_1 - I$. From the discussion following (6.11), we have $\Phi_1 = P\, J\, Q$ where $J = \text{Diag}[\, I_d, \Lambda_r\,]$, and, hence, the process $Z_t = Q\, Y_t$ follows the model $Z_t = J\, Z_{t-1} + a_t$. The forecasts for the Y_{t+l} are $\hat{Y}_t(l) = \Phi_1^l\, Y_t$, and, thus, the forecasts for $Z_t = Q\, Y_t$ are $\hat{Z}_t(l) = J^l\, Z_t$. So $\hat{Z}_{1t}(l) = Q_1' \hat{Y}_t(l) = Z_{1t}$, consistent with a random walk model for the nonstationary or "common trends" components $Z_{1t} = Q_1' Y_t$, whereas $\hat{Z}_{2t}(l) = Q_2' \hat{Y}_t(l) = \Lambda_r^l\, Z_{2t} \approx 0$ for large l, for the stationary co-integrating relations. In addition, from (2.21) of Section 2.5.2, the lead l forecast error covariance matrix for the forecasts $\hat{Z}_t(l)$ is

$$Q\, \Sigma(l)\, Q' = Q \sum_{j=0}^{l-1} \Phi_1^j\, \Sigma\, \Phi_1'^j = \sum_{j=0}^{l-1} J^j\, \Sigma_a\, J'^j ,$$

where $\Sigma_a = Q\, \Sigma\, Q' = \text{Cov}(\, Q\, \varepsilon_t\,)$. Hence, the covariance matrix of the forecasts errors for $\hat{Z}_{2t}(l) = Q_2' \hat{Y}_t(l)$ is $\sum_{j=0}^{l-1} \Lambda_r^j\, \Sigma_{a_2}\, \Lambda_r'^j$ and will remain finite and bounded for large l. Engle and Yoo (1987) established that similar results concerning the behavior of $Q_2' \hat{Y}_t(l)$ and its forecast error covariance matrix for large l will hold in general co-integrated systems such as the error-correction vector AR model (6.11). Thus, the longer-run forecasts based on such co-integrated model systems tend to be tied together and will satisfy the co-integrating relationships $Q_2' \hat{Y}_t(l) \approx 0$ in the long run.

6.3.8 Explicit Unit-Root Structure of the Nonstationary AR Model and Implications

For the nonstationary vector AR model with d unit roots as studied in this section, it is also instructive and useful to examine the explicit nature of the nonstationary factor in the AR operator of the model. For instance, we may want to express the model in a form such that a first-order operator (with d unit roots) applied to Y_t yields a stationary vector AR model. That is, we consider the possibility of factorization of the matrix AR operator in the form $\Phi(B) = \overline{\Phi}(B)\,(\, I - U_1\, B\,)$, where U_1 has d eigenvalues equal to one and $\overline{\Phi}(B)$ is a stationary operator. A related type of factorization problem was considered by Stensholt and Tjostheim (1981) in theory, and from an empirical view by Tjostheim and Paulsen (1982). Under the above factorization, $\boldsymbol{X}_t = (I - U_1 B) \boldsymbol{Y}_t = \boldsymbol{Y}_t - U_1 \boldsymbol{Y}_{t-1}$ is a stationary AR process, and the factor $(I - U_1 B)$ may be viewed as a "generalized differencing operator" that reduces the series $\boldsymbol{Y}_t$ to a stationary AR process $\boldsymbol{X}_t$. We next show that under the model assumptions of this section, such a factorization always exists and that U_1 is given explicitly as $U_1 = P_1 Q_1'$, where P_1 and Q_1 are defined following equation (6.11).

Now considering the error-correction form (6.11), $\Phi^*(B)(1 - B)\boldsymbol{Y}_t = -\Phi(1)\boldsymbol{Y}_{t-1} + \varepsilon_t$, of the AR($p$) model $\Phi(B)\boldsymbol{Y}_t = \varepsilon_t$, the representation $\Phi(B) = \Phi^*(B)(1 - B) + \Phi(1)B$ is seen to exist for the AR operator. Recall that $\Phi(1) = P_2(I_r - \Lambda_r)Q_2'$, where $P = [P_1, P_2]$ and $Q = P^{-1} = [Q_1, Q_2]'$ are such that $Q\,\Phi(1)P = I - J = \text{Diag}[0, I_r - \Lambda_r]$. Hence, we see that $\Phi(1)P_1 = 0$ and also that $(I - U_2 B)(I - U_1 B) = (I - IB) \equiv (1 - B)$, where $U_i = P_i Q_i'$, $i = 1, 2$, with $U_1 + U_2 = I$ and $U_2 U_1 = 0$. Therefore, we have the factorization

$$\Phi(B) = \Phi^*(B)(1 - B) + \Phi(1)B$$

$$= \{\Phi^*(B)(I - U_2 B) + \Phi(1)B\}(I - U_1 B) \equiv \overline{\Phi}(B)(I - U_1 B),$$

where $U_1 = P_1 Q_1'$ and $\overline{\Phi}(B) = \Phi^*(B)(I - U_2 B) + \Phi(1)B$. The matrix U_1 is idempotent of rank d, so that $\det\{I - U_1 B\} = (1 - B)^d = 0$ has d unit roots, and, hence, $\det\{\overline{\Phi}(B)\} = 0$ has all the remaining roots of $\det\{\Phi(B)\} = 0$, which are outside the unit circle by the assumptions on $\Phi(B)$. Thus, we can express the AR(p) model as

$$\overline{\Phi}(B)(I - U_1 B)\boldsymbol{Y}_t = \varepsilon_t, \qquad \text{or} \qquad \overline{\Phi}(B)\boldsymbol{X}_t = \varepsilon_t, \tag{6.19}$$

where $\boldsymbol{X}_t = (I - U_1 B)\boldsymbol{Y}_t = \boldsymbol{Y}_t - U_1 \boldsymbol{Y}_{t-1}$ is a *stationary* vector AR(p) process. Hence, this provides the "minimal differencing" transformation of the process $\boldsymbol{Y}_t$ to induce stationarity. In empirical work, an estimate of the transforming matrix U_1 can be obtained conveniently through the canonical correlation analysis methods discussed previously. This could then be used to construct the stationary process $\boldsymbol{X}_t$, which may be quite useful in practice to study the characteristics and model this stationary process.

In terms of the process $\boldsymbol{Z}_t = Q\,\boldsymbol{Y}_t = (\boldsymbol{Z}_{1t}', \boldsymbol{Z}_{2t}')'$ introduced earlier in this section, we see that the corresponding process related to $\boldsymbol{X}_t$ is simply

$$\boldsymbol{X}_t^* = Q\,\boldsymbol{X}_t = Q\,\boldsymbol{Y}_t - Q\,U_1 P\,Q\,\boldsymbol{Y}_{t-1} = \boldsymbol{Z}_t - J_1^* \boldsymbol{Z}_{t-1} = [(1 - B)\boldsymbol{Z}_{1t}', \boldsymbol{Z}_{2t}']',$$

since $J_1^* = Q\,U_1 P = Q\,P_1 Q_1' P = \text{Diag}[I_d, 0]$. Thus, from (6.19), the process $\boldsymbol{Z}_t$ has the vector AR(p) model representation

$$[Q\,\overline{\Phi}(B)P]\,\boldsymbol{X}_t^* = Q\,\varepsilon_t, \qquad \text{or} \qquad \overline{\Phi}^*(B)\boldsymbol{X}_t^* = \boldsymbol{a}_t,$$

where $\boldsymbol{X}_t^* = (I - J_1^* B)\boldsymbol{Z}_t = [(1 - B)\boldsymbol{Z}_{1t}', \boldsymbol{Z}_{2t}']'$ and

$$\overline{\Phi}^*(B) = Q\,\overline{\Phi}(B)\,P = Q\,\Phi^*(B)\,(\,I - U_2\,B\,)\,P + Q\,\Phi(1)\,P\,B$$

$$= Q\,\Phi^*(B)\,P\,(\,I - J_2^*\,B\,) + (\,I - J\,)\,B$$

with $J_2^* = Q\,U_2\,P = Q\,P_2 Q_2'\,P = \text{Diag}[\,0,\,I_r\,]$. It follows that the stationary process $\boldsymbol{X}_t^*$ has the infinite MA representation as $\boldsymbol{X}_t^* = \{\,\overline{\Phi}^*(B)\,\}^{-1}\,\boldsymbol{a}_t \equiv \overline{\Psi}(B)\,\boldsymbol{a}_t$. We compare this to the representation $\boldsymbol{Z}_t - J\,\boldsymbol{Z}_{t-1} = \boldsymbol{u}_t = \Psi(B)\,\boldsymbol{a}_t$ stated previously following (6.12), where $J = \text{Diag}[\,I_d,\,\Lambda_r\,]$. Since $I - J\,B = (\,I - (J - J_1^*)\,B\,)\,(\,I - J_1^*\,B\,)$, we see that

$$\boldsymbol{u}_t = (\,I - (J - J_1^*)\,B\,)\,\boldsymbol{X}_t^* = (\,I - (J - J_1^*)\,B\,)\,\overline{\Psi}(B)\,\boldsymbol{a}_t\,,$$

so that $\Psi(B) = (\,I - (J - J_1^*)\,B\,)\,\overline{\Psi}(B)$. Hence, in particular, we see that $\Psi(1) = (\,I - (J - J_1^*)\,)\,\{\,\overline{\Phi}^*(1)\,\}^{-1} = \text{Diag}[\,I_d,\,I_r - \Lambda_r\,]\,\{\,\overline{\Phi}^*(1)\,\}^{-1}$, where

$$\overline{\Phi}^*(1) = Q\,\overline{\Phi}(1)\,P = Q\,\Phi^*(1)\,P\,(\,I - J_2^*\,) + (\,I - J\,)$$

$$= Q\,\Phi^*(1)\,[\,P_1,\,0\,] + (\,I - J\,).$$

It then follows that the matrix Ψ_{11} which appears in the asymptotic distribution results, such as the representation for M in (6.14), is given more explicitly by $\Psi_{11} = \{\,Q_1'\,\Phi^*(1)\,P_1\,\}^{-1}$, since $\overline{\Phi}^*(1)$ is seen to be block lower triangular with upper left $(\,d \times d\,)$ block equal to $Q_1'\,\Phi^*(1)\,P_1$.

Another feature that is also obtained rather directly from the above derivations is an explicit form for the infinite MA representation for the first differences $(\,1 - B\,)\,\boldsymbol{Y}_t$. From (6.19) above we have

$$(\,1 - B\,)\,\boldsymbol{Y}_t = (\,I - U_2\,B\,)\,\boldsymbol{X}_t = (\,I - U_2\,B\,)\,\{\,\overline{\Phi}(B)\,\}^{-1}\,\varepsilon_t \equiv \Psi^*(B)\,\varepsilon_t\,. \qquad (6.20)$$

From this we directly see that $\Psi^*(1) = (\,I - U_2\,)\,\{\,\overline{\Phi}(1)\,\}^{-1} = P_1\,Q_1'\,P\,\{\,\overline{\Phi}^*(1)\,\}^{-1}\,Q = P_1\,[\,\Psi_{11},\,0\,]\,Q$. This clearly illustrates that $Q_2'\,\Psi^*(1) = 0$ which readily shows that the infinite MA form for the first differences $(\,1 - B\,)\,\boldsymbol{Y}_t$ is noninvertible with $\Psi^*(1)$ of reduced rank d, reflecting the "overdifferencing" that occurs in the infinite MA model representation for the first differences $\boldsymbol{W}_t = (\,1 - B\,)\,\boldsymbol{Y}_t$.

6.3.9 Further Numerical Examples

Other numerical illustrations of modeling by use of the error-correction "unit-root" vector AR model approach include the trivariate series of U.S. monthly interest rate series for the Federal Fund Rate, the 90-day Treasury Bill Rate, and the one-year Treasury Bill Rate, presented in Reinsel and Ahn (1992), and the bivariate series of U.S. quarterly interest rate data on AAA corporate bonds and on commercial paper. In each case, the vector series were adequately modeled by a vector AR model, and cointegration among the series was indicated such that the AR model for each data set was constrained to contain exactly one unit root. For example, we briefly discuss the model obtained for the quarterly AAA corporate bonds and commercial paper series for 1953–1970.

EXAMPLE 6.5. These quarterly interest rate data are shown in Figures 6.2(a) and (b), and the data were previously analyzed using transfer function modeling by Haugh and Box (1977).

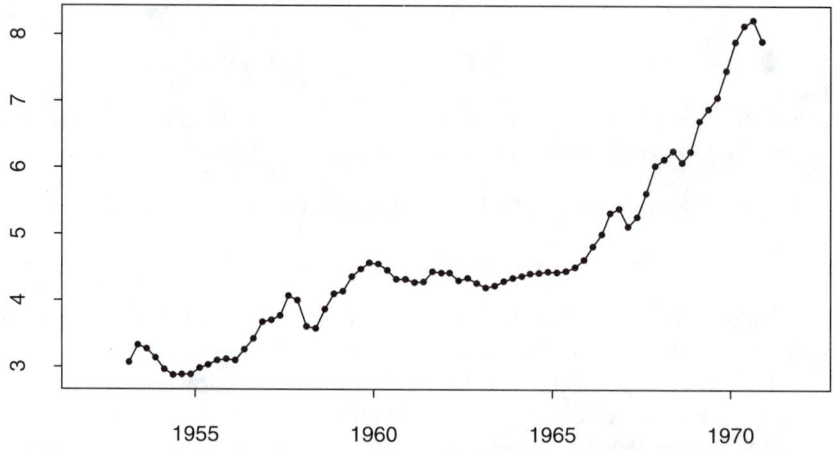

(a) Y_{1t} : AAA Corporate Bond Rate Series

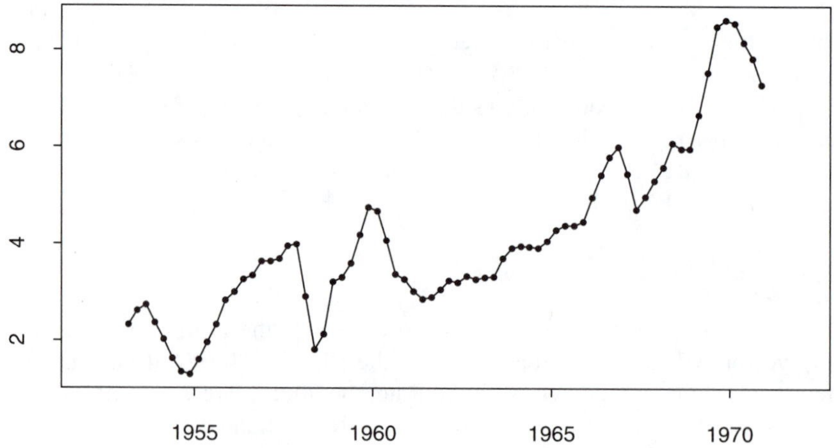

(b) Y_{2t} : Commercial Paper Rate Series

Figure 6.2. Quarterly U.S. Interest Rate Series for the Period 1953 Through 1970

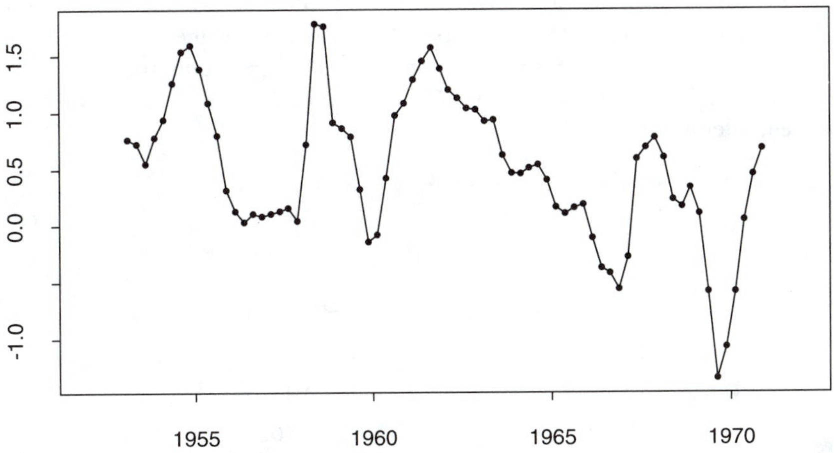

(c) Z_{2t} : Transformed Series, $Z_{2t} = Y_{1t} - 0.991\, Y_{2t}$

Figure 6.2. (continued)

Table 6.4. Summary of Partial Canonical Correlation Analysis Between $\boldsymbol{Y}_t$ and $\boldsymbol{Y}_{t-j}$ and Associated Chi-Squared Test Statistics $C(j, s)$ from (6.3) for the Quarterly Interest Rate Data Example 6.5. (Test statistics are used to determine the appropriate AR order and ranks $k - s$ of coefficient matrices Φ_j.)

j	Squared Correlations	$C(j, s)$	
		$s = 1$	$s = 2$
1	0.684, 0.986	78.96	371.40
2	0.138, 0.562	9.74	63.89
3	0.017, 0.157	1.07	11.72
4	0.022, 0.054	1.31	4.65
5	0.002, 0.085	0.09	5.12
6	0.005, 0.073	0.28	4.34

Using the partial canonical correlation methods of Section 6.1.1, a bivariate AR(3) model was specified, and the testing results from (6.3) also indicated a reduced rank for Φ_3 of rank(Φ_3) = 1 in the AR(3) model. Details of the partial canonical correlation analysis are summarized in Table 6.4. Also, likelihood ratio tests on the rank of C in the error-correction form of the AR(3) model, through the use of partial canonical correlations as discussed in this section, indicated a reduced rank of rank(C) = 1. Specifically, within the AR(3) model context, the (squared) partial canonical correlations between $\boldsymbol{W}_t = \boldsymbol{Y}_t - \boldsymbol{Y}_{t-1}$ and

$\boldsymbol{Y}_{t-1}$, given $\boldsymbol{W}_{t-1}$ and $\boldsymbol{W}_{t-2}$, are obtained as $\hat{\rho}_1^2(p) = 0.241$ and $\hat{\rho}_2^2(p) = 0.011$ (with $p = 3$). Hence, the LR statistic for testing rank(C) = 1 ($d = 1$ unit root) has the value $-(T - p - 1)\log(1 - \hat{\rho}_2^2(p)) = 0.75$, while the value in testing for $d = 2$ unit roots is 19.48, so a model with $d = 1$ unit root ($r = 1$) is selected. The final estimated model for the bivariate series with the unit root imposed, after a little simplification, has the form

$$\boldsymbol{W}_t = \hat{A}\,\hat{B}\,\boldsymbol{Y}_{t-1} + \hat{\Phi}_1^*\,\boldsymbol{W}_{t-1} + \hat{\Phi}_2^*\,\boldsymbol{W}_{t-2} + \varepsilon_t ,$$

with $\boldsymbol{W}_t = \boldsymbol{Y}_t - \boldsymbol{Y}_{t-1}$ and

$$\tilde{C} = \hat{A}\,\hat{B} = \begin{bmatrix} -0.066 \\ 0.180 \end{bmatrix} \begin{bmatrix} 1 & -0.991 \end{bmatrix},$$

$$\hat{\Phi}_1^* = \begin{bmatrix} 0.558 & 0 \\ 1.116 & 0.582 \end{bmatrix}, \qquad \hat{\Phi}_2^* = \begin{bmatrix} -0.441 & 0 \\ -0.863 & 0 \end{bmatrix},$$

with residual covariance matrix estimate $\tilde{\Sigma}$ having diagonal elements 0.01874 and 0.07629, and determinant $|\tilde{\Sigma}| = 0.8021 \times 10^{-3}$. Notice that there is to a certain extent a transfer function structure for the model for the first differences, in the sense that the matrices $\hat{\Phi}_1^*$ and $\hat{\Phi}_2^*$ are lower triangular, and also notice the explicit reduced-rank feature of $\hat{\Phi}_2^*$. The series $Z_{2t} = Y_{1t} - 0.991\,Y_{2t}$ forms a co-integrating linear combination that should be stationary, and this series is plotted in Figure 6.2(c). Relative to the original two series, the series Z_{2t} is certainly much more nearly stationary than the original series. Conversely, $Z_{1t} = 0.180\,Y_{1t} + 0.066\,Y_{2t}$ is the estimate of the purely nonstationary component of the bivariate series $\boldsymbol{Y}_t$, with both Y_{1t} and Y_{2t} being linear combinations of the nonstationary component Z_{1t} and the stationary component Z_{2t}.

6.4 Multiplicative Seasonal Vector ARMA Models

Many time series encountered in economic, meteorological, and environmental studies exhibit strong seasonal behavior, with a period of s. When modeling a seasonal vector time series $\boldsymbol{Y}_t$ in such settings, one may often need to consider the seasonal differences of the series, $\boldsymbol{W}_t = (1 - B^s)\,\boldsymbol{Y}_t = \boldsymbol{Y}_t - \boldsymbol{Y}_{t-s}$. Then, based on the success of seasonal ARIMA modeling in the univariate case, we may consider multiplicative seasonal ARMA models of the general form

$$\boldsymbol{\phi}(B^s)\,\Phi(B)\,\boldsymbol{W}_t = \boldsymbol{\theta}(B^s)\,\Theta(B)\,\varepsilon_t , \tag{6.21}$$

where, for example, $\boldsymbol{\phi}(B^s) = I - \boldsymbol{\phi}_1\,B^s - \cdots - \boldsymbol{\phi}_P\,B^{sP}$ is the seasonal AR operator and $\boldsymbol{\theta}(B^s) = I - \boldsymbol{\theta}_1\,B^s - \cdots - \boldsymbol{\theta}_Q\,B^{sQ}$ is the seasonal MA operator. However, it should be noted immediately that, in general, the matrix seasonal and nonseasonal AR and MA operators do not commute, and, hence, the order in which these operators are presented in the model (6.21) will make a difference. By contrast to the univariate case, where seasonal models have received considerable attention, there has been much less investigation of multiplicative seasonal modeling for vector time series. We will not study vector seasonal models in detail in this section but will just examine a few simple special cases of the general model (6.21).

6.4.1 Some Special Seasonal ARMA Models for Vector Time Series

First consider the simple seasonal MA(1) model, $\boldsymbol{W}_t = (I - \boldsymbol{\theta}_1 B^s) \varepsilon_t = \varepsilon_t - \boldsymbol{\theta}_1 \varepsilon_{t-s}$. It easily follows that the autocovariance matrices for this process are

$$\Gamma(0) = \Sigma + \boldsymbol{\theta}_1 \Sigma \boldsymbol{\theta}_1', \qquad \Gamma(s) = - \Sigma \boldsymbol{\theta}_1' = \Gamma(-s)',$$

and $\Gamma(l) = 0$ otherwise. Hence, this model is rather elementary, and the correlation matrices $\rho(l)$ for such a process exhibit nonzero values only at lags 0 and $\pm s$.

We next consider the first-order multiplicative seasonal MA model, $\boldsymbol{W}_t = (I - \Theta_1 B) (I - \boldsymbol{\theta}_1 B^s) \varepsilon_t$. This model can be written as

$$\boldsymbol{W}_t = (I - \Theta_1 B - \boldsymbol{\theta}_1 B^s + \Theta_1 \boldsymbol{\theta}_1 B^{s+1}) \varepsilon_t .$$

From this we find the autocovariance matrices are given by

$$\Gamma(0) = \Omega + \Theta_1 \Omega \Theta_1', \qquad \Gamma(1) = - \Omega \Theta_1',$$

$$\Gamma(s) = - \Sigma \boldsymbol{\theta}_1' - \Theta_1 \Sigma \boldsymbol{\theta}_1' \Theta_1', \qquad \Gamma(s-1) = \Theta_1 \Sigma \boldsymbol{\theta}_1', \qquad \Gamma(s+1) = \Sigma \boldsymbol{\theta}_1' \Theta_1',$$

and $\Gamma(l) = 0$ for all other $l \geq 0$, where $\Omega = \Sigma + \boldsymbol{\theta}_1 \Sigma \boldsymbol{\theta}_1'$. Thus, the correlation matrices $\rho(l)$, $l \geq 0$, for this process will have nonzero values only at the lags $0, 1, s-1, s, s+1$. However, in this model we notice the dependence of the autocovariance matrix structure on the order in which the nonseasonal and seasonal MA(1) operators are included in the model. The alternate model that might be considered is obtained by reversing the order of the MA(1) operators and is given by $\boldsymbol{W}_t = (I - \boldsymbol{\theta}_1 B^s) (I - \Theta_1 B) \varepsilon_t$. The autocovariance matrices for this model will be of a somewhat similar form as above, in the sense that $\Gamma(l) = 0$ for $l \neq 0, \pm 1, \pm (s-1), \pm s, \pm (s+1)$, but there will be differences due to the order in which the MA operators are included in the model. Both models might be viewed as special cases of the nonmultiplicative MA model given by $\boldsymbol{W}_t = (I - \Theta_1 B - \Theta_s B^s - \Theta_{s+1} B^{s+1}) \varepsilon_t$. In practice, in attempting to determine which (if either) form of the multiplicative MA model is appropriate, one might fit the nonmultiplicative form of the seasonal MA model and then determine to what extent either of the relations $\Theta_{s+1} = - \Theta_1 \boldsymbol{\theta}_1$ or $\Theta_{s+1} = - \boldsymbol{\theta}_1 \Theta_1$ hold.

One further simple seasonal model that may be fairly useful in practice is of the form $(I - \Phi B) \boldsymbol{W}_t = (I - \boldsymbol{\theta}_1 B^s) \varepsilon_t$. From the relations (2.14) in Section 2.3.2, it follows that the autocovariance matrices for this model satisfy

$$\Gamma(l) = \Gamma(l-1)\,\Phi' - \Phi^{s-l}\,\Sigma\,\boldsymbol{\theta}_1', \qquad l = 1, 2, \ldots, s,$$

with $\Gamma(l) = \Gamma(l-1)\,\Phi'$ for $l > s$. Hence, for a typical situation of monthly time series with $s = 12$, if the eigenvalues of Φ are not close to one, the terms Φ^{12-l} may be generally small for low lags l and, hence, the autocovariance matrices for this process may follow a pattern somewhat similar to that of an AR(1) process for low lags. A bivariate model of the form $(I - \Phi B)(1 - B^{12})\,\boldsymbol{Y}_t = (I - \boldsymbol{\theta}_1 B^{12})\,\varepsilon_t$ is, in fact, found to be quite adequate to represent the seasonal time series of monthly housing-starts and housing-sold that is similar to the data considered in Example 6.4 of Section 6.3. We briefly indicate the modeling for this monthly seasonal bivariate time series.

EXAMPLE 6.6. The data consist of monthly U.S. housing-starts and housing-sold (not seasonally adjusted) over the period 1965 through 1974. The original data exhibit strong seasonal behavior, and so the seasonal differences $\boldsymbol{W}_t = (1 - B^{12})\,\boldsymbol{Y}_t = \boldsymbol{Y}_t - \boldsymbol{Y}_{t-12}$ are formed and considered for further modeling. The original series are displayed in Figures 6.3(a) and (b), and the seasonal differences are shown in Figures 6.4(a) and (b). The autocorrelation matrices for the series $\boldsymbol{W}_t$ show high correlations at low lags, and correlation structure is also anticipated in the model for $\boldsymbol{W}_t$ at the seasonal lag of 12. Results of a partial canonical correlation analysis for $\boldsymbol{W}_t$ are given in Table 6.5. These results indicate that, in terms of modeling of the low lag correlation structure, a vector AR(1) model will probably be adequate. (The test statistic results in Table 6.5 are not strictly valid because the series still exhibits correlation at the higher, seasonal, lag of 12, and, hence, the series does not actually follow a low order vector AR model. However, these results still tend to provide useful information on an appropriate model to represent the low order correlation structure for the series $\boldsymbol{W}_t$.)

Table 6.5. Summary of Partial Canonical Correlation Analysis Between Seasonal Differences $\boldsymbol{W}_t = \boldsymbol{Y}_t - \boldsymbol{Y}_{t-12}$ and $\boldsymbol{W}_{t-j}$ and Associated Chi-Squared Test Statistics $C(j, s)$ from (6.3) for the Monthly Seasonal Housing Data Example 6.6.

j	Squared Correlations	$C(j, s)$	
		$s = 1$	$s = 2$
1	0.037, 0.870	3.89	217.45
2	0.003, 0.027	0.29	3.11
3	0.011, 0.024	1.13	3.48
4	0.003, 0.065	0.28	6.70
5	0.024, 0.034	2.21	5.41
6	0.021, 0.070	1.93	8.39

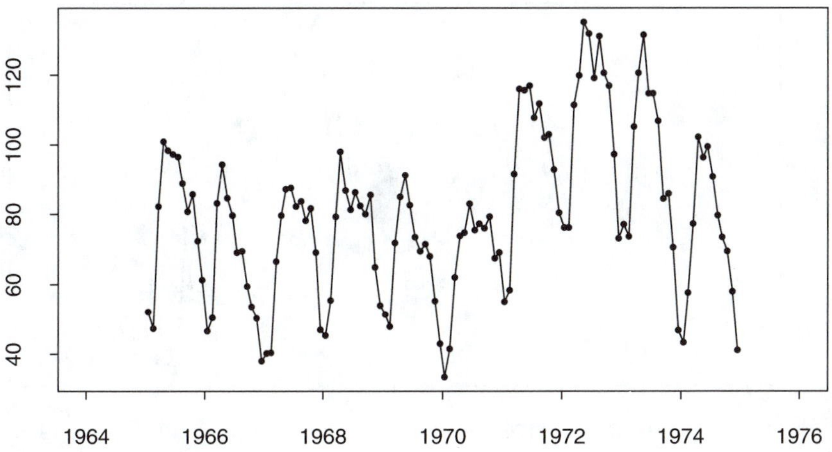

(a) Y_{1t} : Housing Starts Series

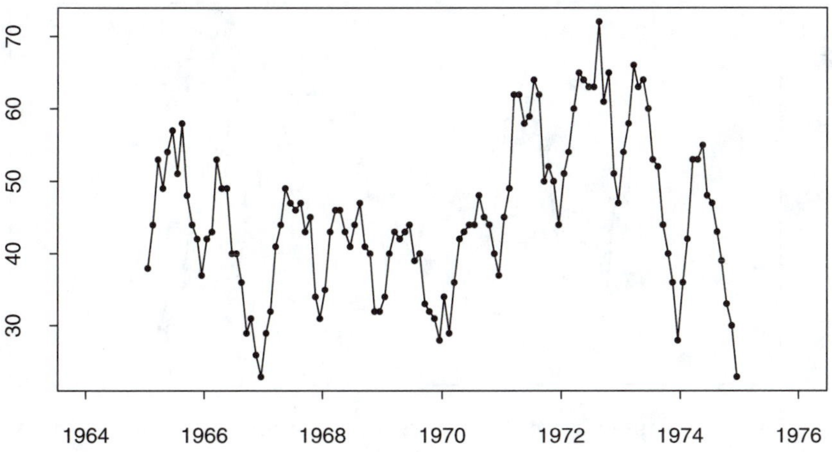

(b) Y_{2t} : Houses Sold Series

Figure 6.3. Unadjusted U.S. Monthly Housing Data (in thousands) for the Period January 1965 Through December 1974

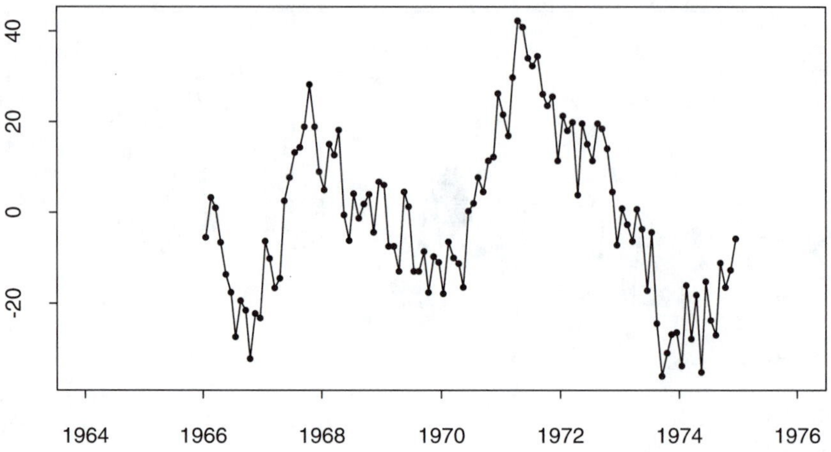

(a) $Y_{1t} - Y_{1,t-12}$: Housing Starts Series

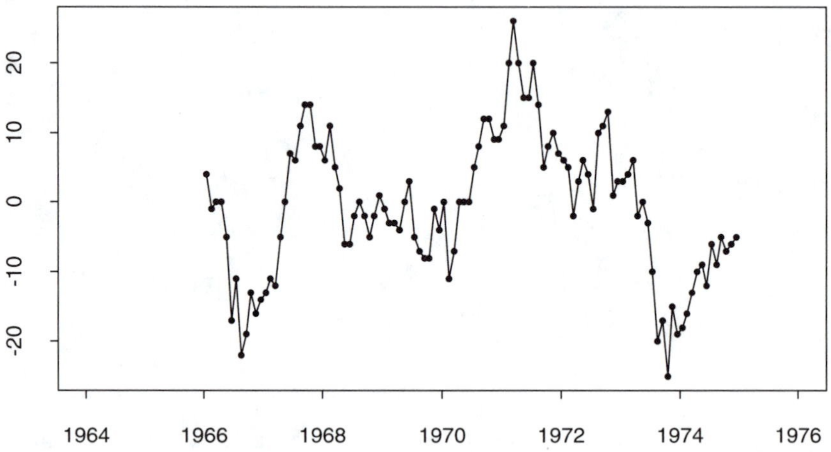

(b) $Y_{2t} - Y_{2,t-12}$: Houses Sold Series

Figure 6.4. Seasonal Differences of U.S. Monthly Housing Data (in thousands) for the Period January 1965 Through December 1974

However, as expected, examination of the autocorrelation matrices of the residuals from an AR(1) model fit to $\boldsymbol{W}_t$ reveals highly significant correlation at the seasonal lag of 12. The residual correlation matrix at lag 12 from the AR(1) model fit is equal to

$$\hat{\rho}_\varepsilon(12) = \begin{bmatrix} -0.329 & -0.176 \\ -0.077 & -0.402 \end{bmatrix}.$$

This strongly suggests the inclusion in the bivariate model of a moving average term at lag 12, a term which would already be anticipated to be appropriate based on the univariate modeling of the individual series [e.g., see Hillmer and Tiao (1979)]. Hence, the model of the form

$$(I - \Phi B) (1 - B^{12}) \boldsymbol{Y}_t = (I - \boldsymbol{\theta}_1 B^{12}) \varepsilon_t$$

was estimated. The maximum likelihood estimators were obtained as

$$\hat{\Phi} = \begin{bmatrix} 0.417 & 0.979 \\ 0.144 & 0.701 \end{bmatrix}, \quad \hat{\boldsymbol{\theta}}_1 = \begin{bmatrix} 1.002 & -0.050 \\ 0.051 & 0.993 \end{bmatrix}, \quad \hat{\Sigma} = \begin{bmatrix} 28.735 & 6.049 \\ 6.049 & 10.314 \end{bmatrix}.$$

The off-diagonal elements of $\hat{\boldsymbol{\theta}}_1$ are not significant and could reasonably be set to zero. Diagnostic checking of residuals from this model do not indicate any inadequacy. Note that the above fitted model (as well as univariate ARIMA models fit to the individual series) suggest the presence of deterministic seasonal components in both series, because the seasonal MA operator is nearly equal to $I - B^{12}$. That is, it suggests the model structure for $\boldsymbol{Y}_t$ of the form $(I - \Phi B) \boldsymbol{Y}_t = \boldsymbol{S}_t + \varepsilon_t$, where $\boldsymbol{S}_t$ is a deterministic (vector) seasonal component such that $(1 - B^{12}) \boldsymbol{S}_t = 0$. [Refer to Abraham and Box (1978) and Bell (1987) for a discussion of the relation between the occurrence of a common seasonal differencing operator $1 - B^{12}$ in both the AR and MA sides of a univariate ARIMA model for a series and the presence of deterministic seasonal components in the series.] In addition, the estimate $\hat{\Phi}$ obtained from this seasonal model is very similar to the estimate obtained in Example 6.4 for the AR(1) model fitted to the seasonally adjusted series. So, as in that example, since the eigenvalues of $\hat{\Phi}$ are equal to 0.960 and 0.158, there is strong indication of a single unit root in the AR(1) operator and, hence, of a single nonstationarity in the seasonally differenced bivariate series $\boldsymbol{W}_t$.

In general, it seems that more research is needed in the application of seasonal time series modeling, especially multiplicative seasonal models, to vector time series. The topic of seasonal cointegration and unit roots at seasonal frequencies in vector models is another interesting area which also requires further research, and some initial examination of this topic for quarterly seasonal vector time series has been performed by Hylleberg et al. (1990), Lee (1992), and Ahn and Reinsel (1993).

CHAPTER 7

State-Space Models, Kalman Filtering, and Related Topics

In this chapter we present and discuss the state-variable (state-space) model and its basic properties. The Kalman filtering and smoothing procedures associated with state-space models will be developed. The relation of time-invariant state-variable models to vector ARMA models, including the state-space representations for vector ARMA models, and the associated topics of minimal dimensionality of the state vector and the relation with Kronecker indices and McMillan degree of a process will also be discussed. The use of the state-space formulation for construction of the exact likelihood function for the vector ARMA model will be presented also. In addition, discussion of results for the classical approach to smoothing and filtering of time series will be presented.

7.1 State-Variable Models and Kalman Filtering

We now consider the basic *state-variable* (*state-space*) *model*, with origins in the engineering control literature (Kalman, 1960), which consists of a *state equation* (or *transition equation*)

$$\mathbf{Z}_t = \Phi_t \, \mathbf{Z}_{t-1} + \boldsymbol{a}_t \tag{7.1}$$

and an *observation equation* (or *measurement equation*)

$$\boldsymbol{Y}_t = H_t \, \mathbf{Z}_t + \boldsymbol{N}_t \,. \tag{7.2}$$

In this model, $\mathbf{Z}_t$ is an (unobservable) $r \times 1$ random vector which describes the state of the dynamic system at time t and, hence, is called the *state vector*, $\boldsymbol{Y}_t$ is a $k \times 1$ random vector of observations related to the state, $\boldsymbol{\Phi}_t$ is an $r \times r$ transition (or system) matrix, H_t is a $k \times r$ observation matrix, and $\boldsymbol{a}_t$ and $\boldsymbol{N}_t$ are random white noise vectors assumed to be independently distributed with zero means and covariance matrices $\Sigma_a = \text{Cov}(\boldsymbol{a}_t)$ and $\Sigma_n = \text{Cov}(\boldsymbol{N}_t)$, respectively. The covariance matrices of the noise vectors $\boldsymbol{a}_t$ and $\boldsymbol{N}_t$ could also be allowed to vary with time t, but for notational convenience this will not be done here. Often, in applications the matrices Φ_t and H_t [as well as $\text{Cov}(\boldsymbol{a}_t)$ and $\text{Cov}(\boldsymbol{N}_t)$] are constant matrices, $\Phi \equiv \Phi_t$ and $H \equiv H_t$ for all t, that do not depend on t, in which case the system or model is said to be *time-invariant*. The state equation (7.1) describes the evolution of the dynamic system in time, while the measurement equation (7.2) indicates that the observations $\boldsymbol{Y}_t$ consist of linear combinations of the (unobservable) state variables corrupted by additive noise. Typically, the dimension r of the state vector $\mathbf{Z}_t$ may be much larger than the dimension k of the observation vector $\boldsymbol{Y}_t$, since it needs to be sufficiently so that the dynamics of the system can be represented by the simple Markovian (first-order) structure as in (7.1). Conceptually, the state vector $\mathbf{Z}_t$ is meant to contain a complete summary of all the past and present information available to describe the dynamic system, and, hence, it contains all information needed to predict future behavior of the system. Also, the basic state-space model (7.1)–(7.2) is readily extended to include the effects of observed exogenous input or control variables $\boldsymbol{X}_t$ on the output $\boldsymbol{Y}_t$, so that we have $\mathbf{Z}_t = \Phi_t \mathbf{Z}_{t-1} + B_t \boldsymbol{X}_{t-1} + \boldsymbol{a}_t$ and $\boldsymbol{Y}_t = H_t \mathbf{Z}_t + F_t \boldsymbol{X}_t + \boldsymbol{N}_t$.

The uses of the state-variable model (7.1)–(7.2) or its extensions have somewhat varied backgrounds. In engineering control and other physical applications, the state variable generally represents a well-defined set of physical variables in a dynamic system such as the coordinates in space of a rocket or satellite, but those variables are not directly observable, and the state equation represents the dynamics that govern the physical system. Typically, in this setting, the coefficients Φ and H may be exactly or at least approximately "well known". In more statistical applications, the state-variable model is used as a convenient form to represent many different types of models, such as vector ARMA models, structural component models of the "signal-plus-noise" form (e.g., economic models for observed series that are represented as the sum of unobservable seasonal, trend, and irregular components), or time-varying parameter models (e.g., regression models with time-varying regression coefficients). In these contexts, some parts of the matrices Φ and H involve unknown parameters which need to be estimated by appropriate statistical methods. The model could also have uses in "empirical model building" without reference to any other model structures, such as in dynamic factor analysis modeling of multivariate time series $\boldsymbol{Y}_t$ where the state vector $\mathbf{Z}_t$ might represent a lower-dimensional ($r < k$) vector of underlying "dynamic factors" (e.g., Engle and Watson, 1981, and Pena and Box, 1987).

7.1.1 The Kalman Filtering Relations

In the context of the state-variable model, one of the traditional uses of the model is to filter or predict current or future values of the unobservable state vector $\mathbf{Z}_t$ (or to predict future values of $\boldsymbol{Y}_t$). This model provides a convenient computational method for recursively obtaining filtered or predicted values. This computation has come to be known as the *Kalman filter*. The basic filter produces the minimum mean squared error (MSE) linear estimator (predictor) of the current state $\mathbf{Z}_t$ given the observations $\boldsymbol{Y}_1, \ldots, \boldsymbol{Y}_t$ through time t, which we denote as $\hat{\mathbf{Z}}_{t|t}$. For convenience of notation, we assume that the minimum MSE linear estimator is also $\hat{\mathbf{Z}}_{t|t} = E[\,\mathbf{Z}_t \mid \boldsymbol{Y}_1, \ldots, \boldsymbol{Y}_t\,]$. More generally, let

$$\hat{\mathbf{Z}}_{t+h|t} = E[\,\mathbf{Z}_{t+h} \mid \boldsymbol{Y}_1, \ldots, \boldsymbol{Y}_t\,]$$

denote the minimum MSE linear estimator (predictor) of $\mathbf{Z}_{t+h}$, with

$$P_{t+h|t} = E[\,(\mathbf{Z}_{t+h} - \hat{\mathbf{Z}}_{t+h|t})(\mathbf{Z}_{t+h} - \hat{\mathbf{Z}}_{t+h|t})'\,],$$

so that under normality the conditional distribution of $\mathbf{Z}_{t+h}$, given $\boldsymbol{Y}_1, \ldots, \boldsymbol{Y}_t$, is normal with (conditional) mean $\hat{\mathbf{Z}}_{t+h|t}$ and (conditional) covariance matrix $P_{t+h|t}$. Starting with some appropriate initial values $\mathbf{Z}_0 \equiv \hat{\mathbf{Z}}_{0|0}$ and $P_0 \equiv P_{0|0}$, it is known that the optimal filtered estimate, $\hat{\mathbf{Z}}_{t|t}$, is given through the following recursive relations:

$$\hat{\mathbf{Z}}_{t|t} = \hat{\mathbf{Z}}_{t|t-1} + K_t\,(\boldsymbol{Y}_t - H_t\,\hat{\mathbf{Z}}_{t|t-1}), \tag{7.3}$$

where

$$K_t = P_{t|t-1}\,H_t'\,[\,H_t\,P_{t|t-1}\,H_t' + \Sigma_n\,]^{-1}, \tag{7.4}$$

with

$$\hat{\mathbf{Z}}_{t|t-1} = \Phi_t\,\hat{\mathbf{Z}}_{t-1|t-1}, \qquad P_{t|t-1} = \Phi_t\,P_{t-1|t-1}\,\Phi_t' + \Sigma_a, \tag{7.5a-b}$$

and

$$\begin{aligned} P_{t|t} &= [\,I - K_t\,H_t\,]\,P_{t|t-1} \\ &= P_{t|t-1} - P_{t|t-1}\,H_t'\,[\,H_t\,P_{t|t-1}\,H_t' + \Sigma_n\,]^{-1}\,H_t\,P_{t|t-1} \end{aligned} \tag{7.6}$$

for $t = 1, 2, \ldots$.

In (7.3), the quantity $\varepsilon_{t|t-1} = \boldsymbol{Y}_t - H_t\,\hat{\mathbf{Z}}_{t|t-1} \equiv \boldsymbol{Y}_t - \hat{\boldsymbol{Y}}_{t|t-1}$ is called the *innovation* at time t, because it is the new information provided by the measurements $\boldsymbol{Y}_t$ which was not available from the previous observed history of the system. The factor K_t is called the "*Kalman gain*" *matrix*. The filtering procedure in (7.3) has the recursive "prediction-correction" or "updating" form, and the validity of these equations as representing the minimum MSE linear predictor can readily be verified through the principles of "updating". For example, verification of (7.3) follows from the principle for linear regression functions [e.g., see relations (A1.11)–(A1.12) of Appendix A1] that

$$\begin{aligned} E[\,\mathbf{Z}_t \mid \boldsymbol{Y}_1, \ldots, \boldsymbol{Y}_t\,] &= E[\,\mathbf{Z}_t \mid \boldsymbol{Y}_1, \ldots, \boldsymbol{Y}_{t-1}, \boldsymbol{Y}_t - \hat{\boldsymbol{Y}}_{t|t-1}] \\ &= E[\,\mathbf{Z}_t \mid \boldsymbol{Y}_1, \ldots, \boldsymbol{Y}_{t-1}] + E[\,\mathbf{Z}_t \mid \boldsymbol{Y}_t - \hat{\boldsymbol{Y}}_{t|t-1}], \end{aligned}$$

since $\varepsilon_{t|t-1} = \boldsymbol{Y}_t - \hat{\boldsymbol{Y}}_{t|t-1}$ is independent of $\boldsymbol{Y}_1, \ldots, \boldsymbol{Y}_{t-1}$. From (7.3) it is seen that the estimate of $\mathbf{Z}_t$ based on observations through time t equals the prediction of $\mathbf{Z}_t$ from observations through time $t-1$ updated by the factor K_t times the innovation $\varepsilon_{t|t-1}$. The equation in (7.4) indicates that K_t can be interpreted as the regression coefficient matrix of $\mathbf{Z}_t$ on the innovation $\varepsilon_{t|t-1}$,

$$K_t = \text{Cov}(\, \mathbf{Z}_t, \varepsilon_{t|t-1} \,) \, \{ \, \text{Cov}(\, \varepsilon_{t|t-1} \,) \, \}^{-1},$$

with $\text{Cov}(\, \varepsilon_{t|t-1} \,) = H_t \, P_{t|t-1} \, H_t' + \Sigma_n$ and $\text{Cov}(\, \mathbf{Z}_t, \varepsilon_{t|t-1} \,) = P_{t|t-1} \, H_t'$ following directly from (7.2) since $\varepsilon_{t|t-1} = H_t \, (\, \mathbf{Z}_t - \hat{\mathbf{Z}}_{t|t-1} \,) + \boldsymbol{N}_t$. Thus, the general *updating relation* in (7.3) is

$$\hat{\mathbf{Z}}_{t|t} = \hat{\mathbf{Z}}_{t|t-1} + \text{Cov}(\, \mathbf{Z}_t, \varepsilon_{t|t-1} \,) \, \{ \, \text{Cov}(\, \varepsilon_{t|t-1} \,) \, \}^{-1} \, \varepsilon_{t|t-1},$$

where $\varepsilon_{t|t-1} = \boldsymbol{Y}_t - \hat{\boldsymbol{Y}}_{t|t-1}$. Note that this equation corresponds to the general relation (A1.12) given in Appendix A1 for the updating of the minimum MSE (linear regression) function. The relation in (7.6) is the usual updating of the error covariance matrix to account for the new information available from the innovation $\varepsilon_{t|t-1}$ [e.g., see relation (A1.13) of Appendix A1], while the *prediction relations* (7.5) follow directly from (7.1). In general, forecasts of future state values are available directly as $\hat{\mathbf{Z}}_{t+h|t} = \Phi_{t+h} \, \hat{\mathbf{Z}}_{t+h-1|t}$ for $h = 1, 2, \ldots$, with the covariance matrix of the forecast errors generated recursively essentially through (7.5) as $P_{t+h|t} = \Phi_{t+h} \, P_{t+h-1|t} \, \Phi_{t+h}' + \Sigma_a$. Forecasts of future observations $\boldsymbol{Y}_{t+h}$ are $\hat{\boldsymbol{Y}}_{t+h|t} = H_{t+h} \, \hat{\mathbf{Z}}_{t+h|t}$, since $\boldsymbol{Y}_{t+h} = H_{t+h} \, \mathbf{Z}_{t+h} + \boldsymbol{N}_{t+h}$, with forecast error covariance matrix $\Sigma_{t+h|t} = E[\, (\, \boldsymbol{Y}_{t+h} - \hat{\boldsymbol{Y}}_{t+h|t} \,) \, (\, \boldsymbol{Y}_{t+h} - \hat{\boldsymbol{Y}}_{t+h|t} \,)' \,]$ $= H_{t+h} \, P_{t+h|t} \, H_{t+h}' + \Sigma_n$.

EXAMPLE 7.1. As a simple example of the state-space model and associated Kalman filtering, consider a basic structural model from econometrics in which an observed vector series $\boldsymbol{Y}_t$ is viewed as the sum of unobserved trend and noise components. To be specific, assume that the observed process can be represented as

$$\boldsymbol{Y}_t = \boldsymbol{\mu}_t + \boldsymbol{N}_t, \qquad \text{where} \qquad \boldsymbol{\mu}_t = \boldsymbol{\mu}_{t-1} + \boldsymbol{a}_t,$$

so that $\boldsymbol{\mu}_t$ is a vector random walk process and $\boldsymbol{N}_t$ is an independent (white) noise process. This is a simple example of a time-invariant state-space model with $\Phi = I$ and $H = I$ in (7.1)–(7.2) and with the state vector $\mathbf{Z}_t = \boldsymbol{\mu}_t$ representing an underlying (unobservable) "trend or level" process (or "permanent" component). For this model, application of the Kalman filter (and associated smoothing algorithm discussed below) can be viewed as estimation of the underlying trend process $\boldsymbol{\mu}_t$ based on the observed process $\boldsymbol{Y}_t$. The Kalman filtering relations (7.3)–(7.6) for this basic model reduce to

$$\hat{\boldsymbol{\mu}}_{t|t} = \hat{\boldsymbol{\mu}}_{t-1|t-1} + K_t \, (\, \boldsymbol{Y}_t - \hat{\boldsymbol{\mu}}_{t-1|t-1} \,) = K_t \, \boldsymbol{Y}_t + (\, I - K_t \,) \, \hat{\boldsymbol{\mu}}_{t-1|t-1},$$

where the gain matrix is $K_t = P_{t|t-1} \, [\, P_{t|t-1} + \Sigma_n \,]^{-1}$, with

$$P_{t+1\,|\,t} = P_{t\,|\,t-1} - P_{t\,|\,t-1}\,[\,P_{t\,|\,t-1} + \Sigma_n\,]^{-1}\,P_{t\,|\,t-1} + \Sigma_a\,.$$

Then $\hat{\boldsymbol{\mu}}_{t\,|\,t}$ represents the current estimate of the trend component $\boldsymbol{\mu}_t$ given the observations $\boldsymbol{Y}_1, \ldots, \boldsymbol{Y}_t$ through time t. The steady-state solution to the Kalman filter relations is obtained as $t \to \infty$ for the matrix P ($P = \lim_{t\to\infty} P_{t+1\,|\,t}$) which satisfies $P = P - P\,[\,P + \Sigma_n\,]^{-1}\,P + \Sigma_a$, that is, $P\,[\,P + \Sigma_n\,]^{-1}\,P = \Sigma_a$, and the corresponding steady-state gain is $K = P\,[\,P + \Sigma_n\,]^{-1}$. More explicit details about the steady-state Kalman filtering results for this example will be presented later in Section 7.4.3.

We remark that, from a computational view, it may be better to modify the basic Kalman filtering recursions in (7.3)–(7.6) so that (7.4), (7.5b), and (7.6) are replaced by other equivalent recursive equations [see Anderson and Moore (1979, Chap. 6) for a general discussion]. These alternate algorithms include the information filter form, which gives an updating recursion for $P_{t\,|\,t-1}^{-1}$ rather than for $P_{t\,|\,t-1}$ in (7.5b)–(7.6), square-root algorithms (e.g., Morf and Kailath, 1975) which provide recursions for the square-root matrices of the covariance matrices $P_{t\,|\,t-1}$ and $P_{t\,|\,t}$ to ensure that the $P_{t\,|\,t-1}$ and $P_{t\,|\,t}$ remain non-negative definite throughout the recursions, and filter equations of the Chandrasekhar-type (for time-invariant coefficient models) as described by Morf, Sidhu, and Kailath (1974).

7.1.2 Smoothing Relations in the State-Variable Model

Another problem of interest within the state-variable model framework, particularly in applications to economics and business, is to obtain "smoothed" estimates of past values of the state vector $\mathbf{Z}_t$ given observations $\boldsymbol{Y}_1, \ldots, \boldsymbol{Y}_T$ through some fixed time T. One convenient method to obtain the desired estimates, known as the *fixed-interval smoothing* algorithm, makes use of the Kalman filter estimates $\hat{\mathbf{Z}}_{t\,|\,t}$ obtainable through (7.3)–(7.6). The smoothing algorithm produces the minimum MSE estimator (predictor) of the state value $\mathbf{Z}_t$ given observations through time T, $\hat{\mathbf{Z}}_{t\,|\,T} = E[\,\mathbf{Z}_t \mid \boldsymbol{Y}_1, \ldots, \boldsymbol{Y}_T\,]$. In general, define $\hat{\mathbf{Z}}_{t\,|\,n} = E[\,\mathbf{Z}_t \mid \boldsymbol{Y}_1, \ldots, \boldsymbol{Y}_n\,]$ and $P_{t\,|\,n} = E[\,(\mathbf{Z}_t - \hat{\mathbf{Z}}_{t\,|\,n})\,(\mathbf{Z}_t - \hat{\mathbf{Z}}_{t\,|\,n})'\,]$. We assume that the filtered estimates $\hat{\mathbf{Z}}_{t\,|\,t}$ and their error covariance matrices $P_{t\,|\,t}$, for $t = 1, \ldots, T$, have already been obtained through the Kalman filter equations. Then the optimal smoothed estimates are obtained by the (backward) recursive relations, in which the filtered estimate $\hat{\mathbf{Z}}_{t\,|\,t}$ is updated, as

$$\hat{\mathbf{Z}}_{t\,|\,T} = \hat{\mathbf{Z}}_{t\,|\,t} + A_t\,(\,\hat{\mathbf{Z}}_{t+1\,|\,T} - \hat{\mathbf{Z}}_{t+1\,|\,t}\,), \tag{7.7}$$

where

$$A_t = P_{t\,|\,t}\,\Phi_{t+1}'\,P_{t+1\,|\,t}^{-1} \equiv \mathrm{Cov}(\,\mathbf{Z}_t, \mathbf{Z}_{t+1} - \hat{\mathbf{Z}}_{t+1\,|\,t}\,)\,\{\,\mathrm{Cov}(\,\mathbf{Z}_{t+1} - \hat{\mathbf{Z}}_{t+1\,|\,t}\,)\,\}^{-1} \tag{7.8}$$

and

$$P_{t\,|\,T} = P_{t\,|\,t} - A_t\,(\,P_{t+1\,|\,t} - P_{t+1\,|\,T}\,)\,A_t'\,. \tag{7.9}$$

The result (7.7) is established from the following argument. First consider $\boldsymbol{u}_t = E\,[\,\mathbf{Z}_t \mid \boldsymbol{Y}_1, \ldots, \boldsymbol{Y}_t, \mathbf{Z}_{t+1} - \hat{\mathbf{Z}}_{t+1\,|\,t}, \boldsymbol{N}_{t+1}, \boldsymbol{a}_{t+2}, \boldsymbol{N}_{t+2}, \ldots, \boldsymbol{a}_T, \boldsymbol{N}_T\,]$. Then because $\{\,\boldsymbol{a}_{t+j}\,, j \geq 2\,\}$ and $\{\,\boldsymbol{N}_{t+j}\,, j \geq 1\,\}$ are independent of the other conditioning variables in the definition of $\boldsymbol{u}_t$ and are independent of $\mathbf{Z}_t$, we have $\boldsymbol{u}_t = \hat{\mathbf{Z}}_{t\,|\,t} + E\,[\,\mathbf{Z}_t \mid \mathbf{Z}_{t+1} - \hat{\mathbf{Z}}_{t+1\,|\,t}\,] = \hat{\mathbf{Z}}_{t\,|\,t} + A_t\,(\,\mathbf{Z}_{t+1} - \hat{\mathbf{Z}}_{t+1\,|\,t}\,)$ where A_t is given by (7.8). Thus, because the conditioning variables in $\boldsymbol{u}_t$ generate $\boldsymbol{Y}_1, \ldots, \boldsymbol{Y}_T$, it follows that

$$\hat{\mathbf{Z}}_{t\,|\,T} = E\,[\,\mathbf{Z}_t \mid \boldsymbol{Y}_1, \ldots, \boldsymbol{Y}_T\,]$$

$$= E\,[\,\boldsymbol{u}_t \mid \boldsymbol{Y}_1, \ldots, \boldsymbol{Y}_T\,] = \hat{\mathbf{Z}}_{t\,|\,t} + A_t\,(\,\hat{\mathbf{Z}}_{t+1\,|\,T} - \hat{\mathbf{Z}}_{t+1\,|\,t}\,)$$

as in (7.7). The relation (7.9) for the error covariance matrix follows from rather straightforward calculations. This derivation of the fixed-interval smoothing relations is given by Ansley and Kohn (1982).

Thus, it is seen from (7.7)–(7.9) that the optimal smoothed estimates $\hat{\mathbf{Z}}_{t\,|\,T}$ are obtained by first obtaining the filtered values $\hat{\mathbf{Z}}_{t\,|\,t}$ through the forward recursion of the Kalman filter relations, followed by the backward recursions of (7.7)–(7.9) for $t = T-1, \ldots, 1$. This type of smoothing procedure has applications for estimation of trend and seasonal components (seasonal adjustment) in economic time series, as will be discussed in Section 7.4. When smoothed estimates $\hat{\mathbf{Z}}_{t\,|\,T}$ are desired only at a fixed time point (or only a few fixed points), for example, in relation to problems that involve the estimation of isolated missing values in a time series, then an alternate "fixed-point" smoothing algorithm may be useful [e.g., see Anderson and Moore (1979) or Brockwell and Davis (1987)].

7.1.3 Innovations Form of State-Space Model and Steady-State for Time-Invariant Models

In general, it must be emphasized that a state-variable model representation is not unique, and there are many ways to express the basic state-space model (7.1)–(7.2). Specifically, if we transform the state vector $\mathbf{Z}_t$ into $\overline{\mathbf{Z}}_t = P\,\mathbf{Z}_t$, where P is an arbitrary nonsingular $r \times r$ matrix, then model (7.1)–(7.2) can be written in a similar form in terms of $\overline{\mathbf{Z}}_t$ with $\overline{\Phi}_t = P\,\Phi_t\,P^{-1}$, $\overline{H}_t = H_t\,P^{-1}$, and $\overline{\boldsymbol{a}}_t = P\,\boldsymbol{a}_t$. One particular form of the state-variable model, referred to as the innovations representation, is worth noting here. If we set $\mathbf{Z}_t^* = \hat{\mathbf{Z}}_{t\,|\,t-1}$ and $\varepsilon_t^* = \varepsilon_{t\,|\,t-1} = \boldsymbol{Y}_t - H_t\,\hat{\mathbf{Z}}_{t\,|\,t-1}$, then from (7.3) and (7.5) we have that $\mathbf{Z}_{t+1}^* = \Phi_{t+1}\,\mathbf{Z}_t^* + \Phi_{t+1}\,K_t\,\varepsilon_t^* \equiv \Phi_{t+1}\,\mathbf{Z}_t^* + B_t\,\varepsilon_t^*$ and $\boldsymbol{Y}_t = H_t\,\mathbf{Z}_t^* + \varepsilon_t^*$, which is also of the general form of a state-space model.

In the "stationary case" (i.e., time-invariant and stable case) of the state-space model, where $\Phi_t \equiv \Phi$ and $H_t \equiv H$ in (7.1)–(7.2) are constant matrices and Φ has all eigenvalues less than one in absolute value, we can obtain the steady-state form of the innovations representation by setting $\mathbf{Z}_t^* = E[\,\mathbf{Z}_t \mid \boldsymbol{Y}_{t-1}, \boldsymbol{Y}_{t-2}, \ldots\,]$, the projection of $\mathbf{Z}_t$ based on the infinite past. In this case, in the Kalman filter relations (7.4)–(7.6) the error covariance matrix $P_{t+1|t}$ approaches the "steady-state" matrix $P = \lim_{t\to\infty} P_{t+1|t}$ as $t \to \infty$ which satisfies the algebraic Riccati equation [e.g., see Harvey (1989, Sec. 3.3)]

$$P = \Phi P \Phi' - \Phi P H' [H P H' + \Sigma_n]^{-1} H P \Phi' + \Sigma_a . \tag{7.10}$$

Then, also, the Kalman gain matrix K_t in (7.4) approaches the "steady-state" matrix, $K_t \to K$ as $t \to \infty$, where $K = P H' [H P H' + \Sigma_n]^{-1}$, $\varepsilon_t^* = \varepsilon_{t|t-1} \to \varepsilon_t = \boldsymbol{Y}_t - H \mathbf{Z}_t^*$, and $\Sigma_{t|t-1} \to \Sigma = \text{Cov}(\varepsilon_t)$, where $\Sigma = H P H' + \Sigma_n$, as t increases. In this situation, the state-variable model can be expressed in the steady-state innovations or prediction error form as $\mathbf{Z}_{t+1}^* = \Phi \mathbf{Z}_t^* + \Phi K \varepsilon_t$ and $\boldsymbol{Y}_t = H \mathbf{Z}_t^* + \varepsilon_t$ (e.g., Hannan and Deistler, 1988, Sec. 1.2). Notice from this form it follows that any time-invariant stable state-space model has a steady-state solution $\{\boldsymbol{Y}_t\}$ which is a stationary process and which possesses a causal convergent infinite MA representation. Since $\mathbf{Z}_t^* = (I - \Phi B)^{-1} \Phi K \varepsilon_{t-1}$ is convergent, this infinite MA representation of the process $\boldsymbol{Y}_t$ is obtained from the above prediction error form as

$$\boldsymbol{Y}_t = H (I - \Phi B)^{-1} \Phi K \varepsilon_{t-1} + \varepsilon_t = H \sum_{j=1}^{\infty} \Phi^j K \varepsilon_{t-j} + \varepsilon_t \equiv \Psi(B) \varepsilon_t , \tag{7.11}$$

with infinite MA coefficient matrices $\Psi_j = H \Phi^j K$, $j \geq 1$.

7.2 State-Variable Representations of the Vector ARMA Model

7.2.1 A State-Space Form Based on the Prediction Space of Future Values

Consider the vector ARMA(p,q) model given by $\boldsymbol{Y}_t - \sum_{j=1}^{p} \Phi_j \boldsymbol{Y}_{t-j} = \varepsilon_t - \sum_{j=1}^{q} \Theta_j \varepsilon_{t-j}$. As in Section 2.5, let us define the predictors $\hat{\boldsymbol{Y}}_t(j) = E(\boldsymbol{Y}_{t+j} \mid \boldsymbol{Y}_t, \boldsymbol{Y}_{t-1}, \ldots)$ for $j = 0, 1, \ldots, r-1$, with $r = \max(p, q+1)$. Of course, we have $\hat{\boldsymbol{Y}}_t(0) = \boldsymbol{Y}_t$. We may recall from the infinite MA representation for forecasts that the following updating equations hold: $\hat{\boldsymbol{Y}}_t(j-1) = \hat{\boldsymbol{Y}}_{t-1}(j) + \Psi_{j-1} \varepsilon_t$, $j = 1, 2, \ldots, r-1$. Also for $j = r > q$ we recall that $\hat{\boldsymbol{Y}}_t(j-1) = \hat{\boldsymbol{Y}}_{t-1}(j) + \Psi_{j-1} \varepsilon_t = \sum_{i=1}^{p} \Phi_i \hat{\boldsymbol{Y}}_{t-1}(j-i) + \Psi_{j-1} \varepsilon_t$. So let us define the "state" vector $\mathbf{Z}_t$ with r vector components as $\mathbf{Z}_t = (\hat{\boldsymbol{Y}}_t(0)', \hat{\boldsymbol{Y}}_t(1)', \ldots, \hat{\boldsymbol{Y}}_t(r-1)')'$. Then from the above relations we have that $\mathbf{Z}_t$ satisfies the state-space (transition) equations

$$\mathbf{Z}_t = \begin{bmatrix} 0 & I & 0 & . & . & 0 \\ 0 & 0 & I & . & . & 0 \\ . & . & . & . & . & . \\ 0 & 0 & . & . & . & I \\ \Phi_r & \Phi_{r-1} & . & . & . & \Phi_1 \end{bmatrix} \mathbf{Z}_{t-1} + \begin{bmatrix} I \\ \Psi_1 \\ . \\ . \\ \Psi_{r-1} \end{bmatrix} \varepsilon_t , \tag{7.12}$$

where $\Phi_i = 0$ if $i > p$. So we have $\mathbf{Z}_t = \Phi\, \mathbf{Z}_{t-1} + \Psi\, \varepsilon_t$, together with the observation equation $\boldsymbol{Y}_t^* = \boldsymbol{Y}_t + \boldsymbol{N}_t = [\, I, 0, \ldots, 0\,]\, \mathbf{Z}_t + \boldsymbol{N}_t = H\, \mathbf{Z}_t + \boldsymbol{N}_t$, where the vector noise $\boldsymbol{N}_t$ would be present only if the process $\boldsymbol{Y}_t$ is observed subject to noise; otherwise we simply have $\boldsymbol{Y}_t = \boldsymbol{Y}_t^* = H\, \mathbf{Z}_t$. (For convenience of presentation, in the remainder of this section we will assume the case where the additional white noise $\boldsymbol{N}_t$ is not present.) This state-space representation for the process $\{\boldsymbol{Y}_t\}$ is exactly of the form discussed previously in Section 7.1 regarding the Kalman filter. Thus, the Kalman filter equations can directly be applied to this representation to produce a set of (one-step) predicted values $\tilde{\boldsymbol{Y}}_{t-1}(1) = E[\, \boldsymbol{Y}_t \mid \underset{\sim}{\boldsymbol{Y}}_{t-1}, \ldots, \boldsymbol{Y}_1\,]$ and their associated error covariance matrices $\Sigma_{t|t-1} = E[\, (\boldsymbol{Y}_t - \tilde{\boldsymbol{Y}}_{t-1}(1))\, (\boldsymbol{Y}_t - \tilde{\boldsymbol{Y}}_{t-1}(1))'\,]$, based on the finite set of past data $\boldsymbol{Y}_1, \boldsymbol{Y}_2, \ldots, \boldsymbol{Y}_{t-1}$. Hence, exact finite sample forecasts and their error covariance matrices, for all longer lead times, can be generated through the state-space approach. [The "steady-state" values of the Kalman filtering lead l forecast error covariance matrices, obtained as t increases, will equal the usual expressions given previously in (2.21) of Section 2.5.2, $\Sigma(l) = \sum_{j=0}^{l-1} \Psi_j\, \Sigma\, \Psi_j'$. That is, $\Sigma_{t+l|t}$ approaches $\Sigma(l)$ as $t \to \infty$.]

7.2.2 Exact Likelihood Function Through the State-Variable Approach

Given a set of T observations $\boldsymbol{Y}_1, \boldsymbol{Y}_2, \ldots, \boldsymbol{Y}_T$ from the vector ARMA(p,q) process, $\Phi(B)\, \boldsymbol{Y}_t = \Theta(B)\, \varepsilon_t$, the Kalman filtering approach can be used as a convenient method for evaluation of the exact likelihood function recursively. That is, if $f(\boldsymbol{Y}_1, \ldots, \boldsymbol{Y}_T)$ denotes the joint p.d.f. of $\mathbf{y} = (\boldsymbol{Y}_1', \boldsymbol{Y}_2', \ldots, \boldsymbol{Y}_T')'$, then

$$f(\boldsymbol{Y}_1, \ldots, \boldsymbol{Y}_T) = \prod_{t=1}^{T} f(\boldsymbol{Y}_t \mid \boldsymbol{Y}_{t-1}, \ldots, \boldsymbol{Y}_1),$$

where $f(\boldsymbol{Y}_t \mid \boldsymbol{Y}_{t-1}, \ldots, \boldsymbol{Y}_1)$ denotes the conditional p.d.f. of $\boldsymbol{Y}_t$, given $\boldsymbol{Y}_{t-1}, \ldots, \boldsymbol{Y}_1$. Now under normality, the conditional distribution of $\boldsymbol{Y}_t$, given $\underset{\sim}{\boldsymbol{Y}}_{t-1}, \ldots, \boldsymbol{Y}_1$, is multivariate normal with (conditional) mean vector $\tilde{\boldsymbol{Y}}_{t-1}(1) = E[\, \boldsymbol{Y}_t \mid \underset{\sim}{\boldsymbol{Y}}_{t-1}, \ldots, \boldsymbol{Y}_1\,]$, and (conditional) covariance matrix $\Sigma_{t|t-1} = E[\, (\boldsymbol{Y}_t - \tilde{\boldsymbol{Y}}_{t-1}(1))\, (\boldsymbol{Y}_t - \tilde{\boldsymbol{Y}}_{t-1}(1))'\,]$. Hence, the joint p.d.f. of $\mathbf{y}$, that is, the *exact likelihood function*, can conveniently be expressed as

$$f(\boldsymbol{Y}_1, \ldots, \boldsymbol{Y}_T) = \Big(\prod_{t=1}^{T} |\Sigma_{t|t-1}|^{-1/2}\Big)$$
$$\times \exp\Big[-(1/2) \sum_{t=1}^{T} (\boldsymbol{Y}_t - \tilde{\boldsymbol{Y}}_{t-1}(1))'\, \Sigma_{t|t-1}^{-1}\, (\boldsymbol{Y}_t - \tilde{\boldsymbol{Y}}_{t-1}(1))\Big]$$

where the quantities $\tilde{\boldsymbol{Y}}_{t-1}(1) = \hat{\boldsymbol{Y}}_{t|t-1}$ and $\Sigma_{t|t-1}$ are determined recursively from the Kalman filter procedure, noting the starting values $\tilde{\boldsymbol{Y}}_0(1) = E(\boldsymbol{Y}_1) = 0$ and $\Sigma_{1|0} = E(\boldsymbol{Y}_1\, \boldsymbol{Y}_1') = \Gamma(0)$ in the stationary case. Specifically, based on the state-space form (7.12), $\mathbf{Z}_t = \Phi\, \mathbf{Z}_{t-1} + \Psi\, \varepsilon_t$, of the vector ARMA model, the updating equations

$$\hat{\mathbf{Z}}_{t|t} = \hat{\mathbf{Z}}_{t|t-1} + K_t \, \varepsilon_{t|t-1} \tag{7.13a}$$

with

$$K_t = P_{t|t-1} H' \, [H P_{t|t-1} H']^{-1} \equiv P_{t|t-1} H' \Sigma_{t|t-1}^{-1}, \tag{7.13b}$$

and the prediction equations

$$\hat{\mathbf{Z}}_{t|t-1} = \Phi \, \hat{\mathbf{Z}}_{t-1|t-1} , \qquad P_{t|t-1} = \Phi \, P_{t-1|t-1} \, \Phi' + \Psi \, \Sigma \, \Psi' \tag{7.14a-b}$$

with $P_{t|t} = [I - K_t H] P_{t|t-1}$ are used for $t = 1, 2, \ldots, T$. From these we obtain $\boldsymbol{Y}_{t|t-1} = H \hat{\mathbf{Z}}_{t|t-1}$ and $\varepsilon_{t|t-1} = \boldsymbol{Y}_t - H \hat{\mathbf{Z}}_{t|t-1} \equiv \boldsymbol{Y}_t - \hat{\boldsymbol{Y}}_{t|t-1}$, and also $\Sigma_{t|t-1} = H \, P_{t|t-1} H'$ to construct the likelihood function recursively.

For the initialization of the above Kalman filtering relations, the unconditional mean vector $\hat{\mathbf{Z}}_{0|0} = E(\mathbf{Z}_0) = 0$ and covariance matrix $P_{0|0} = \mathrm{Cov}(\mathbf{Z}_0) \equiv P_*$ are used in the stationary vector ARMA case. Since the state vector $\mathbf{Z}_t$ follows the stationary AR(1) model $\mathbf{Z}_t = \Phi \, \mathbf{Z}_{t-1} + \Psi \, \varepsilon_t$, its covariance matrix $P_* = \mathrm{Cov}(\mathbf{Z}_t)$ satisfies $P_* = \Phi \, P_* \, \Phi' + \Psi \, \Sigma \, \Psi'$. This can be solved for P_* by the vectorizing method, as first discussed in Section 2.2.3, as $\mathrm{vec}(P_*) = [I - (\Phi \otimes \Phi)]^{-1} \, \mathrm{vec}\{ \Psi \, \Sigma \, \Psi' \}$. To provide a more explicit form for P_*, note that $P_* = \mathrm{Cov}(\mathbf{Z}_t)$ with $\mathbf{Z}_t = (\boldsymbol{Y}_t', \hat{\boldsymbol{Y}}_t(1)', \ldots, \hat{\boldsymbol{Y}}_t(r-1)')'$. Then we use the representation $\boldsymbol{Y}_{t+j} = \hat{\boldsymbol{Y}}_t(j) + \boldsymbol{e}_t(j)$, where $\boldsymbol{e}_t(j) = \sum_{i=0}^{j-1} \Psi_i \, \varepsilon_{t+j-i}$ is the j-step ahead forecast error from Section 2.5, and note that $\boldsymbol{e}_t(j)$ is independent of $\hat{\boldsymbol{Y}}_t(l)$ for any $l \geq 0$. Thus, we find that

$$\Gamma(l-j) = \mathrm{Cov}(\boldsymbol{Y}_{t+j}, \boldsymbol{Y}_{t+l}) = \mathrm{Cov}(\hat{\boldsymbol{Y}}_t(j), \hat{\boldsymbol{Y}}_t(l)) + \mathrm{Cov}(\boldsymbol{e}_t(j), \boldsymbol{e}_t(l)),$$

so that

$$\mathrm{Cov}(\hat{\boldsymbol{Y}}_t(j), \hat{\boldsymbol{Y}}_t(l)) = \Gamma(l-j) - \sum_{i=0}^{j-1} \Psi_i \, \Sigma \, \Psi_{i+l-j}' \qquad \text{for} \qquad 1 \leq j \leq l \leq r-1$$

together with $\mathrm{Cov}(\boldsymbol{Y}_t, \hat{\boldsymbol{Y}}_t(l)) = \mathrm{Cov}(\boldsymbol{Y}_t, \boldsymbol{Y}_{t+l}) = \Gamma(l)$ for $l = 0, 1, \ldots, r-1$ provide the required expressions for the elements of the initial state covariance matrix P_*. In matrix form, we write the equations $\boldsymbol{Y}_{t+j} = \hat{\boldsymbol{Y}}_t(j) + \boldsymbol{e}_t(j) = \hat{\boldsymbol{Y}}_t(j) + \sum_{i=0}^{j-1} \Psi_i \, \varepsilon_{t+j-i}$, $j = 1, \ldots, r-1$, as

$$\mathbf{Y}_t = \mathbf{Z}_t + G_\psi \, \mathbf{e}_{t+1} \equiv \mathbf{Z}_t + \begin{bmatrix} 0 & 0 & . & . & . & 0 \\ I & 0 & . & . & . & 0 \\ \Psi_1 & I & . & . & . & 0 \\ . & . & . & . & . & . \\ \Psi_{r-3} & . & . & . & I & 0 \\ \Psi_{r-2} & \Psi_{r-3} & . & . & \Psi_1 & I \end{bmatrix} \mathbf{e}_{t+1} ,$$

where $\mathbf{Y}_t = (\boldsymbol{Y}_t', \boldsymbol{Y}_{t+1}', \ldots, \boldsymbol{Y}_{t+r-1}')'$, $\mathbf{e}_{t+1} = (\varepsilon_{t+1}', \ldots, \varepsilon_{t+r-1}')'$, and G_ψ is a $kr \times k(r-1)$ matrix. Thus, from the above representation of the state vector as $\mathbf{Z}_t = \mathbf{Y}_t - G_\psi \, \mathbf{e}_{t+1}$, the covariance matrix of the state vector is explicitly given by

$$
\begin{aligned}
P_* = \mathrm{Cov}(\mathbf{Z}_t) &= \mathrm{Cov}(\mathbf{Y}_t) + G_\psi\, \mathrm{Cov}(\mathbf{e}_{t+1})\, G_\psi' \\
&\quad - G_\psi\, \mathrm{Cov}(\mathbf{e}_{t+1}, \mathbf{Y}_t) - \mathrm{Cov}(\mathbf{Y}_t, \mathbf{e}_{t+1})\, G_\psi' \\
&= \Gamma_r - G_\psi\, (I_{r-1} \otimes \Sigma)\, G_\psi' ,
\end{aligned}
\tag{7.15}
$$

since $\mathrm{Cov}(\mathbf{e}_{t+1}) = I_{r-1} \otimes \Sigma$ and $\mathrm{Cov}(\mathbf{Y}_t, \mathbf{e}_{t+1}) = G_\psi\, (I_{r-1} \otimes \Sigma)$ is easily verified, and where $\Gamma_r = \mathrm{Cov}(\mathbf{Y}_t)$ is the $kr \times kr$ matrix with $\Gamma(j-i) = E(Y_{t+i-1}\, Y_{t+j-1}')$ in the (i, j)th block. Notice that the matrix $P_* = \Gamma_r - G_\psi\, (I_{r-1} \otimes \Sigma)\, G_\psi'$ in (7.15) is also $P_* = \mathrm{Cov}(\mathbf{Y}_t \mid \mathbf{e}_{t+1})$ and is essentially the same as the matrix given after equation (5.25) in Section 5.3, $K = \Gamma_p - \Psi'\, (I_q \otimes \Sigma)\, \Psi = \mathrm{Cov}(\mathbf{y}_* \mid \mathbf{e}_*)$ in the notation of Section 5.3. The matrix K is fundamental in the "conventional" procedure for exact likelihood construction presented in Section 5.3, connected with the specification of the covariance matrix $\Omega = \mathrm{Cov}(\mathbf{a}_*)$ of the presample initial values $\mathbf{a}_* = (\mathbf{y}_*', \mathbf{e}_*')'$. Hence, we see that there is a close relationship between the determination of the initial state covariance matrix P_* in the state-space model approach and the specification of the covariance matrix of initial values in the "conventional" approach of Section 5.3.

For nonstationary vector ARIMA processes, additional assumptions need to be specified concerning initialization of the process to determine appropriate starting values for the Kalman filtering calculations, e.g., Ansley and Kohn (1985a), Kohn and Ansley (1986), and Bell and Hillmer (1987, 1991). The approach taken by these authors involves transformation of the data to eliminate the dependence on initial conditions which may not be completely specified and use of a resultant modified Kalman filter algorithm. DeJong (1988, 1991) has also extended the basic Kalman filtering relations, leading to a modified "diffuse" Kalman filtering algorithm, to handle such nonstationary situations where conditions specified for initial values are purposefully vague (i.e., the initial state is diffuse).

The above technique for construction of the exact likelihood function in vector ARMA models using the state-space form and associated Kalman filtering calculations has been presented by several authors, including Ansley and Kohn (1983), Solo (1984b), and Shea (1987). Cooper and Wood (1981) presented some specific details for maximum likelihood estimation of vector ARMA models in the state-space representation using the method of scoring, while Shumway and Stoffer (1982) used an EM algorithm approach for maximum likelihood estimation of parameters in the state-space model (with the unobserved state variables $\mathbf{Z}_0, \mathbf{Z}_1, \ldots, \mathbf{Z}_T$ treated as missing values). The state-space approach to exact likelihood estimation has also been shown to be quite useful in dealing with estimation problems for ARMA models when some values of the series are not observed, that is, there are missing values among the sequence Y_t (Jones, 1980; Ansley and Kohn, 1983; Wincek and Reinsel, 1986), as will be discussed in Section 7.3.

EXAMPLE 7.2. As an illustration of the use of Kalman filter methods for likelihood calculations, consider the vector ARMA(1,1) model $\boldsymbol{Y}_t = \Phi_1 \boldsymbol{Y}_{t-1} + \varepsilon_t - \Theta_1 \varepsilon_{t-1}$. The state-space representation (7.12) for this model is

$$\mathbf{Z}_t = \begin{bmatrix} 0 & I \\ 0 & \Phi_1 \end{bmatrix} \mathbf{Z}_{t-1} + \begin{bmatrix} I \\ \Psi_1 \end{bmatrix} \varepsilon_t ,$$

where $\mathbf{Z}_t = (\boldsymbol{Y}_t', \hat{\boldsymbol{Y}}_t(1)')'$ and $\Psi_1 = \Phi_1 - \Theta_1$ from Section 2.3.3, with $\boldsymbol{Y}_t = H \mathbf{Z}_t = [I, 0] \mathbf{Z}_t$. For this model the filtering equations (7.13a) and (7.14a), $\mathbf{Z}_{t|t} = \Phi \mathbf{Z}_{t-1|t-1} + K_t \varepsilon_{t|t-1}$, reduce to $\boldsymbol{Y}_t = \hat{\boldsymbol{Y}}_{t|t-1} + \varepsilon_{t|t-1}$ and

$$\hat{\boldsymbol{Y}}_{t+1|t} = \Phi_1 \hat{\boldsymbol{Y}}_{t|t-1} + [0, I] K_t \varepsilon_{t|t-1} = \Phi_1 \hat{\boldsymbol{Y}}_{t|t-1} + (\Phi_1 - \Theta_1 \Sigma \Sigma_{t|t-1}^{-1}) \varepsilon_{t|t-1}$$

$$= \Phi_1 \boldsymbol{Y}_t - \Theta_1 \Sigma \Sigma_{t|t-1}^{-1} \varepsilon_{t|t-1} ,$$

where $\Sigma = \text{Cov}(\varepsilon_t)$, using (7.13b) and (7.14b) to establish that

$$[0, I] K_t = [0, I] \{ \Phi P_{t-1|t-1} \Phi' + \Psi \Sigma \Psi' \} H' \Sigma_{t|t-1}^{-1}$$

$$= \{ \Phi_1 (\Sigma_{t|t-1} - \Sigma) + \Psi_1 \Sigma \} \Sigma_{t|t-1}^{-1}$$

$$= \Phi_1 - \Theta_1 \Sigma \Sigma_{t|t-1}^{-1} .$$

Hence, the innovations are explicitly obtained as

$$\varepsilon_{t|t-1} = \boldsymbol{Y}_t - \hat{\boldsymbol{Y}}_{t|t-1} = \boldsymbol{Y}_t - \Phi_1 \boldsymbol{Y}_{t-1} + \Theta_1 \Sigma \Sigma_{t-1|t-2}^{-1} \varepsilon_{t-1|t-2} , \qquad t = 2, \ldots , T ,$$

with $\hat{\boldsymbol{Y}}_{1|0} = 0$ and $\varepsilon_{1|0} = \boldsymbol{Y}_1$. In addition, from (7.14b) and the relation $P_{t|t} = [I - K_t H] P_{t|t-1}$, we can find that the innovations covariance matrix $\Sigma_{t|t-1} = \text{Cov}(\varepsilon_{t|t-1})$ satisfies

$$\Sigma_{t|t-1} = H P_{t|t-1} H' = [0, I] P_{t-1|t-1} [0, I]' + \Sigma$$

$$= \Sigma + \Theta_1 \Sigma \Theta_1' - \Theta_1 \Sigma \Sigma_{t-1|t-2}^{-1} \Sigma \Theta_1' ,$$

for $t > 1$, with $\Sigma_{1|0} = \Gamma(0)$ which can be determined explicitly from the discussion of Section 2.3.3. Note that $\Sigma_{t|t-1}$ approaches Σ as t increases, and, hence, from this we see that the Kalman gain matrix $K_t = [I, (\Phi_1 - \Theta_1 \Sigma \Sigma_{t|t-1}^{-1})']'$ approaches the steady-state value $\Psi = [I, (\Phi_1 - \Theta_1)']' \equiv [I, \Psi_1']'$ and that the coefficient matrix $\Theta_{1,t} \equiv \Theta_1 \Sigma \Sigma_{t|t-1}^{-1}$ in the above ARMA(1,1) model prediction relations, $\hat{\boldsymbol{Y}}_{t+1|t} = \Phi_1 \boldsymbol{Y}_t - \Theta_{1,t} \varepsilon_{t|t-1}$, approaches Θ_1 as well. Also note that these prediction relations give the innovations expressed as $\varepsilon_{t+1|t} = \boldsymbol{Y}_{t+1} - \hat{\boldsymbol{Y}}_{t+1|t} = \boldsymbol{Y}_{t+1} - \Phi_1 \boldsymbol{Y}_t + \Theta_{1,t} \varepsilon_{t|t-1}$, which corresponds to the relation (5.34) of Section 5.4.1 for the innovations algorithm in a vector ARMA(1,1) model.

The exact likelihood function calculated using the Kalman filtering approach via the recursive equations (7.13)–(7.14) can be maximized by using numerical optimization algorithms. These typically require some form of first partial derivatives of the log-likelihood, and it would be more satisfactory numerically to obtain analytical derivatives. From the form of the exact log-likelihood presented in this section, we see that what is needed are partial derivatives of the one-step predictions $\tilde{\boldsymbol{Y}}_{t-1}(1) \equiv \hat{\boldsymbol{Y}}_{t|t-1} = H\,\hat{\mathbf{Z}}_{t|t-1}$ and of their error covariance matrices $\Sigma_{t|t-1}$ with respect to each unknown parameter in the model for each $t = 1, \ldots, T$. These derivatives can be obtained by an additional pass through the Kalman filter type of recursive calculations for each unknown parameter in the model, where the recursion for derivatives is constructed, over $t = 1, \ldots, T$, by differentiation of the Kalman filtering equations [the updating and prediction equations (7.13)–(7.14)] with respect to each unknown model parameter. This same fundamental approach to recursively obtain the gradient vector of first partial derivatives of the exact log-likelihood function also applies when the related innovations form recursive algorithm, as discussed previously in Section 5.4, is used in the calculation of the exact likelihood. Explicit details on this method of recursive calculation of analytical derivatives of the log-likelihood function have been presented by Ansley and Kohn (1985b) and Wincek and Reinsel (1986) for the univariate ARMA model [see also Zhou (1992) for the vector ARMA model case].

7.2.3 Alternate State-Space Forms for the Vector ARMA Model

The fact, noted earlier in Section 7.1.3, that there are many different state-space representations is also true, in particular, for the state-space representation of the vector ARMA model. For example, for a pure AR(p) process an alternate state-space form to (7.12) that is more convenient to use in practice is the representation given in equation (5.39) of Section 5.6, with state vector $\overline{\mathbf{Z}}_t = (\boldsymbol{Y}_t', \boldsymbol{Y}_{t-1}', \ldots, \boldsymbol{Y}_{t-p+1}')'$. One particular popular state-space form of the general vector ARMA(p,q) model, in addition to (7.12), that is worth mentioning is the representation that can be obtained by a specific linear transformation applied to (7.12) using the matrix

$$P = \begin{bmatrix} I & 0 & 0 & \cdot & \cdot & \cdot & 0 \\ -\Phi_1 & I & 0 & \cdot & \cdot & \cdot & 0 \\ -\Phi_2 & -\Phi_1 & I & \cdot & \cdot & \cdot & 0 \\ \cdot & \cdot & \cdot & \cdot & \cdot & \cdot & \cdot \\ -\Phi_{r-2} & -\Phi_{r-3} & \cdot & \cdot & \cdot & I & 0 \\ -\Phi_{r-1} & -\Phi_{r-2} & -\Phi_{r-3} & \cdot & \cdot & -\Phi_1 & I \end{bmatrix},$$

with $\overline{\Phi} = P\,\Phi\,P^{-1}$, $\overline{\Psi} = P\,\Psi$, $\overline{H} = H\,P^{-1} = H = [\,I, 0, \ldots, 0\,]$, and state vector $\overline{\mathbf{Z}}_t = P\,\mathbf{Z}_t$. This leads to the state-space representation [equivalent to (7.12)] given by

$$\bar{\mathbf{Z}}_t = \bar{\Phi}\,\bar{\mathbf{Z}}_{t-1} + \bar{\Psi}\,\varepsilon_t = \begin{bmatrix} \Phi_1 & I & 0 & . & . & 0 \\ \Phi_2 & 0 & I & . & . & 0 \\ . & . & . & . & . & . \\ \Phi_{r-1} & 0 & . & . & . & I \\ \Phi_r & 0 & 0 & . & . & 0 \end{bmatrix} \bar{\mathbf{Z}}_{t-1} + \begin{bmatrix} I \\ -\Theta_1 \\ . \\ . \\ -\Theta_{r-1} \end{bmatrix} \varepsilon_t\,, \quad (7.16)$$

with $\boldsymbol{Y}_t = \bar{H}\,\bar{\mathbf{Z}}_t$, and the state vector $\bar{\mathbf{Z}}_t = P\,\mathbf{Z}_t = (\,\bar{Z}'_{1,t}, \ldots, \bar{Z}'_{r,t}\,)'$ has its first $k \times 1$ block component equal to $\bar{Z}_{1,t} = \boldsymbol{Y}_t$ and the remaining block components equal to $\bar{Z}_{j+1,t} = \hat{\boldsymbol{Y}}_t(j) - \sum_{i=1}^{j} \Phi_i\,\hat{\boldsymbol{Y}}_t(j-i)$, $j = 1, \ldots, r-1$. This state-space form, which we might call the "left-companion matrix form", has been used in particular for exact likelihood evaluations by Harvey and Phillips (1979), Gardner, Harvey, and Phillips (1980), Pearlman (1980), Melard (1984), and Harvey and Pierse (1984) for the univariate ARMA model, and by Ansley and Kohn (1983), Shea (1987, 1989), and Mittnik (1991) in the multivariate ARMA case. Note from the relation $\mathbf{Z}_t = \mathbf{Y}_t - G_\psi\,\mathbf{e}_{t+1}$ satisfied by the state vector in the representation (7.12), and used in (7.15), that the state vector $\bar{\mathbf{Z}}_t$ in the alternate representation (7.16) satisfies $\bar{\mathbf{Z}}_t = P\,\mathbf{Z}_t = P\,\mathbf{Y}_t - P\,G_\psi\,\mathbf{e}_{t+1} \equiv P\,\mathbf{Y}_t - G_\theta\,\mathbf{e}_{t+1}$, where $G_\theta = P\,G_\psi$ has exactly the same form as the matrix G_ψ in (7.15) but with the matrices $-\Theta_i$ in place of the Ψ_i. That is, the components of the state vector $\bar{\mathbf{Z}}_t$ satisfy

$$\bar{Z}_{j+1,t} = \boldsymbol{Y}_{t+j} - \sum_{i=1}^{j} \Phi_i\,\boldsymbol{Y}_{t+j-i} - \varepsilon_{t+j} + \sum_{i=1}^{j-1} \Theta_i\,\varepsilon_{t+j-i}$$

($\bar{Z}_{j+1,t} \equiv \sum_{i=j+1}^{r} \Phi_i\,\boldsymbol{Y}_{t+j-i} - \sum_{i=j}^{r-1} \Theta_i\,\varepsilon_{t+j-i}$, also), for $j = 1, \ldots, r-1$. Note that this fact, as well as the specific form of $\bar{\Psi} = P\,\Psi = (\,I, -\Theta'_1, \ldots, -\Theta'_{r-1}\,)'$ obtained above in (7.16) and of $G_\theta = P\,G_\psi$, uses the relations $\Psi_j - \sum_{i=1}^{p} \Phi_i\,\Psi_{j-i} = -\Theta_j$ from (2.12) of Section 2.3.1. Hence, when using the state-space representation (7.16) for likelihood calculations, the initial state covariance matrix $\bar{P}_* = \mathrm{Cov}(\,\bar{\mathbf{Z}}_t\,)$ can be obtained in a similar manner to (7.15) explicitly as

$$\bar{P}_* = P\,P_*\,P' = P\,\Gamma_r\,P' - P\,G_\psi\,(\,I_{r-1} \otimes \Sigma\,)\,G'_\psi\,P'$$

$$\equiv P\,\Gamma_r\,P' - G_\theta\,(\,I_{r-1} \otimes \Sigma\,)\,G'_\theta\,.$$

Remark (*Kalman filtering in the vector ARMA model innovations form*). When state-space likelihood evaluations for the vector ARMA model are performed using the model representation (7.16), then updating and prediction equations of the same form as in (7.13)–(7.14) apply,

$$\bar{\mathbf{Z}}_{t\,|\,t} = \bar{\mathbf{Z}}_{t\,|\,t-1} + \bar{K}_t\,\varepsilon_{t\,|\,t-1} \quad (7.13a')$$

with

$$\overline{K}_t = \overline{P}_{t|t-1}\,\overline{H}'\,[\,\overline{H}\,\overline{P}_{t|t-1}\,\overline{H}'\,]^{-1} \equiv \overline{P}_{t|t-1}\,\overline{H}'\,\Sigma_{t|t-1}^{-1}\,, \tag{7.13b$'$}$$

$$\overline{\mathbf{Z}}_{t|t-1} = \overline{\Phi}\,\overline{\mathbf{Z}}_{t-1|t-1}\,, \qquad \overline{P}_{t|t-1} = \overline{\Phi}\,\overline{P}_{t-1|t-1}\,\overline{\Phi}' + \overline{\Psi}\,\Sigma\,\overline{\Psi}'. \tag{7.14a$'$-b$'$}$$

From the relations between the model representations (7.12) and (7.16), that is, $\overline{\Phi} = P\,\Phi\,P^{-1}$, $\overline{\Psi} = P\,\Psi$, and $\overline{H} = H\,P^{-1} = H$, it follows that the relations $\overline{\mathbf{Z}}_{t|t} = P\,\hat{\mathbf{Z}}_{t|t}$, $\overline{K}_t = P\,K_t$, $\overline{P}_{t|t} = P\,P_{t|t}\,P'$, and so on in the Kalman filtering equations exist between the two representations. Define the $k \times kr$ matrices $H_i = [\,0, \ldots, I, \ldots, 0\,]$, $i = 1, \ldots, r$, with $H_1 \equiv H$, where H_i has the $k \times k$ identity matrix in the ith block and zeros otherwise. Then from (7.13a′) and (7.14a′) we have $\overline{\mathbf{Z}}_{t|t} = \overline{\Phi}\,\overline{\mathbf{Z}}_{t-1|t-1} + \overline{K}_t\,\varepsilon_{t|t-1}$. Making use of this relation successively for times t through $t-r+1$, and using the easily verified relation $H_i\,\overline{\Phi} = \Phi_i\,H + H_{i+1}$, $i = 1, \ldots, r-1$, $H_r\,\overline{\Phi} = \Phi_r\,H$, at each step, we obtain

$$\boldsymbol{Y}_t = H\,\overline{\mathbf{Z}}_{t|t} = H\,\overline{\Phi}\,\overline{\mathbf{Z}}_{t-1|t-1} + H\,\overline{K}_t\,\varepsilon_{t|t-1} = \Phi_1\,\boldsymbol{Y}_{t-1} + H_2\,\overline{\mathbf{Z}}_{t-1|t-1} + \varepsilon_{t|t-1}$$

$$= \Phi_1\,\boldsymbol{Y}_{t-1} + H_2\,(\,\overline{\Phi}\,\overline{\mathbf{Z}}_{t-2|t-2} + \overline{K}_{t-1}\,\varepsilon_{t-1|t-2}\,) + \varepsilon_{t|t-1} = \ldots$$

$$= \sum_{i=1}^{r} \Phi_i\,\boldsymbol{Y}_{t-i} + \varepsilon_{t|t-1} + \sum_{i=1}^{r-1} H_{i+1}\,\overline{K}_{t-i}\,\varepsilon_{t-i|t-i-1} \equiv \hat{\boldsymbol{Y}}_{t|t-1} + \varepsilon_{t|t-1}\,, \tag{7.17}$$

where $\hat{\boldsymbol{Y}}_{t|t-1} = \sum_{i=1}^{r} \Phi_i\,\boldsymbol{Y}_{t-i} + \sum_{i=1}^{r-1} H_{i+1}\,\overline{K}_{t-i}\,\varepsilon_{t-i|t-i-1}$. So, we find that (7.17) provides the ARMA model representation for the innovations obtained through the state-space Kalman filtering equations as

$$\varepsilon_{t|t-1} = \boldsymbol{Y}_t - \sum_{i=1}^{r} \Phi_i\,\boldsymbol{Y}_{t-i} + \sum_{i=1}^{r-1} \Theta_{i,t-1}\,\varepsilon_{t-i|t-i-1}\,, \tag{7.17$'$}$$

where $\Theta_{i,t-1} = -\,H_{i+1}\,\overline{K}_{t-i} \equiv \overline{K}_{i+1,t-i}$, $i = 1, \ldots, r-1$, is the $(i+1)$st block component of $\overline{K}_{t-i}$. Note that (7.17′) corresponds to the (unique) ARMA model innovations algorithm representation for the $\varepsilon_{t|t-1}$ given in equation (5.34) of Section 5.4.1. These relations also imply that $\overline{K}_{i+1,t-i} \equiv \Theta_{i,t-1} = 0$ if $i > q$, for $t > r$, which can occur when $p > q+1$, and note that the $\Theta_{i,t-1}$ can also be expressed directly in terms of the elements of K_t in the Kalman filtering equations (7.13)–(7.14) that use the representation (7.12), because of the relation $\overline{K}_t = P\,K_t$. Hence, as already illustrated in Example 7.2 for the ARMA(1,1) model, the Kalman filtering calculations for the likelihood of the ARMA model are closely related to the innovations algorithm calculations presented in Section 5.4.1; the two methods both calculate the innovations $\varepsilon_{t|t-1}$ and their corresponding covariance matrices $\Sigma_{t|t-1}$ but using different numerical recursion algorithms. Actually, from a numerical computation point of view, the Kalman filtering recursions to obtain the ARMA model innovations are generally less efficient than the innovations algorithm recursions, unless the covariance matrix filtering and updating equations such as in (7.13b′) and (7.14b′) are replaced by the more efficient Chandrasekhar-type equation recursions, as has been done by Pearlman (1980), Melard (1984), and Shea (1987, 1989).

In fact, concerning alternate state-space forms, in the vector ARMA model setting it may be possible to obtain a state-space representation with a state vector $\mathbf{Z}_t^*$ of smaller dimension than $k\,r$ as given in (7.12). As a simple illustration of this, note that the vector ARMA(p,q) model can always be put in an alternate state-space representation that is similar to (7.12), that is,

$$\mathbf{Z}_{t+1}^* = \Phi\,\mathbf{Z}_t^* + \Psi^*\varepsilon_t\,, \qquad \text{with} \qquad Y_t = H\,\mathbf{Z}_t^* + \varepsilon_t\,, \tag{7.18a-b}$$

but using the state vector $\mathbf{Z}_{t+1}^* = (\,\hat{Y}_t(1)', \ldots, \hat{Y}_t(r^*)'\,)'$ and with $\Psi^* = (\,\Psi_1', \Psi_2', \ldots, \Psi_{r^*}'\,)'$, where $r^* = \max(\,p, q\,)$. Note that this representation has smaller dimension than (7.12) whenever $p \le q$. The form (7.18) is, in fact, just the innovations or prediction error form, as discussed in Section 7.1.3, of the state-space model representation (7.12), with state vector $E[\,\mathbf{Z}_{t+1} \mid Y_t, Y_{t-1}, \ldots\,] = \Phi\,\mathbf{Z}_t = (\,\hat{Y}_t(1)', \ldots, \hat{Y}_t(r)'\,)'$ in this case. Except that when $p \le q$, $r = \max(\,p, q+1\,) = q+1 > p$ so that $\Phi_r = 0$ and, hence, $\hat{Y}_t(r) = \Phi_1\,\hat{Y}_t(r-1) + \cdots + \Phi_{r-1}\,\hat{Y}_t(1)$ is a linear combination of the remaining components of the state vector. Thus, when $p \le q$, $\hat{Y}_t(r)$ is redundant and can be eliminated from the state vector in the prediction error form to obtain the representation (7.18). For example, in the vector ARMA(1,1) model, $Y_t = \Phi_1\,Y_{t-1} + \varepsilon_t - \Theta_1\,\varepsilon_{t-1}$, we have $\mathbf{Z}_{t+1}^* = \hat{Y}_t(1) = \Phi_1\,\hat{Y}_{t-1}(1) + \Psi_1\,\varepsilon_t \equiv \Phi\,\mathbf{Z}_t^* + \Psi^*\varepsilon_t$, with $\Phi = \Phi_1$ and $\Psi^* = \Psi_1 = \Phi_1 - \Theta_1$. This state equation together with the observation equation, $Y_t = \hat{Y}_{t-1}(1) + \varepsilon_t \equiv H\,\mathbf{Z}_t^* + \varepsilon_t$, gives a representation with state vector of dimension $k\,r^* = k$ rather than $k\,r = 2k$ which would be the dimension in (7.12). However, note that in this representation (7.18) the transition equation errors ($\Psi^*\varepsilon_t$) and the measurement equation errors (ε_t) are now correlated, so that the Kalman filtering and smoothing equations would need to be modified to account for this, for example, see Harvey (1989, Sec. 3.2.4). Of course, the alternate "left-companion matrix" state-space form mentioned in (7.16) can also be expressed in the prediction error form analogous to (7.18), $\overline{\mathbf{Z}}_{t+1}^* = \overline{\Phi}\,\overline{\mathbf{Z}}_t^* + \overline{\Psi}^*\varepsilon_t$, $Y_t = \overline{H}\,\overline{\mathbf{Z}}_t^* + \varepsilon_t$, with $\overline{\mathbf{Z}}_{t+1}^*$ obtained from $E[\,\overline{\mathbf{Z}}_{t+1} \mid Y_t, Y_{t-1}, \ldots\,] = \overline{\Phi}\,\overline{\mathbf{Z}}_t$, and $\overline{\Psi}^*$ having the form $\overline{\Psi}^* = [\,(\Phi_1 - \Theta_1)', \ldots, (\Phi_{r^*} - \Theta_{r^*})'\,]'$ in this representation since $\overline{\Phi}\,\overline{\Psi} = [\,(\Phi_1 - \Theta_1)', \ldots, (\Phi_r - \Theta_r)'\,]'$.

In general, the state-space representation with the smallest possible dimension for the state vector is called a *minimal realization*, and it is desirable to have methods which determine such minimal dimension representations for vector ARMA models. Such representations can, in principle, be determined through the use of canonical correlation analysis between the "present and past variables" $\mathbf{P}_t = (\,Y_t', Y_{t-1}', \ldots\,)'$ and the "future variables" $\mathbf{F}_{t+1} = (\,Y_{t+1}', Y_{t+2}', \ldots\,)'$ of the process, as was discussed in Section 4.5.2. This leads to state-space representations of the prediction error form (7.18) with the state vector consisting of a subset of the components of $(\,\hat{Y}_t(1)', \ldots, \hat{Y}_t(r^*)'\,)'$, which forms a basis for the (linear) predictor space of all future values (Akaike, 1974a). [Some procedures also include Y_t among the set of "future variables" as well, which leads to state-space representations analogous to the form (7.12) with the state vector consisting of $\boldsymbol{Y}_t$ together with a subset of the components of $(\hat{\boldsymbol{Y}}_t(1)', \ldots, \hat{\boldsymbol{Y}}_t(r-1)')'$.] These notions and techniques have been explored by Akaike (1974c, 1976), Tuan (1978), Cooper and Wood (1982), Tsay (1989a), and Swift (1990) and have been described and compared with other viewpoints by Tsay (1989b). We illustrate the possibility of reduced dimension in the state-space representation (7.12) with the following simple example.

EXAMPLE 7.3. Consider a bivariate $(k=2)$ process $\{\boldsymbol{Y}_t\}$ that follows an ARMA(2,1) model, $\boldsymbol{Y}_t = \Phi_1 \boldsymbol{Y}_{t-1} + \Phi_2 \boldsymbol{Y}_{t-2} + \varepsilon_t - \Theta_1 \varepsilon_{t-1}$. The "usual" state-space representation for this model, based on (7.12), is given by

$$\mathbf{Z}_t = \begin{bmatrix} 0 & I \\ \Phi_2 & \Phi_1 \end{bmatrix} \mathbf{Z}_{t-1} + \begin{bmatrix} I \\ \Psi_1 \end{bmatrix} \varepsilon_t ,$$

where $\mathbf{Z}_t = (\boldsymbol{Y}_t', \hat{\boldsymbol{Y}}_t(1)')'$ and $\Psi_1 = \Phi_1 - \Theta_1$ from Section 2.3.1. But now suppose that $\operatorname{rank}(\Phi_2, \Theta_1) = 1$, so that the system has a reduced-rank structure. Then there exists a vector F such that $F'\Phi_2 = 0$ and $F'\Theta_1 = 0$. Hence, from Section 2.5.3, or directly from the above representation, we find that $F'(\hat{\boldsymbol{Y}}_t(1) - \Phi_1 \boldsymbol{Y}_t) = 0$ or $[-F'\Phi_1, F']\,\mathbf{Z}_t = 0$. Thus, one of the components of $\hat{\boldsymbol{Y}}_t(1)$ of the state vector $\mathbf{Z}_t$, say $\hat{Y}_{2t}(1)$, can be expressed as a linear combination of the other components of $\mathbf{Z}_t$. So by substitution of this linear combination expression for $\hat{Y}_{2t}(1)$ into the right side of the first three equations of the above state-space representation, we can obtain an equivalent representation of the form $\mathbf{Z}_t^* = \Phi^* \mathbf{Z}_{t-1}^* + \Psi^* \varepsilon_t$, where the state vector $\mathbf{Z}_t^* = (Y_{1t}, Y_{2t}, \hat{Y}_{1t}(1))'$ is of a reduced and minimal dimension of three.

7.2.4 Minimal Dimension State-Variable Representation and Kronecker Indices

Notice in the above example the close connection between reduced-rank structure for the ARMA model and reduced dimensionality in the minimal state-variable representation. In the terminology of scalar component models introduced by Tiao and Tsay (1989), for the above example an AR(1) scalar component would be said to exist in the sense that there is a linear combination $F'\boldsymbol{Y}_t$ that satisfies $F'\boldsymbol{Y}_t - F'\Phi_1 \boldsymbol{Y}_{t-1} = F'\varepsilon_t$. In general, some of the components of the state vector $\mathbf{Z}_t$ or $\mathbf{Z}_t^*$ in the representations (7.12) or (7.18) for the vector ARMA model may be linearly dependent on other components, in which case an equivalent representation is possible with a reduced dimension for the state vector. This situation will correspond to the presence of certain forms of reduced-rank structure in the coefficient matrices Φ_i and Θ_i in the ARMA model or, equivalently, to the presence of certain lower-order scalar component models for the vector process.

We can summarize these notions of minimal dimension more formally through the concepts of Kronecker indices and McMillan degree of a vector ARMA(p,q) process $\boldsymbol{Y}_t$. As discussed in Section 3.1 and in Section 4.5.2, the ith Kronecker index of a process is the minimum index or lag K_i such that the prediction $\hat{Y}_{it}(K_i + 1)$ is *linearly dependent* on the collection of its predecessors,

$$\{ \hat{\boldsymbol{Y}}_t(1)', \hat{\boldsymbol{Y}}_t(2)', \ldots, \hat{\boldsymbol{Y}}_t(K_i)', \hat{Y}_{1t}(K_i + 1), \ldots, \hat{Y}_{i-1,t}(K_i + 1) \},$$

for $i = 1, 2, \ldots, k$. The set $\{ K_i \}$ of Kronecker indices for an ARMA process is unique and does not depend on the particular ordering of the components Y_{it} in $\boldsymbol{Y}_t$. Hence, K_1 is not necessarily associated with Y_{1t}, and so on. Then the McMillan degree M is given by $M = \sum_{i=1}^{k} K_i$, and clearly represents the minimal dimension of the prediction space of all future values of the process. Thus, a state-space representation, of the prediction error form (7.18), with state vector $\mathbf{Z}_{t+1}^*$ of minimal dimension M can be constructed for any vector ARMA model, with $\mathbf{Z}_{t+1}^*$ consisting of a basis for the prediction space of all future values. As indicated above, the structure of this representation can be determined through the use of canonical correlation analysis between the present and past variables and the future variables. This canonical correlation analysis was discussed in detail in Section 4.5.2 and leads to a determination of the Kronecker indices $K_1, \ldots, K_k$ of the process, which, in turn, determines a minimal dimension state-space form.

7.2.5 (Minimal Dimension) Echelon Canonical State-Space Representation

Recall that a minimal state-space representation, of minimal dimension M, is not unique for a given ARMA process $\{\boldsymbol{Y}_t\}$. However, it is known that any two such minimal representations must be related by a nonsingular linear transformation as indicated in Section 7.1.3. That is, $\mathbf{Z}_{t+1}^* = \Phi\, \mathbf{Z}_t^* + \Psi^* \varepsilon_t$, $\boldsymbol{Y}_t = H\, \mathbf{Z}_t^* + \varepsilon_t$ and $\overline{\mathbf{Z}}_{t+1}^* = \overline{\Phi}\, \overline{\mathbf{Z}}_t^* + \overline{\Psi}^* \varepsilon_t$, $\boldsymbol{Y}_t = \overline{H}\, \overline{\mathbf{Z}}_t^* + \varepsilon_t$ are two (equivalent) minimal state-space representations for the process $\boldsymbol{Y}_t$ if and only if there exists an $M \times M$ nonsingular matrix P such that $\overline{\Phi} = P\, \Phi\, P^{-1}$, $\overline{H} = H\, P^{-1}$, and $\overline{\Psi}^* = P\, \Psi^*$ [e.g., see Hannan and Deistler (1988, Sec. 2.3); also Tuan (1978)], so that $\overline{\mathbf{Z}}_t^* = P\, \mathbf{Z}_t^*$. One specific minimal state-space form is the (unique) *echelon canonical state-space representation* that is in correspondence with the (unique) echelon canonical form of the vector ARMA model, as discussed in Section 3.1 for the model of the form (3.4). In this canonical minimal state-space form, the state vector $\mathbf{Z}_{t+1}^*$ in (7.18) is chosen to contain the elements

$$\{ \hat{Y}_{1t}(1), \ldots, \hat{Y}_{1t}(K_1), \hat{Y}_{2t}(1), \ldots, \hat{Y}_{2t}(K_2), \ldots, \hat{Y}_{kt}(1), \ldots, \hat{Y}_{kt}(K_k) \}, \quad (7.19)$$

which clearly constitutes a basis for the prediction space of all future values, of minimal dimension $M = \sum_{i=1}^{k} K_i$. So, suppose the process $\boldsymbol{Y}_t$ has specified Kronecker indices $K_1, \ldots, K_k$ and has echelon canonical ARMA model of the form

$$\Phi_0^{\#}\, \boldsymbol{Y}_t - \sum_{i=1}^{p} \Phi_j^{\#}\, \boldsymbol{Y}_{t-j} = \Theta_0^{\#}\, \varepsilon_t - \sum_{i=1}^{q} \Theta_j^{\#}\, \varepsilon_{t-j}, \qquad \Phi_0^{\#} = \Theta_0^{\#}.$$

As in Section 3.1.3 associated with the reduced-rank form of the model given in

(3.8), we introduce the $r_j \times k$ matrix D_j' which selects the nonzero rows of $(\Phi_j^{\#}, \Theta_j^{\#})$, where $r_j = \text{rank}(\Phi_j^{\#}, \Theta_j^{\#})$ is equal to the number of Kronecker indices that are greater than or equal to the lag j. Then we define the subvectors of predictors as $\hat{\boldsymbol{Y}}_t^*(j) = D_j' \hat{\boldsymbol{Y}}_t(j)$, for $j = 1, \ldots, r^*$, where $r^* = \max(p, q)$. Note that $\hat{\boldsymbol{Y}}_t^*(j)$ consists of those components of $\hat{\boldsymbol{Y}}_t(j)$ whose corresponding Kronecker indices are greater than or equal to j and that, for instance, all elements of $D_{j+1}' \hat{\boldsymbol{Y}}_t(j)$ are contained in $D_j' \hat{\boldsymbol{Y}}_t(j)$.

The minimal dimension state vector in the echelon canonical form is then taken as

$$\mathbf{Z}_{t+1}^* = (\hat{\boldsymbol{Y}}_t^*(1)', \hat{\boldsymbol{Y}}_t^*(2)', \ldots, \hat{\boldsymbol{Y}}_t^*(r^*)')' = D^*(\hat{\boldsymbol{Y}}_t(1)', \hat{\boldsymbol{Y}}_t(2)', \ldots, \hat{\boldsymbol{Y}}_t(r^*)')', \quad (7.19')$$

where $D^* = \text{Diag}(D_1', D_2', \ldots, D_{r^*}')$ is $M \times kr^*$. Note that the elements of $\mathbf{Z}_{t+1}^*$ in (7.19′) are the same as those in (7.19), just with a different ordering. From the relations that are contained in (7.18), $\hat{\boldsymbol{Y}}_t(j) = \hat{\boldsymbol{Y}}_{t-1}(j+1) + \Psi_j \varepsilon_t$, $j = 1, \ldots, r^* - 1$, we obtain that $\sum_{j=2}^{r^*} r_j = M - r_1$ of the state equations in the echelon canonical form are

$$D_{j+1}' \hat{\boldsymbol{Y}}_t(j) = D_{j+1}' \hat{\boldsymbol{Y}}_{t-1}(j+1) + D_{j+1}' \Psi_j \varepsilon_t, \qquad j = 1, \ldots, r^* - 1. \quad (7.20)$$

For the remaining state equations, we recall from (4.17) of Section 4.5.2 that the forecasts from the echelon ARMA model form satisfy

$$\phi_0(i)' \hat{\boldsymbol{Y}}_t(K_i+1) = \sum_{j=1}^{K_i} \phi_j(i)' \hat{\boldsymbol{Y}}_t(K_i+1-j), \qquad \text{for} \quad i = 1, \ldots, k,$$

where $\phi_j(i)'$ denotes the ith row of the matrix $\Phi_j^{\#}$. Hence, we have

$$\begin{aligned}\phi_0(i)' \hat{\boldsymbol{Y}}_t(K_i) &= \phi_0(i)' \hat{\boldsymbol{Y}}_{t-1}(K_i+1) + \phi_0(i)' \Psi_{K_i} \varepsilon_t \\ &= \sum_{j=1}^{K_i} \phi_j(i)' \hat{\boldsymbol{Y}}_{t-1}(K_i+1-j) + \phi_0(i)' \Psi_{K_i} \varepsilon_t, \qquad i = 1, \ldots, k,\end{aligned}$$

and recall that the lth element of $\phi_j(i)'$ can be specified to equal zero in the echelon ARMA model form whenever $j + K_l \le K_i$, i.e., when $K_l < K_i + 1 - j$. Therefore, if we let $\phi_j^*(i)'$ denote the row vector consisting of the elements of $\phi_j(i)'$ that are not specified to equal zero, then it follows that the above equation can be reexpressed as

$$\phi_0^*(i)' \hat{\boldsymbol{Y}}_t^*(K_i) = \sum_{j=1}^{K_i} \phi_j^*(i)' \hat{\boldsymbol{Y}}_{t-1}^*(K_i+1-j) + \phi_0(i)' \Psi_{K_i} \varepsilon_t, \qquad i = 1, \ldots, k. \quad (7.21)$$

In (7.21), if the ith Kronecker index is equal to $K_i = r^*$ then the vector $\phi_0(i)'$ contains a one in the ith position and zeros elsewhere, and the left side of (7.21) is simply equal to $\hat{Y}_{it}(K_i)$. If $K_i < r^*$, then the remaining elements on the left side of (7.21) other than $\hat{Y}_{it}(K_i)$ (that is, elements with corresponding Kronecker indices that are greater that K_i) are replaced using the identity in

(7.20),

$$D_{K_i+1}' \hat{\boldsymbol{Y}}_t(K_i) = D_{K_i+1}' \hat{\boldsymbol{Y}}_{t-1}(K_i+1) + D_{K_i+1}' \Psi_{K_i} \varepsilon_t = \hat{\boldsymbol{Y}}^*_{t-1}(K_i+1) + D_{K_i+1}' \Psi_{K_i} \varepsilon_t .$$

Then equation (7.21) can be written as

$$\hat{Y}_{it}(K_i) = -\phi_0^*(i)' \hat{\boldsymbol{Y}}^*_{t-1}(K_i+1) + \sum_{j=1}^{K_i} \phi_j^*(i)' \hat{\boldsymbol{Y}}^*_{t-1}(K_i+1-j)$$

$$+ (\, 0 \; \cdots \; 0 \;\; 1 \;\; 0 \; \cdots \; 0 \,) \Psi_{K_i} \varepsilon_t , \qquad i = 1, \ldots, k, \qquad (7.21')$$

where now $\phi_0^*(i)'$ denotes the row vector consisting of the elements of $\phi_0(i)'$ that are not specified to equal *zero or one*, and $(\, 0 \; \cdots \; 0 \;\; 1 \;\; 0 \; \cdots \; 0 \,)$ has a one in the ith position and zeros elsewhere. Thus, with the state vector $\mathbf{Z}^*_{t+1}$ given in (7.19′), the M equations of (7.20) and (7.21′) constitute the state equations $\mathbf{Z}^*_{t+1} = \Phi^* \mathbf{Z}^*_t + \Psi^* \varepsilon_t$ in the echelon canonical state-space representation, and the observation equations $\boldsymbol{Y}_t = H \mathbf{Z}^*_t + \varepsilon_t$ are simply $\boldsymbol{Y}_t = \hat{\boldsymbol{Y}}_{t-1}(1) + \varepsilon_t$. A particular instance of this result includes the nested reduced-rank autoregressive model of Section 6.1, which, as we see, possesses a minimal dimension state-space form with dimension $M = \sum_{i=1}^{k} K_i = \sum_{l=1}^{p} r_l$.

For an illustrative example of construction of the echelon canonical state-space representation, consider a bivariate $(k = 2)$ process $\boldsymbol{Y}_t$ with Kronecker indices $K_1 = 2$ and $K_2 = 1$. Then $\boldsymbol{Y}_t$ has an echelon canonical ARMA(2,2) model form as in (3.4) of Section 3.1.2,

$$\Phi_0^{\#} \boldsymbol{Y}_t - \Phi_1^{\#} \boldsymbol{Y}_{t-1} - \Phi_2^{\#} \boldsymbol{Y}_{t-2} = \Theta_0^{\#} \varepsilon_t - \Theta_1^{\#} \varepsilon_{t-1} - \Theta_2^{\#} \varepsilon_{t-2} , \quad \Phi_0^{\#} = \Theta_0^{\#} ,$$

with

$$\Phi_0^{\#} = \begin{bmatrix} 1 & 0 \\ \phi_0(2,1) & 1 \end{bmatrix}, \quad \Phi_1^{\#} = \begin{bmatrix} \phi_1(1,1) & 0 \\ \phi_1(2,1) & \phi_1(2,2) \end{bmatrix}, \quad \Phi_2^{\#} = \begin{bmatrix} \phi_2(1,1) & \phi_2(1,2) \\ 0 & 0 \end{bmatrix},$$

where $\phi_j(i,l)$ denotes the $(i,\, l\,)$th element of $\Phi_j^{\#}$. In the notation of the previous paragraphs, we have $D_1' = I_2$ and $D_2' = (\, 1 \;\; 0 \,)$, and the minimal dimension state vector is $\mathbf{Z}^*_{t+1} = (\, \hat{Y}_{1t}(1), \hat{Y}_{2t}(1), \hat{Y}_{1t}(2) \,)' \equiv (\, \hat{\boldsymbol{Y}}^*_t(1)', \hat{\boldsymbol{Y}}^*_t(2)' \,)'$, with $\hat{\boldsymbol{Y}}^*_t(1) = D_1' \hat{\boldsymbol{Y}}_t(1) = \hat{\boldsymbol{Y}}_t(1)$ and $\hat{\boldsymbol{Y}}^*_t(2) = D_2' \hat{\boldsymbol{Y}}_t(1) = \hat{Y}_{1t}(2)$. The state equation corresponding to (7.20) with $j = 1$ is $\hat{Y}_{1t}(1) = \hat{Y}_{1,t-1}(2) + D_2' \Psi_1 \varepsilon_t = \hat{Y}_{1,t-1}(2) + (\, 1 \;\; 0 \,) \Psi_1 \varepsilon_t$. The equations corresponding to (7.21)–(7.21′) are

$$\hat{Y}_{1t}(2) = \phi_1^*(1)' \hat{\boldsymbol{Y}}^*_{t-1}(2) + \phi_2(1)' \hat{\boldsymbol{Y}}_{t-1}(1) + \phi_0(1)' \Psi_2 \varepsilon_t$$

$$= \phi_1(1,1) \hat{Y}_{1,t-1}(2) + \phi_2(1,1) \hat{Y}_{1,t-1}(1) + \phi_2(1,2) \hat{Y}_{2,t-1}(1)$$

$$+ (\, 1 \;\; 0 \,) \Psi_2 \varepsilon_t$$

and $\phi_0(2)' \hat{\boldsymbol{Y}}_t(1) = \phi_1(2)' \hat{\boldsymbol{Y}}_{t-1}(1) + \phi_0(2)' \Psi_1 \varepsilon_t$, or

$$\hat{Y}_{2t}(1) = -\phi_0(2,1)\,\hat{Y}_{1t}(1) + \phi_1(2,1)\,\hat{Y}_{1,t-1}(1) + \phi_1(2,2)\,\hat{Y}_{2,t-1}(1) + \phi_0(2)'\,\Psi_1\,\varepsilon_t$$

$$= -\phi_0(2,1)\,\hat{Y}_{1,t-1}(2) + \phi_1(2,1)\,\hat{Y}_{1,t-1}(1) + \phi_1(2,2)\,\hat{Y}_{2,t-1}(1)$$

$$+ (\,0 \quad 1\,)\,\Psi_1\,\varepsilon_t\,.$$

Therefore, the corresponding echelon canonical state-space representation for the process $\boldsymbol{Y}_t$ has state equation $\mathbf{Z}^*_{t+1} = \Phi^*\,\mathbf{Z}^*_t + \Psi^*\varepsilon_t$ given by

$$\begin{bmatrix} \hat{Y}_{1t}(1) \\ \hat{Y}_{2t}(1) \\ \hat{Y}_{1t}(2) \end{bmatrix} = \begin{bmatrix} 0 & 0 & 1 \\ \phi_1(2,1) & \phi_1(2,2) & -\phi_0(2,1) \\ \phi_2(1,1) & \phi_2(1,2) & \phi_1(1,1) \end{bmatrix} \begin{bmatrix} \hat{Y}_{1,t-1}(1) \\ \hat{Y}_{2,t-1}(1) \\ \hat{Y}_{1,t-1}(2) \end{bmatrix} + \begin{bmatrix} (\,1 \quad 0\,)\,\Psi_1 \\ (\,0 \quad 1\,)\,\Psi_1 \\ (\,1 \quad 0\,)\,\Psi_2 \end{bmatrix} \varepsilon_t\,,$$

with observation equation $\boldsymbol{Y}_t = H\,\mathbf{Z}^*_t + \varepsilon_t$. Similar to the reduced-rank feature in Example 7.3, a reduction in dimension of the state vector over the standard form $(\,\hat{\boldsymbol{Y}}_t(1)',\,\hat{\boldsymbol{Y}}_t(2)'\,)'$ is possible in this example. The reason is that the ARMA model with Kronecker indices $K_1 = 2$ and $K_2 = 1$ implies the following linear dependency in forecasts, $\hat{Y}_{2t}(2) + \phi_0(2,1)\,\hat{Y}_{1t}(2) = \phi_1(2,1)\,\hat{Y}_{1t}(1) + \phi_1(2,2)\,\hat{Y}_{2t}(1)$, so that the component $\hat{Y}_{2t}(2)$ is redundant and can be eliminated from the state vector to obtain minimal dimension.

Example 7.4. As an illustration of the (minimal) state-space formulation of the vector ARMA model, we further consider the bivariate quarterly time series on the first differences of U.S. fixed investment and U.S. changes in business inventories from Example 4.2. We first consider the determination of the echelon canonical ARMA model form for these data, by use of the canonical correlation analyses methods of Akaike (1976) and Cooper and Wood (1982) as discussed in Section 4.5.2. For the vector of present and past values, we use $m = 4$ in the notation of Section 4.5.2, and set $\mathbf{U}_t = (\,\boldsymbol{Y}_t',\,\boldsymbol{Y}_{t-1}',\,\ldots,\,\boldsymbol{Y}_{t-4}'\,)'$. Then, in an attempt to determine the Kronecker indices and, hence, the overall (minimal) dimension or McMillan degree of the process, the squared sample canonical correlations between $\mathbf{V}^*_t$ and $\mathbf{U}_t$ are determined for various vectors $\mathbf{V}^*_t$ of the future variables. The resulting squared sample canonical correlations are presented in Table 7.1. These results rather clearly indicate the occurrence of a first small (squared) sample canonical correlation, equal to 0.076, when $\mathbf{V}^*_t = (\,Y_{1,t+1},\,Y_{2,t+1},\,Y_{1,t+2}\,)'$. This implies a first Kronecker index value of $K_1 = 1$. We might interpret that an additional small (squared) sample canonical correlation value $(\,0.124\,)$ occurs when $\mathbf{V}^*_t = (\,Y_{1,t+1},\,Y_{2,t+1},\,Y_{1,t+2},\,Y_{2,t+2}\,)'$, and this would imply that $K_2 = 1$ also. Hence this would lead to a standard ARMA(1,1) model formulation, as considered previously in Example 4.2 of Section 4.5 for these data, and ML estimation results for that model were presented there. As an alternative, suppose we consider the value 0.124 at this stage as not being sufficiently small, and we continue on with the canonical correlation analysis procedure. At the next step, a second small (squared) sample canonical correlation $(\,0.092$ in addition to $0.054\,)$ is introduced, as expected, due to the inclusion of the (redundant) variable $Y_{1,t+3}$ in the future vector $\mathbf{V}_t^*$. Finally, with $\mathbf{V}_t^* = (\boldsymbol{Y}_{t+1}', \boldsymbol{Y}_{t+2}', \boldsymbol{Y}_{t+3}')'$, we interpret that one further small (squared) sample canonical correlation (0.119) is introduced and this implies a Kronecker index of $K_2 = 2$.

Table 7.1 Results of the Canonical Correlation Analysis, Based on the Kronecker Indices Approach of Akaike (1976) and Cooper and Wood (1982), for the U.S. Business Investment and Inventories Data. Values are the squared sample canonical correlations between the present and past vector $\mathbf{U}_t = (\boldsymbol{Y}_t', \ldots, \boldsymbol{Y}_{t-4}')'$ and various future vectors $\mathbf{V}_t^*$.

Future Vector $\mathbf{V}_t^*$	Squared Canonical Correlations
$Y_{1,t+1}$	0.422
$Y_{1,t+1}, Y_{2,t+1}$	0.587, 0.420
$Y_{1,t+1}, Y_{2,t+1}, Y_{1,t+2}$	0.664, 0.428, 0.076
$Y_{1,t+1}, Y_{2,t+1}, Y_{1,t+2}, Y_{2,t+2}$	0.674, 0.430, 0.124, 0.069
$Y_{1,t+1}, Y_{2,t+1}, Y_{1,t+2}, Y_{2,t+2}, Y_{1,t+3}$	0.686, 0.426, 0.140, 0.092, 0.054
$Y_{1,t+1}, Y_{2,t+1}, Y_{1,t+2}, Y_{2,t+2}, Y_{1,t+3}, Y_{2,t+3}$	0.688, 0.430, 0.145, 0.119, 0.068, 0.049

This analysis would thus result in the specification of an ARMA(2,2) model in the echelon form (3.4) of Section 3.1.2, with Kronecker indices equal to $K_1 = 1$ and $K_2 = 2$. Notice that for this set of Kronecker indices, for the model in (3.4) we have $\Phi_0^\# = \Theta_0^\# = I$, the first rows of both $\Phi_2^\#$ and $\Theta_2^\#$ have values specified to be zero, and a zero is also specified for the value in the (2, 1) position of $\Phi_1^\#$ for parameter identifiability. This model form was estimated by conditional ML, and after eliminating a few parameters whose estimates were clearly nonsignificant, the ML estimates were obtained as

$$\hat{\Phi}_1^\# = \begin{bmatrix} 0.491 & -0.201 \\ (0.085) & (0.060) \\ & \\ 0 & 0.000 \\ & (\text{---}) \end{bmatrix}, \qquad \hat{\Phi}_2^\# = \begin{bmatrix} 0 & 0 \\ & \\ & \\ 0.843 & 0.438 \\ (0.224) & (0.093) \end{bmatrix},$$

$$\hat{\Theta}_1^\# = \begin{bmatrix} 0.000 & -0.348 \\ (\text{---}) & (0.075) \\ & \\ -0.300 & -0.644 \\ (0.161) & (0.075) \end{bmatrix}, \qquad \hat{\Theta}_2^\# = \begin{bmatrix} 0 & 0 \\ & \\ & \\ 0.468 & 0.000 \\ (0.241) & (\text{---}) \end{bmatrix},$$

with

$$\tilde{\Sigma} = \begin{bmatrix} 4.7490 & 1.3582 \\ 1.3582 & 15.3811 \end{bmatrix} \qquad \text{and} \qquad \det(\tilde{\Sigma}) = 71.2007\,.$$

Compared to the estimation results for the ARMA(1,1) model given in Example 4.2, this ARMA(2,2) model seems to offer a little improvement, and residual checks do not indicate any significant correlations in the residuals $\hat{\varepsilon}_t$ from this model.

This ARMA model result then gives rise to a minimal dimension state-space model formulation, in a representation of the prediction error form of (7.18), with state vector equal to $\mathbf{Z}_{t+1}^* = (\hat{Y}_{1t}(1), \hat{Y}_{2t}(1), \hat{Y}_{2t}(2))'$. From the developments in (7.20)–(7.21), the explicit form of the state equation implied is

$$\mathbf{Z}_{t+1}^* = \begin{bmatrix} 0.491 & -0.201 & 0 \\ 0 & 0 & 1 \\ 0.842 & 0.438 & 0.000 \end{bmatrix} \mathbf{Z}_t^* + \begin{bmatrix} (1 \quad 0)\,\Psi_1 \\ (0 \quad 1)\,\Psi_1 \\ (0 \quad 1)\,\Psi_2 \end{bmatrix} \varepsilon_t$$

with

$$\Psi_1 = \Phi_1^{\#} - \Theta_1^{\#} = \begin{bmatrix} 0.491 & 0.146 \\ 0.300 & 0.644 \end{bmatrix},$$

$$\Psi_2 = \Phi_1^{\#}\Psi_1 + \Phi_2^{\#} - \Theta_2^{\#} = \begin{bmatrix} 0.180 & -0.058 \\ 0.375 & 0.438 \end{bmatrix}.$$

The observation equation is $\boldsymbol{Y}_t = [\, I \quad 0\,]\,\mathbf{Z}_t^* + \varepsilon_t$. A state-space representation analogous to the form in (7.12) could also be obtained, by including these last two (observation) equations in the state equations, with state vector then equal to $\mathbf{Z}_t = (\boldsymbol{Y}_t', \hat{\boldsymbol{Y}}_t(1)', \hat{Y}_{2t}(2))'$, and observation equation simply $\boldsymbol{Y}_t = [\, I \quad 0\,]\,\mathbf{Z}_t$.

As an alternative approach to formulation of a state-space model, we could also consider a modification, as originally proposed by Akaike (1976), to the preceding canonical correlation analysis. In this approach, the present values $\boldsymbol{Y}_t$ are also included in the vector of "future" variables $\mathbf{V}_t^*$. This approach leads, in general, to specification of an ARMA model with order of the form $(p, p-1)$, and corresponding state-space representation analogous to the form (7.12). Results from performing this canonical correlation analysis between $\mathbf{U}_t = (\boldsymbol{Y}_t', \ldots, \boldsymbol{Y}_{t-4}')'$ and various vectors $\mathbf{V}_t^*$ of present and future variables are presented in Table 7.2. From these results, we may interpret that a first small (squared) sample canonical correlation occurs when $\mathbf{V}_t^* = (Y_{1t}, Y_{2t}, Y_{1,t+1}, Y_{2,t+1})'$, that is, when $Y_{2,t+1}$ is introduced into $\mathbf{V}_t^*$, and a second small (squared) sample canonical correlation is obtained when the variable $Y_{1,t+2}$ is further included.

Table 7.2 Results of the Canonical Correlation Analysis, Based on the Original Approach of Akaike (1976), for the U.S. Business Investment and Inventories Data. Values are the squared sample canonical correlations between the present and past vector $\mathbf{U}_t = (\boldsymbol{Y}_t', \ldots, \boldsymbol{Y}_{t-4}')'$ and various "future" vectors $\mathbf{V}_t^*$.

Future Vector $\mathbf{V}_t^*$	Squared Canonical Correlations
$Y_{1t}, Y_{2t}, Y_{1,t+1}$	1.000, 1.000, 0.235
$Y_{1t}, Y_{2t}, Y_{1,t+1}, Y_{2,t+1}$	1.000, 1.000, 0.277, 0.091
$Y_{1t}, Y_{2t}, Y_{1,t+1}, Y_{2,t+1}, Y_{1,t+2}$	1.000, 1.000, 0.302, 0.096, 0.074
$Y_{1t}, Y_{2t}, Y_{1,t+1}, Y_{2,t+1}, Y_{1,t+2}, Y_{2,t+2}$	1.000, 1.000, 0.313, 0.116, 0.087, 0.059

These results imply the existence of a linear combination of $Y_{2,t+1}$, in terms of $(Y_{1t}, Y_{2t}, Y_{1,t+1})'$, which is uncorrelated with the present and past $\mathbf{U}_t = (\boldsymbol{Y}_t', \boldsymbol{Y}_{t-1}', \ldots)'$, and similarly for $Y_{1,t+2}$. Hence, this gives rise to the specification of an ARMA(2,1) model for $\boldsymbol{Y}_t$, in the form $\Phi_0^\# \boldsymbol{Y}_t - \Phi_1^\# \boldsymbol{Y}_{t-1} - \Phi_2^\# \boldsymbol{Y}_{t-2} = \Theta_0^\# \varepsilon_t - \Theta_1^\# \varepsilon_{t-1}$ with $\Phi_0^\# = \Theta_0^\#$, where the structure of the coefficient matrices is such that

$$\Phi_0^\# = \begin{bmatrix} 1 & 0 \\ X & 1 \end{bmatrix}, \quad \Phi_1^\# = \begin{bmatrix} X & 0 \\ X & X \end{bmatrix}, \quad \Phi_2^\# = \begin{bmatrix} X & X \\ 0 & 0 \end{bmatrix}, \quad \Theta_1^\# = \begin{bmatrix} X & X \\ 0 & 0 \end{bmatrix}.$$

This model form was estimated by conditional ML, leading to an error covariance matrix estimate $\tilde{\Sigma}$ with $\det(\tilde{\Sigma}) = 76.3272$. After eliminating one parameter from $\Theta_1^\#$ whose estimate was quite nonsignificant, the final ML estimation results were obtained as

$$\hat{\Phi}_0^\# = \begin{bmatrix} 1 & 0 \\ & \\ 0.396 & 1 \\ (0.424) & \end{bmatrix}, \quad \hat{\Phi}_1^\# = \begin{bmatrix} 0.512 & 0 \\ (0.093) & \\ 0.523 & 0.656 \\ (0.263) & (0.077) \end{bmatrix},$$

$$\hat{\Phi}_2^\# = \begin{bmatrix} -0.133 & -0.129 \\ (0.090) & (0.040) \\ 0 & 0 \end{bmatrix}, \quad \hat{\Theta}_1^\# = \begin{bmatrix} 0.000 & -0.132 \\ (\text{---}) & (0.057) \\ 0 & 0 \end{bmatrix},$$

with

$$\tilde{\Sigma} = \begin{bmatrix} 4.9677 & 1.6874 \\ 1.6874 & 15.9577 \end{bmatrix} \qquad \text{and} \qquad \det(\tilde{\Sigma}) = 76.4258 .$$

Notice that this model has similar form to the reduced-rank AR(2) model, as obtained and presented in Section 6.1.4 for these data. In fact, the two models are nearly identical in terms of fit. [Note that the parameter in the $(1, 2)$ position of the MA matrix $\hat{\Theta}_1^{\#}$ in the above ARMA(2,1) model is essentially exchangeable with the parameter in the $(1, 2)$ position of the AR matrix $\hat{\Phi}_1^{\#}$ in the reduced-rank AR(2) model.] In comparison to the ARMA(2,2) model obtained earlier, it seems that the second approach which leads to the ARMA(2,1) model does not provide quite as good fitting model as the first approach.

The above ARMA(2,1) model has the corresponding state-space model formulation, analogous to the form in (7.12), with state vector $\mathbf{Z}_t = (Y_{1t}, Y_{2t}, \hat{Y}_{1t}(1))'$. The explicit form of the state equation implied by the ARMA(2,1) model result is

$$\mathbf{Z}_t = \begin{bmatrix} 0 & 0 & 1 \\ 0.523 & 0.656 & -0.396 \\ -0.133 & -0.129 & 0.512 \end{bmatrix} \mathbf{Z}_{t-1} + \begin{bmatrix} 1 & 0 \\ 0 & 1 \\ (1 & 0)\Psi_1 \end{bmatrix} \varepsilon_t$$

with $(1 \quad 0)\Psi_1 = (1 \quad 0)(\Phi_1^{\#} - \Theta_1^{\#}) = (0.512 \quad 0.132)$, and the observation equation is simply $\boldsymbol{Y}_t = [\, I \quad 0 \,]\, \mathbf{Z}_t$.

Remark. We note that a state-space model represented as in the latter forms in the above example, analogous to the form in (7.12), can be estimated using the Statespace procedure in the SAS/ETS computer system (SAS, 1988). Results obtained from this procedure differ substantially from the conditional ML estimates given above, however, since the estimation of model parameters in the SAS procedure uses an approximate likelihood function that is based on sample autocovariance matrices of the $\boldsymbol{Y}_t$ and truncation of an infinite AR form.

7.3 Exact Likelihood Estimation for Vector ARMA Processes with Missing Values

In some situations, in practice, the vector time series $\boldsymbol{Y}_t$ will not be observed at equally spaced times because of failure to observe the series at some time points, or some components Y_{it} may not be observed at certain times. For such "missing data" situations, we briefly discuss the calculation of the exact Gaussian likelihood function of the vector ARMA model for the observed data. It is shown that in missing data situations, the likelihood function can readily and conveniently be constructed using the state-space form of the model and associated Kalman filtering procedures, as discussed in general terms in Section 7.1 and for ARMA models previously in Section 7.2, but modified to accommodate the missing data. These methods for evaluation of the exact likelihood in cases of irregularly spaced observations have been examined by Jones (1980), Harvey and Pierse (1984), Ansley and Kohn (1983, 1985a), and Wincek and Reinsel (1986).

We suppose that observations on at least some components of the vector series $\{ \boldsymbol{Y}_t \}$ are available at the n integer times $t_1 < t_2 < \cdots < t_n$, not equally spaced, and that $\boldsymbol{Y}_t$ follows the vector ARMA(p,q) model $\Phi(B)\, \boldsymbol{Y}_t = \Theta(B)\, \varepsilon_t$. Thus, at time t_i we observe $\boldsymbol{Y}^*_{t_i} = M_i\, \boldsymbol{Y}_{t_i}$, where M_i is a known incidence matrix of dimension $k_i \times k$ ($k_i \le k$), whose elements are equal to 1 or 0 to indicate the occurrence of an observation for each given component. In particular, $M_i = I_k$ when all components of $\boldsymbol{Y}_{t_i}$ are observed at time t_i. From (7.12), the process $\boldsymbol{Y}_t$ has the state-space representation given by

$$\mathbf{Z}_t = \Phi\, \mathbf{Z}_{t-1} + \Psi\, \varepsilon_t , \tag{7.22}$$

with $\boldsymbol{Y}_t = H\, \mathbf{Z}_t = [\, I, 0, \dots, 0\,]\, \mathbf{Z}_t$, where $\mathbf{Z}_t$ is the kr-dimensional state vector and $r = \max(\, p, q+1\,)$. Let $\Delta_i = t_i - t_{i-1}$ denote the time difference between successive observations $\boldsymbol{Y}^*_{t_i-1}$ and $\boldsymbol{Y}^*_{t_i}$, $i = 2, \dots, n$. By successive substitutions, Δ_i times, in the right-hand side of (7.22), we obtain

$$\mathbf{Z}_{t_i} = \Phi^{\Delta_i}\, \mathbf{Z}_{t_i-1} + \sum_{j=0}^{\Delta_i-1} \Phi^j\, \Psi\, \varepsilon_{t_i-j} \equiv \Phi^*_i\, \mathbf{Z}_{t_i-1} + \varepsilon^*_{t_i} , \tag{7.23}$$

where $\Phi^*_i = \Phi^{\Delta_i}$ and $\varepsilon^*_{t_i} = \sum_{j=0}^{\Delta_i-1} \Phi^j\, \Psi\, \varepsilon_{t_i-j}$, with $\mathrm{Cov}(\, \varepsilon^*_{t_i}\,) = \Sigma^*_i = \sum_{j=0}^{\Delta_i-1} \Phi^j \Psi\, \Sigma\, \Psi' \Phi'^j$. Thus, (7.23) together with the observation equation $\boldsymbol{Y}^*_{t_i} = M_i\, \boldsymbol{Y}_{t_i} = M_i\, H\, \mathbf{Z}_{t_i} \equiv H_i\, \mathbf{Z}_{t_i}$ constitute a state-space model form for the *observed* time series data $\boldsymbol{Y}^*_{t_1}, \boldsymbol{Y}^*_{t_2}, \dots, \boldsymbol{Y}^*_{t_n}$.

Therefore, the Kalman filter recursive equations as in (7.3)–(7.6) can be directly employed to obtain the state predictors $\hat{\mathbf{Z}}_{t_i | t_i-1}$ and their error covariance matrices $P_{t_i | t_i-1}$. So we can readily obtain the predictors

$$\hat{\boldsymbol{Y}}^*_{t_i | t_i-1} = E[\, \boldsymbol{Y}^*_{t_i} \mid \boldsymbol{Y}^*_{t_i-1}, \dots, \boldsymbol{Y}^*_{t_1}\,] = H_i\, \hat{\mathbf{Z}}_{t_i | t_i-1} \tag{7.24}$$

for the observations $\boldsymbol{Y}^*_{t_i}$, based on the previous *observed* data, and their error covariance matrices

$$\Sigma_{t_i | t_i-1} = H_i\, P_{t_i | t_i-1}\, H_i' = E[\, (\, \boldsymbol{Y}^*_{t_i} - \hat{\boldsymbol{Y}}^*_{t_i | t_{i-1}})\, (\, \boldsymbol{Y}^*_{t_i} - \hat{\boldsymbol{Y}}^*_{t_i | t_{i-1}})'\,] \tag{7.25}$$

from the recursive Kalman filtering procedure. More specifically, the updating equations (7.3) and (7.4) in this missing data setting take the form

$$\hat{\mathbf{Z}}_{t_i | t_i} = \hat{\mathbf{Z}}_{t_i | t_{i-1}} + K_i\, (\, \boldsymbol{Y}^*_{t_i} - H_i\, \hat{\mathbf{Z}}_{t_i | t_{i-1}}) \tag{7.26a}$$

with

$$K_i = P_{t_i | t_{i-1}}\, H_i'\, [\, H_i\, P_{t_i | t_{i-1}}\, H_i'\,]^{-1} \tag{7.26b}$$

while the prediction equations (7.5) are given by

$$\hat{\mathbf{Z}}_{t_i \mid t_{i-1}} = \Phi_i^* \hat{\mathbf{Z}}_{t_{i-1} \mid t_{i-1}} = \Phi^{\Delta_i} \hat{\mathbf{Z}}_{t_{i-1} \mid t_{i-1}} , \qquad P_{t_i \mid t_{i-1}} = \Phi_i^* P_{t_{i-1} \mid t_{i-1}} \Phi_i^{*\prime} + \Sigma_i^* \quad \text{(7.27a-b)}$$

with $P_{t_i \mid t_i} = [\, I - K_i H_i \,] P_{t_i \mid t_{i-1}}$. Notice that the calculation of the prediction equations (7.27) can be interpreted as computation of the successive "one-step ahead" predictions

$$\hat{\mathbf{Z}}_{t_{i-1}+j \mid t_{i-1}} = \Phi \hat{\mathbf{Z}}_{t_{i-1}+j-1 \mid t_{i-1}} ,$$

$$P_{t_{i-1}+j \mid t_{i-1}} = \Phi P_{t_{i-1}+j-1 \mid t_{i-1}} \Phi' + \Psi \Sigma \Psi'$$

for $j = 1, \ldots, \Delta_i$, without any updating since there are no observations available between the time points t_{i-1} and t_i to provide any additional information for updating.

The exact likelihood for the vector of observations $\mathbf{y} = (Y_{t_1}^{*\prime}, Y_{t_2}^{*\prime}, \ldots, Y_{t_n}^{*\prime})'$ is obtained directly from the quantities in (7.24) and (7.25) because the joint density of $\mathbf{y}$ can be expressed as the product of the conditional densities of the $Y_{t_i}^*$, given $Y_{t_{i-1}}^*, \ldots, Y_{t_1}^*$, for $i = 2, \ldots, n$, which are Gaussian with (conditional) mean vectors and covariance matrices given by (7.24)–(7.25). Hence, the joint density of the observations $\mathbf{y}$ can be expressed as

$$f(\mathbf{y}) = \Big(\prod_{i=1}^{n} | \Sigma_{t_i \mid t_{i-1}} |^{-1/2} \Big)$$
$$\times \exp\Big[-(1/2) \sum_{i=1}^{n} (Y_{t_i}^* - \hat{Y}_{t_i \mid t_{i-1}}^*)' \, \Sigma_{t_i \mid t_{i-1}}^{-1} \, (Y_{t_i}^* - \hat{Y}_{t_i \mid t_{i-1}}^*) \Big] . \quad (7.28)$$

In (7.28), the quantities $\hat{Y}_{t_i \mid t_i - 1}^*$ and $\Sigma_{t_i \mid t_{i-1}}$ are directly determined from the recursive filtering calculations (7.26)–(7.27). In the case of a stationary ARMA(p,q) model, the initial conditions required to start the filtering procedure can be determined readily as discussed in Sections 7.2.2 and 7.2.3 (e.g., Akaike, 1978; Jones, 1980; Ansley and Kohn, 1983). However, for the nonstationary ARIMA model situation, some additional assumptions need to be specified concerning the process and initial conditions, and appropriate modified methods for such cases have been examined by Ansley and Kohn (1985a, 1990), for example.

A related problem of interest that often arises in the context of missing values for time series is that of estimation of the missing values. Within the framework of the state-space formulation, estimates of missing values and their corresponding error covariance matrices can be derived conveniently through the use of recursive smoothing methods associated with the Kalman filter, which were discussed briefly in Section 7.1.2. Such smoothing methods are described in general terms in Anderson and Moore (1979), for example. These methods have been considered more specifically for the ARIMA model missing data situation by Harvey and Pierse (1984) and by Kohn and Ansley (1986).

7.4 Classical Approach to Smoothing and Filtering of Time Series

Let $(\boldsymbol{S}_t', \boldsymbol{Y}_t')'$ denote a vector stationary process with zero mean and covariance matrices $\Gamma_s(l) = E(\boldsymbol{S}_t \boldsymbol{S}_{t+l}')$, $\Gamma_y(l) = E(\boldsymbol{Y}_t \boldsymbol{Y}_{t+l}')$, and $\Gamma_{sy}(l) = E(\boldsymbol{S}_t \boldsymbol{Y}_{t+l}')$. The problem to be considered is the estimation of values of the series $\boldsymbol{S}_t$ which is not observable, based on values of the observed series $\boldsymbol{Y}_t$. Specifically, suppose we observe the values $\boldsymbol{Y}_t$, $t \leq \tau$. Then we want to determine the linear filter

$$\hat{\boldsymbol{S}}_t = \sum_{u=0}^{\infty} A_u^{(\tau)} \boldsymbol{Y}_{\tau-u} \equiv A^{(\tau)}(B) \boldsymbol{Y}_\tau \tag{7.29}$$

of $\{ \boldsymbol{Y}_t \}$ such that the value $\hat{\boldsymbol{S}}_t$ is close to $\boldsymbol{S}_t$ in the sense that the mean square estimation error matrix, $E[(\boldsymbol{S}_t - \hat{\boldsymbol{S}}_t) (\boldsymbol{S}_t - \hat{\boldsymbol{S}}_t)']$, is a minimum among all possible linear filters. A common model formulation for which this problem arises is the signal extraction model, in which there is a "signal" $\boldsymbol{S}_t$ of interest but what is observed is assumed to be a noise-corrupted version of the signal, that is, we observe $\boldsymbol{Y}_t$ where we assume $\boldsymbol{Y}_t = \boldsymbol{S}_t + \boldsymbol{N}_t$, and $\boldsymbol{N}_t$ is a noise component. The problem then is to estimate values of the signal series $\boldsymbol{S}_t$ given values on the observed series $\boldsymbol{Y}_t$. Often, the filtering and smoothing algorithms for the state-space model, as discussed in Section 7.1, can be applied to this situation. However, it is not always direct to obtain explicit expressions from the state-space algorithms for the coefficients or weights $A_u^{(\tau)}$ in (7.29) applied to the observations in the formation of the estimate $\hat{\boldsymbol{S}}_t$. These expressions can be derived more readily in the classical approach, which assumes that an infinite extent of observations are available for filtering or smoothing. In this section we will first develop some classical filtering and smoothing results and indicate relationships of these with the state-space results. Typically, from a practical point of view, the classical results provide a good approximation to exact filtering and smoothing results that are based on a finite sample of observations $\boldsymbol{Y}_1, \ldots, \boldsymbol{Y}_T$.

7.4.1 Smoothing for Univariate Time Series

We first consider the scalar case where S_t and Y_t are both univariate time series. We suppose that $\{Y_t\}$ has the infinite MA representation, $Y_t = \psi(B)\, \varepsilon_t = \sum_{j=0}^{\infty} \psi_j\, \varepsilon_{t-j}$, where the ε_t are white noise with variance σ_ε^2. Also let $g_{ys}(B) = \sum_{j=-\infty}^{\infty} \gamma_{ys}(j)\, B^j$ be the cross-covariance generating function between Y_t and S_t. Then it can be derived (e.g., Whittle, 1963b, Chaps. 5–6; Priestley, 1981, Chap. 10) that the optimal linear filter for the estimate $\hat{S}_t = \sum_{u=0}^{\infty} a_u^{(\tau)}\, Y_{\tau-u} = A^{(\tau)}(B)\, Y_\tau$, where $A^{(\tau)}(B) = \sum_{u=0}^{\infty} a_u^{(\tau)}\, B^u$, is given by

$$A^{(\tau)}(B) = \frac{1}{\sigma_\varepsilon^2\, \psi(B)} \left[\frac{B^{\tau-t}\, g_{ys}(B)}{\psi(B^{-1})} \right]_+ , \tag{7.30}$$

where, for a general operator $v(B) = \sum_{j=-\infty}^{\infty} v_j B^j$, the notation $[\, v(B)\,]_+$ denotes $\sum_{j=0}^{\infty} v_j B^j$.

To derive the result (7.30) for the optimal linear filter, note that since $Y_t = \psi(B)\, \varepsilon_t$, the linear filter can be expressed as $\hat{S}_t = A^{(\tau)}(B)\, Y_\tau = A^{(\tau)}(B)\, \psi(B)\, \varepsilon_\tau = H^{(\tau)}(B)\, \varepsilon_\tau$, where $H^{(\tau)}(B) = A^{(\tau)}(B)\, \psi(B) = \sum_{j=0}^{\infty} h_j^{(\tau)} B^j$. So we can determine the coefficients $h_j^{(\tau)}$ to minimize the mean squared error $E[\, (S_t - \hat{S}_t)^2\,] = E[\, (S_t - \sum_{j=0}^{\infty} h_j^{(\tau)} \varepsilon_{\tau-j})^2\,]$. Since the $\{\varepsilon_t\}$ are mutually uncorrelated, it follows by standard linear least squares arguments that the values of the coefficients that minimize the above mean squared error are

$$h_j^{(\tau)} = \mathrm{Cov}(\, \varepsilon_{\tau-j},\, S_t\,) \,/\, \mathrm{Var}(\, \varepsilon_{\tau-j}\,) = \gamma_{\varepsilon s}(j+t-\tau) \,/\, \sigma_\varepsilon^2\,, \qquad j \ge 0\,.$$

Hence, the optimal linear filter is

$$H^{(\tau)}(B) = \frac{1}{\sigma_\varepsilon^2} \sum_{j=0}^{\infty} \gamma_{\varepsilon s}(j+t-\tau)\, B^j = \frac{1}{\sigma_\varepsilon^2}\, [\, B^{\tau-t}\, g_{\varepsilon s}(B)\,]_+\,, \tag{7.31}$$

where $g_{\varepsilon s}(B)$ denotes the cross-covariance generating function between ε_t and S_t. Also, note that $\gamma_{ys}(j) = \mathrm{Cov}(\, \sum_{i=0}^{\infty} \psi_i\, \varepsilon_{t-i},\, S_{t+j}\,) = \sum_{i=0}^{\infty} \psi_i\, \gamma_{\varepsilon s}(i+j)$ so that

$$g_{ys}(B) = \sum_{j=-\infty}^{\infty} \gamma_{ys}(j)\, B^j = \sum_{j=-\infty}^{\infty} \left[\, \sum_{i=0}^{\infty} \psi_i\, \gamma_{\varepsilon s}(i+j) \right] B^j$$

$$= \sum_{i=0}^{\infty} \psi_i\, B^{-i} \sum_{u=-\infty}^{\infty} \gamma_{\varepsilon s}(u)\, B^u = \psi(B^{-1})\, g_{\varepsilon s}(B)\,.$$

Therefore, the optimal linear filter in (7.31) is $H^{(\tau)}(B) = (\, 1/\sigma_\varepsilon^2\,)\, [\, B^{\tau-t}\, g_{ys}(B)/\psi(B^{-1})\,]_+$ and, hence, the optimal filter in terms of $\hat{S}_t = A^{(\tau)}(B)\, Y_\tau$ is $A^{(\tau)}(B) = H^{(\tau)}(B)/\psi(B)$ which yields the result (7.30). The mean squared error (MSE) of the optimal linear filter, since $\hat{S}_t = \sum_{j=0}^{\infty} h_j^{(\tau)}\, \varepsilon_{\tau-j}$, is easily seen from the above derivation to be equal to $E[\, (S_t - \hat{S}_t)^2\,] = \mathrm{Var}(\, S_t\,) - \sigma_\varepsilon^2 \sum_{j=0}^{\infty} \{\, h_j^{(\tau)}\, \}^2 = \mathrm{Var}(\, S_t\,) - \mathrm{Var}(\, \hat{S}_t\,)$.

In the smoothing case where $\tau = +\infty$, that is, we estimate S_t based on the infinite record of observations Y_u, $-\infty < u < \infty$, by a linear filter $\hat{S}_t = \sum_{u=-\infty}^{\infty} a_u\, Y_{t-u} = A(B)\, Y_t$, then the above result (7.30) for the optimal filter reduces to

$$A(B) = g_{ys}(B) \,/\, g_{yy}(B) = g_{ys}(B) \,/\, [\, \sigma_\varepsilon^2\, \psi(B)\, \psi(B^{-1})\,]\,. \tag{7.32}$$

For the signal extraction problem, we have $Y_t = S_t + N_t$ where it is usually assumed that the signal $\{S_t\}$ and the noise process $\{N_t\}$ are independent. Thus, in this case we have $g_{ys}(B) = g_{ss}(B)$, and so in the smoothing case $\tau = +\infty$, for example, we have $A(B) = g_{ss}(B)/g_{yy}(B)$ or $A(B) = g_{ss}(B)/[\, g_{ss}(B) + g_{nn}(B)\,]$. It can also be shown that the MSE of the optimal estimate $\hat{S}_t$ is given by $E[\, (S_t - \hat{S}_t)^2\,] = \mathrm{Var}(\, S_t\,) - \mathrm{Var}(\, \hat{S}_t\,)$, since $S_t = \hat{S}_t + (\, S_t - \hat{S}_t\,)$ and the error $S_t - \hat{S}_t$ of the optimal estimate is uncorrelated with $\hat{S}_t$.

EXAMPLE 7.5. Suppose $Y_t = S_t + N_t$, where $(1 - \phi B) S_t = a_t$ is AR(1) and N_t is white noise with variance σ_n^2. Then we have $(1 - \phi B) Y_t = a_t + (1 - \phi B) N_t$. Hence, we know that the observed process Y_t satisfies an ARMA(1,1) model, $(1 - \phi B) Y_t = (1 - \theta B) \varepsilon_t$, where θ and σ_ε^2 are determined from the relations

$$\sigma_a^2 + (1 + \phi^2) \sigma_n^2 = \sigma_\varepsilon^2 (1 + \theta^2), \qquad -\phi \sigma_n^2 = -\theta \sigma_\varepsilon^2, \qquad |\theta| < 1,$$

so that

$$\theta = [-1 + \sqrt{1 - 4\rho(1)^2}\,] / (2\rho(1)) \qquad \text{where} \qquad \rho(1) = \frac{-\phi \sigma_n^2}{\sigma_a^2 + (1 + \phi^2) \sigma_n^2}$$

with $\sigma_\varepsilon^2 = (\phi \sigma_n^2) / \theta$. Thus, $\psi(B) = (1 - \phi B)^{-1} (1 - \theta B)$, and $g_{ys}(B) = g_{ss}(B) = \sigma_a^2 (1 - \phi B)^{-1} (1 - \phi B^{-1})^{-1}$. Hence, for example, from (7.32) the optimal filter in the smoothing case $\tau = +\infty$ becomes $A(B) = g_{ss}(B) / g_{yy}(B) = \sigma_a^2 / [\sigma_\varepsilon^2 (1 - \theta B)(1 - \theta B^{-1})]$. So the optimal smoothing estimate can be written explicitly as

$$\hat{S}_t = \frac{\sigma_a^2}{\sigma_\varepsilon^2 (1 - \theta^2)} \sum_{u=-\infty}^{\infty} \theta^{|u|} Y_{t-u},$$

with coefficients or weights of the linear filter for the observations Y_{t-u} that are symmetric about $u = 0$ and decline exponentially with rate θ as $|u|$ moves away from zero. In practice, when one computes the estimate $\hat{S}_t = A(B) Y_t$, the computations can proceed as follows. Since

$$\hat{S}_t = c (1 - \theta B)^{-1} (1 - \theta B^{-1})^{-1} Y_t = c (1 - \theta B^{-1})^{-1} X_t,$$

where $c = \sigma_a^2 / \sigma_\varepsilon^2$ and $X_t = (1 - \theta B)^{-1} Y_t$, the X_t are first computed in the forward recursion from the relation $(1 - \theta B) X_t = Y_t$ as $X_u = \theta X_{u-1} + Y_u$, $u = -\infty, \ldots, +\infty$. Then we compute the "backward recursion" $\hat{S}_t = c (1 - \theta B^{-1})^{-1} X_t$ to obtain $\hat{S}_t$ from $\hat{S}_u = \theta \hat{S}_{u+1} + c X_u$, $u = +\infty, \ldots, t$. In the more general smoothing case with finite $\tau \geq t$, we are estimating S_t from Y_u, $u \leq \tau$. Then the general results (7.30) give the optimal linear filter as

$$A^{(\tau)}(B) = c^* (1 - \theta B)^{-1} \left[B^{\tau-t} + \theta \frac{(B^{\tau-t} - \theta^{\tau-t})}{B - \theta} (1 - \phi B) \right], \quad \tau - t \geq 0, \quad (7.33)$$

where $c^* = \sigma_a^2 / [\sigma_\varepsilon^2 (1 - \phi\theta)] = 1 - (\theta/\phi)$, since

$$[B^{\tau-t} g_{ss}(B) / \psi(B^{-1})]_+ = \sigma_a^2 [B^{\tau-t} (1 - \phi B)^{-1} / (1 - \theta B^{-1})]_+$$

$$= [\sigma_a^2 / (1 - \phi\theta)] \left[\frac{B^{\tau-t}}{1 - \phi B} + \frac{\theta (B^{\tau-t} - \theta^{\tau-t})}{B - \theta} \right].$$

A special case of interest is that for which $\tau = t$, so we are estimating the "current" S_t from Y_u, $u \leq t$. Then the result (7.33) with $\tau - t = 0$ reduces to give the optimal filter as $A^{(\tau)}(B) = c^* \, (1 - \theta B)^{-1}$. Thus the filtered estimate explicitly in this case is $\hat{S}_t = c^* \, (1 - \theta B)^{-1} \, Y_t = c^* \sum_{u=0}^{\infty} \theta^u \, Y_{t-u}$. We compute $\hat{S}_t = c^* \, (1 - \theta B)^{-1} \, Y_t$ recursively from $(1 - \theta B) \, \hat{S}_t = c^* \, Y_t$ as $\hat{S}_u = \theta \, \hat{S}_{u-1} + c^* \, Y_u$, $u = -\infty, \ldots, t$. Since $\hat{S}_t = c^* \, (1 - \theta B)^{-1} \, Y_t = c^* \, (1 - \phi B)^{-1} \, \varepsilon_t$, the MSE of the filter in this case is

$$E[\,(S_t - \hat{S}_t)^2\,] = \mathrm{Var}(S_t) - \mathrm{Var}(\hat{S}_t) = \frac{\sigma_a^2}{1 - \phi^2} \left[1 - \frac{c^*}{1 - \phi\,\theta} \right] = \frac{\theta\, \sigma_a^2}{\phi\,(1 - \phi\,\theta)} .$$

We note that this example is a simple case of a state-space model, $S_t = \phi \, S_{t-1} + a_t$ and $Y_t = S_t + N_t$, and the Kalman filter produces the optimal filtered estimate $\hat{S}_t$ based on the finite set of data $Y_t, \ldots, Y_1$. As t increases, it can be seen that the "steady-state" form of the Kalman filter equations will reduce to the classical filtering results. That is, the "steady-state" form of the Kalman filtering equation (7.3) will be

$$\hat{S}_t = \phi \, \hat{S}_{t-1} + K\,(Y_t - \phi \, \hat{S}_{t-1}) = \phi\,(1 - K)\,\hat{S}_{t-1} + K \, Y_t ,$$

and it can be established from (7.4)–(7.6) that the "steady-state" form of the Kalman gain K is

$$K = c^* = \sigma_a^2 \,/\, [\,\sigma_\varepsilon^2 (1 - \phi\,\theta)\,] = 1 - (\theta/\phi) \qquad \text{with} \qquad \phi\,(1 - K) = \theta ,$$

which is thus seen to correspond with the classical filtering equation given above. In addition, the "steady-state" form of the MSE $P_{t|t}$ in (7.6) is the MSE given above, $\theta\,\sigma_a^2 \,/\, [\,\phi\,(1 - \phi\,\theta)\,]$.

7.4.2 *Smoothing Relations for the Signal Plus Noise or Structural Components Model*

The preceding results can be generalized to the situation where we observe $Y_t = S_t + N_t$, where we assume the signal process $\{S_t\}$ and the noise process $\{N_t\}$ are independent and satisfy ARMA models, $\phi_s(B)\, S_t = \eta_s(B)\, a_t$, $\phi_n(B)\, N_t = \eta_n(B)\, b_t$, where a_t and b_t are independent white noise processes with variances σ_a^2 and σ_b^2. It follows that the observed process Y_t also satisfies an ARMA model $\phi(B)\, Y_t = \theta(B)\, \varepsilon_t$, where $\phi(B) = \phi_s(B)\, \phi_n(B)$ assuming no common factors in the AR operators. It then follows that the optimal linear "smoother" $\hat{S}_t = \sum_{u=-\infty}^{\infty} a_u \, Y_{t-u} = A(B)\, Y_t$ of S_t, based on the infinite set of values Y_u, $-\infty < u < \infty$, has filter given by

$$A(B) = \frac{g_{ss}(B)}{g_{yy}(B)} = \frac{\sigma_a^2 \, \phi(B)\, \phi(B^{-1})\, \eta_s(B)\, \eta_s(B^{-1})}{\sigma_\varepsilon^2 \, \theta(B)\, \theta(B^{-1})\, \phi_s(B)\, \phi_s(B^{-1})} .$$

In practice, the series S_t and N_t are not observable, and, hence, the models for S_t and N_t would usually not be known. Thus, the optimal filter would not be known in practice. However, an accurate estimate of the model for the observed series Y_t can be obtained, and, hence, by placing certain restrictions on the form of the models for S_t and N_t beyond those implied by the model for Y_t (e.g., by assuming N_t is white noise with the largest possible variance), one may obtain reasonable approximations to the optimal linear filter $A(B)$. These filtering and smoothing methods have applications to seasonal and trend adjustment (extracting seasonal and trend components) for economic time series, and approaches of the type just mentioned have been explored by Hillmer and Tiao (1982). Although the optimal smoothing results, such as (7.32), have been derived for the case where S_t and N_t are stationary processes, Bell (1984) has established that the results extend to the nonstationary case under reasonable assumptions for the nonstationary signal S_t and noise N_t processes. As suggested above, an alternative to the classical approach to filtering in the structural components models is to formulate the model in the state-space form and use the Kalman filtering and smoothing techniques, for example, as illustrated by Kitagawa and Gersch (1984). We now illustrate these ideas using the following simple example of a (nonstationary) seasonal plus trend decomposition model.

EXAMPLE 7.6. Consider a seasonal time series $\{Y_t\}$ of quarterly data and assume that $Y_t = S_t + T_t + N_t$, where the (unobservable) components S_t, T_t, and N_t correspond to the seasonal, trend, and stationary noise features of the series. We suppose models for the component series are

$$U(B)\, S_t = b_t\,, \qquad (1-B)\, T_t = (1+B)\, c_t\,, \qquad N_t = d_t\,,$$

where $U(B) = 1 + B + B^2 + B^3$, and b_t, c_t, and d_t are independent white noise series with variances σ_b^2, σ_c^2, and σ_d^2, respectively. Then the model for the observable series Y_t is

$$\begin{aligned}(1-B)\, U(B)\, Y_t &= (1-B^4)\, Y_t \\ &= (1-B)\, b_t + U(B)\,(1+B)\, c_t + (1-B^4)\, d_t \equiv \theta_4(B)\, \varepsilon_t\,,\end{aligned}$$

where $\theta_4(B) = 1 - \sum_{i=1}^4 \theta_i B^i$ is a certain MA(4) operator whose coefficients θ_i can be determined by solving equations that involve the autocovariances at lags zero through four of the series $W_t = (1-B^4)\, Y_t$. Optimal smoothers for the seasonal and trend components $\{S_t\}$ and $\{T_t\}$ from the classical approach based on the observations $\{Y_t\}$ are given by $\hat{S}_t = W_S(B)\, Y_t$ and $\hat{T}_t = W_T(B)\, Y_t$, where

$$W_S(B) = \frac{\sigma_b^2\,(1-B^4)\,(1-F^4)}{\sigma_\varepsilon^2\,\theta_4(B)\,\theta_4(F)\,U(B)\,U(F)} = \frac{\sigma_b^2\,(1-B)\,(1-F)}{\sigma_\varepsilon^2\,\theta_4(B)\,\theta_4(F)}$$

and

$$W_T(B) = \frac{\sigma_c^2\,(1-B^4)(1-F^4)(1+B)(1+F)}{\sigma_\varepsilon^2\,\theta_4(B)\theta_4(F)(1-B)(1-F)} = \frac{\sigma_c^2\,U(B)(1+B)U(F)(1+F)}{\sigma_\varepsilon^2\,\theta_4(B)\theta_4(F)}\,,$$

where $F = B^{-1}$ denotes the forward shift operator and $U(B)\,(1+B) = 1 + 2B + 2B^2 + 2B^3 + B^4$. Considering computation of the smoothed seasonal component $\hat{S}_t$, for example, we have

$$\hat{S}_t = b\,[\,(1-B)\,(1-F)\,/\,(\,\theta_4(B)\,\theta_4(F)\,)\,]\,Y_t \equiv b\,[\,(1-F)\,/\,\theta_4(F)\,]\,X_t^s\,,$$

say, where $b = \sigma_b^2 / \sigma_\varepsilon^2$, and $X_t^s = [\,(1-B)\,/\,\theta_4(B)\,]\,Y_t$. This implies that $\theta_4(B)\,X_t^s = (1-B)\,Y_t$, so that X_t^s is computed recursively from $X_t^s = \sum_{i=1}^{4} \theta_i\, X_{t-i}^s + (\,Y_t - Y_{t-1}\,)$. Then $\hat{S}_t$ is computed in a backward recursion from the relation $\theta_4(F)\,\hat{S}_t = b\,(1-F)\,X_t^s$ as $\hat{S}_t = \sum_{i=1}^{4} \theta_i\, \hat{S}_{t+i} + b\,(\,X_t^s - X_{t+1}^s\,)$. Similar computations are involved in forming the smoothed estimate $\hat{T}_t$ of the trend component. However, for the computations of both $\hat{S}_t$ and $\hat{T}_t$ the smoothing calculations need to be specialized to account for the initializing values at both ends of the finite data series Y_t, possibly using methods of back-casting and forecasting of the observed series Y_t.

As an alternative to the above approach, one can formulate the components model $Y_t = S_t + T_t + N_t$ as a state-variable form, and then utilize the Kalman filtering and smoothing techniques from Section 7.1 to obtain the desired smoothed estimates of the seasonal and trend components. In Example 7.6, for example, we define the state vector $\mathbf{Z}_t = (\,S_t,\, S_{t-1},\, S_{t-2},\, T_t,\, \hat{T}_t(1),\, N_t\,)'$, and let $\boldsymbol{e}_t = (\,b_t,\, c_t,\, d_t\,)'$, with $\Sigma_e = \mathrm{Cov}(\,\boldsymbol{e}_t\,) = \mathrm{Diag}(\,\sigma_b^2,\, \sigma_c^2,\, \sigma_d^2\,)$. Then assuming the models for the components as given above, we have the state-variable equation representation $\mathbf{Z}_t = \Phi\,\mathbf{Z}_{t-1} + \Psi\,\boldsymbol{e}_t$ given by

$$\mathbf{Z}_t = \begin{bmatrix} -1 & -1 & -1 & 0 & 0 & 0 \\ 1 & 0 & 0 & 0 & 0 & 0 \\ 0 & 1 & 0 & 0 & 0 & 0 \\ 0 & 0 & 0 & 0 & 1 & 0 \\ 0 & 0 & 0 & 0 & 1 & 0 \\ 0 & 0 & 0 & 0 & 0 & 0 \end{bmatrix} \begin{bmatrix} S_{t-1} \\ S_{t-2} \\ S_{t-3} \\ T_{t-1} \\ \hat{T}_{t-1}(1) \\ N_{t-1} \end{bmatrix} + \begin{bmatrix} 1 & 0 & 0 \\ 0 & 0 & 0 \\ 0 & 0 & 0 \\ 0 & 1 & 0 \\ 0 & \psi_1 & 0 \\ 0 & 0 & 1 \end{bmatrix} \begin{bmatrix} b_t \\ c_t \\ d_t \end{bmatrix},$$

where $\psi_1 = 2$, together with the observation equation $Y_t = [\,1,\, 0,\, 0,\, 1,\, 0,\, 1\,]\,\mathbf{Z}_t$. Hence, by applying the Kalman filtering and smoothing techniques to this state-variable representation, the desired optimal smoothed estimates $\hat{S}_{t|T}$ and $\hat{T}_{t|T}$ of the seasonal and trend components based on the observed data $Y_1, \ldots, Y_T$ can be readily obtained.

The above considerations concerning optimal filtering and smoothing can also be generalized to the vector case. For example, in the vector smoothing problem where we wish to estimate $\boldsymbol{S}_t$ from $\boldsymbol{Y}_u$, $-\infty < u < \infty$, it can be proven that the optimal filter $\hat{\boldsymbol{S}}_t = A(B)\,\boldsymbol{Y}_t = \sum_{u=-\infty}^{\infty} A_u\, \boldsymbol{Y}_{t-u}$ is obtained with $A(B) = g_{SY}(B^{-1})\, g_{YY}(B^{-1})^{-1}$, where $g_{SY}(B) = \sum_{l=-\infty}^{\infty} \Gamma_{sy}(l)\, B^l$ and $g_{YY}(B) = \sum_{l=-\infty}^{\infty} \Gamma_y(l)\, B^l$. When $\boldsymbol{S}_t$ and $\boldsymbol{Y}_t$ are generated by vector ARMA models, $\Phi_S(B)\,\boldsymbol{S}_t = \Theta_S(B)\,\boldsymbol{a}_t$ and $\Phi(B)\,\boldsymbol{Y}_t = \Theta(B)\,\varepsilon_t$, respectively, then it is known from Section 2.3.1 that the generating functions are given by

$$g_{YY}(B) = \Phi(B^{-1})^{-1}\, \Theta(B^{-1})\, \Sigma\, \Theta(B)'\, \Phi(B)'^{-1} = \Psi(B^{-1})\, \Sigma\, \Psi(B)',$$

where $\Psi(B) = \Phi(B)^{-1}\Theta(B)$, and similarly for $g_{SS}(B)$. Hence, in the signal extraction model $\boldsymbol{Y}_t = \boldsymbol{S}_t + \boldsymbol{N}_t$, the filter

$$A(B) = g_{SY}(B^{-1})\, g_{YY}(B^{-1})^{-1} = g_{SS}(B^{-1})\, g_{YY}(B^{-1})^{-1}$$

can be determined in terms of the parameters of the ARMA models for $\boldsymbol{Y}_t$ and $\boldsymbol{S}_t$.

7.4.3 A Simple Vector Structural Component Model for Trend

We close this section with consideration of another notable example of the structural components model formulation, that of the "vector exponential smoothing model". In this model, the observed vector time series $\boldsymbol{Y}_t$ is represented as $\boldsymbol{Y}_t = \boldsymbol{\mu}_t + \boldsymbol{N}_t$, where $\boldsymbol{\mu}_t$ is an unobserved vector of trend components assumed to follow a vector random walk model, possibly with drift, $\boldsymbol{\mu}_t = \boldsymbol{\mu}_{t-1} + \boldsymbol{\delta} + \boldsymbol{a}_t$, and $\boldsymbol{a}_t$ and $\boldsymbol{N}_t$ are independent vector white noise processes. Such special forms of structural models have been studied by many authors, including Jones (1966), Harvey (1986, 1989), and Fernandez (1990). This model is clearly a special form of the state-variable model discussed in Section 7.1, with $\Phi = I$ and $H = I$ in (7.1)–(7.2) and with state vector $\mathbf{Z}_t = \boldsymbol{\mu}_t$. Hence, the Kalman filtering and smoothing techniques can be directly applied to the analysis of this model, including the Gaussian estimation of unknown covariance matrix parameters Σ_a and Σ_n for $\boldsymbol{a}_t$ and $\boldsymbol{N}_t$ and the estimation of the trend component $\boldsymbol{\mu}_t$. This leads, in particular, to the usual "exponential smoothing formula" for the filtered estimate $\hat{\boldsymbol{\mu}}_{t|t} = E[\, \boldsymbol{\mu}_t \mid \boldsymbol{Y}_1, \ldots, \boldsymbol{Y}_t \,]$ of the trend component $\boldsymbol{\mu}_t$. That is, applying (7.3)–(7.6) to this model (with $\boldsymbol{\delta} = 0$) we obtain

$$\hat{\boldsymbol{\mu}}_{t|t} = \hat{\boldsymbol{\mu}}_{t-1|t-1} + K_t\, (\, \boldsymbol{Y}_t - \hat{\boldsymbol{\mu}}_{t-1|t-1}\,) = K_t\, \boldsymbol{Y}_t + (\, I - K_t\,)\, \hat{\boldsymbol{\mu}}_{t-1|t-1}\,, \tag{7.34}$$

with

$$K_t = P_{t|t-1}\, [\, P_{t|t-1} + \Sigma_n\,]^{-1} \tag{7.35a}$$

and

$$P_{t|t-1} = [\, I - K_{t-1}\,]\, P_{t-1|t-2} + \Sigma_a\,. \tag{7.35b}$$

The "steady-state" value of $K_t \to K$ as $t \to \infty$ is given by $K = I - \Theta$ with $P_{t|t-1} \to P = \Sigma_\varepsilon - \Sigma_n = \Sigma_\varepsilon\, (\, I - \Theta\,)'$, where $\Theta = \Sigma_n\, \Sigma_\varepsilon^{-1}$ and Σ_ε are as defined in the next paragraph. From (7.10) of Section 7.1.3, the steady-state value $P = \Sigma_\varepsilon - \Sigma_n$ is determined so as to satisfy the equation $P\, [\, P + \Sigma_n\,]^{-1}\, P = \Sigma_a$, and the steady-state value $K = I - \Theta$ is determined by $K = P\, [\, P + \Sigma_n\,]^{-1}$. Hence, the steady-state form of the equation for the filtered estimate of $\boldsymbol{\mu}_t$ is $\hat{\boldsymbol{\mu}}_{t|t} = (\, I - \Theta\,)\, \boldsymbol{Y}_t + \Theta\, \hat{\boldsymbol{\mu}}_{t-1|t-1}$, which can also be expressed as $\hat{\boldsymbol{\mu}}_{t|t} = (\, I - \Theta\,) \sum_{j=0}^{\infty} \Theta^j\, \boldsymbol{Y}_{t-j}$, and has the "vector exponential smoothing" form.

An interesting modification of the basic model above is to represent $\boldsymbol{Y}_t$ as $\boldsymbol{Y}_t = H\,\boldsymbol{\mu}_t + \boldsymbol{N}_t$, where H is $k \times d$ with $d < k$ and now $\boldsymbol{\mu}_t$ is a lower-dimensional $d \times 1$ vector random walk, which represents a reduced number of "common trends" among the components of the series $\boldsymbol{Y}_t$. Such a model has direct connections with the concept of cointegration among the components of $\boldsymbol{Y}_t$ as considered in Section 6.3, and we see that such a model implies the existence of $r = k - d$ co-integrating relationships among the components of $\boldsymbol{Y}_t$. In fact, r linear combinations, $\boldsymbol{Z}_{2t} = Q_2^{'}\boldsymbol{Y}_t$ such that $Q_2^{'}H = 0$, would form a (stationary) r-dimensional white noise process. This model is a special case of more general common trends representations considered by Stock and Watson (1988) and Engle and Yoo (1987). By considering

$$\boldsymbol{W}_t = (1 - B)\,\boldsymbol{Y}_t = H\,(\,\boldsymbol{\mu}_t - \boldsymbol{\mu}_{t-1}\,) + (\,\boldsymbol{N}_t - \boldsymbol{N}_{t-1}\,) = H\,\boldsymbol{\delta} + H\,\boldsymbol{a}_t + (\,\boldsymbol{N}_t - \boldsymbol{N}_{t-1}\,),$$

we see that $\boldsymbol{W}_t = (\,1 - B\,)\,\boldsymbol{Y}_t$ has the covariance structure that implies a representation as a vector MA(1) model, $\boldsymbol{W}_t = (\,1 - B\,)\,\boldsymbol{Y}_t = \boldsymbol{\delta}^* + (\,I - \Theta B\,)\,\varepsilon_t$ where ε_t is a vector white noise with $\mathrm{Cov}(\,\varepsilon_t\,) = \Sigma_\varepsilon$. That is, the process $\boldsymbol{Y}_t$ follows the familiar IMA(1,1) model considered in Sections 2.4.3 and 2.5.4. The relations between parameters in the two model forms are

$$\Gamma_w(0) = H\,\Sigma_a\,H^{'} + 2\,\Sigma_n \equiv \Sigma_\varepsilon + \Theta\,\Sigma_\varepsilon\,\Theta^{'}$$

and

$$\Gamma_w(1) = -\,\Sigma_n \equiv -\,\Sigma_\varepsilon\,\Theta^{'} = -\,\Theta\,\Sigma_\varepsilon\,,$$

which imply that $H\,\Sigma_a\,H^{'} = (\,I - \Theta\,)\,\Sigma_\varepsilon\,(\,I - \Theta\,)^{'}$. These relations uniquely determine the parameters Θ and Σ_ε in terms of Σ_a and Σ_n. Thus, in the case of a reduced number d of common trends, the matrix Θ in the IMA(1,1) representation is seen to correspond to a noninvertible MA operator and Θ possesses $r = k - d$ eigenvalues equal to one, since $Q_2^{'}\,H = 0$ implies that $Q_2^{'}\,(\,I - \Theta\,) = 0$. Hence, this situation results from a form of "overdifferencing" in the IMA(1,1) model structure. Generally, in this situation we can write $\Theta = P\,\Lambda\,Q$ where $\Lambda = \mathrm{Diag}[\,\Lambda_d,\,I_r\,]$. Then by "cancellation" of the r "common differencing factors" $(\,1 - B\,)$ from both sides of the IMA(1,1) model equation, the process $\boldsymbol{Y}_t$ can also be brought into a more appropriate ARMA(1,1) representation, $(\,I - \Phi^*\,B\,)\,\boldsymbol{Y}_t = \boldsymbol{\mu}^* + (\,I - \Theta^*\,B\,)\,\varepsilon_t$, in which the autoregressive matrix $\Phi^* = P\,\mathrm{Diag}[\,I_d,\,0\,]\,Q$ has d unit eigenvalues, while the MA matrix $\Theta^* = P\,\mathrm{Diag}[\,\Lambda_d,\,0\,]\,Q$ yields an invertible MA operator (see, for instance, Example 2.7 in Section 2.4.3). Thus, the model will also have an error-correction form as an ARMA(1,1) model,

$$(\,1 - B\,)\,\boldsymbol{Y}_t = A\,B\,\boldsymbol{Y}_{t-1} + \boldsymbol{\mu}^* + (\,I - \Theta^*\,B\,)\,\varepsilon_t\,,$$

with $C = A\,B = \Phi^* - I = -\,P\,\mathrm{Diag}[\,0,\,I_r\,]\,Q$ and Θ^* of reduced ranks r and d, respectively.

Appendix. Time Series Data Sets

Table A.1. Natural Logarithms of the Annual Sales of Mink Furs, Y_{1t}, and Muskrat Furs, Y_{2t}, by the Hudson's Bay Company for the Years 1850–1911.

Year	Log Mink	Log Muskrat	Year	Log Mink	Log Muskrat
1850	10.2962	12.0752	1881	10.4957	13.6280
1851	9.9594	12.1791	1882	10.7277	13.8444
1852	10.1210	12.5863	1883	10.7687	13.8824
1853	10.1327	13.1102	1884	10.8646	13.8953
1854	10.6543	13.1466	1885	11.4167	13.6134
1855	10.8364	12.7531	1886	11.2451	12.7572
1856	11.0281	12.4638	1887	11.0714	12.8483
1857	11.0341	12.6191	1888	11.3269	12.7509
1858	11.2415	12.6556	1889	10.6152	12.3177
1859	11.0551	12.4461	1890	10.4800	12.6828
1860	10.7084	12.0855	1891	10.2914	13.2617
1861	10.3448	12.2357	1892	10.6517	13.6000
1862	10.8088	12.7230	1893	10.9711	13.7479
1863	10.6911	12.7857	1894	10.8359	13.3827
1864	11.0305	13.1417	1895	10.8452	13.4222
1865	11.0077	12.9441	1896	11.1595	13.6087
1866	10.8475	12.6786	1897	11.2433	13.2208
1867	10.9759	12.9292	1898	11.1620	13.2515
1868	11.2061	13.3344	1899	10.6416	13.4610
1869	11.2164	12.9096	1900	10.7359	13.5512
1870	10.2295	12.3556	1901	10.7751	13.7410
1871	10.3730	13.0036	1902	10.9616	14.3164
1872	10.5781	13.4657	1903	11.1057	14.2131
1873	10.7086	13.5514	1904	10.9091	13.7369
1874	11.0092	13.4180	1905	10.9330	13.8702
1875	11.1882	13.1689	1906	11.0030	13.4518
1876	11.2799	13.2765	1907	10.5756	12.9177
1877	11.2780	12.9880	1908	9.9774	12.5151
1878	11.3415	13.0940	1909	9.7902	12.6188
1879	11.0444	13.1218	1910	9.9891	13.5267
1880	10.4652	13.0775	1911	10.4045	13.7784

Table A.2. Quarterly, Seasonally Adjusted, U.S. Fixed Investment, Y^*_{1t}, and Changes in Business Inventories, Y_{2t}, for the Period 1947–1971 (from Lutkepohl, 1991).

Year	Quarter	Investment	Changes in Inventories	Year	Quarter	Investment	Changes in Inventories
1947	1	69.6	0.1	1957	1	96.2	2.5
	2	67.6	−0.9		2	95.3	2.9
	3	69.5	−2.9		3	96.4	3.7
	4	74.7	2.7		4	94.9	−3.0
1948	1	77.1	4.1	1958	1	90.0	−6.8
	2	77.4	5.6		2	87.2	−6.2
	3	76.6	6.9		3	88.0	0.3
	4	76.1	5.3		4	93.0	5.3
1949	1	71.8	−0.3	1959	1	98.3	5.0
	2	68.9	−7.1		2	101.6	13.0
	3	68.5	−2.5		3	102.6	−0.4
	4	70.6	−7.7		4	101.4	8.2
1950	1	75.4	4.4	1960	1	104.9	13.5
	2	82.3	7.7		2	101.8	4.9
	3	88.2	8.0		3	98.8	3.0
	4	86.9	22.1		4	98.6	−3.9
1951	1	83.4	13.4	1961	1	97.7	−3.8
	2	80.3	19.9		2	99.2	1.9
	3	79.4	14.6		3	101.3	6.6
	4	78.6	7.0		4	104.6	6.7
1952	1	79.3	7.3	1962	1	106.1	10.6
	2	80.3	−2.7		2	109.9	9.2
	3	75.3	5.4		3	111.1	8.0
	4	80.6	7.2		4	110.1	4.7
1953	1	83.9	3.9	1963	1	110.7	7.6
	2	84.2	5.1		2	116.0	7.0
	3	84.4	1.9		3	118.5	9.3
	4	83.8	−5.0		4	122.0	7.1
1954	1	82.8	−3.4	1964	1	124.0	6.1
	2	84.1	−4.1		2	124.0	8.0
	3	87.0	−2.7		3	124.9	7.3
	4	88.5	1.5		4	126.4	7.9
1955	1	92.1	5.9	1965	1	133.4	13.4
	2	96.1	8.0		2	137.9	10.6
	3	98.3	7.8		3	140.1	12.4
	4	98.8	9.2		4	143.8	8.8
1956	1	96.6	7.5	1966	1	147.5	13.5
	2	97.4	5.5		2	146.2	17.8
	3	97.6	4.9		3	145.0	15.1
	4	96.6	5.4		4	139.7	20.5

Table A.2. (continued)

Year	Quarter	Investment	Changes in Inventories	Year	Quarter	Investment	Changes in Inventories
1967	1	136.4	14.6	1970	1	151.8	2.9
	2	139.6	7.5		2	150.0	4.8
	3	141.1	12.2		3	150.4	6.3
	4	145.5	13.8		4	149.5	3.3
1968	1	148.9	6.3	1971	1	154.3	7.9
	2	148.9	11.8		2	158.4	10.0
	3	150.7	9.2		3	162.1	5.0
	4	155.0	7.6		4	166.0	3.7
1969	1	159.1	9.8				
	2	158.4	12.2				
	3	158.1	13.4				
	4	154.3	6.8				

Table A.3. Weekly Production Schedule Figures, Y_{1t} (in thousands of units), and Weekly Billing Figures, Y_{2t} (in millions of dollars), of a Company (from Makridakis and Wheelwright, 1978).

Period	Production	Billing	Period	Production	Billing	Period	Production	Billing
1	50.900	103.000	35	50.533	103.838	68	48.454	104.410
2	49.112	103.916	36	50.628	110.873	69	49.372	111.124
3	48.791	100.115	37	51.531	110.924	70	46.517	104.956
4	50.114	101.577	38	48.488	108.998	71	47.724	95.525
5	52.127	100.812	39	46.747	105.282	72	49.661	92.163
6	50.706	96.455	40	49.301	105.015	73	48.638	83.464
7	51.100	98.459	41	51.849	94.262	74	52.285	82.083
8	50.164	101.856	42	49.635	85.900	75	51.087	89.417
9	49.998	98.898	43	50.549	86.393	76	48.851	100.043
10	51.269	99.189	44	48.722	97.544	77	47.761	111.950
11	48.894	99.455	45	49.824	101.861	78	45.251	115.786
12	52.673	98.775	46	51.045	106.499	79	49.121	109.128
13	53.406	103.087	47	50.943	104.903	80	48.933	102.147
14	51.192	100.569	48	50.249	104.643	81	47.942	88.770
15	50.114	105.557	49	50.538	103.684	82	48.715	90.126
16	49.968	109.592	50	50.569	103.667	83	48.746	92.812
17	53.321	105.883	51	50.671	103.087	84	49.058	92.246
18	51.683	102.970	52	51.360	96.766	85	51.504	95.298
19	49.843	94.108	53	49.646	95.956	86	48.549	100.044
20	51.464	97.793	54	49.415	95.719	87	49.037	102.806
21	51.671	99.117	55	48.940	96.644	88	49.361	112.437
22	50.456	101.337	56	50.746	93.025	89	50.291	106.934
23	50.395	101.926	57	48.912	92.576	90	47.061	106.283
24	52.591	103.925	58	49.358	95.205	91	47.768	102.776
25	51.916	103.674	59	47.812	100.016	92	47.585	100.181
26	49.967	100.041	60	49.362	101.049	93	48.311	92.541
27	51.100	101.488	61	50.630	104.115	94	48.049	86.056
28	50.214	97.475	62	52.009	101.299	95	45.933	85.623
29	47.217	88.786	63	48.907	98.391	96	49.883	89.016
30	48.172	88.079	64	49.114	99.020	97	46.565	96.537
31	50.618	90.521	65	52.602	103.847	98	44.731	95.561
32	51.055	92.665	66	53.808	99.947	99	47.420	106.057
33	52.299	92.890	67	49.900	97.395	100	49.694	103.590
34	51.282	98.968						

Table A.4. Monthly U.S. Grain Price Data (in dollars per 100-pound sack for wheat flour, and per bushel for corn, wheat, and rye) for the Period January 1961 – October 1972.

Year	Month	Wheat Flour	Corn	Wheat	Rye	Year	Month	Wheat Flour	Corn	Wheat
1961	1	5.58	1.12	1.62	1.075	1964	1	6.45	1.26	1.78
	2	5.46	1.15	1.58	1.064		2	6.90	1.23	1.75
	3	5.50	1.18	1.56	1.158		3	6.64	1.24	1.48
	4	5.53	1.09	1.39	1.073		4	6.68	1.26	1.66
	5	5.63	1.15	1.28	1.125		5	6.70	1.29	1.54
	6	5.65	1.13	1.28	1.070		6	6.53	1.26	1.43
	7	5.70	1.18	1.36	1.298		7	7.09	1.23	1.46
	8	5.68	1.15	1.46	1.168		8	6.70	1.25	1.44
	9	5.75	1.11	1.45	1.205		9	6.73	1.30	1.52
	10	5.66	1.11	1.47	1.298		10	6.84	1.25	1.50
	11	5.70	1.14	1.51	1.320		11	6.90	1.19	1.56
	12	5.75	1.12	1.53	1.295		12	6.90	1.28	1.59
1962	1	5.75	1.09	1.52	1.298	1965	1	6.68	1.29	1.57
	2	5.78	1.11	1.52	1.293		2	6.70	1.31	1.58
	3	5.72	1.11	1.54	1.253		3	6.70	1.34	1.56
	4	5.80	1.12	1.59	1.228		4	6.70	1.35	1.56
	5	5.95	1.16	1.65	1.230		5	6.70	1.37	1.46
	6	6.09	1.15	1.63	1.278		6	6.90	1.36	1.45
	7	6.25	1.12	1.62	1.125		7	7.07	1.33	1.43
	8	6.37	1.14	1.57	1.145		8	7.10	1.29	1.53
	9	6.45	1.13	1.59	1.120		9	7.10	1.31	1.54
	10	6.44	1.12	1.55	1.148		10	7.20	1.27	1.53
	11	6.46	1.10	1.59	1.164		11	7.28	1.17	1.66
	12	6.39	1.17	1.58	1.210		12	7.21	1.24	1.74
1963	1	6.03	1.21	1.60	1.258	1966	1	7.22	1.32	1.71
	2	6.05	1.21	1.63	1.228		2	7.22	1.32	1.75
	3	6.02	1.20	1.64	1.195		3	7.22	1.26	1.67
	4	6.15	1.21	1.68	1.233		4	7.19	1.30	1.65
	5	6.09	1.23	1.63	1.196		5	7.20	1.32	1.67
	6	6.05	1.30	1.38	1.220		6	7.63	1.31	1.77
	7	5.70	1.34	1.30	1.205		7	8.07	1.42	1.90
	8	5.28	1.34	1.30	1.185		8	8.09	1.45	1.89
	9	5.76	1.36	1.47	1.355		9	7.92	1.46	1.92
	10	5.98	1.22	1.64	1.433		10	7.72	1.40	1.69
	11	6.41	1.18	1.69	1.420		11	7.73	1.35	1.75
	12	6.83	1.23	1.69	1.423		12	7.62	1.44	1.83

Table A.4. (continued)

Year	Month	Wheat Flour	Corn	Wheat	Rye	Year	Month	Wheat Flour	Corn	Wheat
1967	1	7.35	1.44	1.80	1.203	1970	1	7.14	1.26	1.50
	2	7.48	1.39	1.68	1.163		2	7.17	1.26	1.52
	3	7.55	1.42	1.83	1.208		3	7.19	1.25	1.54
	4	7.27	1.38	1.80	1.188		4	7.19	1.29	1.54
	5	7.43	1.39	1.67	1.180		5	7.19	1.32	1.55
	6	7.34	1.38	1.57	1.160		6	7.20	1.36	1.40
	7	7.40	1.33	1.48	1.191		7	7.16	1.39	1.43
	8	7.29	1.21	1.42	1.099		8	7.21	1.37	1.45
	9	7.27	1.20	1.45	1.173		9	7.20	1.45	1.68
	10	7.24	1.20	1.49	1.140		10	7.35	1.44	1.67
	11	7.18	1.08	1.44	1.118		11	7.43	1.42	1.74
	12	7.21	1.18	1.50	1.148		12	7.50	1.33	1.63
1968	1	7.18	1.15	1.52	1.154	1971	1	7.45	1.39	1.72
	2	7.04	1.17	1.56	1.190		2	7.45	1.40	1.74
	3	6.94	1.14	1.51	1.179		3	7.45	1.36	1.68
	4	6.76	1.16	1.46	1.145		4	7.51	1.33	1.66
	5	6.78	1.21	1.47	1.130		5	7.58	1.31	1.66
	6	6.77	1.16	1.26	1.073		6	7.55	1.38	1.63
	7	6.73	1.15	1.29	1.125		7	7.52	1.30	1.44
	8	6.72	1.08	1.19	1.046		8	7.48	1.29	1.31
	9	6.85	1.08	1.18	1.119		9	7.48	1.17	1.33
	10	6.89	1.10	1.28	1.145		10	7.48	1.12	1.39
	11	6.90	1.17	1.36	1.169		11	7.48	1.09	1.49
	12	6.85	1.15	1.34	1.168		12	7.48	1.26	1.60
1969	1	6.85	1.19	1.43	1.205	1972	1	7.28	1.23	1.58
	2	6.83	1.17	1.40	1.201		2	7.15	1.21	1.52
	3	6.76	1.17	1.35	1.215		3	7.15	1.23	1.58
	4	6.75	1.22	1.36	1.215		4	7.15	1.25	1.68
	5	6.75	1.31	1.40	1.240		5	7.07	1.28	1.67
	6	6.75	1.32	1.33	1.205		6	7.07	1.28	1.36
	7	6.92	1.30	1.30	1.185		7	7.07	1.30	1.47
	8	6.86	1.29	1.28	1.035		8	7.66	1.29	1.60
	9	6.78	1.27	1.33	1.055		9	8.39	1.39	1.86
	10	6.89	1.27	1.33	1.105		10	8.57	1.31	1.92
	11	6.95	1.16	1.42	1.115					
	12	7.14	1.18	1.48	1.135					

Note: The wheat grain price series has been adjusted over the period January 1961 through May 1964 for a price level shift effect, by subtracting 0.55 from the original series.

Table A.5. Monthly U.S. Housing-Starts, Y_{1t}, and Housing-Sold, Y_{2t}, (in thousands) for the Period January 1965 Through December 1974.

		Housing				Housing				Housing	
Year	Month	Starts	Sold	Year	Month	Starts	Sold	Year	Month	Starts	Sold
1965	1	52.1	38		5	86.8	43		9	102.1	50
	2	47.2	44		6	81.4	41		10	102.9	52
	3	82.2	53		7	86.4	44		11	92.9	50
	4	100.9	49		8	82.5	47		12	80.4	44
	5	98.4	54		9	80.1	41	1972	1	76.2	51
	6	97.4	57		10	85.6	40		2	76.3	54
	7	96.5	51		11	64.8	32		3	111.4	60
	8	88.8	58		12	53.8	32		4	119.8	65
	9	80.9	48	1969	1	51.3	34		5	135.2	64
	10	85.8	44		2	47.9	40		6	131.9	63
	11	72.4	42		3	71.9	43		7	119.1	63
	12	61.2	37		4	85.0	42		8	131.3	72
1966	1	46.6	42		5	91.3	43		9	120.5	61
	2	50.4	43		6	82.7	44		10	117.0	65
	3	83.2	53		7	73.5	39		11	97.4	51
	4	94.3	49		8	69.5	40		12	73.2	47
	5	84.7	49		9	71.5	33	1973	1	77.1	54
	6	79.8	40		10	68.0	32		2	73.6	58
	7	69.1	40		11	55.1	31		3	105.1	66
	8	69.4	36		12	42.8	28		4	120.5	63
	9	59.4	29	1970	1	33.4	34		5	131.6	64
	10	53.5	31		2	41.4	29		6	114.8	60
	11	50.2	26		3	61.9	36		7	114.7	53
	12	38.0	23		4	73.8	42		8	106.8	52
1967	1	40.2	29		5	74.8	43		9	84.5	44
	2	40.3	32		6	83.0	44		10	86.0	40
	3	66.6	41		7	75.5	44		11	70.5	36
	4	79.8	44		8	77.3	48		12	46.8	28
	5	87.3	49		9	76.0	45	1974	1	43.3	36
	6	87.6	47		10	79.4	44		2	57.6	42
	7	82.3	46		11	67.4	40		3	77.3	53
	8	83.7	47		12	69.0	37		4	102.3	53
	9	78.2	43	1971	1	54.9	45		5	96.4	55
	10	81.7	45		2	58.3	49		6	99.6	48
	11	69.1	34		3	91.6	62		7	90.9	47
	12	47.0	31		4	116.0	62		8	79.8	43
1968	1	45.2	35		5	115.6	58		9	73.4	39
	2	55.4	43		6	116.9	59		10	69.5	33
	3	79.3	46		7	107.7	64		11	57.9	30
	4	98.0	46		8	111.7	62		12	41.0	23

Table A.6. Quarterly, Seasonally Adjusted, U.S. AAA Corporate Bonds and Commercial Paper Interest Rates, 1953–1970.

Year	Quarter	AAA Bond Rate	Commercial Paper Rate	Year	Quarter	AAA Bond Rate	Commercial Paper Rate
1953	1	3.070	2.327	1962	1	4.410	3.243
	2	3.323	2.623		2	4.297	3.203
	3	3.270	2.747		3	4.337	3.333
	4	3.133	2.373		4	4.257	3.263
1954	1	2.957	2.037	1963	1	4.197	3.310
	2	2.877	1.633		2	4.220	3.317
	3	2.883	1.363		3	4.287	3.697
	4	2.887	1.310		4	4.333	3.907
1955	1	2.980	1.613	1964	1	4.370	3.950
	2	3.033	1.967		2	4.407	3.933
	3	3.100	2.327		3	4.410	3.910
	4	3.117	2.833		4	4.430	4.063
1956	1	3.097	3.000	1965	1	4.420	4.300
	2	3.263	3.263		2	4.443	4.380
	3	3.423	3.350		3	4.497	4.380
	4	3.677	3.630		4	4.613	4.470
1957	1	3.700	3.630	1966	1	4.813	4.970
	2	3.773	3.683		2	5.003	5.427
	3	4.070	3.953		3	5.320	5.790
	4	3.997	3.993		4	5.383	6.000
1958	1	3.607	2.917	1967	1	5.120	5.450
	2	3.580	1.817		2	5.263	4.717
	3	3.870	2.130		3	5.617	4.973
	4	4.093	3.213		4	6.027	5.303
1959	1	4.130	3.303	1968	1	6.127	5.580
	2	4.353	3.603		2	6.253	6.080
	3	4.473	4.193		3	6.077	5.963
	4	4.570	4.760		4	6.243	5.963
1960	1	4.553	4.678	1969	1	6.700	6.657
	2	4.453	4.073		2	6.887	7.540
	3	4.313	3.373		3	7.063	8.487
	4	4.320	3.270		4	7.467	8.620
1961	1	4.270	3.013	1970	1	7.893	8.553
	2	4.283	2.860		2	8.140	8.167
	3	4.437	2.897		3	8.220	7.837
	4	4.410	3.057		4	7.907	7.293

Table A.7. Monthly U.S. Interest Rate Series for the Federal Fund Rate, 90-Day Treasury Bill Rate, and the One-Year Treasury Bill Rate, for the Period January 1960 Through December 1979.

Year	Month	Federal Fund	90-Day	One-Year	Year	Month	Federal Fund	90-Day	One-Year
1960	1	3.99	4.35	4.95	1963	1	2.91	2.91	3.00
	2	3.97	3.96	4.45		2	3.00	2.92	3.00
	3	3.84	3.31	3.68		3	2.98	2.89	2.97
	4	3.92	3.23	3.83		4	2.90	2.90	3.03
	5	3.85	3.29	4.01		5	3.00	2.92	3.06
	6	3.32	2.46	3.10		6	2.99	2.99	3.11
	7	3.23	2.30	3.03		7	3.02	3.18	3.40
	8	2.98	2.30	2.82		8	3.49	3.32	3.50
	9	2.60	2.48	2.86		9	3.48	3.38	3.57
	10	2.47	2.30	2.92		10	3.50	3.45	3.61
	11	2.44	2.37	2.87		11	3.48	3.52	3.67
	12	1.98	2.25	2.64		12	3.38	3.52	3.69
1961	1	1.85	2.24	2.63	1964	1	3.48	3.52	3.68
	2	2.14	2.42	2.75		2	3.48	3.53	3.71
	3	2.02	2.39	2.76		3	3.43	3.54	3.78
	4	1.50	2.29	2.74		4	3.47	3.47	3.75
	5	1.98	2.29	2.72		5	3.50	3.48	3.71
	6	1.73	2.33	2.80		6	3.50	3.48	3.70
	7	1.56	2.24	2.79		7	3.42	3.46	3.64
	8	2.00	2.39	2.91		8	3.50	3.50	3.67
	9	1.88	2.28	2.88		9	3.45	3.53	3.73
	10	2.26	2.30	2.90		10	3.36	3.57	3.79
	11	2.62	2.48	2.90		11	3.52	3.64	3.86
	12	2.33	2.60	2.97		12	3.85	3.84	3.96
1962	1	2.14	2.72	3.19	1965	1	3.90	3.81	3.91
	2	2.37	2.73	3.21		2	3.98	3.93	4.00
	3	2.70	2.72	2.98		3	4.04	3.93	4.02
	4	2.69	2.73	2.90		4	4.09	3.93	4.00
	5	2.29	2.68	2.91		5	4.10	3.89	3.96
	6	2.68	2.73	2.89		6	4.04	3.80	3.89
	7	2.71	2.92	3.17		7	4.09	3.84	3.89
	8	2.93	2.82	3.10		8	4.12	3.84	3.96
	9	2.90	2.78	2.99		9	4.01	3.92	4.09
	10	2.90	2.74	2.90		10	4.08	4.03	4.16
	11	2.94	2.83	2.94		11	4.10	4.09	4.23
	12	2.93	2.87	2.94		12	4.32	4.38	4.56

Table A.7. (continued)

Year	Month	Federal Fund	90-Day	One-Year	Year	Month	Federal Fund	90-Day	One-Year
1966	1	4.42	4.59	4.69	1969	1	6.30	6.14	6.07
	2	4.60	4.65	4.81		2	6.64	6.12	6.21
	3	4.65	4.59	4.81		3	6.79	6.02	6.20
	4	4.67	4.62	4.76		4	7.41	6.11	6.04
	5	4.90	4.64	4.85		5	8.67	6.04	6.11
	6	5.17	4.50	4.78		6	8.90	6.44	6.88
	7	5.30	4.80	4.94		7	8.61	7.00	7.19
	8	5.53	4.96	5.34		8	9.19	6.98	7.28
	9	5.40	5.37	5.80		9	9.15	7.09	7.35
	10	5.53	5.35	5.52		10	9.00	7.00	7.21
	11	5.77	5.32	5.49		11	8.85	7.24	7.36
	12	5.40	4.96	5.00		12	8.97	7.82	7.62
1967	1	4.94	4.72	4.61	1970	1	8.98	7.87	7.51
	2	5.00	4.56	4.57		2	8.98	7.13	7.05
	3	4.53	4.26	4.18		3	7.76	6.63	6.50
	4	4.05	3.84	3.90		4	8.10	6.51	6.53
	5	3.94	3.60	3.88		5	7.94	6.84	7.12
	6	3.98	3.54	4.16		6	7.60	6.68	7.07
	7	3.79	4.21	4.90		7	7.21	6.45	6.62
	8	3.89	4.27	5.05		8	6.61	6.41	6.55
	9	4.00	4.42	5.10		9	6.29	6.13	6.39
	10	3.88	4.56	5.22		10	6.20	5.91	6.23
	11	4.12	4.73	5.39		11	5.60	5.28	5.39
	12	4.51	4.97	5.59		12	4.90	4.87	4.84
1968	1	4.60	5.00	5.30	1971	1	4.14	4.44	4.40
	2	4.72	4.98	5.24		2	3.72	3.70	3.84
	3	5.05	5.17	5.42		3	3.71	3.38	3.60
	4	5.76	5.38	5.47		4	4.15	3.86	4.09
	5	6.12	5.66	5.84		5	4.63	4.14	4.64
	6	6.07	5.52	5.68		6	4.91	4.75	5.33
	7	6.02	5.31	5.39		7	5.31	5.40	5.74
	8	6.03	5.09	5.15		8	5.57	4.94	5.52
	9	5.78	5.19	5.18		9	5.55	4.69	5.19
	10	5.92	5.35	5.33		10	5.20	4.46	4.75
	11	5.81	5.45	5.51		11	4.91	4.22	4.49
	12	6.02	5.96	5.97		12	4.14	4.01	4.40

Table A.7. (continued)

Year	Month	Federal Fund	90-Day	One-Year	Year	Month	Federal Fund	90-Day	One-Year
1972	1	3.50	3.38	3.82	1975	1	7.13	6.26	6.27
	2	3.29	3.20	4.06		2	6.24	5.50	5.56
	3	3.83	3.73	4.43		3	5.54	5.49	5.70
	4	4.17	3.71	4.65		4	5.49	5.61	6.40
	5	4.27	3.69	4.46		5	5.22	5.23	5.91
	6	4.46	3.91	4.71		6	5.55	5.34	5.86
	7	4.55	3.98	4.90		7	6.10	6.13	6.64
	8	4.80	4.02	4.90		8	6.14	6.44	7.16
	9	4.87	4.66	5.44		9	6.24	6.42	7.20
	10	5.04	4.74	5.39		10	5.82	5.96	6.48
	11	5.06	4.78	5.20		11	5.22	5.48	6.07
	12	5.33	5.07	5.28		12	5.20	5.44	6.16
1973	1	5.94	5.41	5.58	1976	1	4.87	4.87	5.44
	2	6.58	5.60	5.93		2	4.77	4.88	5.53
	3	7.09	6.09	6.53		3	4.84	5.00	5.82
	4	7.12	6.26	6.51		4	4.82	4.86	5.54
	5	7.84	6.36	6.63		5	5.29	5.20	5.98
	6	8.49	7.19	7.05		6	5.48	5.41	6.12
	7	10.40	8.01	7.97		7	5.31	5.23	5.82
	8	10.50	8.67	8.32		8	5.29	5.14	5.64
	9	10.78	8.29	8.07		9	5.25	5.08	5.50
	10	10.01	7.22	7.17		10	5.03	4.92	5.19
	11	10.03	7.83	7.40		11	4.95	4.75	5.00
	12	9.95	7.45	7.01		12	4.65	4.35	4.64
1974	1	9.65	7.77	7.01	1977	1	4.61	4.62	5.00
	2	8.97	7.12	6.51		2	4.68	4.67	5.16
	3	9.35	7.96	7.34		3	4.69	4.60	5.19
	4	10.51	8.33	8.08		4	4.73	4.54	5.10
	5	11.31	8.23	8.21		5	5.35	4.96	5.43
	6	11.93	7.90	8.16		6	5.39	5.02	5.41
	7	12.92	7.55	8.04		7	5.42	5.19	5.57
	8	12.01	8.96	8.88		8	5.90	5.49	5.97
	9	11.34	8.06	8.52		9	6.14	5.81	6.13
	10	10.06	7.46	7.59		10	6.47	6.16	6.52
	11	9.45	7.47	7.29		11	6.51	6.10	6.52
	12	8.53	7.15	6.79		12	6.56	6.07	6.52

Table A.7. (continued)

Year	Month	Federal Fund	90-Day	One-Year	Year	Month	Federal Fund	90-Day	One-Year
1978	1	6.70	6.44	6.80	1979	1	10.07	9.35	9.54
	2	6.78	6.45	6.86		2	10.06	9.32	9.39
	3	6.79	6.29	6.82		3	10.09	9.48	9.38
	4	6.89	6.29	6.96		4	10.01	9.46	9.28
	5	7.36	6.41	7.28		5	10.24	9.61	9.27
	6	7.60	6.73	7.53		6	10.29	9.06	8.81
	7	7.81	7.01	7.79		7	10.47	9.24	8.87
	8	8.04	7.08	7.73		8	10.94	9.52	9.16
	9	8.45	7.85	8.01		9	11.43	10.26	9.89
	10	8.96	7.99	8.45		10	13.77	11.70	11.23
	11	9.76	8.64	9.20		11	13.18	11.79	11.22
	12	10.03	9.08	9.44		12	13.78	12.04	10.92

Exercises and Problems

Chapter 2

2.1 Let $\varepsilon = (\varepsilon_1', \varepsilon_2')'$ be a k-dimensional random vector distributed as multivariate normal $N(0, \Sigma)$, where Σ is partitioned as

$$\Sigma = \begin{bmatrix} \Sigma_{11} & \Sigma_{12} \\ \Sigma_{12}' & \Sigma_{22} \end{bmatrix}.$$

Find the joint distribution of $a_1 = \varepsilon_1$, $a_2 = \varepsilon_2 - \Sigma_{12}' \Sigma_{11}^{-1} \varepsilon_1$, and in particular show that a_2 and $\varepsilon_1 \equiv a_1$ are independent. Hence, one can always write $\varepsilon_2 = B' \varepsilon_1 + a_2$, where $B' = \Sigma_{12}' \Sigma_{11}^{-1}$ is the regression coefficient matrix of ε_2 on ε_1, and ε_1 and a_2 are independent. (Note that this is a generalization of a bivariate result discussed in Examples 2.2 and 2.4.)

2.2 Consider the bivariate MA(1) process $Y_t = (I - \Theta B)\varepsilon_t$, with

$$\Theta = \begin{bmatrix} 0.4 & 0.3 \\ -0.5 & 0.8 \end{bmatrix}, \qquad \Sigma = \begin{bmatrix} 4 & 1 \\ 1 & 2 \end{bmatrix}.$$

(a) Find the lag 0 and lag 1 autocorrelations and cross-correlations of Y_t, i.e., find the matrices $\Gamma(0)$, $\Gamma(1)$, and $\rho(0)$, $\rho(1)$.

(b) Find the individual univariate MA(1) models for Y_{1t} and Y_{2t}, i.e., in the models $Y_{it} = (1 - \eta_i B) a_{it}$, $i = 1, 2$, find the values of the parameters η_i and $\sigma_{a_i}^2 = \text{Var}(a_{it})$, from $\Gamma(0)$ and $\Gamma(1)$.

(c) State the matrix difference equation satisfied by the matrix weights Π_j in the infinite AR form of the above bivariate MA(1) model, and explicitly evaluate the Π_j for $j = 1, 2, 3, 4$.

(d) It follows from Section 2.5 that the diagonal elements of Σ represent the one-step ahead forecast error variances for the two series when each series is forecast from the past history of both series, i.e., when each series is forecast based on the bivariate model. Compare these one-step forecast error variances in the bivariate model with the one-step forecast error variances $\sigma_{a_i}^2$ based on the individual univariate models in (b).

(e) Reexpress the above MA(1) model in the form $Y_t = \Theta_0^+ a_t - \Theta_1^+ a_{t-1}$, where Θ_0^+

is lower triangular with ones on the diagonal, and $\boldsymbol{a}_t$ is a vector white noise process with *diagonal* covariance matrix $\text{Cov}(\boldsymbol{a}_t) = \Sigma^{\#}$.

(f) For a general bivariate MA(1) process, $\boldsymbol{Y}_t = (I - \Theta B)\,\varepsilon_t$, verify that the lag-1 autocorrelations $\rho_{ii}(1) = \text{Corr}(Y_{it}, Y_{i,t+1})$ satisfy $|\rho_{ii}(1)| \le 0.5$, for $i = 1, 2$.

2.3 For the multivariate AR(1) model $(I - \Phi B)\,\boldsymbol{Y}_t = \varepsilon_t$, it is known that $\Gamma(0) - \Phi\,\Gamma(0)\,\Phi' = \Sigma$. Hence, if the model parameters Φ and Σ are given, the above matrix equation may be solved to determine $\Gamma(0)$. In the bivariate case, this leads to three linear equations in the unknowns $\gamma_{11}(0)$, $\gamma_{12}(0)$, and $\gamma_{22}(0)$. If these equations are expressed in matrix form as $A\,[\,\gamma_{11}(0), \gamma_{12}(0), \gamma_{22}(0)\,]' = \boldsymbol{b}$, give explicitly the expressions for A and $\boldsymbol{b}$. Consider the specific case

$$\Phi = \begin{bmatrix} 0.2 & 0.3 \\ -0.6 & 1.1 \end{bmatrix}, \qquad \Sigma = \begin{bmatrix} 4 & 1 \\ 1 & 1 \end{bmatrix}.$$

(a) Show that

$$\Gamma(0) = \begin{bmatrix} 5.667 & 4.00 \\ 4.00 & 10.667 \end{bmatrix}.$$

Also, determine the stationarity of the above AR(1) model, state the difference equation satisfied by the $\Gamma(j)$, $j \ge 1$, and find the values of $\Gamma(1)$, $\Gamma(2)$, and $\Gamma(3)$. In addition, compute the cross-correlation matrices $\rho(0)$, $\rho(1)$, $\rho(2)$, and $\rho(3)$.

(b) For the above specific case, determine the forms of the individual univariate models for the Y_{it} as ARMA(2,1) models and determine the values of the variances $\sigma^2_{a_i}$ of the random shocks a_{it} in these univariate models. Compare these univariate one-step forecast error variances with the diagonal elements of Σ.

(c) Find the matrix coefficients Ψ_1, Ψ_2, and Ψ_3 in the infinite MA representation for $\boldsymbol{Y}_t$, and, hence, compute the covariance matrix of the bivariate lead-l forecast errors from the bivariate model using the formula $\Sigma(l) = \sum_{j=0}^{l-1} \Psi_j\,\Sigma\,\Psi_j'$, for $l = 1, 2, 3$. Compare the variances with the corresponding variances of the forecast errors from the individual univariate models in (b).

(d) For the general bivariate AR(1) model, indicate what simplifications occur in the individual univariate models for the Y_{it} when Φ is lower triangular, when Φ is diagonal, and when $\det(\Phi) = 0$ (i.e., Φ is singular). Show that the case $\det(\Phi) = 0$ implies that there exists a linear combination of Y_{1t} and Y_{2t}, $Z_{1t} = c_{11}\,Y_{1t} + c_{12}\,Y_{2t}$, which is a white noise series, and a second linear combination $Z_{2t} = c_{21}\,Y_{1t} + c_{22}\,Y_{2t}$ which is again a univariate AR(1) process.
Hint: If $\det(\Phi) = 0$, then Φ has rank at most one and can be written as

$$\Phi = \begin{bmatrix} \phi_{11} & \phi_{12} \\ \alpha\,\phi_{11} & \alpha\,\phi_{12} \end{bmatrix} = \begin{bmatrix} 1 \\ \alpha \end{bmatrix} \begin{bmatrix} \phi_{11} & \phi_{12} \end{bmatrix}.$$

2.4 Consider the vector AR(p) model

$$\boldsymbol{Y}_t = \Phi_1\,\boldsymbol{Y}_{t-1} + \Phi_2\,\boldsymbol{Y}_{t-2} + \cdots + \Phi_p\,\boldsymbol{Y}_{t-p} + \varepsilon_t.$$

Verify that the model can be expressed as a vector AR(1) model in terms of the kp-dimensional vector $\mathbf{Y}_t = (\boldsymbol{Y}_t', \boldsymbol{Y}_{t-1}', \ldots, \boldsymbol{Y}_{t-p+1}')'$, $\mathbf{Y}_t = \mathbf{\Phi}\,\mathbf{Y}_{t-1} + \mathbf{e}_t$, using the $kp \times kp$ *companion matrix* $\mathbf{\Phi}$ for the AR(p) operator $\Phi(B) = I - \Phi_1 B - \cdots - \Phi_p B^p$,

$$\Phi = \begin{bmatrix} \Phi_1 & \Phi_2 & . & . & . & \Phi_p \\ I & 0 & . & . & . & 0 \\ 0 & I & . & . & . & 0 \\ . & . & . & . & . & . \\ . & . & . & . & . & . \\ 0 & 0 & . & . & I & 0 \end{bmatrix}.$$

In addition, show that $\det\{ I - \Phi B \} = \det\{ I - \Phi_1 B - \cdots - \Phi_p B^p \}$, and hence the stationarity condition for the AR(p) process is equivalent to the condition that all eigenvalues of the companion matrix Φ be less than one in absolute value. [Hint: To evaluate $\det\{ I - \Phi B \}$, multiply the ith column of $I - \Phi B$ by B and add to the $(i-1)$st column, successively, for $i = p, p-1, \ldots, 2$.]

2.5 Consider the vector ARMA(1,1) model $\boldsymbol{Y}_t - \Phi \boldsymbol{Y}_{t-1} = \varepsilon_t - \Theta \varepsilon_{t-1}$, and assume the covariance matrices $\Gamma(l)$ of the process are given, with $\Gamma(1)$ nonsingular (for purposes of unique identifiability). We want to determine the appropriate equations to obtain the parameters Φ, Θ, and Σ from the $\Gamma(l)$. First, Φ can be obtained from $\Gamma(2) = \Gamma(1)\, \Phi'$. Then Σ and Θ can be obtained from the equations (see Section 2.3.3)

$$\Gamma(0) - \Gamma(1)'\, \Phi' = \Sigma - (\Phi - \Theta)\, \Sigma\, \Theta',$$

$$\Gamma(1) - \Gamma(0)\, \Phi' = - \Sigma\, \Theta'.$$

(a) From the above two equations, solve for Σ given Θ.

(b) Show that Θ can then be determined to be the solution of a matrix equation of the form $\Theta^2 A_1 + \Theta A_2 + A_3 = 0$. Find the expressions for the matrices A_1, A_2, and A_3, as functions of $\Gamma(0)$, $\Gamma(1)$, and $\Gamma(2)$.

2.6 Let $\mathbf{X} = (\mathbf{X}_1', \mathbf{X}_2')'$ be distributed such that $\boldsymbol{\mu} = E(\mathbf{X}) = (\boldsymbol{\mu}_1', \boldsymbol{\mu}_2')'$, and

$$\mathrm{Cov}(\mathbf{X}) = \Sigma = \begin{bmatrix} \Sigma_{11} & \Sigma_{12} \\ \Sigma_{12}' & \Sigma_{22} \end{bmatrix}.$$

Suppose we want to predict $\mathbf{X}_1$ by a linear function of $\mathbf{X}_2$ of the form $\hat{\mathbf{X}}_1 = \boldsymbol{a} + B\, \mathbf{X}_2$, where $\boldsymbol{a}$ is a vector of constants and B is a matrix of constants. Show that for any arbitrary linear combination $\boldsymbol{c}'\mathbf{X}_1$, the values of $\boldsymbol{a}$ and B which give the minimum mean square error (MSE) of prediction,

$$E[\,(\boldsymbol{c}'\mathbf{X}_1 - \boldsymbol{c}'\hat{\mathbf{X}}_1)^2\,] = \boldsymbol{c}'\, E[\,(\mathbf{X}_1 - \hat{\mathbf{X}}_1)(\mathbf{X}_1 - \hat{\mathbf{X}}_1)'\,]\, \boldsymbol{c},$$

are $B = \Sigma_{12}\Sigma_{22}^{-1}$ and $\boldsymbol{a} = \boldsymbol{\mu}_1 - B\, \boldsymbol{\mu}_2$, so that the minimum MSE predictor of $\mathbf{X}_1$ is $\hat{\mathbf{X}}_1 = \boldsymbol{\mu}_1 + \Sigma_{12}\Sigma_{22}^{-1}(\mathbf{X}_2 - \boldsymbol{\mu}_2)$. Also show that for this predictor, the minimum MSE matrix is

$$E[\,(\mathbf{X}_1 - \hat{\mathbf{X}}_1)(\mathbf{X}_1 - \hat{\mathbf{X}}_1)'\,] = \Sigma_{11} - \Sigma_{12}\Sigma_{22}^{-1}\Sigma_{12}',$$

and, hence, $\mathrm{Cov}(\mathbf{X}_1 - \hat{\mathbf{X}}_1) = \mathrm{Cov}(\mathbf{X}_1) - \mathrm{Cov}(\hat{\mathbf{X}}_1)$. Also note that $\mathbf{X}_2$, and hence $\hat{\mathbf{X}}_1$, is uncorrelated with the prediction error $\boldsymbol{e} = \mathbf{X}_1 - \hat{\mathbf{X}}_1$, so that we have $\mathbf{X}_1 = \hat{\mathbf{X}}_1 + \boldsymbol{e}$ and $\mathrm{Cov}(\mathbf{X}_1) = \mathrm{Cov}(\hat{\mathbf{X}}_1) + \mathrm{Cov}(\boldsymbol{e})$.

Remark: If $\mathbf{X}$ is normally distributed, then $\hat{\mathbf{X}}_1 = E(\mathbf{X}_1 \mid \mathbf{X}_2)$ and

$\mathrm{Cov}(\mathbf{X}_1 - \hat{\mathbf{X}}_1) = \mathrm{Cov}(\mathbf{X}_1 \mid \mathbf{X}_2)$.

Relate the above results in the context of prediction of stationary vector time series $\boldsymbol{Y}_t = (Y_{1t}, \ldots, Y_{kt})'$ based on AR model fitting and solutions to Yule-Walker equations, with the correspondences that

$$\mathbf{X}_1 = \boldsymbol{Y}_t, \qquad \mathbf{X}_2 = \mathbf{Y}_{p,t-1} = (\boldsymbol{Y}'_{t-1}, \ldots, \boldsymbol{Y}'_{t-p})', \qquad \Sigma_{11} = \Gamma(0) = \mathrm{Cov}(\boldsymbol{Y}_t),$$

$$\Sigma_{22} = \Gamma_p = \mathrm{Cov}(\mathbf{Y}_{p,t-1}), \qquad \text{and} \qquad \Sigma'_{12} = \Gamma_{(p)} = (\Gamma(1)', \ldots, \Gamma(p)')'.$$

2.7 For a bivariate AR(2) model $\boldsymbol{Y}_t = \Phi_1 \boldsymbol{Y}_{t-1} + \Phi_2 \boldsymbol{Y}_{t-2} + \varepsilon_t$, with

$$\Phi_1 = \begin{bmatrix} 1.5 & -0.6 \\ 0.3 & 0.2 \end{bmatrix}, \qquad \Phi_2 = \begin{bmatrix} -0.5 & 0.3 \\ 0.7 & -0.2 \end{bmatrix}, \qquad \Sigma = \begin{bmatrix} 4 & 1 \\ 1 & 2 \end{bmatrix},$$

(a) calculate forecasts $\hat{\boldsymbol{Y}}_T(l)$ for $l = 1, \ldots, 5$ steps ahead, given that $\boldsymbol{Y}_T = (1.2, 0.6)'$ and $\boldsymbol{Y}_{T-1} = (0.5, 0.9)'$.

(b) Find the coefficient matrices Ψ_j, $j = 1, \ldots, 4$, in the infinite MA representation of the process, and find the forecast error covariance matrices $\Sigma(l)$ for $l = 1, \ldots, 5$.

2.8 Consider the simple transfer function model

$$(1 - B) Y_{1t} = a_{1t} - \theta a_{1t-1}, \qquad Y_{2t} = \omega Y_{1t} + a_{2t},$$

where a_{1t} and a_{2t} are independent white noise series.

(a) Determine the univariate model for Y_{2t}, and note that Y_{2t} is nonstationary.

(b) Express the bivariate model for $\boldsymbol{Y}_t = (Y_{1t}, Y_{2t})'$ in the general form of a "generalized" ARMA(1,1) model, $(I - \Phi_1 B) \boldsymbol{Y}_t = (I - \Theta_1 B) \varepsilon_t$, and determine that one of the eigenvalues of Φ_1 is equal to one.

(c) Determine the bivariate model for the first differences $(1 - B) \boldsymbol{Y}_t$, and show that it has the form of an IMA(1,1) model, $(1 - B) \boldsymbol{Y}_t = (I - \Theta^* B) \varepsilon_t$, where the MA operator $(I - \Theta^* B)$ is not invertible. Hence, this model represents an "overdifferencing" of the bivariate series $\boldsymbol{Y}_t$.

2.9 For the bivariate IMA(1,1) series $(1 - B) \boldsymbol{Y}_t = (I - \Theta B) \varepsilon_t$, with

$$\Theta = \begin{bmatrix} 0.4 & 0.3 \\ -0.5 & 0.8 \end{bmatrix}, \qquad \Sigma = \begin{bmatrix} 4 & 1 \\ 1 & 2 \end{bmatrix},$$

you may deduce the univariate IMA(1,1) models for the individual series directly from the results of Exercise 2.2.

(a) Let $\boldsymbol{e}_t(l) = \boldsymbol{Y}_{t+l} - \hat{\boldsymbol{Y}}_t(l)$ denote the l-step ahead forecast error from the bivariate model. Find a general (explicit) expression for the covariance matrix of $\boldsymbol{e}_t(l)$, $l = 1, 2, \ldots$, and, in particular, show that the variances of the forecast errors are of the form

$$\mathrm{Var}(e_{it}(l)) = a_{1i} + b_{1i}(l - 1), \qquad i = 1, 2,$$

and determine the exact values of the coefficients a_{1i} and b_{1i}.

(b) Let $\boldsymbol{a}_t(l) = \boldsymbol{Y}_{t+l} - \tilde{\boldsymbol{Y}}_t(l)$ denote the l-step ahead forecast error vector obtained from forecasting each series individually from their univariate models. Show that the variances of these forecast errors are of the form

$$\text{Var}(\, a_{it}(l)\,) = a_{2i} + b_{2i}\,(\, l-1\,)\,, \qquad i = 1, 2\,,$$

and determine the exact values of the a_{2i} and b_{2i}. Notice that the b_{2i} are the same as the b_{1i} from (a), but that $a_{2i} \geq a_{1i}$, $i = 1, 2$. [This type of result is true, in general, for any multivariate IMA(1,1) model; attempt to prove it in general.]

2.10 Suppose the stationary vector process $\boldsymbol{Y}_t$ has infinite MA representation $\boldsymbol{Y}_t = \sum_{j=0}^{\infty} \Psi_j\, \varepsilon_{t-j} = \Psi(B)\, \varepsilon_t$, where $\Psi(B) = \sum_{j=0}^{\infty} \Psi_j\, B^j$, and $E(\, \varepsilon_t\, \varepsilon_t'\,) = \Sigma$. Define the autocovariance matrix generating function of $\boldsymbol{Y}_t$ as $g(z) = \sum_{j=-\infty}^{\infty} \Gamma(j)\, z^j$, where $\Gamma(j) = E(\, \boldsymbol{Y}_t\, \boldsymbol{Y}_{t+j}'\,)$, $j = 0, \pm 1, \pm 2, \ldots$. Show that $g(z)$ can be expressed as $g(z) = \Psi(\, z^{-1}\,)\, \Sigma\, \Psi(\, z\,)'$.

2.11 Let $\mathbf{u} = (\, \boldsymbol{e}_t(1)', \boldsymbol{e}_t(2)', \ldots, \boldsymbol{e}_t(L)'\,)'$ denote the kL-dimensional vector of forecast errors made at origin t from a vector ARMA process for lead times $l = 1, 2, \ldots, L$, so that $\boldsymbol{e}_t(l) = \sum_{j=0}^{l-1} \Psi_j\, \varepsilon_{t+l-j}$.

(a) Then, with $\boldsymbol{\varepsilon} = (\, \varepsilon_{t+1}', \varepsilon_{t+2}', \ldots, \varepsilon_{t+L}'\,)'$ denoting the vector of corresponding random shocks, determine that $\mathbf{u} = M\, \boldsymbol{\varepsilon}$, where

$$M = \begin{bmatrix} I & 0 & 0 & \cdot & \cdot & \cdot & 0 \\ \Psi_1 & I & 0 & \cdot & \cdot & \cdot & 0 \\ \Psi_2 & \Psi_1 & I & \cdot & \cdot & \cdot & 0 \\ \cdot & \cdot & \cdot & \cdot & \cdot & \cdot & \cdot \\ \cdot & \cdot & \cdot & \cdot & \cdot & \cdot & \cdot \\ \Psi_{L-1} & \Psi_{L-2} & \Psi_{L-3} & \cdot & \cdot & \Psi_1 & I \end{bmatrix},$$

and, hence, show that the covariance matrix of the forecast errors $\mathbf{u}$ is $E(\, \mathbf{u}\, \mathbf{u}'\,) = M\,(\, I_L \otimes \Sigma\,)\, M'$. Thus, we see that forecast errors for different lead times, based on the same forecast origin t, are correlated.

(b) Deduce from (a) that the r.v. defined by $Q = \mathbf{u}'\, M'^{-1}\,(\, I_L \otimes \Sigma^{-1}\,)\, M^{-1}\, \mathbf{u}$ has a chi-squared distribution with $k^2 L$ degrees of freedom, and note that Q can also be expressed as $\boldsymbol{\varepsilon}'\,(\, I_L \otimes \Sigma^{-1}\,)\, \boldsymbol{\varepsilon} = \sum_{l=1}^{L} \varepsilon_{t+l}'\, \Sigma^{-1}\, \varepsilon_{t+l}$. Hence, once "future" values $(\, \boldsymbol{Y}_{t+1}', \ldots, \boldsymbol{Y}_{t+L}'\,)'$ have been observed, describe how the statistic $Q = \sum_{l=1}^{L} \varepsilon_{t+l}'\, \Sigma^{-1}\, \varepsilon_{t+l}$ [with $\varepsilon_{t+l} = \boldsymbol{Y}_{t+l} - \hat{\boldsymbol{Y}}_{t+l-1}(1)$] may be used to compare forecasts generated from the ARMA model with the subsequent actual data values, and, hence, it may be useful in checking the adequacy of the model over the forecast period or for detecting changes (e.g., interventions) in the process model over the forecast period.

Chapter 3

3.1 Consider the vector ARMA(1,1) model $\boldsymbol{Y}_t - \Phi\, \boldsymbol{Y}_{t-1} = \varepsilon_t - \Theta\, \varepsilon_{t-1}$. By referring to Exercise 2.5 of Chapter 2, establish that the lag-1 partial autoregression matrix Φ_{11} is such that $\Phi_{11} \neq \Phi$ in the ARMA(1,1) model, and that the lag-2 partial AR matrix is such that $\Phi_{22} \neq 0$.

3.2 Consider the bivariate MA(1) model $Y_t = (I - \Theta B)\,\varepsilon_t$, with

$$\Theta = \begin{bmatrix} 0.4 & 0.3 \\ -0.5 & 0.8 \end{bmatrix}, \qquad \Sigma = \begin{bmatrix} 4 & 1 \\ 1 & 2 \end{bmatrix},$$

of Exercise 2.2.

(a) Find the roots of $\det(I - \Theta B) = 0$ and check for invertibility.

(b) Use the matrix Yule-Walker equations to "fit" (approximating) bivariate AR models of orders $p = 1, 2, 3$, and 4 to the autocovariances of this MA(1) process, computing both the AR matrices Φ_{jp}, $j = 1, \ldots, p$, and the error covariance matrix Σ_p, for each p. Compare your results with the theoretical Π_j weights from Exercise 2.2, and with the theoretical error covariance matrix Σ.

3.3 Suppose a 3-dimensional vector ARMA process Y_t has Kronecker indices $K_1 = 3$, $K_2 = 1$, $K_3 = 2$. Write the form of the coefficient matrices $\Phi_j^{\#}$ and $\Theta_j^{\#}$ in the echelon canonical ARMA model structure of equation (3.4) for this process.

3.4 Consider the stationary vector ARMA(1,1) process, $(I - \Phi B)\,Y_t = (I - \Theta B)\,\varepsilon_t$, where the ε_t are distributed independently with mean zero and covariance matrix Σ. For this vector process $\{Y_t\}$, state what vector variables of the form $(Y_{t-l-1}', Y_{t-l-2}')'$ (i.e., what values of l) will have zero canonical correlations with the vector $(Y_t', Y_{t-1}')'$ and explain how many zero canonical correlations and why. Also, consider the possibility of zero canonical correlations for any special cases where the matrices Φ and/or Θ may have reduced ranks.

3.5 Suppose that Y_t is a trivariate ARMA(1,1) process, $(I - \Phi_1 B)\,Y_t = (I - \Theta_1 B)\,\varepsilon_t$, and that $\operatorname{rank}(\Phi_1) = 2$, $\operatorname{rank}(\Theta_1) = 1$, and $\operatorname{rank}(\Phi_1, \Theta_1) = 3$.

(a) Establish that there are two linearly independent linear combinations of Y_t which possess AR(1) scalar component models and that there is one linear combination of Y_t which has MA(1) scalar component model.

(b) Construct a table for the number of zero canonical correlations between $\mathbf{Y}_{m,t}$ and $\mathbf{Y}_{m,t-l-1}$ for $m, l = 0, 1, 2, 3$, where $\mathbf{Y}_{m,t} = (Y_t', Y_{t-1}', \ldots, Y_{t-m}')'$.

Chapter 4

4.1 Suppose $Y_1, \ldots, Y_T$, with $T = 60$, is a sample from a bivariate AR(1) process, with sample covariance matrices obtained as

$$\hat{\Gamma}(0) = \begin{bmatrix} 1.0 & 1.0 \\ 1.0 & 2.0 \end{bmatrix}, \qquad \hat{\Gamma}(1) = \begin{bmatrix} 0.6 & 0.4 \\ 0.7 & 1.2 \end{bmatrix}, \qquad \hat{\Gamma}(2) = \begin{bmatrix} 0.30 & 0.10 \\ 0.42 & 0.64 \end{bmatrix}.$$

(a) Find the sample Yule-Walker estimates for Φ and Σ in the AR(1) model, and find an estimate for the approximate covariance matrix of the estimator $\hat{\Phi}$, that is, for the covariance matrix of $\operatorname{vec}(\hat{\Phi}')$.

(b) Based on the results in (a), test whether the matrix Φ has a lower triangular structure, i.e., test whether $\phi_{12} = 0$, using a Wald statistic.

(c) Based on YW-type estimation results, construct a (Wald) statistic as in equation (4.6) of Section 4.2.1 for the test of $H_0 : \Phi_2 = 0$ in an AR(2) model.

4.2 Consider the MA(1) process from Exercise 3.2 of Chapter 3, $\boldsymbol{Y}_t = (I - \Theta B)\, \varepsilon_t$, with

$$\Theta = \begin{bmatrix} 0.4 & 0.3 \\ -0.5 & 0.8 \end{bmatrix}, \qquad \Sigma = \begin{bmatrix} 4 & 1 \\ 1 & 2 \end{bmatrix}.$$

Suppose one theoretically "fits" an AR(1) model to the process $\{\boldsymbol{Y}_t\}$ and, hence, obtains the following value for the AR(1) parameter, $\Phi_{11} = \Gamma(1)' \, \Gamma(0)^{-1}$. We then consider the residual series $\tilde{\varepsilon}_t = \boldsymbol{Y}_t - \Phi_{11} \boldsymbol{Y}_{t-1}$ after the AR(1) fit.

For the above series, using previous results first determine the value of Φ_{11}, and then the autocovariances $\Gamma_{\tilde{\varepsilon}}(l)$, $l = 0, 1, 2, 3$, and hence the autocorrelation matrices at lags $0, 1, 2$, and 3, for the residual series $\tilde{\varepsilon}_t$. Use these results to discuss the extent to which use of the residual correlations after fitting the AR(1) model might lead to the correct identification of the MA(1) model, or at least to a conclusion that the AR(1) model is not appropriate.

Note: The autocovariance matrices of the residuals $\tilde{\varepsilon}_t = \boldsymbol{Y}_t - \Phi_{11} \boldsymbol{Y}_{t-1}$ may be derived directly in terms of those of the original series $\boldsymbol{Y}_t$, for example,

$$\Gamma_{\tilde{\varepsilon}}(1) = E[\, (\boldsymbol{Y}_t - \Phi_{11} \boldsymbol{Y}_{t-1}) (\boldsymbol{Y}_{t+1} - \Phi_{11} \boldsymbol{Y}_t)' \,]$$

$$= \Gamma(1) - \Phi_{11} \Gamma(2) - \Gamma(0) \Phi_{11}' + \Phi_{11} \Gamma(1) \Phi_{11}'.$$

4.3 For the 3-dimensional ARMA process $\boldsymbol{Y}_t$ with Kronecker indices as given in Exercise 3.3 of Chapter 3, describe the nature of the zero canonical correlations that occur in the canonical correlation analysis of the past vector $\mathbf{P}_t$ and various future vectors $\mathbf{F}^*_{t+1}$.

4.4 Consider properties of the least squares (conditional ML) estimator $\hat{\phi}_T$ of ϕ in the simple univariate AR(1) model $Y_t = \phi Y_{t-1} + \varepsilon_t$, $t = 0, 1, \ldots, T$, given by $\hat{\phi}_T = \sum_{t=1}^T Y_{t-1} Y_t \,/\, \sum_{t=1}^T Y_{t-1}^2$, where we assume $|\phi| < 1$.

(a) Show that $\hat{\phi}_T - \phi = \sum_{t=1}^T Y_{t-1} \varepsilon_t \,/\, \sum_{t=1}^T Y_{t-1}^2$; then since $\sum_{t=1}^T Y_{t-1}^2 / T \rightarrow \gamma(0) = \sigma^2 / (1 - \phi^2)$ in probability as $T \rightarrow \infty$, standard large sample results imply that

$$\sqrt{T}\, (\hat{\phi}_T - \phi) = \frac{1}{\sqrt{T}} \sum_{t=1}^T Y_{t-1} \varepsilon_t \,/\, \frac{1}{T} \sum_{t=1}^T Y_{t-1}^2$$

has the same limiting distribution as $T^{-1/2} \sum_{t=1}^T Y_{t-1} \varepsilon_t \,/\, \gamma(0)$.

To establish the limiting distribution for $T^{-1/2} \sum_{t=1}^T Y_{t-1} \varepsilon_t$, we introduce the concept of *martingales*.

Definition. Let $Z_1, Z_2, \ldots$ be a sequence of r.v.'s and define the partial sums $S_T = \sum_{i=1}^T Z_i$, $T = 1, 2, \ldots$. The sequence $\{S_T\}$ is called a martingale if

(i) $E(| S_T |)$ is finite for all T, and

(ii) $E(S_{T+1} \mid Z_T, Z_{T-1}, \ldots, Z_1) = S_T$, for $T = 1, 2, \ldots,$ [or, equivalently, $E(Z_{T+1} \mid Z_T, Z_{T-1}, \ldots, Z_1) = 0$].

(b) Use the relation $E(Z_{T+1}) = E[E(Z_{T+1} \mid Z_T, \ldots, Z_1)]$ to verify that condition (ii) above implies that $E(S_T) = 0$ for all T, with $E(Z_1) = 0$, and that if second moments exist, then the Z_i are uncorrelated, $\mathrm{Cov}(Z_i, Z_j) = 0$, $i \neq j$, so that $\mathrm{Var}(S_T) = E(S_T^2) = \sum_{i=1}^T \mathrm{Var}(Z_i)$.

(c) Show that the sequence $S_T = \sum_{t=1}^{T} Y_{t-1}\, \varepsilon_t$ from part (a) forms a martingale, and find that $\mathrm{Var}(S_T) = T\, \gamma(0)\, \sigma^2$.

The following *Martingale Central Limit Theorem* (CLT) is due to Billingsley (1968) :
Let $S_T = \sum_{i=1}^{T} Z_i$ be a martingale, where the Z_i are stationary with finite (common) variance. Then,

$$S_T \,/\, [\, \mathrm{Var}(S_T)\,]^{1/2} = S_T \,/\, [\, \sqrt{T}\, (\mathrm{Var}(Z_i))^{1/2}\,]$$

converges in distribution to $N(0, 1)$ as $T \to \infty$.

(d) Use the above martingale CLT result to prove that $\sum_{t=1}^{T} Y_{t-1}\, \varepsilon_t \,/\, \sqrt{T}\, \gamma(0)$, and, hence, from comments in part (a), $\sqrt{T}\, (\hat{\phi}_T - \phi)$ converges in distribution to $N(0, 1 - \phi^2)$ as $T \to \infty$. Hence, for large T, $\mathrm{Var}(\hat{\phi}_T) \approx (1 - \phi^2) / T$.
Briefly comment on reasons why you might think the above result does not hold for the nonstationary case where $\phi = 1$, that is, for the model $Y_t = Y_{t-1} + \varepsilon_t$ with starting value $Y_0 = 0$, say. [Hint: Note that in the nonstationary case we have $Y_t = \sum_{i=1}^{t} \varepsilon_i$. Consider the behavior of $\sum_{t=1}^{T} Y_{t-1}^2$ as $T \to \infty$, for example.]

Chapter 5

5.1 Consider the vector MA(1) model, $\boldsymbol{Y}_t = (I - \Theta B)\, \varepsilon_t$. Based on T observations from this model, write down the equation explicitly for one-step of the modified Newton-Raphson iteration for obtaining the conditional MLE $\hat{\Theta}$.

5.2 Consider the vector MA(1) model $\boldsymbol{Y}_t = (I - \Theta_1 B)\, \varepsilon_t$, and let $\hat{\Theta}_1$ denote the MLE of Θ_1 based on T observations $\boldsymbol{Y}_1, \ldots, \boldsymbol{Y}_T$. Letting $\hat{\boldsymbol{\theta}} = \mathrm{vec}(\hat{\Theta}_1)$ and $\boldsymbol{\theta} = \mathrm{vec}(\Theta_1)$, it can be established from the results of Section 5.1.4 that the asymptotic distribution of $T^{1/2} (\hat{\boldsymbol{\theta}} - \boldsymbol{\theta})$ as $T \to \infty$ is $N(0, \Sigma^{-1} \otimes V^{-1})$, where $V = \sum_{j=0}^{\infty} \Theta_1^{\prime j}\, \Sigma^{-1}\, \Theta_1^{j}$. Compare this result with the asymptotic distribution of $T^{1/2}\, [\, \mathrm{vec}(\hat{\Phi}_1^{\prime}) - \mathrm{vec}(\Phi_1^{\prime})\,]$ for the vector AR(1) model, as described in Section 4.3.1, and comment on similarities, if any, in these two asymptotic distributional results.

5.3 Consider the simple seasonal MA model for monthly data, $\boldsymbol{Y}_t = \varepsilon_t - \Theta\, \varepsilon_{t-12}$, $t = 1, \ldots, T$, and assume there are $T = 12\, N$ observations available, i.e., N years of monthly data. Use general (exact) likelihood results from Section 5.3 to obtain explicit expressions for the exact likelihood function for this model, including the matrix D and the backcasted values for $\mathbf{e}_* = (\varepsilon_{-11}^{\prime}, \varepsilon_{-10}^{\prime}, \ldots, \varepsilon_{-0}^{\prime})^{\prime}$.
When N is relatively small (the typical case) and the matrix Θ is close to the invertible boundary (i.e., eigenvalues close to the unit circle), briefly discuss why the conditional and exact likelihood functions may show substantial differences, and try to refer to the simple MA(1) model situation in this discussion.
[Hint: Note the correspondences $\Omega = \mathrm{Cov}(\mathbf{e}_*) = I_{12} \otimes \Sigma$, $F = (B_{12}^{\prime}, 0^{\prime})^{\prime}$ where $B_{12} = -(I_{12} \otimes \Theta)$, and so on. Also, eventually try to interpret the likelihood function for all $T = 12\, N$ vector observations as equal to the product of 12 likelihood functions, one each for the data $\{ \boldsymbol{Y}_m, \boldsymbol{Y}_{12+m}, \ldots, \boldsymbol{Y}_{12(N-1)+m} \}$ in month m for months $m = 1, 2, \ldots, 12$.]

5.4 For the bivariate MA(1) process $\boldsymbol{Y}_t$ given in Exercise 2.2, perform the innovations algorithm calculations of Section 5.4 for 10 time points $t = 1, 2, \ldots, 10$, to determine the coefficient matrices $\Theta_{1,t-1}$ and the innovations covariance matrices $\Sigma_{t|t-1} = \text{Cov}(\varepsilon_{t|t-1})$. Examine the nature of the convergence of the $\Theta_{1,t-1}$ and $\Sigma_{t|t-1}$ to the corresponding values Θ and Σ.

5.5 Investigate the explicit nature and form of the score test for testing $\Phi_2 = 0$ in a vector AR(2) model, after having fitted an AR(1) model to the time series, and examine the asymptotic equivalence of this test statistic with the alternate test statistic forms (e.g., Wald statistic) for testing $\Phi_2 = 0$ as discussed in Chapter 4.

5.6 Consider the monthly data on U.S. housing starts and houses sold covering the period January 1965 – December 1974 listed in the Appendix. Denote the series as $\boldsymbol{Y}_t = (Y_{1t}, Y_{2t})'$, $t = 1, \ldots, 120$, where Y_{1t} is the monthly U.S. housing starts and Y_{2t} is the monthly U.S. houses sold. Identify and fit an appropriate vector ARIMA model for this bivariate series. In doing so, obtain sample autocorrelation matrices, partial AR matrices, canonical correlations, and so on, of suitably transformed versions of the series. Experience gained from univariate ARIMA modeling of the two series separately will also be useful to help your bivariate model specification. In particular, it is probably reasonable to consider seasonal differences $(Y_{it} - Y_{i,t-12})$ of both series in the analysis, due to the strong seasonal nature of these series. Be sure to check your model for adequacy, using residual correlation matrices and so on. Also, forecast 12 periods ahead for both series using the bivariate model obtained and determine standard deviations of the forecast errors.

Chapter 6

6.1 Express the model for the nonstationary bivariate process $\boldsymbol{Y}_t$ given in Exercise 2.8 of Chapter 2 in an error-correction form, similar to equation (6.11) of Section 6.3, as $\boldsymbol{W}_t = C\,\boldsymbol{Y}_{t-1} + \varepsilon_t - \Theta\,\varepsilon_{t-1}$, where $\boldsymbol{W}_t = (1 - B)\,\boldsymbol{Y}_t$. Determine the structure (and the ranks) of the matrices C and Θ explicitly.

6.2 Consider the bivariate IMA(1,1) process $(1 - B)\,\boldsymbol{Y}_t = (I - \Theta B)\,\varepsilon_t$, where

$$\Theta = \begin{bmatrix} 1.6 & -0.6 \\ 1.2 & -0.2 \end{bmatrix}.$$

(a) Show that the process $\boldsymbol{Y}_t$ is not invertible and determine a matrix P such that $P^{-1}\Theta P = \text{Diag}(\lambda_1, \lambda_2)$, where λ_1 and λ_2 are the eigenvalues of Θ.

(b) Let $Z_t = P^{-1}\boldsymbol{Y}_t$, and determine the bivariate model for Z_t. Simplify this model and show that the model for Z_t can be written as $(I - \phi^* B)\,Z_t = \mu^* + (I - \Lambda^* B)\,\boldsymbol{a}_t$, where

$$\phi^* = \begin{bmatrix} 0.0 & 0.0 \\ 0.0 & 1.0 \end{bmatrix}, \qquad \Lambda^* = \begin{bmatrix} 0.0 & 0.0 \\ 0.0 & 0.4 \end{bmatrix}.$$

(c) By multiplying the above model in (b) on the left by P, noting that $\boldsymbol{Y}_t = P\,Z_t$, obtain a model representation for $\boldsymbol{Y}_t$ of the form

$$(I - \Phi_1 B) \boldsymbol{Y}_t = \boldsymbol{\mu} + (I - \Theta_1 B) \varepsilon_t ,$$

and obtain the values of Φ_1 and Θ_1 .

(d) Verify that the model representation in (c) is invertible, but nonstationary. Furthermore, show from the model representation in (c) that the process $\boldsymbol{Y}_t$ is co-integrated, and find a co-integrating vector.

(e) Find an error-correction representation of the model for $\boldsymbol{Y}_t$ in (c).

Chapter 7

7.1 For the 3-dimensional vector ARMA process of Exercise 3.3, with Kronecker indices as specified there, write down the form of the minimal dimension echelon state-space model corresponding to the echelon canonical ARMA model for the process.

7.2 For the log mink and muskrat furs data, write down explicitly the minimal dimension echelon state-space model that corresponds to the fitted echelon canonical form ARMA(2,2) model which was presented in Example 5.1 of Section 5.7.

7.3 Consider the simple univariate signal extraction model, $Y_t = S_t + N_t$, where the signal S_t follows an AR(1) model $S_t = \phi S_{t-1} + a_t$, and $\{N_t\}$ and $\{a_t\}$ are independent white noise processes with variances σ_n^2 and σ_a^2 , respectively.

(a) Obtain the Kalman filtering equations explicitly in this case for estimating S_t based on Y_u , $0 \le u \le t$. Also, determine the "steady-state" equilibrium values (as $t \to \infty$) of the gain K_t and the error variances $\Omega_t = E [(S_t - \hat{S}_{t|t-1})^2]$ and $P_t = E [(S_t - \hat{S}_{t|t})^2]$, and show (verify) that these are equal to $K = \sigma_a^2 / [\sigma_\varepsilon^2 (1 - \phi \theta)] \equiv (1 - \theta / \phi)$, $\Omega = \sigma_a^2 / (1 - \phi \theta)$, and $(1 - K) \Omega = \theta \sigma_a^2 / [\phi (1 - \phi \theta)]$, respectively.

(b) Compare the Kalman filtering technique of estimating S_t with the (classical) method given for computing $\hat{S}_t$ based on the infinite past data Y_u , $u \le t$, and find and compare the variance of the filtering error $S_t - \hat{S}_t$ using this approach with that of the Kalman filtering approach (especially compare results to the "steady-state" results for the Kalman filtering method).

[Note: Refer to Example 7.5 given in Section 7.4.1.]

7.4 Consider the structural components model for quarterly time series $\{Y_t\}$ given by $Y_t = S_t + T_t + N_t$, where it is supposed that the models for the component seasonal $\{S_t\}$, trend $\{T_t\}$, and noise $\{N_t\}$ series are $U(B) S_t = b_t$, $(1 - B) T_t = (1 + B) c_t$, and $N_t = d_t$, with $U(B) = 1 + B + B^2 + B^3$, where b_t , c_t , and d_t are independent white noise series with variances σ_b^2, σ_c^2, and σ_d^2, respectively. Express this model in a state-variable form.

7.5 Consider the simple regression model with a random regression coefficient, $Y_t = \alpha + \beta_t x_t + N_t$, where β_t is a random process following the AR(1) model $\beta_t = \phi \beta_{t-1} + a_t$, with N_t distributed as $N(0, \sigma_n^2)$ and a_t distributed as $N(0, \sigma_a^2)$. Assuming the x_t are known values, determine the conditional distribution of β_t , given $Y_1 , \dots , Y_t$. Based on T observations $Y_1 , \dots , Y_T$, find an expression for the log-likelihood for the model and provide as much detail as possible.

References

Aasnaes, H. B., and Kailath, T. (1973). An innovations approach to least-squares estimation–Part VII: Some applications of vector autoregressive-moving average models. *IEEE Transactions on Automatic Control*, AC-18, 601–607.

Abraham, B. (1980). Intervention analysis and multiple time series. *Biometrika*, 67, 73–78.

Abraham, B., and Box, G. E. P. (1978). Deterministic and forecast-adaptive time-dependent models. *Applied Statistics*, 27, 120–130.

Ahn, S. K., and Reinsel, G. C. (1988). Nested reduced-rank autoregressive models for multiple time series. *Journal of the American Statistical Association*, 83, 849–856.

Ahn, S. K., and Reinsel, G. C. (1990). Estimation for partially nonstationary multivariate autoregressive models. *Journal of the American Statistical Association*, 85, 813–823.

Ahn, S. K., and Reinsel, G. C. (1993). Estimation of partially nonstationary vector autoregressive models with seasonal behavior. To appear in the *Journal of Econometrics*.

Akaike, H. (1971). Autoregressive model fitting for control. *Annals of the Institute of Statistical Mathematics*, 23, 163–180.

Akaike, H. (1974a). Stochastic theory of minimal realization. *IEEE Transactions on Automatic Control*, AC-19, 667–674.

Akaike, H. (1974b). A new look at the statistical model identification. *IEEE Transactions on Automatic Control*, AC-19, 716–723.

Akaike, H. (1974c). Markovian representation of stochastic processes and its application to the analysis of autoregressive moving average processes. *Annals of the Institute of Statistical Mathematics*, 26, 363–387.

Akaike, H. (1976). Canonical correlation analysis of time series and the use of an information criterion. In *Systems Identification: Advances and Case Studies*, eds. R. K. Mehra and D. G. Lainiotis, New York: Academic Press, pp. 27–96.

Akaike, H. (1978). Covariance matrix computation of the state variable of a stationary Gaussian process. *Annals of the Institute of Statistical Mathematics*, 30, 499–504.

An Hong-Zhi, Chen Zhao-Guo, and Hannan, E. J. (1983). A note on ARMA estimation. *Journal of Time Series Analysis*, 4, 9–17.

Anderson, B. D. O., and Moore, J. B. (1979). *Optimal Filtering*. Englewood Cliffs, NJ: Prentice-Hall.

Anderson, T. W. (1951). Estimating linear restrictions on regression coefficients for multivariate normal distributions. *Annals of Mathematical Statistics*, 22, 327–351.

Anderson, T. W. (1971). *The Statistical Analysis of Time Series*. New York: John Wiley.

Anderson, T. W. (1980). Maximum likelihood estimation for vector autoregressive moving average models. In *Directions in Time Series*, eds. D. R. Brillinger and G. C. Tiao, Institute of Mathematical Statistics, 49–59.

Anderson, T. W. (1984). *An Introduction to Multivariate Statistical Analysis*, 2nd edition. New York: John Wiley.

Ansley, C. F. (1979). An algorithm for the exact likelihood of a mixed autoregressive-moving average process. *Biometrika*, 66, 59–65.

Ansley, C. F. (1980). Computation of the theoretical autocovariance function for a vector ARMA process. *Journal of Statistical Computation and Simulation*, 12, 15–24.

Ansley, C. F., and Kohn, R. (1982). A geometrical derivation of the fixed interval smoothing algorithm. *Biometrika*, 69, 486–487.

Ansley, C. F., and Kohn, R. (1983). Exact likelihood of vector autoregressive-moving average process with missing or aggregated data. *Biometrika*, 70, 275–278.

Ansley, C. F., and Kohn, R. (1985a). Estimation, filtering, and smoothing in state space models with incompletely specified initial conditions. *Annals of Statistics*, 13, 1286–1316.

Ansley, C. F., and Kohn, R. (1985b). A structured state space approach to computing the likelihood of an ARIMA process and its derivatives. *Journal of Statistical Computation and Simulation*, 21, 135–169.

Ansley, C. F., and Kohn, R. (1985c). On the rate of convergence of the innovation representation of a moving average process. *Biometrika*, 72, 325–330.

Ansley, C. F., and Kohn, R. (1990). Filtering and smoothing in state space models with partially diffuse initial conditions. *Journal of Time Series Analysis*, 11, 275–293.

Ansley, C. F., and Newbold, P. (1979). Multivariate partial autocorrelations. *ASA Proceedings of the Business and Economic Statistics Section*, 349–353.

Baillie, R. T. (1979). Asymptotic prediction mean squared error for vector autoregressive models. *Biometrika*, 66, 675–678.

Bartlett, M. S. (1955). *An Introduction to Stochastic Processes*. Cambridge: Cambridge University Press.

Bell, W. (1984). Signal extraction for nonstationary time series. *Annals of Statistics*, 12, 646–664. Correction (1991), 19, 2280.

Bell, W. (1987). A note on overdifferencing and the equivalence of seasonal time series models with monthly means and models with $(0, 1, 1)_{12}$ seasonal parts when $\Theta = 1$. *Journal of Business and Economic Statistics*, 5, 383–387.

Bell, W. R., and Hillmer, S. (1987). Initializing the Kalman filter in the nonstationary case. *ASA Proceedings of the Business and Economic Statistics Section*, 693–697.

Bell, W., and Hillmer, S. (1991). Initializing the Kalman filter for nonstationary time series models. *Journal of Time Series Analysis*, 12, 283–300. Correction (1992), 13, 281–282.

Billingsley, P. (1968). *Convergence of Probability Measures*. New York: John Wiley.

Box, G. E. P., and Jenkins, G. M. (1976). *Time Series Analysis: Forecasting and Control*, Revised Edition. San Francisco: Holden-Day.

Box, G. E. P., and Tiao, G. C. (1977). A canonical analysis of multiple time series. *Biometrika*, 64, 355–365.

Brockwell, P. J., and Davis, R. A. (1987). *Time Series: Theory and Methods*. New York: Springer-Verlag.

Brown, B. M. (1971). Martingale central limit theorems. *Annals of Mathematical*

Statistics, 42, 59–66.

Caines, P. E., and Rissanen, J. (1974). Maximum likelihood estimation of parameters in multivariate Gaussian stochastic processes. *IEEE Transactions on Information Theory*, IT-20, 102–104.

Chan, N. H., and Wei, C. Z. (1988). Limiting distributions of least squares estimates of unstable autoregressive processes. *Annals of Statistics*, 16, 367–401.

Chan, W. Y. T., and Wallis, K. F. (1978). Multiple time series modelling: another look at the mink-muskrat interaction. *Applied Statistics*, 27, 168–175.

Cooper, D. M., and Wood, E. F. (1981). Estimation of the parameters of the Markovian representation of the autoregressive-moving average model. *Biometrika*, 68, 320–322.

Cooper, D. M., and Wood, E. F. (1982). Identifying multivariate time series models. *Journal of Time Series Analysis*, 3, 153–164.

Deistler, M., Dunsmuir, W., and Hannan, E. J. (1978). Vector linear time series models: corrections and extensions. *Advances in Applied Probability*, 10, 360–372.

DeJong, P. (1988). The likelihood for a state space model. *Biometrika*, 75, 165–169.

DeJong, P. (1991). The diffuse Kalman filter. *Annals of Statistics*, 19, 1073–1083.

Dickey, D. A., Bell, W. R., and Miller, R. B. (1986). Unit roots in time series models: tests and implications. *American Statistician*, 40, 12–26.

Dickey, D. A., and Fuller, W. A. (1979). Distribution of the estimates for autoregressive time series with a unit root. *Journal of the American Statistical Association*, 74, 427–431.

Dickinson, B. W., Kailath, T., and Morf, M. (1974). Canonical matrix fraction and state-space descriptions for deterministic and stochastic linear systems. *IEEE Transactions on Automatic Control*, AC-19, 656–667.

Dunsmuir, W., and Hannan, E. J. (1976). Vector linear time series models. *Advances in Applied Probability*, 8, 339–364.

Engle, R. F., and Granger, C. W. J. (1987). Co-integration and error correction: Representation, estimation, and testing. *Econometrica*, 55, 251–276.

Engle, R., and Watson, M. (1981). A one-factor multivariate time series model of metropolitan wage rates. *Journal of the American Statistical Association*, 76, 774–781.

Engle, R. F., and Yoo, B. S. (1987). Forecasting and testing in co-integrated systems. *Journal of Econometrics*, 35, 143–159.

Fernandez, F. J. (1990). Estimation and testing of a multivariate exponential smoothing model. *Journal of Time Series Analysis*, 11, 89–105.

Forney, G. D. (1975). Minimal bases of rational vector spaces, with applications to multivariable linear systems. *SIAM Journal on Control*, 13, 493–520.

Fountis, N. G., and Dickey, D. A. (1989). Testing for a unit root nonstationarity in multivariate autoregressive time series. *Annals of Statistics*, 17, 419–428.

Fuller, W. A. (1976). *Introduction to Statistical Time Series*. New York: John Wiley.

Gardner, G., Harvey, A. C., and Phillips, G. D. A. (1980). An algorithm for exact maximum likelihood estimation of autoregressive-moving average models by means of Kalman filtering. *Applied Statistics*, 29, 311–322.

Godfrey, L. G. (1979). Testing the adequacy of a time series model. *Biometrika*, 66, 67–72.

Granger, C. W. J., and Newbold, P. (1986). *Forecasting Economic Time Series*, Second Edition. New York: Academic Press.

Graybill, F. A. (1969). *Introduction to Matrices With Applications in Statistics*. Belmont, CA: Wadsworth.

Hall, A. D., and Nicholls, D. F. (1980). The evaluation of exact maximum likelihood estimates for VARMA models. *Journal of Statistical Computation and Simulation*, 10, 251–262.

Hannan, E. J. (1969). The identification of vector mixed autoregressive-moving average systems. *Biometrika*, 56, 223–225.

Hannan, E. J. (1970). *Multiple Time Series*. New York: John Wiley.

Hannan, E. J. (1971). The identification problem for multiple equation systems with moving average errors. *Econometrica*, 39, 751–766.

Hannan, E. J. (1975). The estimation of ARMA models. *Annals of Statistics*, 3, 975–981. Correction (1981), 9, 233.

Hannan, E. J. (1976). The identification and parameterisation of ARMAX and state space forms. *Econometrica*, 44, 713–723.

Hannan, E. J. (1981). Estimating the dimension of a linear system. *Journal of Multivariate Analysis*, 11, 459–473.

Hannan, E. J., and Deistler, M. (1988). *The Statistical Theory of Linear Systems*. New York: John Wiley.

Hannan, E. J., and Kavalieris, L. (1984). Multivariate linear time series models. *Advances in Applied Probability*, 16, 492–561.

Hannan, E. J., and Quinn, B. G. (1979). The determination of the order of an autoregression. *Journal of the Royal Statistical Society*, **B**, 41, 190–195.

Hannan, E. J., and Rissanen, J. (1982). Recursive estimation of mixed autoregressive-moving average order. *Biometrika*, 69, 81–94. Correction (1983), 70, 303.

Harvey, A. C. (1986). Analysis and generalisation of a multivariate exponential smoothing model. *Management Science*, 32, 374–380.

Harvey, A. C. (1989). *Forecasting, Structural Time Series Models and the Kalman Filter*. Cambridge: Cambridge University Press.

Harvey, A. C., and Phillips, G. D. A. (1979). Maximum likelihood estimation of regression models with autoregressive-moving average disturbances. *Biometrika*, 66, 49–58.

Harvey, A. C., and Pierse, R. G. (1984). Estimating missing observations in economic time series. *Journal of the American Statistical Association*, 79, 125–131.

Haugh, L. D., and Box, G. E. P. (1977). Identification of dynamic regression (distributed lag) models connecting two time series. *Journal of the American Statistical Association*, 72, 121–130.

Hillmer, S. C., and Tiao, G. C. (1979). Likelihood function of stationary multiple autoregressive moving average models. *Journal of the American Statistical Association*, 74, 652–660.

Hillmer, S. C., and Tiao, G. C. (1982). An ARIMA-model-based approach to seasonal adjustment. *Journal of the American Statistical Association*, 77, 63–70.

Hosking, J. R. M. (1980). The multivariate portmanteau statistic. *Journal of the American Statistical Association*, 75, 602–608.

Hosking, J. R. M. (1981a). Lagrange multiplier tests of multivariate time series models. *Journal of the Royal Statistical Society*, **B**, 43, 219–230.

Hosking, J. R. M. (1981b). Equivalent forms of the multivariate portmanteau statistics. *Journal of the Royal Statistical Society*, **B**, 43, 261–262.

Hylleberg, S., Engle, R. F., Granger, C. W. J., and Yoo, B. S. (1990). Seasonal integration and co-integration. *Journal of Econometrics*, 44, 215–238.

Jenkins, G. M. (1979). *Practical Experiences With Modelling and Forecasting Time Series*. Jersey, Channel Islands: Gwilym Jenkins & Partners Ltd.

Jenkins, G. M., and Alavi, A. S. (1981). Some aspects of modeling and forecasting multivariate time series. *Journal of Time Series Analysis*, 2, 1–47.

Jenkins, G. M., and Watts, D. G. (1968). *Spectral Analysis and Its Applications*. San Francisco: Holden-Day.

Johansen, S. (1988). Statistical analysis of cointegration vectors. *Journal of Economic Dynamics and Control*, 12, 231–254.

Johansen, S. (1991). Estimation and hypothesis testing of cointegration vectors in Gaussian vector autoregressive models. *Econometrica*, 59, 1551–1580.

Johansen, S., and Juselius, K. (1990). Maximum likelihood estimation and inference on cointegration–with applications to the demand for money. *Oxford Bulletin of Economics and Statistics*, 52, 169–210.

Jones, R. H. (1966). Exponential smoothing for multivariate time series. *Journal of the Royal Statistical Society*, **B**, 28, 241–251.

Jones, R. H. (1980). Maximum likelihood fitting of ARMA models to time series with missing observations. *Technometrics*, 22, 389–395.

Kalman, R. E. (1960). A new approach to linear filtering and prediction problems. *Journal of Basic Engineering*, 82, 35–45.

Kashyap, R. L., and Nasburg, R. E. (1974). Parameter estimation in multivariate stochastic difference equations. *IEEE Transactions on Automatic Control*, AC-19, 784–797.

Kavalieris, L. (1991). A note on estimating autoregressive-moving average order. *Biometrika*, 78, 920–922.

Kitagawa, G., and Gersch, W. (1984). A smoothness priors-state space modeling of time series with trend and seasonality. *Journal of the American Statistical Association*, 79, 378–389.

Kohn, R. (1979). Asymptotic estimation and hypothesis testing results for vector linear time series models. *Econometrica*, 47, 1005–1030.

Kohn, R., and Ansley, C. F. (1982). A note on obtaining the theoretical autocovariances of an ARMA process. *Journal of Statistical Computation and Simulation*, 15, 273–283.

Kohn, R., and Ansley, C. F. (1986). Estimation, prediction, and interpolation for ARIMA models with missing data. *Journal of the American Statistical Association*, 81, 751–761.

Koreisha, S. G., and Pukkila, T. (1993). Determining the order of a vector autoregression when the number of component series is large. *Journal of Time Series Analysis*, 14, 47–69.

LeCam, L. (1986). *Asymptotic Methods in Statistical Decision Theory*. New York: Springer-Verlag.

Lee, H. S. (1992). Maximum likelihood inference on cointegration and seasonal cointegration. *Journal of Econometrics*, 54, 1–47.

Lewis, R., and Reinsel, G. C. (1985). Prediction of multivariate time series by autoregressive model fitting. *Journal of Multivariate Analysis*, 16, 393–411.

Lewis, R., and Reinsel, G. C. (1988). Prediction error of multivariate time series with mis-specified models. *Journal of Time Series Analysis*, 9, 43–57.

Li, W. K., and McLeod, A. I. (1981). Distribution of the residual autocorrelations in multivariate ARMA time series models. *Journal of the Royal Statistical Society*, **B**, 43, 231–239.

Liu, L. M., and Hudak, G. B. (1986). *The SCA Statistical System, Version III*. DeKalb, IL: Scientific Computing Associates.

Lutkepohl, H. (1984). Linear transformations of vector ARMA processes. *Journal of Econometrics*, 26, 283–293.

Lutkepohl, H. (1985). Comparison of criteria for estimating the order of a vector autoregressive process. *Journal of Time Series Analysis*, 6, 35–52. Correction (1987), 8, 373.

Lutkepohl, H. (1991). *Introduction to Multiple Time Series Analysis*. Berlin: Springer-Verlag.

Makridakis, S., and Wheelwright, S. C. (1978). *Forecasting Methods and Applications*. Santa Barbara, Calif.: John Wiley.

Melard, G. (1984). A fast algorithm for the exact likelihood of autoregressive-moving average models. *Applied Statistics*, 33, 104–114.

Mittnik, S. (1990). Computation of theoretical autocovariance matrices of multivariate autoregressive moving average time series. *Journal of the Royal Statistical Society*, **B**, 52, 151–155.

Mittnik, S. (1991). Derivation of the unconditional state-covariance matrix for exact maximum-likelihood estimation of ARMA models. *Journal of Economic Dynamics and Control*, 15, 731–740.

Mittnik, S. (1993). Computing theoretical autocovariances of multivariate autoregressive moving average models by using a block Levinson method. *Journal of the Royal Statistical Society*, **B**, 55, 435–440.

Morf, M., and Kailath, T. (1975). Square-root algorithms for least-squares estimation. *IEEE Transactions on Automatic Control*, AC-20, 487–497.

Morf, M., Sidhu, G. S., and Kailath, T. (1974). Some new algorithms for recursive estimation in constant, linear, discrete-time systems. *IEEE Transactions on Automatic Control*, AC-19, 315–323.

Morf, M., Vieira, A., and Kailath, T. (1978). Covariance characterization by partial autocorrelation matrices. *Annals of Statistics*, 6, 643–648.

Newton, H. J. (1978). The information matrices of the parameters of multiple mixed time series. *Journal of Multivariate Analysis*, 8, 317–323.

Newton, H. J. (1980). Efficient estimation of multivariate moving average autocovariances. *Biometrika*, 67, 227–231.

Nicholls, D. F., and Hall, A. D. (1979). The exact likelihood function of multivariate autoregressive moving average models. *Biometrika*, 66, 259–264.

Osborn, D. R. (1977). Exact and approximate maximum likelihood estimators for vector moving average processes. *Journal of the Royal Statistical Society*, **B**, 39, 114–118.

Osborn, D. R. (1982). On the criteria functions used for the estimation of moving average processes. *Journal of the American Statistical Association*, 77, 388–392.

Paulsen, J. (1984). Order determination of multivariate autoregressive time series with unit roots. *Journal of Time Series Analysis*, 5, 115–127.

Pearlman, J. G. (1980). An algorithm for the exact likelihood of a high-order autoregressive-moving average process. *Biometrika*, 67, 232–233.

Pena, D., and Box, G. E. P. (1987). Identifying a simplifying structure in time series. *Journal of the American Statistical Association*, 82, 836–843.

Phadke, M. S., and Kedem, G. (1978). Computation of the exact likelihood function of multivariate moving average models. *Biometrika*, 65, 511–519.

Phillips, P. C. B. (1991). Optimal inference in cointegrated systems. *Econometrica*, 59, 283–306.

Phillips, P. C. B., and Durlauf, S. N. (1986). Multiple time series with integrated process. *Review of Economic Studies*, 53, 473–495.

Phillips, P. C. B., and Ouliaris, S. (1986). Testing for cointegration. Discussion Paper No. 809, Cowles Foundation for Research in Economics, Yale University.

Phillips, P. C. B., and Perron, P. (1988). Testing for a unit root in time series regression. *Biometrika*, 75, 335–346.

Poskitt, D. S. (1990). Estimation and structure determination of multivariate input output systems. *Journal of Multivariate Analysis*, 33, 157–182.

Poskitt, D. S. (1992). Identification of echelon canonical forms for vector linear processes using least squares. *Annals of Statistics*, 20, 195–215.

Poskitt, D. S., and Tremayne, A. R. (1982). Diagnostic tests for multiple time series models. *Annals of Statistics*, 10, 114–120.

Priestley, M. B. (1981). *Spectral Analysis and Time Series*. New York: Academic Press.

Pukkila, T. M., and Krishnaiah, P. R. (1988). On the use of autoregressive order determination criteria in multivariate white noise tests. *IEEE Transactions on Acoustics, Speech, and Signal Processing*, ASSP-36, 1396–1403.

Quinn, B. G. (1980). Order determination for a multivariate autoregression. *Journal of the Royal Statistical Society*, **B**, 42, 182–185.

Rao, C. R. (1973). *Linear Statistical Inference and Its Applications*, Second Edition. New York: John Wiley.

Reinsel, G. (1979). FIML estimation of the dynamic simultaneous equations model with ARMA disturbances. *Journal of Econometrics*, 9, 263–281.

Reinsel, G. C. (1980). Asymptotic properties of prediction errors for the multivariate autoregressive model using estimated parameters. *Journal of the Royal Statistical Society*, **B**, 42, 328–333.

Reinsel, G. C., and Ahn, S. K. (1992). Vector autoregressive models with unit roots and reduced rank structure: Estimation, likelihood ratio test, and forecasting. *Journal of Time Series Analysis*, 13, 353–375.

Reinsel, G. C., Basu, S., and Yap, S. F. (1992). Maximum likelihood estimators in the multivariate autoregressive moving-average model from a generalized least squares viewpoint. *Journal of Time Series Analysis*, 13, 133–145.

Reinsel, G. C., and Lewis, R. A. (1987). Prediction mean square error for non-stationary multivariate time series using estimated parameters. *Economics Letters*, 24, 57–61.

Rissanen, J. (1973a). A fast algorithm for optimum linear predictors. *IEEE Transactions on Automatic Control*, AC-18, 555.

Rissanen, J. (1973b). Algorithms for triangular decomposition of block Hankel and Toeplitz matrices with application to factoring positive matrix polynomials. *Mathematics of Computation*, 27, 147–154.

Rissanen, J., and Barbosa, L. (1969). Properties of infinite covariance matrices and stability of optimum predictors. *Information Sciences*, 1, 221–236.

Rissanen, J., and Caines, P. E. (1979). The strong consistency of maximum likelihood estimators for ARMA processes. *Annals of Statistics*, 7, 297–315.

Roy, R. (1989). Asymptotic covariance structure of serial correlations in multivariate time series. *Biometrika*, 76, 824–827.

SAS Institute Inc. (1988). *SAS/ETS User's Guide, Version 6, First Edition*. Cary, NC: SAS Institute Inc.

Said, S. E., and Dickey, D. A. (1984). Testing for unit roots in autoregressive-moving average models of unknown order. *Biometrika*, 71, 599–607.

Said, S. E., and Dickey, D. A. (1985). Hypothesis testing in ARIMA(p, 1, q) models. *Journal of the American Statistical Association*, 80, 369–374.

Saikkonen, P. (1992). Estimation and testing of cointegrated systems by an autoregressive approximation. *Econometric Theory*, 8, 1–27.

Samaranayake, V. A., and Hasza, D. P. (1988). Properties of predictors for multivariate autoregressive models with estimated parameters. *Journal of Time Series Analysis*, 9, 361–383.

Schwarz, G. (1978). Estimating the dimension of a model. *Annals of Statistics*, 6, 461–464.

Searle, S. R. (1982). *Matrix Algebra Useful for Statistics*. New York: John Wiley.

Serfling, R. J. (1980). *Approximation Theorems of Mathematical Statistics*. New York: John Wiley.

Shaman, P., and Stine, R. A. (1988). The bias of autoregressive coefficient estimators. *Journal of the American Statistical Association*, 83, 842–848.

Shea, B. L. (1987). Estimation of multivariate time series. *Journal of Time Series Analysis*, 8, 95–109.

Shea, B. L. (1989). The exact likelihood of a vector autoregressive moving average model. *Applied Statistics*, 38, 161–184.

Shumway, R. H., and Stoffer, D. S. (1982). An approach to time series smoothing and forecasting using the EM algorithm. *Journal of Time Series Analysis*, 3, 253–264.

Silvey, S. D. (1959). The Lagrangian multiplier test. *Annals of Mathematical Statistics*, 30, 389–407.

Sims, C. A., Stock, J. H., and Watson, M. W. (1990). Inference in linear time series models with some unit roots. *Econometrica*, 58, 113–144.

Solo, V. (1984a). The order of differencing in ARIMA models. *Journal of the American Statistical Association*, 79, 916–921.

Solo, V. (1984b). The exact likelihood for a multivariate ARMA model. *Journal of Multivariate Analysis*, 15, 164–173.

Solo, V. (1986). Topics in advanced time series analysis. In *Lectures in Probability and Statistics*, eds. G. del Pino and R. Rebolledo, New York: Springer-Verlag, pp. 165–328.

Spliid, H. (1983). A fast estimation method for the vector autoregressive moving average model with exogenous variables. *Journal of the American Statistical Association*, 78, 843–849.

Srivastava, M. S., and Khatri, C. G. (1979). *An Introduction to Multivariate Statistics*. New York: North Holland.

Stensholt, E., and Tjostheim, D. (1981). Factorizing multivariate time series operators. *Journal of Multivariate Analysis*, 11, 244–249.

Stock, J. H., and Watson, M. W. (1988). Testing for common trends. *Journal of the American Statistical Association*, 83, 1097–1107.

Swift, A. L. (1990). Orders and initial values of non-stationary multivariate ARMA models. *Journal of Time Series Analysis*, 11, 349–359.

Terasvirta, T. (1985). Mink and muskrat interaction: A structural analysis. *Journal of Time Series Analysis*, 6, 171–180.

Tiao, G. C., and Box, G. E. P. (1981). Modeling multiple time series with applications. *Journal of the American Statistical Association*, 76, 802–816.

Tiao, G. C., and Tsay, R. S. (1989). Model specification in multivariate time series (with Discussion). *Journal of the Royal Statistical Society*, **B**, 51, 157–213.

Tjostheim, D., and Paulsen, J. (1982). Empirical identification of multiple time series. *Journal of Time Series Analysis*, 3, 265–282.

Tjostheim, D., and Paulsen, J. (1983). Bias of some commonly-used time series estimates. *Biometrika*, 70, 389–399; Corrections (1984), 71, 656.

Tsay, R. S. (1989a). Identifying multivariate time series models. *Journal of Time Series Analysis*, 10, 357–372.

Tsay, R. S. (1989b). Parsimonious parameterization of vector autoregressive moving average models. *Journal of Business and Economic Statistics*, 7, 327–341.

Tsay, R. S. (1991). Two canonical forms for vector ARMA processes. *Statistica Sinica*, 1, 247–269.

Tsay, R. S., and Tiao, G. C. (1990). Asymptotic properties of multivariate nonstationary processes with applications to autoregressions. *Annals of Statistics*, 18, 220–250.

Tuan, P.-D. (1978). On the fitting of multivariate processes of the autoregressive-moving average type. *Biometrika*, 65, 99–107.

Tunnicliffe Wilson, G. (1972). The factorization of matricial spectral densities. *SIAM Journal on Applied Mathematics*, 23, 420–426.

Tunnicliffe Wilson, G. (1973). The estimation of parameters in multivariate time series models. *Journal of the Royal Statistical Society*, **B**, 35, 76–85.

Velu, R. P., Reinsel, G. C., and Wichern, D. W. (1986). Reduced rank models for multiple time series. *Biometrika*, 73, 105–118.

Wei, W. W. S. (1990). *Time Series Analysis: Univariate and Multivariate Methods*. Reading, MA: Addison-Wesley.

Whittle, P. (1963a). On the fitting of multivariate autoregressions, and the approximate canonical factorization of a spectral density matrix. *Biometrika*, 50, 129–134.

Whittle, P. (1963b). *Prediction and Regulation by Linear Least-Square Methods*. New York: Van Nostrand.

Wincek, M. A., and Reinsel, G. C. (1986). An exact maximum likelihood estimation procedure for regression-ARMA time series models with possibly nonconsecutive data. *Journal of the Royal Statistical Society*, **B**, 48, 303–313.

Yamamoto, T. (1981). Prediction of multivariate autoregressive-moving average models. *Biometrika*, 68, 485–492.

Yamamoto, T., and Kunitomo, N. (1984). Asymptotic bias of the least squares estimator for multivariate autoregressive models. *Annals of the Institute of Statistical Mathematics*, 36, 419–430.

Yap, S. F., and Reinsel, G. C. (1992). Estimation and testing for unit roots in a partially nonstationary vector autoregressive moving average model. Technical Report, Department of Statistics, University of Wisconsin, Madison.

Zhou, G. (1992). Algorithms for estimation of possibly nonstationary vector time series. *Journal of Time Series Analysis*, 13, 171–188.

Index

AAA corporate bond rate data, 183-186, 233
Adjoint of a matrix, 12
AIC, 87, 92
Akaike's information criterion (*see* AIC)
Approximations by AR models, 64-65,
ARMAX models, 10-11
Asymptotic distribution
  of Gaussian estimator in error-correction AR model, 169-172
  of least squares (LS) estimator in AR models, 81-84
  of least squares estimator in error-correction AR model, 166-168, 176
  of likelihood ratio (LR) test statistic, 85, 107-109,
  of maximum likelihood (ML) estimator in ARMA models, 117-118
  of portmanteau model checking test statistic, 133-134
  of residual correlations, 132, 134
  of sample correlations, 76-77
  of Wald statistic, 79, 86, 109, 120
Asymptotic information matrix, 117
Asymptotic normality
  of least squares (LS) estimator in AR model, 81-84
  of maximum likelihood (ML) estimator in ARMA model, 117-118
Autocorrelations (*see* Correlations)
Autocovariances (*see* Covariances)
Autoregressive (AR(p)) process, 26-30
  AR(1) representation of, 27, 138
  covariances matrices of, 28
  examples of, 30-33
  infinite MA representation, 26-27
  LS estimation of, 81-84
  nested reduced-rank, 154-161
  nonstationary unit-root, 165-176
  specification of the order, 79-80, 85-86, 92-93
  stationarity condition, 26
  univariate models for components, 29-30
  Yule-Walker equations for, 28, 65
  Yule-Walker estimation of, 89-90
Autoregressive process of infinite order (AR(∞)), 8, 141
Autoregressive moving average (ARMA(p,q)) process, 8-9, 33-38
  conditional likelihood function for, 112-113
  covariance matrices of, 34
  echelon canonical form, 37, 54-56
  estimation of, 113-116, 119-120,

121-122
exact likelihood function for, 123-126, 129-131,
identifiability conditions, 36-37
infinite AR representation coefficients, 33
infinite MA representation coefficients, 33
invertibility condition, 8, 33
spectral density matrix of, 34
state-space representation of, 198-199, 203-210
stationarity condition, 8, 33

Backshift operator, 7
BIC, 92
Bayesian information criterion (*see* BIC)
Billing figures data, 149-153, 229
Brownian motion process
in unit-root asymptotic distributions, 163-164, 167-168, 172, 176

Canonical correlations, 58-59
for ARMA processes, 60-61, 64
relation to Kronecker indices, 64
sample estimates of, 94
use for ARMA model specification, 94-96
use for error-correction AR model, 173, 175
Cholesky decomposition of a symmetric positive-definite matrix, 13
Cofactor of element of a square matrix, 12
Coherency (*see* Squared coherency)
Co-integrated processes, 42-43,
error-correction form of AR model, 165-166
estimation of parameters in AR model, 167, 169-171, 175
vector AR processes, 165-177
Cointegration rank, 42, 166
likelihood ratio (LR) test for, 173-176
Commercial paper rate data, 183-186, 233
Companion matrix, 27
Conditional likelihood function
for ARMA model, 112-113
Conditional maximum likelihood estimator
asymptotic distribution of, 117-118
computation for ARMA model parameters, 114-116, 119, 121-122
Conditional least squares estimator
for AR model, 80-84
Convergence
in distribution, 19
in probability, 18-19
Coprime (left), 36
Covariance-generating function, 4
Covariance matrix, 13
for stationary process, 3
for white noise error, estimation of, 81, 84, 105, 113
in asymptotic distribution of LSE in AR model, 81-82, 84
in asymptotic distribution of MLE in ARMA model, 117-118, 119
in conditional multivariate normal distribution, 14, 16
of errors in Kalman filtering, 194-195
of initial state vector in ARMA model, 200-201, 204
of l-step ahead forecast error, 47
of LS estimator in multivariate linear model, 106
of presample values in ARMA model, 123
Correlations
asymptotic properties of sample estimates, 76-77
of a stationary process, 3
of residuals, 132
sample estimate of, 75

Covariances
of a stationary process, 2-3
of an AR(p) model, 28-29
of an ARMA(p,q) model, 34-35
of an MA(q) model, 22-23
of residuals, 132, 134
sample estimate of, 75

Diagnostic checking
of vector ARMA models, 132-137
use of portmanteau test statistic, 133-134
use of residual correlations, 132
use of score test statistic, 135-137
Differencing operator, 41
Dimension of prediction space, 53, 94-96, 206-209

Echelon (canonical) form of ARMA model, 37, 54-56
examples of, 57-58, 63, 148, 212
ML estimation for, 121-122
specification of, 95-96
Effects of parameter estimation on forecasting properties, 138-142
Eigenvalue of a square matrix, 12
Eigenvector of a square matrix, 12
Error-correction model
asymptotic distribution of least squares estimator, 166-168, 176
asymptotic distribution of Gaussian reduced-rank estimator, 169-170, 172
asymptotic distribution of two-step reduced-rank estimator, 170
estimation of parameters in AR model, 166-172
rank of cointegration in, 166
use of partial canonical correlation analysis in, 174-175
Estimation
least squares, 81-84, 105-107
maximum likelihood, 105-106, 112-117, 119-121
of autoregressive models, 81-84, 89-90
of autoregressive moving average models, 113-116, 119-120, 121-122
of co-integrated AR models, 166-172
of correlations, 75
of covariance matrices, 75
of process mean vector, 74-75
of reduced-rank AR models, 158-159
of white noise covariance matrix, 81, 84, 105, 113
under linear constraints in ARMA model, 119-120
Yule-Walker, 78, 89-90
Exact backcasts, 49, 124-126
Exact likelihood function, 123-126, 129-131
innovations form, 129-131
of ARMA process, 123-126
using state-space representation, 199-202, 205, 216-217
with missing observations, 216-217
Exchangeable ARMA models, 39-40
Exponential smoothing model, 43-44
forecasting in the IMA(1,1) model, 50-51
noninvertible example, 44-45

Filter (*see* Linear filter)
Final prediction error (FPE), 71, 92, 140
Forecasting for ARMA processes, 46-51,
computation of forecasts, 48-49,
infinite MA representation of forecasts, 47
l-step ahead forecasts, 47
prediction intervals, 47-48
updating, 50
use of finite order AR approximation, 141-142
use of innovations algorithm for

forecasting, 131-132
use of state-space model for forecasting, 194-195
Forecast errors, 47
covariance matrix of, 47
effects of parameter estimation on, 139-142
impulse response weights of, 47-48

Generalized least squares estimation
for MLE in ARMA model, 115-116
in multivariate linear model, 106
Grain price data, 161-162, 230-231

Hankel matrix, 53
Hannan-Kavalieris procedure, 97
Hannan-Quinn (HQ) criterion, 92
Hannan-Rissanen procedure, 96-97
Hessian matrix, 115, 119
Housing sold data, 177-180, 188-191, 232, 246
Housing starts data, 165, 177-180, 188-191, 232, 246

Idempotent matrix, 18
Identifiability
block, 37
conditions for ARMA model, 36-38
examples of, 38-39
Impulse response matrices of infinite MA, 33, 47
Innovations representation, 67-68
algorithm for recursive computations of, 68
Innovations form of exact likelihood function for ARMA model, 129-131
Integrated moving average (IMA) model, 43-45
Inventories data, 98-104, 156-157, 160, 211-215, 227-228
Invertible, 8, 21-22, 33
Investment data, 98-104, 156-157, 160, 211-215, 227-228

Irreducible factorization
of infinite MA operator in ARMA model, 36

Kalman filtering, 194-196
innovations, 194
Kalman gain matrix, 194
prediction relations, 194-195
updating relations, 194-195
Kronecker indices
of vector ARMA process, 37, 54, 95-96
relation to echelon (canonical) ARMA form, 54-56
relation with minimal dimension state-space form, 207-211
relation with reduced-rank ARMA model, 56-57, 160
relation with scalar component ARMA models, 64
specification of, 64, 94-96
Kronecker product, 29, 106

Lag matrix, 112, 122
Lagrange multiplier (LM) test (*see* Score statistic)
Least squares (LS) estimation
asymptotic properties of, for AR model, 81-83, 84
of co-integrated AR model, 166-168, 176
of multivariate linear model, 105-107
of vector AR model, 80-84
Likelihood function (*see* Conditional and Exact)
Likelihood ratio (LR) test
for linear hypothesis in multivariate linear model, 107-108
for linear restrictions in ARMA models, 120-121
for number of unit roots in nonstationary AR model, 173-174
for order of AR model, 79, 85
for reduced rank in multivariate linear model, 109-110

Linear filter, 5-6
causal, 5
stable, 5

Martingale process, 83
Maximum likelihood (ML) estimation
conditional, 112-122
exact, 123-131
in ARMA models, with missing data, 216-217
of co-integrated AR model, 169-172
of multivariate linear model, 105-106
of reduced-rank AR model, 158-159
of vector ARMA models, 112-131
under linear restrictions, in ARMA model, 119-120
using state-space model, 199-202
McMillan degree of vector ARMA process, 53, 96, 208
Mean
conditional, 14
estimation of, 74-75
of stationary process, 2
Mean squared error matrix, 15-16, 46, 47
Measurement equation of state-space model, 192
Mink furs data, 86-88, 143-149, 226, 247
Model checking
using portmanteau test statistic, 133
using residual correlations, 132-133
Model selection
AIC criterion, 87, 92, 96
BIC criterion, 92, 97
FPE criterion, 92
HQ criterion, 92, 143
of vector AR models, 79-80, 85-86, 92-93
of vector ARMA models, 92-98
use of canonical correlation analysis, 94-96
use of linear estimation methods, 96-98
use of partial canonical correlation analysis for AR models, 155-156
Moving average (MA(q)) process, 21-24
covariance matrices of, 22-23
examples of, 24-26
infinite AR representation, 22
invertibility condition, 22
models for subsets of components, 24
Moving average process of infinite order, 7
Multivariate normal distribution, 14-18
conditional distributions, 14, 16-17
distribution of quadratic forms, 18
linear regression matrix, 14
relation of conditional mean to best linear prediction, 15-16
Muskrat furs data, 86-88, 143-149, 226, 247

Newton-Raphson algorithm, 114, 119
Non-negative definite sequences, 3
Nonstationary ARMA processes (ARIMA), 41-44
co-integrated processes, 42-43, 165-166, 225

Observation equation of state-space model, 192
Observationally equivalent ARMA models, 36
Order determination (*see also* Model selection)
for AR model, 79-80, 85-86, 92-93
criteria, 92-93
Order selection criteria (*see also* Model selection), 92-93

Partial autoregressive matrices, 65
example of, 71-73

recursive computation of, 66-67
sample estimate of, 78-79, 90
tests based on sample estimate, 79
Partial correlation matrices, 69-70
sample estimate of, 79-80
tests based on sample estimate, 80
Partial canonical correlations, 59
for stationary processes, 71
sample estimate of, 80, 109
use of sample estimates for testing in reduced-rank AR models, 155-156
use in reduced-rank estimation, 174-175
Portmanteau lack-of-fit test, 133-134
approximate distribution of, 134
Positive definiteness of a matrix, 13
Prediction (*see also* Forecasting)
minimum mean squared error prediction, 15-16, 46
Production schedule figures data, 149-153, 229

Random walk process, 41-42
Recursive calculation
in Kalman filtering and smoothing, 194-197
of AR matrices, 66-67
of forecasts, 48, 131-132, 195
of innovations, 67-68, 130-131
of partial derivatives, 114-116, 135, 203
of residuals, 115, 125-126
Reduced-rank AR models, 154-161
canonical form of model, 157-158
maximum likelihood estimation of parameters, 158-159
relation of ranks to Kronecker indices, 159-160
use of sample partial canonical correlations for testing rank, 155-156
Reduced-rank form of ARMA model, 56-57
Relationship
between ARMA model and state-space model, 198-199, 203-204, 206-211
between conditional LS and Yule-Walker estimation, 89-90
between conditional mean and best prediction, 15-16, 46
Residual correlations
asymptotic properties of, 134
Residuals
correlations of, 132
of estimated vector ARMA model, 115, 125-126, 132
use for model checking, 132-134
Restrictions
ML estimation of ARMA model under, 119-120
LR test for, 120-121

Scalar component models
examples of, 62-63
for ARMA processes, 61-62, 94, 160
relation with Kronecker indices, 64
Score statistic, 134-137
Seasonal multiplicative ARMA models, 186-188
Smoothing
classical methods based on infinite data, 218-222
fixed interval, 196
in the IMA(1,1) model, 44, 50, 195-196, 224-225
in the state-space model, 196-197
Spectral density matrix, 4
for vector ARMA processes, 34
Spectral representation
of a stationary process, 4, 6
of covariances from a stationary process, 4
Squared coherency, 4
Stability, 5, 11, 26, 27-28
State equation of state-space model, 192
State-space model

canonical correlation in, 206
Kalman filtering for, 194-196
minimal dimension of state vector, 206-207
prediction error form of, 197, 206
prediction space, 94, 206
representation for vector ARMA models, 198-199, 202-204, 206-210
steady-state form, 198
time-invariant, 193, 197
use for exact likelihood of ARMA model, 199-201
use for exact likelihood with missing data, 216-217
State vector
covariance matrix of, in ARMA model, 200-201, 204
in state-space model, 192-193
in state-space representation of vector ARMA model, 198, 203-204, 206-209
minimal dimension, 206-210
Stationarity, 2-3
of AR process, 26-27
of ARMA process, 8, 33
Stationary process, 2

Test for
linear restrictions in ARMA model, 120-121
order of vector AR model, 78-80, 85-86
rank of cointegration, 173-174
reduced ranks in AR model, 155-156
unit root in univariate model, 163-164
white noise, 93, 133
Transfer function
examples of form for ARMA models, 25, 32
form for ARMA models, 10
of a linear filter, 5
Transition equation of state-space model, 192

Unimodular matrix, 36, 40
Unit roots
in vector ARMA processes, 41-42
in vector AR processes, 165-166
likelihood ratio test for number of, 173-174
testing in univariate AR models, 163-164
Updating forecasts, 50
in multivariate normal distribution, 17
in state-space model, 194-195
U.S. interest rate data, 183, 234-237

Vec operator, 29, 105-106

Wald test statistic
for linear hypothesis in multivariate linear model, 109
for linear restrictions in ARMA model, 120-121
for order of AR model, 79, 85-86
White noise process, 6
Whittle algorithm
for AR model coefficients, 66-67
Wold representation of stationary process, 7

Yule-Walker equations, 28-29, 65
recursive calculations, 66-67
Yule-Walker estimators for AR model, 79, 89-90

Springer Series in Statistics

(continued from p. ii)

Sachs: Applied Statistics: A Handbook of Techniques, 2nd edition.
Salsburg: The Use of Restricted Significance Tests in Clinical Trials.
Särndal/Swensson/Wretman: Model Assisted Survey Sampling.
Seneta: Non-Negative Matrices and Markov Chains, 2nd edition.
Shedler: Regeneration and Networks of Queues.
Siegmund: Sequential Analysis: Tests and Confidence Intervals.
Tanner: Tools for Statistical Inference: Methods for the Exploration of Posterior Distributions and Likelihood Functions
Todorovic: An Introduction to Stochastic Processes and Their Applications.
Tong: The Multivariate Normal Distribution.
Vapnik: Estimation of Dependences Based on Empirical Data.
West/Harrison: Bayesian Forecasting and Dynamic Models.
Wolter: Introduction to Variance Estimation.
Yaglom: Correlation Theory of Stationary and Related Random Functions I: Basic Results.
Yaglom: Correlation Theory of Stationary and Related Random Functions II: Supplementary Notes and References.